P9-DTW-572

SHIPS AND SEAMANSHIP IN
THE ANCIENT WORLD

BY LIONEL CASSON

Ships and Seamanship in
THE ANCIENT WORLD

THE JOHNS HOPKINS UNIVERSITY PRESS

BALTIMORE AND LONDON

Copyright © 1971 by Princeton University Press
New material copyright © 1995 by The Johns Hopkins University Press
All rights reserved
Printed in the United States of America on acid-free paper

Originally published in a hardcover edition by Princeton University Press, 1971
Johns Hopkins Paperbacks edition, 1995
9 8 7 6 5 4 3

The Johns Hopkins University Press
2715 North Charles Street
Baltimore, Maryland 21218-4363
www.press.jhu.edu

Library of Congress Cataloging-in-Publication Data will be found
at the end of this book.

A catalog record for this book is available from the British Library.

ISBN 0-8018-5130-0 (pbk.)

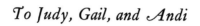
To Judy, Gail, and Andi

PREFACE

MEN OF THE ancient world, throughout the span of their history, were
loth to stray far from the sea. It was woven into the fabric of their
lives, and among their great contributions to later ages was their mastering
of this superbly useful but tricky and dangerous way of communication.

The ancient mariners of the Mediterranean can claim credit for most
of the major discoveries in ships and sailing that the western world was
to know until the age of steam. The details of this achievement—the arrangements they hit upon for rowing war galleys, the rigs they devised
for merchantmen, the ways they worked out for assembling a hull, and
the like—make up a highly technical and specialized subject, yet one that
has an intimate connection with ancient man's day-to-day experience. It
is no accident that the west's first epic poet chose to sing of a storm-tossed
captain, its first historian and first dramatist to highlight a crucial naval
battle.

Despite the manifest importance of the subject, no scholar came near
to doing it justice until the very end of the last century. In 1895 Cecil Torr
published his *Ancient Ships*, a short but admirable summary of what was
then known about the design, construction, and equipment of Greek and
Roman craft. This filled much of the need, but hardly all; and, though
Torr promised studies of the other phases of shipping, he never carried
them to completion. Moreover, because of the growing mass of additional
material archaeologists were steadily unearthing, even what he did provide soon became outdated.

Then, after World War II, underwater archaeology sprang into being
and began to tap a source of totally new evidence. This meant that Torr's
book was now hopelessly obsolete. At the same time, by a curious irony,
the very ones who had cut short its usefulness guaranteed it a new lease
on life: the ever increasing number of divers who took up marine archaeology, and ipso facto turned into students of Mediterranean nautical antiquities, made *Ancient Ships* their vade mecum—it was, after all, the sole
reference work available. The stepped-up demand brought about a new
printing in 1964; though called a second edition, it reproduced Torr's text
without alteration.

I started to gather material for the present work in 1953 when, with the
help of a grant from the Guggenheim Foundation, I visited the pioneer
excavation of an ancient wreck that Jacques Cousteau had opened that

year off an islet near Marseilles. I completed the manuscript almost a decade and a half later, with the help this time of a grant from the National Endowment for the Humanities, which, by giving me a year away from classroom duties, enabled me to assemble the evidence I had gathered into a coherent presentation.

My aim has been to replace Torr's monograph with a comprehensive and up-to-date work of reference. I have broadened the scope to include all makes of craft, from tiny fishing smacks to monster grain-carriers, from an admiral's gig to a catapult-carrying supergalley. I have tried to cover them in all aspects: the wood that went into them, the oars and sails that drove them, the officers and men that manned them, the names that were given to them, the speed, the handling, and so on. The task was somewhat lightened by the timely appearance of Morrison and Williams' *Greek Oared Ships*, an authoritative study of Greek warships down to ca. 300 B.C.; the lion's share of the subject, however, required an independent review of the material for the most part. Though I begin the story with the remote moment when men first went down to the sea and carry it as far as the ninth century A.D., the heyday of the fleets of the Byzantine Empire, my protagonists have necessarily been the shipwrights and sailors of Greece and Rome, whose contribution was far and away the most significant.

The notes bulk as large as the text. I did not want merely to formulate what we know—or think we know—on any given point; I was equally concerned to put before the reader precise indications of what our knowledge rests on. And so the text is everywhere supported by footnotes that cite whatever evidence is available—ancient writings (inscriptions and papyrus documents as well as the works of ancient authors), paintings and sculptures and models, the reports of archaeologists and divers. The notes treat as well the many controversial points that are unavoidable in a study such as this. Those that required extended discussion, as well as certain topics too specialized for the text and too voluminous for a note, are dealt with in appendices.

Unlike Torr or any of the others who have involved themselves with some phase of this field, I have designed my work for the full spectrum of readers, from the specialist in the history of technology or in ancient history and literature to those with but a casual interest in the subject. The text is free of Greek characters or similar esoterica; citations from the Greek or Latin—generally quoted in extenso for handy reference—are everywhere given in translation as well as the original; the publications of nautical experts are referred to as freely as those of Classical scholars and archaeologists; wherever possible, parallels have been adduced from the ships and practices of later ages.

I have had much welcome help.

I have already mentioned the grants I received from the National Endowment for the Humanities and the Guggenheim Foundation. I must thank in particular the former director of the Guggenheim Foundation, Henry Allen Moe, who, a professional seaman himself at one time, has followed my work with gratifying interest. New York University on two occasions awarded me grants for travel and the acquiring of photographs.

I owe an enormous debt to the underwater archaeologists. George Bass, Gerhard Kapitän, Michael Katzev, Peter Throckmorton, Frederick van Doorninck, Jr.—all have unstintingly supplied photographs of, and information about, their latest findings, while Frédéric Dumas, M. Y. Girault, Anna Marguerite McCann kindly allowed me to reproduce photographs they had taken. Indeed, I am grateful to many who generously provided photographs: A. and A.-M. Bon, Michael Eisman, Alison Frantz, Ch. Makaronas, Mario Moretti, John Morrison, Ernest Nash, Josephine Powell, G. M. Richter, Otello Testaguzza. The drawings for Figs. 171 and 172 were made by Joseph Ascherl, for Fig. 173 by Milton Brown. John Morrison put at my disposal a set of proofs of his *Greek Oared Ships* six months before it saw publication, and Roberto Peliti, publisher of O. Testaguzza's *Portus*, rushed the first copy off the press into my hands. I have had invaluable advice on naval matters from Ole Crumlin-Pedersen, Manlio Guberti, above all, R. C. Anderson; on Greek naval tactics from Herman Wallinga; on points of Roman law from Arthur Schiller; on papyrological problems from Naphtali Lewis; on numismatic points from Bluma Trell; on matters relating to ancient art from John Ward-Perkins and Blanche Brown; on Egyptian ship-names from Alan Schulman.

A difficult and involved manuscript was put in superb shape for printer and plate-maker by the skill of Polly Hanford and Jan Lilly. The manuscript and the endless drafts that preceded it were typed by my wife; without her patient and able help the completion of this book would have taken years more. And Rae D. Michelman contributed long and arduous hours to the preparation of the indexes.

July 1970, Rome

CONTENTS

CONTENTS

ILLUSTRATIONS

The Aegean Merchantman

Cretal Vessels Pictured on Seals

Round-Hulled Cretan Vessels

(Polyphemus and Galatea) in the House of the Priest Amandus in Pompeii. Photo German Archaeological Institute, Rome.

125. Roman trireme. Relief at Ostia (second half of 1st B.C.). The prominent rectangle is not the *epotis*, but a plaque bearing the ship's device.

126. Graffito (end of 1st B.C. or beginning of 1st A.D.) of a Roman quadrireme (*navis tetreris longa*). Found at Alba Fucens. The low hull would fit a single-banked, at most a double-banked, arrangement of rowers. After *Not. Sc.* (1953) 120.

127. The prow of Trajan's trireme with *artemon* raised. Above, a liburnian with sailing rig furled and resting on crutches. From Trajan's column (Cichorius 80.211). Photo German Archaeological Institute, Rome.

128. Emperor Trajan at the helm of a trireme (note the square *vexillum* alongside the cabin and the commander's light hanging from the *aplustre*). Below, a liburnian carrying legionary standards. From Trajan's Column (Cichorius 79.209). Photo German Archaeological Institute, Rome.

129. Roman trireme. Relief (1st B.C. to 1st A.D.) found at Pozzuoli and now in the Naples National Museum. Photo Anderson.

130. Two-banked Roman galley, probably a quadrireme or larger. Relief (second half of 1st B.C.) found at Praeneste and now in the Vatican Museum. Photo courtesy of the museum.

131. Roman trireme. Relief (1st B.C. to 1st A.D.) found at Pozzuoli and now in the Naples National Museum. Photo Anderson.

132. Another view of galley shown in Fig. 130.

133. Roman galleys, probably larger than triremes, with a full complement of marines. Wall-painting from the Temple of Isis at Pompeii (1st A.D.) and now in the Naples National Museum.

134. Dromon using Greek fire. Illustration in a manuscript of Ioannes Scylitzes (14th A.D.) in the Biblioteca Nacional, Madrid. Photo courtesy of the Biblioteca.

135. Small well-fitted galley for carrying passengers, perhaps an admiral's barge. Renaissance copy of an ancient votive model found at Rome near the site of a barracks that included the temple of Juppiter Redux, where the original may have been dedicated; cf. *Memorie della pontificia accademia romana d'archeologia 7* (1944) 232. It now decorates the fountain in front of the church of S. Maria in Domnica. Photo Anderson.

Imperial period) in the Archaeological Museum, Saloniki. Photo courtesy of Dr. Ch. Makaronas.

179. Sprit-rigged craft. Detail of Fig. 147.

180. Lateen-rigged craft. Graffito on a piece of roof tile found on Thasos. Date indeterminate. See A.-M. Bon and A. Bon, *Les timbres amphoriques de Thasos* (École française d'Athènes, Études thasiennes IV, Paris 1957) no. 2274. Photo courtesy of M. and Mme. Bon.

181. Lateen-rigged craft. Relief on the tombstone of Alexander of Miletus (2nd A.D.). Found near Piraeus and now in the National Archaeological Museum, Athens. Photo courtesy of the museum.

182. Lateen-rigged craft and small boats off a seaside villa. Mosaic (probably 4th A.D.) in the Museo Correr, Venice. Photo courtesy of the museum.

183. Iron anchor, with removable stock, cased in wood. From one of the Nemi barges. After Ucelli, fig. 270.

184. Wooden anchor with lead stock. From one of the Nemi barges. Behind, the hull of one of the barges covered to the gunwales with a lead sheathing. Photo Fototeca Unione, Rome.

185. Reconstruction of an anchor with arms and shank held together by a lead collar. After Benoit, fig. 91.

186. Lead core of a wooden anchor stock. The piece has accidentally been bent double; it originally was flat, the short, rounded part in the center passing through a hole in the shank. Found in Syracuse Harbor. Photo courtesy of G. Kapitän.

187. Stone anchors recovered from a wreck of ca. 350 B.C. off Taranto. Photo courtesy of Dr. A. McCann.

188. Sketch showing method of sailing against the wind with sail brailed up; see TWELVE, note 24.

189. A rivercraft of the Rhine or Moselle. Relief on the tombstone of the sailor Blussus (mid-1st A.D.) in the Mittelrheinisches Landesmuseum, Mainz. Espérandieu 5815. Photo courtesy of the museum.

190. Plan of a harbor barge. Same provenience as Fig. 194.

191. Coastal craft unloading bars of lead onto a beach. Mosaic from a tomb chamber near Sousse (3rd A.D.), now in the Bardo Museum, Tunis. Photo Fototeca Unione, Rome.

SHIPS AND SEAMANSHIP IN
THE ANCIENT WORLD

CHAPTER ONE

Floats, Rafts, and the Earliest Boats

MEN FIRST WENT down to the sea not in boats but on whatever they could find that would keep them afloat. A New Zealand aborigine today paddles over lakes astride a bundle of reeds, an Iraqi herdsman crosses streams on an inflated goatskin, a Tamil native does his fishing drifting with a log under his arms while a Sindhi does his lying prone over an openmouthed pot. Such devices must reach back in an unbroken line to man's earliest forays on the water. Available, simple, cheap, and convenient, they retained a certain usefulness even after far superior and more sophisticated devices had made their appearance.[1]

I RAFTS

ONE early step must have been from float to raft, from the single log or bundle of reeds that would support one person to a platform that would support several. In wooded areas, it must have been the raft most of us know best, of bound logs.[2] Along the Nile or amid

[1] Cf. Hornell 1-20. The best known examples of inflated bladders used as floats are those in the Assyrian reliefs with fishermen astride them (Fig. 4) or with Assyrian soldiers lying on them and paddling across water (*IH* ill. 1). Other armies besides the Assyrian carried them as standard equipment, e.g., the Spanish troops that fought against Caesar in 49 B.C. (Caesar, *Bell. Civ.* 1.48.7; cf. Suetonius, *Iul.* 57.2). After Julian's death in A.D. 363, some of his frantically retreating soldiers recrossed the Tigris on inflated skins (Ammianus 25.8.2). In 74 B.C., Mithridates had Cyzicus, which sits on the tip of a peninsula, under siege; the Romans got a message through by means of a strong swimmer using a pair of inflated skins (Frontinus, *Strateg.* 3.13.6).

When soldiers had no bladders as part of their equipment, they could improvise by taking their leather tent covers, stuffing them with hay or the like, and sewing them tight. This is the way some of Xenophon's Greeks once crossed the Euphrates (*Anab.* 1.5.10) and the way Alexander transported whole armies over the Danube (Arrian, *Anab.* 1.3.6) and the Oxus (Arrian, *Anab.* 3.29.4). Cf. Hornell 21-22.

[2] Hornell (61-90) discusses the manifold forms of log rafts in use today from the

3

the marshy lower stretches of the Tigris and Euphrates, regions of few trees but thick with reeds, rafts of reed bundles early came into use and, in the course of time, served as stepping stone to an important form of boat, the reed canoe (11 below).

Where particular geographical conditions or special requirements demanded something better, a more sophisticated form of raft came into being, the buoyed raft. In Mesopotamia, for example, the upper reaches of the Tigris and Euphrates with their swift waters and stony rapids would be death on any ordinary type. Here, in remote times, someone with imagination, observing his fellows crossing the river on inflated skins, figured out that, if one float could hold up one rider, a number of floats should be able to hold up a platform carrying several, and thereby invented the *kelek*, the raft made up of a wooden frame resting on multiple bladders. Rock-strewn rapids which would splinter a log raft to bits, merely give a *kelek* a few blowouts, which can be repaired in short order. Moreover, when the raftsmen want to get home, instead of fighting their way upstream, they simply sell the wood of the platform, deflate the skins, pile them on some donkeys which they had mindfully loaded aboard at the outset, and leisurely walk back. We see such rafts pictured in Assyrian reliefs of the eighth century b.c. (Fig. 1); until a few years ago they were still in use on the twin rivers, including some that were all of 50 feet square and kept afloat by as many as a thousand skin bladders.[3]

---

crudest types to the highly sophisticated catamarans, or shaped sailing rafts, of India's Coromandel coast.

Ancient authors, understandably finding little occasion to mention ordinary rafts, speak only of exceptional ones. E.g., we hear of the raft 100 feet long by 50 wide that Hannibal built to ferry his war elephants across the Rhone in 218 b.c. (Livy 21.28.8-9; cf. Polybius 3.46, who does not give the size) or the ones Cassander used to raft his from Megara to Epidaurus in 316 b.c. (Diodorus 19.54.3). Theophrastus (*Hist. Plant.* 5.8.2) tells how the Romans, in order to transport Corsican timber, made an enormous raft propelled by no fewer than 50 masts and sails.

[3] See Salonen, *Wasserfahrzeuge* 66-69 and Hornell 26.30. *Kelek* is the modern Arabic version of the Akkadian word *kalakku*; see Salonen, *Wasserfahrzeuge* 66. On Herodotus' confusion of the *kelek* and *quffa*, see note 10 below. Xenophon, *Anab.* 2.4.28 mentions the natives' use of σχεδίαι διφθέριναι "leather rafts" on the

Where there was no danger from rocks, an equally efficient and far cheaper way to buoy a raft was with a line of pots instead of inflated skins. Pot rafts are common in, e.g., modern India, where the natives use them in the mud-bottomed alluvial areas.[4] Their earliest appearance is in connection with a myth of Hercules; a series of Etruscan gems picture him comfortably floating along on such a raft (Fig. 2).[5] From historical times we have reports of versions large enough to ferry substantial numbers of soldiers and even war elephants across considerable distances.[6]

II THE EARLIEST BOATS

ALONG with rafts men created true boats, craft that would not only keep a user afloat but enable him to stay dry in the process.

One of the earliest forms must have been the skin boat, made of sewn hides stretched over a light frame of branches and laced together with withes, cords, or thongs. Such craft can be built with the

---

Tigris; these must be *keleks*. Buoyed rafts have also been reported off the coast of Ethiopia (Pliny, *NH* 6.176) and the southern coast of Arabia (*Per. Mar. Eryth.* 27). They could be improvised. E.g., in A.D. 374 the Armenians crossed the Euphrates by mounting beds on wineskins (Ammianus 30.1.9). Hannibal's Spanish soldiers improvised individual buoyed rafts to get across the Rhone in 218 B.C.: each stuffed a skin with his clothes (thereby keeping them dry), set his shield on top, and crossed on the shield (Livy 21.27.5-6).

Buoyed rafts were sometimes hitched together to form a pontoon bridge. Xenophon reports that such a bridge was suggested as a way for the Greek forces to cross the Tigris (*Anab.* 3.5.9-11), while Julian's engineers in 363 actually constructed some for his army to use (Ammianus 24.3.11, 25.6.15).

During the Roman Imperial period towns near rivers, particularly along the Rhone but also in Dacia, had guilds of *utricularii* "bladdermen." The very reasonable suggestion has been made that these groups operated landing floats, pontoon bridges, buoyed-raft ferries, or other apparatus involving the use of inflated skins; see J. Rougé, "Utricularii," *Cahiers d'histoire* 4 (1959) 285-306.

[4] Hornell 34-37.

[5] The material has been collected by R. Stiglitz, "Herakles auf dem Amphorenfloss," *Jahreshefte des oesterreichischen Instituts* 44 (1959) 112-41.

[6] When Spartacus and his men were trapped at the toe of the boot in 72 B.C., they attempted to cross to Sicily on *rates ex trabibus et dolia conexa virgultis* "rafts of lengths of wood and pots tied together by means of withes" (Florus 3.20.13). Metellus in 252 B.C. transported 140 or so elephants over the same stretch of water on *ratibus, quas doliorum consertis ordinibus imposuerat* "rafts laid over lines of lashed pots" (Pliny, *NH* 8.16. Not "casks" as Rackham translates it in the Loeb edition, vol. III, p. 15).

simplest of tools—a flint knife and bone needle—and in a very short time. They can be made light and small enough to be packed on the back for use when needed, or large and commodious enough to carry up to four and five tons of cargo.[7] The ancient version that we know best is the *quffa*, the round coracle of the lower Euphrates.[8] They are depicted in detail in Assyrian reliefs of the ninth to the seventh centuries B.C. (Fig. 4),[9] they were seen by Herodotus,[10] and the modern versions are still an essential means of river transport. The reliefs show us fairly good-sized coracles, driven by four oarsmen and big enough to carry a chariot or a load of massive stones (modern *quffas* run 13 feet in diameter and 7½ feet deep).[11]

[7] Cf. Hornell 91-133.

[8] *Quffa* is the modern Arabic version of the Akkadian word *quppu*; see Salonen, *Wasserfahrzeuge* 71.

[9] Salonen, *Wasserfahrzeuge* pl. 21.1 (9th B.C.); E. Budge, *Assyrian Sculptures in the British Museum* (London 1914) pl. 21.2 (*quffa* transporting a chariot, 9th B.C.); Fig. 4 (*quffa* transporting building stone, 7th B.C.).

[10] 1.194. Herodotus has apparently confused the *quffa* with the *kelek* (cf. Hornell 29-30, 106; Salonen, *Wasserfahrzeuge* 72-74). The physical description he supplies is of a *quffa* (1.194.2): ἐόντα κυκλοτερέα πάντα σκύτινα . . . νομέας εἰτέης ταμόμενοι ποιήσωνται, περιτείνουσι τούτοισι διφθέρας στεγαστρίδας ἔξωθεν ἐδάφεος τρόπον, οὔτε πρύμνην ἀποκρίνοντες οὔτε πρῴρην συνάγοντες, ἀλλ' ἀσπίδος τρόπον κυκλοτερέα ποιήσαντες "They are rounded and all of hides. . . . They are made by cutting willow branches for frames and stretching over these a covering of hides on the outside to form a sort of hull in which they do not make a distinctive stern nor bring it to a point to form a prow but make it rounded like a shield." However, the uses he assigns to it belong to the *kelek* (1.194.3-5): τὰ δὲ μέγιστα αὐτῶν καὶ πεντακισχιλίων ταλάντων γόμον ἔχει. ἐν ἑκάστῳ δὲ πλοίῳ ὄνος ζὼς ἔνεστι, ἐν δὲ τοῖσι μέζοσι πλεῦνες. ἐπεὰν ὦν ἀπίκωνται πλέοντες ἐς τὴν Βαβυλῶνα καὶ διαθέωνται τὸν φόρτον, νομέας μὲν τοῦ πλοίου . . . ἀπ' ὦν ἐκήρυξαν, τὰς δὲ διφθέρας ἐπισάξαντες ἐπὶ τοὺς ὄνους ἀπελαύνουσι ἐς τοὺς Ἀρμενίους "The biggest can carry a good 5,000 talents [*sic*; at least 125 tons, a tall order even for a big *kelek*!]. Each has aboard a live donkey, the larger ones several. After arriving at Babylon and disposing of the cargo, the frames of the boat . . . they auction off, load the hides on the donkeys, and walk back to Armenia." Hornell (29-30) offers the plausible suggestion that Herodotus' visit to Babylon did not coincide with the season when the *keleks* came downstream and that the only craft he actually saw were the *quffas* busily ferrying passengers and cargo across the river then as now (cf. Salonen, *Wasserfahrzeuge* pl. 40 and Hornell, pl. IVB).

The *quffas* pictured in the Assyrian reliefs and those Herodotus saw were of leather. The 500 "boats of skins" (πλοῖα ἀπὸ δερμάτων) used by Julian to ferry his army over the Euphrates during the campaign of A.D. 363 (Zosimus 3.13) were probably leather *quffas*. The type lasted into the last century (Hornell 104). Modern versions are of basketry coated with asphalt (Hornell 101-103). The split reeds, which Assyrian documents record as building material for *quffas* (Salonen, *Wasserfahrzeuge* 71 and 144), may have served for the frame, being then covered with skins.

[11] For dimensions of modern *quffas*, see Hornell 103.

Another region where coracles and skin boats have flourished is
the British Isles; Julius Caesar was the first to report seeing them
there,[12] and they are frequently mentioned by later writers.[13] Other
areas, too, found them of service, for they have also been reported in
the Po Valley, along the north coast of Spain, in the Red Sea, on
Lake Maeotis in the Crimea.[14]

In waters free of rocks, a form of boat that serves the same purpose
as a small coracle but is far cheaper to make is the clay tub. These
are still in use in India, where the natives get about in them during
the rainy season when their villages are flooded for long periods.[15]
In ancient times they were certainly in use in the Nile Delta,[16] and,
to judge from another of the myths of Hercules, one which has him
voyaging in a pot, the Greek world probably knew of them as well.[17]

Where wood was available, men inevitably turned to the bark
canoe and the dugout. Perhaps the bark canoe came first—indeed,
it may even be the earliest form of boat devised, for it can be made
without tools: all that is needed is a troughlike strip of bark and two
lumps of clay to stop up the ends.[18] The dugout itself requires little

[12] *Bell. Civ.* 1.54: *imperat militibus Caesar ut naves faciant, cuius generis eum
superioribus annis usus Britanniae docuerat. carinae primum ac statumina ex levi
materia fiebant: reliquum corpus navium viminibus contextum coriis integebatur*
"Caesar orders his men to make boats of the kind his experience in Britain the
years before had taught him. The keel and ribs were made first, out of light wood;
then the rest of the body of the craft was made of wickerwork and covered with
skins." Probably Caesar built curraghs, skin boats, rather than coracles; cf. Hornell
112.

[13] Antiphilus (1st A.D.) in *Anth. Pal.* 9.306; Pliny, *NH* 4.104, 7.206, 34.156; Dio
Cassius 48.18, 19 (building of coracles of the British type to use on the Strait of
Messina in 40 B.C.); Solinus (3rd A.D.) 22.6-9 (see critical note in Mommsen's edi-
tion); Avienus (4th A.D.), *Ora maritima* 103-107.

[14] Po Valley: Lucan 4.131-32 (*cana salix madefacto vimine parvam / texitur in
puppim, caesoque inducta iuvenco* "The white willow, its branches soaked [to render
them pliable], is woven into a small craft and covered [with the skin of] a slaugh-
tered bullock"); north coast of Spain: Strabo 3.155 (διφθερίνοις πλοίοις "hide boats");
Red Sea: Agatharchides 101 = Müller, *GGM* I, p. 189 (δερματίνοις πορείοις "craft
of skins"), repeated in Strabo 16.778 (δερματίνοις πλοίοις "skin boats"); the Crimea:
Strabo 7.308 (ῥαπτοῖς πλοίοις "sewn boats"; these, specially designed for marshy
shallows, should be coracles rather than boats of sewn planks).

[15] Hornell 98.

[16] Strabo 17.788 (ὀστράκινα πορθμεῖα "earthenware ferries"); Juvenal 15.127 (*ficti-
libus phaselis* "phaseli [see 167 below] of clay").

[17] See C. Haspels, *Attic Black-Figured Lekythoi* (Paris 1936) 121-22 and pl. 17.3.

[18] Cf. Hornell 182.

more: a stone cutting-tool (or even just a hard shell) or the controlled use of fire, and infinite patience.[19] In the ancient world we can trace the dugout chronologically from the Stone Age to the fifth century A.D.,[20] and geographically from Spain to India, wherever there were forests to supply the logs: on the Guadalquiver, Rhone, Elbe, Danube, Nile, Phasis, Euphrates; along the coasts of Spain, Germany, the eastern Mediterranean, the Black Sea, East Africa, India.[21]

In many parts of the world dugout-builders have fitted their craft for open water by raising the sides with planks and inserting frames to strengthen the complex. Here we have in embryo the fundamental elements of the planked boat—keel, ribs, and strakes. There is good reason to think that at least one avenue which led to the wooden boat proper was by way of such dugouts, and that, in the course of time, so many planks had been added to the sides that they came to form what we may properly call a hull, and the original dugout had shrunk in the process to the dimensions of a keel.[22] Just when this took place no one can say. Our earliest evidence of planked boats is from Egypt, but the Egyptian type seems to have come into being via a very different route (13 below).

[19] Cf. Hornell 189.

[20] A dugout uncovered in Holland has been dated by carbon-14 tests to ca. 6300 B.C.; see J. Hawkes, *History of Mankind: Cultural and Scientific Development.* I, pt. one, *Prehistory* (London 1963) 159-60. The National Archaeological Museum of Ferrara, Italy, has two big dugouts (14.50 and 12.75 m.) dating to the 4th or 5th century A.D.; see *IH* ill. 8, and *Fasti Archaeologici* 3 (1948) 2712. For other dugouts found in the Po Valley, see N. Alfieri, "Tipi navali nel delta antico del Po," *Atti del Convegno Internazionale di Studi sulle Antichità di Classe, Ravenna, 14-17 ottobre 1967* (Ravenna 1968) 187-208, esp. 190-92.

[21] Guadalquivir: Strabo 3.142; Rhone: Polybius 3.42.2, Livy 21.26.8; Elbe: Velleius 2.107.1; Danube: Arrian, *Anab.* 1.3.6, Theophylactus, *Hist.* 6.9; Nile: Heliodorus, *Aeth.* 1.31.2; Phasis: Hippocrates, *Aer.* 15; Euphrates: Ammianus 24.4.8; Spain: Strabo 3.155; Germany, Pliny, *NH* 16.203; eastern Mediterranean: Aristotle, *Hist. anim.* 533b; Black Sea: Xenophon, *Anab.* 5.4.11, Polyaenus 5.23; eastern Africa: *Per. Mar. Eryth.* 15; India: Pliny, *NH* 6.105 (cf. *Per. Mar. Eryth.* 60 and Schoff's note on p. 243). The technical term for dugout throughout these passages is the Greek *monoxylon,* used Latinized in form by Roman writers.

One of the reliefs on Trajan's Column shows army artificers carving log canoes probably to use as pontoons for a floating bridge; see I. Richmond, "Trajan's Army on Trajan's Column," *Papers of the British School at Rome* 13 (1935) 1-40, esp. 28.

[22] Cf. Hornell 187, 192.

SEWN BOATS

With the planked boat we reach a key question in the history of boat-building: how were the planks put together? The sole area for which we have sure information that dates to an early age is Egypt, but the rather special technique devised there had best be delayed for the next chapter. For other places the only evidence available are some straws in the wind to indicate that at least one early way of fastening planks was by sewing them with fibers, cord, or thongs.

The boat of sewn planks is no anomaly but a well-authenticated type. Though rapidly succumbing to twentieth century technology, it is still widely used in East Africa, India, and Ceylon, and even lingered on late in northern Europe. Its particular home has been the Indian Ocean, where it was the boatbuilding technique par excellence right up to the end of the fifteenth century, when the arrival of the Portuguese brought in contemporary European methods.[23] We can trace sewn boats in this region as far back as the first century A.D.,[24] and they were unquestionably on the scene much earlier. Moreover, there are strong clues that they were a primitive form of craft throughout the Mediterranean. When Vergil describes Aeneas' descent into the underworld, he has him cross the Styx in a "sewn skiff."[25] The Roman dramatist Pacuvius (second century B.C.) in a play about Odysseus has the hero make a craft of sewn planks to effect his escape from the charms of Calypso.[26] Both authors must have chosen the sewn boat because they knew their audiences would

---

[23] Hornell 192-93, 234-37; Hourani 87-98. The Boat Hall of the Norsk Sjøfarts-musee in Oslo has on display a *schnjaka*, an open boat used until at least the end of the last century for codfishing off the Murmansk coast, whose planks are sewn together.

[24] πλοιάρια ῥαπτά "sewn boats," *Per. Mar. Eryth.* 15 and 16 (probably Zanzibar), 36 (Arab side of the Persian Gulf).

[25] *cumba sutilis, Aen.* 6.413-14. Jerome, in comparing chastity to a *fragilem et sutilem ratem* "a frail, sewn boat" (*Epist.* 128.3, A.D. 413), very likely is just echoing Vergil's phrase.

[26] *Nec ulla subscus cohibet compagem alvei, / sed suta lino et sparteis serilibus* "Nor did any mortises hold together the joints of the hull, but [the boat] was sewn with flax cord and grass fibers," Pacuvius ap. Festus 508.33 = *ROL* II, p. 268. On boatbuilding with mortises and tenons, see 202 below.

connect it with the days of long ago, that it would conjure up the associations which, say, "galleon" or "ark" does today. Varro (first century B.C.), followed by Pliny the Elder (first century A.D.), both sober encyclopedists and not creative poets, were convinced that the galleys which in remote Homeric times brought the Greek warriors to Troy were of sewn planks,[27] and Aeschylus attributes sewn boats to the heroic age his characters lived in.[28]

All the above gives the impression that, when the dwellers around the Mediterranean turned to making a boat of keel and planks, one of their earliest techniques—if not the earliest—for fastening the members was the quick, cheap, and effective one of binding them with twine.

[27] Varro ap. Gellius 17.3.4: *Liburni . . . plerasque naves loris suebant, Graeci magis cannabo et stuppa ceterisque sativis rebus* "The Liburnians . . . sewed most of their craft with thongs, while the Greeks preferred hemp and tow and other plant products"; Pliny, *NH* 24.65: *et cum fierent sutiles naves, lino tamen, non sparto umquam sutas* "And when sewn boats were made [by the early Greeks], they were sewn with linen, never with esparto grass fiber." Both authors are commenting on *Il.* 2.135 (καὶ δὴ δοῦρα σέσηπε νεῶν καὶ σπάρτα λέλυνται "And the ships' planks were rotten and the cords all loose"), and assume that Homer is talking not of the rigging but of the twine that bound plank to plank; cf. W. Jones' note on the passage in the Loeb edition of Pliny (vol. VII, p. 50).

[28] Thus, in *Suppl.* 134-35, the chorus talks of crossing the sea in a λινορραφής . . . δόμος . . . δορὸς "a linen-sewn abode of wood." The scholiast remarks that the phrase stands for "the ship, inasmuch as they used to bore vessels and sew them together with cords" (ἡ ναῦς, παρόσον τρυπῶντες τὰς ναῦς σπάρτοις αὐτὰς συνέρραπτον). Cf. *GOS* 199 and 211.

CHAPTER TWO

Egypt and Mesopotamia

THE CREATION OF the planked boat from the dugout as suggested in the previous chapter would have taken place by the shores of the Mediterranean or other great bodies of water, where trees were plentiful. Along the three great rivers, the scene of so much of man's pioneering efforts, Egypt's Nile and Mesopotamia's Tigris and Euphrates, special circumstances and needs resulted in a different line of development.

I EGYPT: THE PREDYNASTIC AGE

FROM Aswan at the First Cataract a boatman can ride the Nile's current without obstruction right to the river's mouth, 750 miles away. And, when ready to return, a prevailing wind that blows from the north will waft him back upstream. With so useful a means of communication at their disposal, it was only to be expected that the Egyptians would make swift strides in the development of waterborne transport.

The Valley of the Nile, as it happens, is short of timber. The one tree useful for boatbuilding is the acacia, and its hard brittle wood comes only in short lengths.[1] However, reeds—the famous bulrushes that sheltered the baby Moses—were to be had for the cutting all along the banks, and they were a perfectly adequate material for building craft to ride the river's placid waters. And so, the first vessels the Egyptians created were simple rafts of bundles of reeds lashed together.[2] By the second half of the fourth millennium B.C., they had learned to shape their rafts, making them long and slender and bringing them to a point at each end, had learned to propel them with paddles and to direct them with steering oars slung on the quarters.

[1] A. Lucas, *Ancient Egyptian Materials and Industries* (London 1962⁴) 429, 442; Hornell 215.
[2] Boreux 3.

11

In a word, the reed raft had been transformed into a reed boat. These earliest Egyptian craft are frequently pictured (Figs. 3, 5) but in so sketchy and stylized a fashion that we cannot be sure of the precise features.[3] They boast cabins and a goodly number of oarsmen.[4] In form some are bowed or sickle-shaped, some squarish with prow or stern or both rising in a high vertical line.[5] Most significant of all, some were powered by a sail, a square sail set well forward of amidships (Fig. 6).[6] To use the wind instead of muscle for driving a boat was an epoch-making idea; the Valley of the Nile, with its so very convenient prevailing wind, would seem the logical birthplace.

A reed boat, despite its name, is actually a kind of raft, a solid platform of bundles so lashed as to take the external form of a boat. As Egypt's boatwrights grew more adept, they advanced from the primitive long slender hull to a graceful spoonlike shape (Figs. 7-9) while the plain vertical extremities often gave way to elegant curved stem- and sternposts ending in an adornment resembling a lotus bud.

[3] See W. Flinders Petrie, *Prehistoric Egypt* (London 1920) pls. 19-22 (vase-paintings); H. Winkler, *Rock-Drawings of Southern Upper Egypt* I (London 1938) pls. 33-41. There is a selection of line drawings in Bowen 117-31. For a boat (2nd half of 4th millennium B.C.) painted on linen, see E. Scamuzzi, *Museo Egizio di Torino* (Turin 1964) pl. 5.

[4] Boreux 25. Boreux is not fully convinced that the deck structures are cabins; cf. his remarks on pp. 20-21.

[5] Since boats of the squarish type with high extremities appear in contemporary Mesopotamia (see 22 below), a controversy has long raged about their origin: did Egypt owe the introduction of such craft to Mesopotamia (or other peoples living to the east)? If so, were they brought in by traders or invaders? The argument for introduction from outside and probably by invasion has been developed at length by Boreux (28-56, 80-118; cf. his résumé on 517-18). It received new support from subsequently discovered rock-drawings (see Winkler, *op. cit.* note 3 above, 35-39) and is the prevailing view today. See, e.g., R. Barnett, "Early Shipping in the Near East," *Antiquity* 32 (1958) 220-30, esp. 222; A. Arkell, "Early Shipping in Egypt," *Antiquity* 33 (1959) 52-53; W. Emery, *Archaic Egypt* (Penguin 1961) 39; A. Gardiner, *Egypt of the Pharaohs* (Oxford 1961) 396; I. Edwards in *CAH* I², Chapter XI, section VI, "Southern Mesopotamia"; H. Müller-Karpe, *Handbuch der Vorgeschichte* II (Munich 1968) 279. It has by no means gone unchallenged; e.g., A. Scharff, "Die Frühkulturen Ägyptens und Mesopotamiens," *Der alte Orient* 41 (1941) 48, note 97, argues for a native Egyptian origin.

[6] The earliest examples to date are on pots of the late Gerzean period, ca. 3200-3100 B.C.; see Bowen 117-18. For reproductions in color, see *IH* ills. 14, 15. On a model from Mesopotamia which has been offered as evidence to support Mesopotamia's claim to the discovery, see 22 below.

Light, shallow in draft, easily maneuverable, such craft were particularly useful along Egypt's multitudinous canals and in the reed-choked marshes. With little change in form, they continued to be used throughout the whole of Egyptian history.[7]

Egypt's primitive reed raft had, however, another, far more significant, line of development. By about 2700 B.C., the Egyptians had developed a stone architecture. This meant that some form of cargo vessel was needed to ferry massive blocks of rock from the quarries to the building sites. Egyptian boatwrights had already equipped their fragile floats for harder use by paving them with a layer of planks.[8] They met the new challenge by taking the all-important forward step of building boats entirely of planks.[9]

Three features make it fairly clear that the Nile's first true boats were replicas in wood of her reed craft. First, in shape they often followed the lines of the reed boat, reproducing the latter's distinctive spoonlike design and often its distinctive lotus-bud ends (Fig. 10).[10] Second, they were often rigged with a bipod mast, one leg planted on each gunwale; this was surely a borrowing from reed boats, which, with no point strong enough to socket a pole mast properly, found the two-legged type much more serviceable.[11] Lastly, they paralleled the reed boat in construction: they had no internal frame of keel and ribs; they were a shell of planks edge-joined to

[7] Boreux 182-231. Reed canoes are pictured in a mosaic of a Nile scene dating from the 1st century B.C. (Fig. 116). On the possible method of construction, see A. Servin, "Constructions navales égyptiennes: Les barques de papyrus," *Annales du service des antiquités de l'Égypte* 48 (1948) 55-88. Servin argues cogently (82-88) against Boreux that papyrus rafts were not caulked, that the workers in canoebuilding scenes whom Boreux (184) claims are caulking are actually tightening the bindings of the bundles. Servin's theories about the origin of the sail (76-81) are curiously farfetched and take no account of the predynastic evidence.

[8] The platform of planks is easily discerned in the pictures of small reed boats used in hunting marsh fowl. We see the hunter standing with his feet planted on a solid line that, by its color and shape, is distinguished from the reed hull; cf., e.g., the color reproduction in *IH* ill. 12. This line represents the pavement of planking, short lengths laid from side to side and held in place by a raised border of braided reeds; cf. Servin, *op. cit.* (see previous note) 64-67, Boreux 229-30.

[9] Cf. A. Servin, "Les constructions navales sous l'ancien empire: Le navire en bois," *Annales du service des antiquités de l'Égypte* 43 (1943) 157-71, esp. 158.

[10] Reisner xviii-xix, Boreux 235, 272-75.   [11] Hornell 225-29.

13

each other, even as a reed boat is of bundles lashed to one another.

The earliest planked boats seem to have been square-ended and flat-bottomed (Fig. 12), more barge than boat, a form that might have been chosen because it involved simpler carpentry.[12] But boat-wrights learned quickly, and within a few centuries were giving prow and stern the traditional rounded or pointed shape and were rounding the bottom.[13]

For the mode of construction we have the best possible evidence, that of a number of actual hulls.[14] They date from a somewhat later period than the one under discussion—roughly 2000 B.C.—but this means little: the technique they reveal is identical with that Herodotus observed in the fifth century B.C.[15] and that still is in use along the Upper Nile;[16] we can confidently assume it reflects earliest Egyptian practice. Herodotus likened Egyptian boatbuilding, based on the use of short lengths of acacia, to the laying of bricks, and no more apt

[12] Reisner v-vi; C. Firth and B. Gunn, *Excavations at Saqqara: Teti Pyramid Cemeteries* (Cairo 1926) 33 and pl. 49 (11 models, of which 6 are square-ended); G. Jéquier, *Fouilles à Saqqarah. Les Pyramides des Reines Neit et Apouit* (Cairo 1933) 33-40 and pls. 33-35 (16 models, of which 10 are square-ended). All the above are Old Kingdom, i.e., roughly 2700-2100 B.C. Servin (*op. cit.* note 7 above, 71-74) suggests that certain crudely drawn pictures of tablets dating from Dynasties I-III (roughly 3100-2700 B.C.) show boats of this type.

[13] Reisner v-vi (vessels of his Type I with round bottom and rounded ends).

[14] There are two boats of the XII Dynasty found at Dahshur and now in the Cairo Museum; see Reisner nos. 4925 and 4926. They are modest rivercraft: one measures 10.20 m. long, 2.24 wide, .84 deep amidships (33½′ 7⅓, 2¾); the other measures 9.90 m. long, 2.28 wide, .74 deep amidships (32½′, 7½, 2½). A third is the impressive solar bark of Cheops (IV Dynasty) found dismantled alongside the Great Pyramid and now being assembled; see Mohammad Zaki Nour, Zaky Iskander, Mohammad Salah Osman, Ahmad Youssof Moustafa, *The Cheops Boats*, pt. 1 (Cairo 1960). The dimensions (p. 10) are 43.40 m. long, 5.53 wide, ca. 7.90 deep (142 3/8′, 18 1/8, 26).

[15] 2.96: ἐκ ταύτης ὦν τῆς ἀκάνθης κοψάμενοι ξύλα ὅσον τε διπήχεα πλινθηδὸν συντιθεῖσι, ναυπηγεύμενοι τρόπον τοιόνδε· περὶ γόμφους πυκνοὺς καὶ μακροὺς περιείρουσι τὰ διπήχεα ξύλα· ἐπεὰν δὲ τῷ τρόπῳ τούτῳ ναυπηγήσωνται, ζυγὰ ἐπιπολῆς τείνουσι αὐτῶν· νομεῦσι δὲ οὐδὲν χρέωνται· ἔσωθεν δὲ τὰς ἁρμονίας ἐν ὦν ἐπάκτωσαν τῇ βύβλῳ "From this acacia tree they cut planks 3 feet long, which they put together like courses of brick, building up the hull as follows: they join these 3-foot lengths together with long, close-set dowels; when they have built up a hull in this fashion [out of the planks], they stretch crossbeams over them. They use no ribs, and they caulk seams from the inside, using papyrus fibers." For a discussion of some puzzling points in this passage, see Boreux 236-39.

[16] Hornell 215-17.

14

comparison can be found (Figs. 11, 13). The boatwright began his hull with a sort of keel plank, a plank made of a number of lengths pieced together, which, in keel-like fashion, formed the centerline of the hull.[17] To either side of it he added short planks, all fastened edge to edge by means of dowels or mortises and tenons or wooden clamps or combinations of these.[18] When the shell of planks built up in this fashion had reached the height desired, he finished it off with gunwales and inserted a series of crossbeams at gunwale level;[19] these, besides carrying the deck planking, provided lateral stiffening, keeping the sides from sagging outward. Possibly, for further strength, there was a stringer from stem to stern just under the crossbeams supported by stanchions that stood on the keel plank.[20] Herodotus adds that the seams were caulked with papyrus from the inside; so are they still on some of the craft of the Upper Nile, save that rags take the place of papyrus fibers.[21]

This mode of construction should cause no surprise. We are so used to the modern western style of boatbuilding, which starts with a sturdy skeleton of keel and frames and wraps about this a skin of planks, that we tend to forget the existence in many areas of a method which does things in precisely the reverse order. In Arabia, Persia, India, and other parts of Asia, as well as in certain areas of northern Europe, shipwrights, like the ancient Egyptians, have traditionally

[17] In the larger of the two boats from Dahshur (Reisner no. 4925), the keel plank is of three lengths pieced together by means of dovetailed wooden clamps.

[18] In the solar bark the planking seems to have been joined with mortises and tenons; see *The Cheops Boat* (*op. cit.* note 14 above) 55. In the Dahshur boats mortises and tenons were used along with dovetailed wooden clamps; the latter were laid in shallow slots cut into the interior surface (Reisner no. 4925). The planks were 9 cm. (3½″) thick in the larger boat, 7 cm. (2¾″) in the smaller. A sample tenon measured 2 cm. broad, 1.5 deep, and .08 thick (13/16″, 9/16, 5/16); see S. Clarke, "Nile Boats and Other Matters," *Ancient Egypt* 5 (1920) 2-9 and 40-51, esp. 9, 41. How far apart they were spaced cannot be determined without taking the hulls completely apart.

[19] In the larger of the two boats from Dahshur there are 11, and they extend through the side planking, which seems to be a later practice not used before the Middle Kingdom; cf. 21 below.

[20] Absent in the Dahshur boats but possibly to be inferred from the markings on the numerous ancient models that have been preserved; see Boreux 296-98.

[21] Hornell 217.

begun with the skin of planks.[22] Having no frames to pin these to, they make each fast to its neighbors by means of pegs, joints, staples, nails, or—a technique touched on in the previous chapter—by sewing them together with twine.[23] Only after the shell has been completely built up in this fashion is any internal stiffening provided, generally a certain number of frames inserted as a last step. What the ancient Egyptians used was, in effect, just one variation of the technique that prevailed in the Asian world. It differed from that of Persia and Arabia by using wood joinery instead of sewing—and, as we shall see (202 below), was in this respect more akin to the practice in vogue throughout the Mediterranean. And, as the Mediterranean shipwright was to do during the whole of ancient history, they set the planks "carvel" fashion, i.e., edge to edge, never "lapstrake."[24]

II EGYPT: 3000-1000 B.C.

FROM Old Kingdom (ca. 2700 B.C.) times on, there is a wealth of evidence in the form of paintings, reliefs, and models. These show primarily the multifarious rivercraft of the Nile, but a few precious documents record what the Egyptians produced for use on open water.[25]

RIVERCRAFT

The graceful and shallow spoon-shaped hull of the earliest planked craft remained characteristic throughout the Pharaonic period (Fig. 10).[26] Vessels of any size were generally decked, often with movable planks amidships permitting use as an open boat when circumstances required.[27] Cabins took the form of deckhouses, most often of a frame

---

[22] See 201 below.

[23] See TEN, notes 2-4. [24] See 201, below and TEN, note 2.

[25] For ancient models, see Reisner. For ancient reliefs, see L. Klebs, *Die Reliefs des alten Reiches* (*Abh. der Heidelberger Akad. der Wiss.* Phil.-hist. Kl., 3. Abh., 1915) 100-107; *Die Reliefs und Malereien des mittleren Reiches* (6. Abh., 1922) 136-40; *Die Reliefs und Malereien des neuen Reiches* (9. Abh., 1934) 193-206.

[26] Reisner x, xviii-xix; Boreux 272-75.

[27] Reisner x, Boreux 308-309.

covered with mats.[28] On passenger craft these could extend almost from stem to stern (Fig. 19); cargo boats generally made do with a shelter cabin aft.[29] From early in the third millennium B.C. on, Egypt imported excellent ship timber from Syria (where the famed cedars of Lebanon then grew in profusion) and Asia Minor;[30] still using their traditional technique,[31] they now turned out vessels running up to 150 feet in length and requiring a rowing complement of 40 or better.[32] A thousand years later they were able to build some of the largest vessels known from the ancient world, the brobdingnagian barges that hauled obelisks downriver from the quarries at Aswan (Fig. 14), vessels that must have measured well over 200 feet long and 70 wide and needed a fleet of 30 oar-powered tugs to tow them. Such craft, to be sure, had to be stiffened internally by the insertion of frames and multiple levels of crossbeams and by the addition of rope trusses.[33]

Egyptian man-driven rivercraft were paddled, rowed, or poled. For paddling, a strenuous stroke was used in which the paddler first

[28] Reisner viii, xv, xxi.　　　　　[29] Boreux 258-61.

[30] "Bringing forty ships filled (with) cedar logs" is listed among the accomplishments of Pharaoh Snefru, ca. 2600 B.C.; see J. Pritchard, *Ancient Near Eastern Texts* (Princeton 1955²) 227. For other sources of foreign timber, see Lucas (*op. cit.* note 1 above) 429-39. For woods identified in Cheops' solar bark, see *The Cheops Boats* (note 14 above) 45-46; some were local, some came from Asia Minor and Syria.

[31] The earliest pictorial evidence is a relief of ca. 2400 B.C. (Klebs, *Alten Reiches* [note 25 above] 102, fig. 84 = *IH* ill. 21). In one scene workmen hammer down a plank, driving the joints home. The boat seems to be about 7 or so meters long—some 23 feet—and the plank is a solid timber extending almost the whole length; with such good material at their disposal, the boatwrights were spared much piecing. Some workmen are shown adzing smooth the finished underbody, and Hornell (48) has suggested that this part is a dugout to which the plank is being added. I take the whole hull to be made out of planks; the men may be smoothing some seams which had not come out perfectly evenly.

[32] Records of ca. 2600 B.C. mention ships 100 cubits (ca. 150 feet) long; see J. Breasted, *Ancient Records of Egypt* I (Chicago 1906) §146, 147. "The Tale of the Shipwrecked Sailor" refers to a vessel 120 cubits long and 40 wide and manned by a crew of 120; see A. Erman, *The Literature of the Ancient Egyptians* (London 1927) 30. The largest riverboats pictured seem to have 40 or so rowers or paddlers (Reisner xvi; cf. Boreux 315-17).

[33] See the articles by C. Sølver and G. Ballard in *MM* 26 (1940) 237-56; 27 (1941) 290-306; 33 (1947) 39-43, 158-64. Sølver reckons that one barge, carrying some 750 tons, had three levels of crossbeams with vertical stanchions between to distribute the weight, and an intricate series of antisagging trusses (on these, see 20 below).

raised the blade high over the head and then leaned over the side
to reach the water; moreover, the stroke was not simultaneous but
progressive, each man dipping his blade a fraction of a second behind
the man in front (Fig. 15).[34] By about the V Dynasty (ca. 2400 B.C.),
oars had replaced paddles on all larger craft.[35] Oarlocks were simply
loops of rope passed through holes in the gunwale.[36] The rowing
stroke was as strenuous as that used in paddling. Since the oars were
relatively short and were worked at a steep angle, the oarsmen, like
those who, millennia later, pulled the great sweeps of the galleys of
France or Italy in the sixteenth to eighteenth centuries, had to rise
to their feet to dip the blade in the water and delivered the stroke
by falling back on the bench (rowers had a square patch of stout
leather sewn to the seat of their kilts as chafing gear).[37] Steering
from the outset was taken care of by one or more steering oars, depending on the size of the craft, slung on the quarters (Fig. 19).
From about the V Dynasty on, tiller bars, socketed into the loom
of the steering oars, made the helmsman's job vastly easier.[38] On
smaller vessels the oars on the quarters were sometimes replaced by
a single oar fixed on a vertical pole in the center of the counter or a
pair set on vertical poles on the quarters; beginning at the end of the
Old Kingdom, the single oar at the stern became increasingly common.[39]

Some boats, both passenger and cargo, were equipped only to be
rowed or poled or towed; others carried sail as well, or only sail.[40]

[34] W. Smith, *The Art and Architecture of Ancient Egypt* (Penguin 1958) fig. 32.
[35] Reisner viii, Boreux 314.
[36] Reisner xiv, Boreux 338-39.
[37] Boreux 319-24 and, for the seat-patch, T. Säve-Söderbergh, *The Navy of the Eighteenth Egyptian Dynasty* (Uppsala Universitets Årsskrift 1946:6, Uppsala 1946) 75-76. For an illustration, see N. Davies and A. Gardiner, *The Tomb of Huy* (London 1926) pl. 18.
[38] Boreux 399-400.
[39] Reisner vii-viii, xiii, xx; Boreux 387-403, 494-96. For good examples, see the plates cited in note 50 below. A single sweep mounted in this fashion on the stern is attested elsewhere; see Barnett-Forman cited in App. 2 (Assyria), FOURTEEN, note 3 and Fig. 189 (Sarre or Moselle).
[40] And sailing craft, when there was little wind or when in the stretch between Coptos and Abydos, which is oriented roughly east-west, then as now had to be

18

The earliest type of mast was perhaps the pole mast. Two-legged masts (Fig. 19), better suited for a reed hull (13 above) soon came into use, and the two types existed side by side until the end of the VI Dynasty (ca. 2200 B.C.);[41] thereafter the pole mast took over completely.[42] Down to the end of the third millennium B.C., masts were lofty—the highest, relative to length of hull, in all antiquity—carrying an immensely tall oblong sail (Fig. 19);[43] such a rig, able to catch upper breezes, makes excellent sense on the Nile, which in many places flows between high-rising cliffs.[44] Mast and sail, despite their height, were made to be unstepped and carried on deck when not in use (Fig. 17).[45] The mast was supported by multiple lines running aft which served as both shrouds and backstays. By the V Dynasty (2400 B.C.) forestays make their appearance.[46] At the same time, the position of the mast shifted: well forward of amidships in earlier times, it gradually moved aft until, by ca. 1500 B.C., it reached the center of the vessel, where it stayed; the change reflects the Egyptians' increasing ability to sail on winds from other directions than dead aft.[47] The sail was hung from a yard and, during the whole of the period under discussion, was stretched by a boom along the foot (Figs. 16, 19)—the only sail in antiquity to have this feature. Sails were generally of papyrus. Apparently this material was too frail unaided to carry the weight of the boom, which must have been a heavy spar since men are often shown standing on it (Figs. 18, 57), for the booms invariably hang from a network of lifts. The boom remained fixed on these lifts, and sail was raised or furled by hoisting or lowering the yard (Figs. 16, 18, 57). A single sheet at the clew was not enough to control so tall a sail, so multiple sheets were

---

towed upstream; cf. Herodotus 2.96: ταῦτα τὰ πλοῖα ἀνὰ μὲν τὸν ποταμὸν οὐ δύναται πλέειν, ἢν μὴ λαμπρὸς ἄνεμος ἐπέχῃ, ἐκ γῆς δὲ παρέλκεται "These boats [the standard] Egyptian rivercraft] cannot sail upstream without a good steady wind, so they are towed from the bank." See App. 2.

[41] Reisner vii, xx.   [42] Boreux 348.   [43] Boreux 349.

[44] The lateens of modern Nile craft are tall and narrow for the same reason; see *IH* ill. 234.

[45] *IH* ills. 16, 18.   [46] Boreux 357-58.   [47] Bowen 127-31.

used, made fast to the leech.[48] All lines were handled by sheer muscle, since ancient Egypt did not know the use of the block and tackle; halyards, for example, simply passed through a hole near the mast truck. Lines were of palm fiber, papyrus, or various grass fibers.[49] The tall narrow sail must have proved clumsy to handle for, by about 2000 B.C., it was replaced by its opposite, a low wide type (Fig. 16); this was so rigged, however, that it could be carried quite high above the deck.[50]

SEAGOING SHIPS

Information about Egypt's seagoing ships is far scantier than about her rivercraft. The first solid evidence is a carefully done relief of about 2450 B.C. (Fig. 17) portraying the return of an overseas military expedition. The hulls depicted, long and slender and with considerable overhang fore and aft, reflect the spoonlike shape of the rivercraft. They were built in the same manner, i.e., of edge-joined planks with a minimum of internal stiffening, for, to fit such a lightly built hull for use on the open water, Egyptian shipwrights had to add two distinctive features. The first was a rope truss—a heavy cable which, starting with a loop about the bows, was carried the length of the deck on upright forked sticks to the poop, and ended in a loop about the stern; a lever thrust between the strands enabled the crew to twist the cable like a tourniquet, and, by twisting and twisting, reach just the tension to keep the overhanging bow and stern from sagging. The second feature was a narrow rope netting that girdled the upper part of the hull at deck level. It probably served to keep the planking there from starting under the pressure

[48] Boreux 385-87 and fig. 158. They are not bowlines, as Boreux calls them, for bowlines are to hold the leading edge of a sail flat when sailing into the wind, and Egyptian craft could only sail before the wind. Similar lines are found, e.g., on Japanese junks of the last century, which also carried a tall oblong sail; see *IH* ill. 200.

[49] Lucas (*op. cit.* note 1 above) 134-36. The cordage of Cheops' solar bark was of halfa grass (*The Cheops Boat* [note 14 above] 42-45).

[50] See also *IH* ill. 22; H. Winlock, *Models of Daily Life in Ancient Egypt from the Tomb of Meket-Rē' at Thebes* (Cambridge, Mass. 1955) pls. 33, 34, 42.

exerted by the deck beams.[51] The rig, again like contemporary river-craft, consisted of a two-legged mast set forward of amidships, with a tall narrow square sail; despite its size it was unstepped when not needed. Steering was done by triple steering oars on the quarters.

The next evidence is a relief (Fig. 18) that dates almost a thousand years later. It portrays the fleet Queen Hatshepsut sent down the Red Sea to reopen trade with the east African coast. The hulls, of the same general shape as those in the earlier relief, have cleaner and more graceful lines. There is a foredeck and afterdeck, both protected by screens, and decking over the centerline of the vessel; along the gunwales an open space has been left to allow the rowers to sit to their oars. The stempost is straight, and the sternpost smoothly curved and finished off in the shape of a lotus bud. An important innovation is the extending of the beams through the planking on the sides[52] thereby increasing the lateral strength and enabling these ships to do without the girdle of rope netting. They still have, however, the rope truss to support their long overhangs. As on the river-craft of the period, the tall narrow sail has been replaced by a low wide sail hung on a pole mast set amidships. The width is such that the yards are made of two tapering spars with their thick ends fished together. The shorter mast has far fewer stays than its predecessor— two forestays, one made fast to either bow, and a single backstay— but the lifts that support the boom are as numerous as ever. The yard is raised by twin halyards, which, being secured to the quarters, reinforce the backstay. Neither on these nor any earlier vessel is there any visible gear for shortening sail; perhaps this was done by stripping off the working sail and bending a smaller one, as was the practice on lateen-rigged galleys three millennia later.[53] The multiple

---

[51] Cf. R. Faulkner, "Egyptian Seagoing Ships," *JEA* 26 (1940) 3-9, esp. 5. Boreux (299-304) argues cogently that the netting is not the artist's way of indicating the stitching on a sewn boat, as some have interpreted it (cf. Hornell 220). However, his own suggestion (304-306, 470) that it represents a lacing for detachable bulwarks is not convincing.

[52] Cf. note 19 above. [53] Cf. Lane 22-23.

steering oars on each quarter have given way to a single one of massive size. Fifteen rowers a side are portrayed; since these would occupy a space of about 45 feet, the ship, with its ample fore and poop decks, may have been some 90 feet long.

Eventually, even tradition-bound Egypt introduced radical changes in both hull and rig. These are visible in a relief (Fig. 61) showing the vessels she used to counter an amphibious attack on the Nile delta about 1200 B.C. (36-37 below). From that date on, Egypt ceases to have an independent tradition and joins the Mediterranean mainstream.

III MESOPOTAMIA

THE earliest evidence of Mesopotamia's watercraft so far discovered is a clay model (Fig. 20) of a bowl-like boat dating to about 3400 B.C. To judge from the shape, the craft may have been some form of skin boat; it closely resembles a *quffa* (6 above) in all but the slight elongation that gives it an identifiable prow and stern. A socket in the floor has been explained as being intended for a mast, and three holes at gunwale level for a stay and shrouds.[54] This interpretation is by no means certain, particularly since sails are otherwise unattested in Mesopotamia until much later. The socket could very well have been to hold some ceremonial staff, and the holes for cords by which the model could be suspended.[55]

The only other examples of boats from this early period are small, sketchy representations on seals. These show vessels very much like some of the early Egyptian reed craft, squarish with distinctive high prow and stern.[56] It has been suggested that they were made of wood,[57] but this is far from certain. In the rare instances where details are given, the hulls are marked by frequent parallel oblique

[54] S. Lloyd and F. Safar, "Eridu," *Sumer* 4 (1948) 118 and pl. v (reproduced in *Antiquity* 32, 1958, pl. 21b) shows it with reconstructed mast and stays.

[55] Cf. Bowen 124.

[56] E.g., H. Frankfort, *Cylinder Seals* (London 1939) pls. 3, 11, 19, 20, 24; Salonen, *Wasserfahrzeuge* pls. 3-5. For the possible relation between these and the Egyptian craft resembling them, see note 5 above.

[57] Hornell 49-50.

and vertical lines (Fig. 21), which seem most easily interpreted as the lashings of reed bundles; and reed craft are amply attested in later times, particularly in the marshes of the delta area.[58] None are shown with sails.[59]

Such light craft, propelled by paddles or oars or punts, along with *keleks* and *quffas* (4-6 above), very likely were enough to serve a good part of Mesopotamia's needs, since the Tigris and Euphrates were by no means as useful as Egypt's Nile. Rapids in the upper reaches, shallows, above all a prevailing wind from the north, which meant that towing was the only way to get boats upstream,[60] all militated against the introduction of shipping of any great size. By the second half of the third millennium B.C., local boatwrights were making craft of wood rigged with sail as well as oars, but these were of no great size. The largest class carried no more than 11 tons of cargo, and the normal class only half that much. It has been asserted that they were made with an elaborate pre-erected skeleton, but this is based on a very hazardous interpretation of the evidence; the indications are, rather, that, like craft in contemporary Egypt, they were of edge-joined planks with some inserted framing.[61] There is even reason to think that, like the earliest Egyptian planked boats, they were square-ended.[62] They carried a single sail, no doubt a square sail, apparently of cloth, possibly at times of reed matting.[63] The largest rowing complement recorded included 11 oarsmen.[64]

Contemporary records prove without a doubt that during the third millennium B.C. Babylon carried on extensive overseas trade through the Persian Gulf southward to the east African coast and eastward

---

[58] Salonen, *Wasserfahrzeuge* 68, 70. Reed floats very much like the Egyptian type appear frequently in reliefs (early 7th B.C.) depicting an Assyrian campaign carried out in the marshes of the delta; see A. Paterson, *Assyrian Sculptures. Palace of Sinacherib* (The Hague 1915) pls. 92, 93; *IH* ill. 6.

[59] Salonen repeatedly offers (*Wasserfahrzeuge* 110, 113, 116) a seal of ca. 3200 B.C. as evidence for an early Mesopotamian sail, taking a pair of enigmatic lines as representing a mast and yard. Whatever they stand for (ceremonial pole? punting pole and oar?) it is clearly not a mast and yard; cf. Bowen 124, note 28.

[60] See App. 2.   [61] See App. 1, pt. 1.   [62] See App. 1, pt. 2.

[63] Salonen, *Wasserfahrzeuge* 115-16.

[64] Salonen 1942, 48.

to India. Hardly anything is known about the vessels used on these ambitious runs other than that they were very small; the largest mentioned has a capacity of some 28 tons.[65]

[65] Salonen, *Wasserfahrzeuge* 53-55; *AM* 5-9, 251-52. A "seagoing boat" of 300 gur is mentioned in a document of ca. 2000 B.C.; see A. Oppenheim, "The Seafaring Merchants of Ur," *Journal of the American Oriental Society* 74 (1954) 6-17, esp. 8, note 8. For the size of the gur, see App. 1, note 5.

APPENDIX 1

RIVERBOATS OF MESOPOTAMIA

PART 1: CONSTRUCTION

ARMAS SALONEN, specialist in ancient Near Eastern languages, has published three important studies in which he extracted all the nautical information he could from Sumerian, Akkadian, and Assyrian sources.[1] In these he describes the craft used on the Tigris and Euphrates during the first two and a half millennia B.C. as being built with an elaborate inner structure made up of members that he identifies as keel, close-set frames, and supplementary floor timbers.[2] Careful restudy of these identifications in the light of what we now know of ancient shipbuilding quickly reveals, even to one without special competence in the languages, that they are almost totally based on a most questionable assumption, and, once this is removed, are left with scant independent support.

In interpreting his material, Salonen let himself be guided throughout by the boatbuilding practices of modern Iraq, taking it for granted that the tradition in the area continued from ancient times in an unbroken line. Today an Iraqi boatwright, like his European contemporary, starts with an elaborate skeleton of keel and frames. That this was always so, however, is extremely doubtful. Such a procedure cannot be proved until after the arrival of the Portuguese in India at the end of the fifteenth century; before that time, whatever information we have points to the use of the shell-of-planks technique.[3] Egypt offers an illuminating parallel. There the shell-of-planks technique maintained itself without a break from earliest antiquity to the present all along the Upper Nile; however, to the north of Aswan, the area that felt western influence the earliest and strongest, it was replaced by the skeleton-first technique, introduced presumably from Europe.[4] The Tigris and Euphrates must have shared the fate of the Lower Nile.

The crucial evidence that Salonen interprets in the light of modern

---

[1] Salonen 1938, 1939, 1942.

[2] The identifications are presented in detail in Salonen 1938, 16-21, and listed, with references, in Salonen 1939, 83-88.

[3] Cf. Hornell 229-41.

[4] Hornell, unaware of what we now know of Greco-Roman shipbuilding practices (202-204 below), assumed (223) that the use of a pre-erected skeleton came in with the Greek occupation of Egypt. It must have come in over a millennium later at the earliest.

analogies are three technical terms which occur in inventories of materials for building boats of various classes, most importantly the largest-sized class, which ran some 11 tons in burden (120-gur vessel),[5] and the standard-sized class of half that (60-gur vessel). The terms are as follows:[6]

me-re-za (meaning of elements doubtful)	195 units listed for a boat of the largest class, 150 for a boat of the standard class
eme-sig (eme "tongue" + sig "lower")	180 units listed for a boat of the largest class, 138 for a boat of the standard class
dubbin ("fingernail," "claw")	40 units listed for a boat of the largest class, 20 for the standard; listings in conjunction with the preceding term

The first term he identified as "upper ribs," assigning 98 to each side of a ship of the largest class—that the total number of 195 as given in the sources was not divisible by two he brushed aside not only in this instance but in several others as well.[7] The second term, since one of its elements means "lower," he identified as "lower ribs," and reckoned 90 to a side. His identification in both cases was largely based on the quantities listed for the terms: since there were far more of them than of any other structural element, Salonen equated them with the nearest counterpart he could find in a modern Iraqi riverboat, the ribs. The third term, since it is found in conjunction with the second, the so-called lower ribs, he identified as supplementary floor timbers and reckoned from their number that they were intercalated every fifth frame or so. Salonen went so far as to take the distance between frames on modern Iraqi boats—12-20 cms.—and, on the basis of this, to estimate the length of ancient craft; e.g., he arrived at the 14 meters he gives for the largest class of boat by multiplying the 90 "lower ribs" by 16 cms., the latter representing the average between 12 and 20.[8]

[5] Since the *gur* in the period under discussion contained ca. 121 liters (Salonen 1938, 11), a 120-gur vessel had a capacity of 14,520 liters, or some 11 tons of grain (a liter of grain weighs about 1.7 lbs).

[6] See Salonen 1938, 16-19; 1939, 85-88.

[7] E.g., the number given for the "lower ribs" of a 10-gur vessel is 45, which he reckons as 22 a side (1938, 16); the number given for the "floor timbers" of a 30-gur vessel is 15, which he reckons as 7 a side (1938, 19).

[8] Salonen 1938, 17; cf. 1939, 155.

All the above, it is immediately apparent, is not merely guesswork based on a doubtful assumption, but guesswork that takes surprisingly little cognizance of the only real clues we have—the prime meaning of the terms. The prime meaning of the word *dubbin*, which Salonen renders "floor timbers," is "fingernail," "claw";[9] the determinative sign accompanying it shows that it is an object of wood. If we keep in mind that the craft we are dealing with were most likely made of edge-joined planks, a "claw" of wood suggests not floor timbers but rather something like the dovetailed clamps that hold together the planks of the Nile boats of 2000 B.C. found at Dahshur (15 above). And the word *eme-sig*, occurring regularly in conjunction with *dubbin* and meaning literally "lower tongue,"[10] suggests not "lower ribs" but rather the tonguelike tenon of a mortise and tenon joint, which, as in the Dahshur boats, can be used along with dovetailed clamps. The "tongues" are "lower" possibly because the boatwrights' procedure in building up the shell of planks was to keep the tenons as the lower part of the joint. The quantities offer no difficulty. The largest class of boat, one with a hold that perhaps measured in the neighborhood of 7m. x 2 x 1,[11] was made up of some 59 to 66 planks (see below). These could easily be edge-joined with 180 mortise-and-tenons and 40 dovetailed clamps, along with a generous number of treenails.[12] There were 6 crossbeams, a number that correlates rather well with that found in the Dahshur boats.[13]

In any event, whether the suggestions given above have validity or not, Salonen's equations must be abandoned. With them disappears any evidence from ancient Mesopotamia for the building of hulls with a pre-erected inner structure. Documents other than the inventories of materials indicate that Mesopotamian rivercraft had a "backbone" and "ribs."[14] This I take to mean that they had a prominent keel plank, as in ancient and some modern Nile boats,[15] and a limited number of inserted frames.[16]

[9] A. Deimel, *Sumerisches Lexicon* (Rome 1925-50) III.1.77.

[10] Deimel III.1.91 and III.2.440 (s.v. *šaplu*).

[11] 14,520 liters (see note 5 above) = ca. 14½ cubic meters.

[12] The inventories allot 7,200 treenails for the overall construction of a boat of the largest size (Salonen 1938, 20-21).

[13] Salonen 1939, 94-95 (the standard size had 3-4, the 30-gur size 2, and the 10-gur size 1-2). The Dahshur boat, 10.20 m. in length, had 11 (15 above and Two, note 19).

[14] Salonen 1939, 83-84.

[15] Cf. Hornell 216, 221; Reisner, no. 4925.

[16] Cf. Hornell 221-23.

PART 2: SHAPE

Salonen demonstrates[17] that the number of planks used for vessels of the various sizes was as follows:

Size of Ship	Side Planks	Bow Planks	Stern Planks
largest (120-gur)	43, 46	12, 8	8
normal (60-gur)	26-28	8	8
small (30-gur)	21	8	8
smallest (10-gur)	15, 16	6	4

The nomenclature—"bow planks," "stern planks"—and the quantities give the strong impression that these craft were made with square bow and stern. Such a shape, as we have seen (14 above), is well attested in Egypt of the third millennium B.C.

[17] Salonen 1938, 13-14; 1939, 89-90. He makes the reasonable suggestion that the "water-striking" planks are those for the bow and the "water-limiting" planks those for the stern.

APPENDIX 2

TOWING

THE DOCUMENTS studied by Salonen provide some rare and welcome details about towing (Salonen 1939, 117-19; 1942, 45-46).

The boats, which vary in size from 1 ton burden to 11, were pulled by crews of 2 to 18 men; there is no occurrence of the use of animals. The standard-sized boat of 6 tons averaged 9-10 kms. a day. Big teams of 18 could make 20 kms. a day. Downstream, 30-35 kms. was possible.

There are a number of representations of towing teams at work. For Egypt, see Davies and Gardiner, *loc. cit.* (Two, note 37); cf. Klebs, *Neuen Reiches* (Two, note 25) 206 and fig. 129. For Mesopotamia, see L. W. King, *Bronze Reliefs from the Gates of Shalmaneser* (London 1915) pl. 63 = Salonen 1939, pl. 16.2 (9th B.C.); E. Budge, *op. cit.* (ONE, note 9) pl. 22.1 = R. Barnett and W. Forman, *Assyrian Palace Reliefs* (London 1960) no. 20 (9th B.C.). Towing teams still operate today: for Egypt, see *National Geographic* (Dec. 1966) 776; for Mesopotamia, see Salonen 1939, pl. 34.2, 3.

CHAPTER THREE

The Eastern Mediterranean:
3000-1000 B.C.

THE MAINSTREAM of nautical development that was to flow
throughout ancient history arose not in the river-oriented civili-
zations of Mesopotamia or Egypt, but in the open waters of the eastern Mediterranean: on Crete, on the Aegean islands, along the coasts
of Greece.

The Minoans of Crete, whose heyday spanned, in round numbers,
the half millennium between 2000 and 1500 B.C., included among their
many accomplishments an impressive navy and merchant marine;
the first can be deduced from the total absence of fortifications about
their cities, and the second is attested by the abundant archaeological
traces of their contact with surrounding lands. Their hegemony was
eventually challenged by another race of skilled seafarers, the Mycenaeans or Aegeans, as the Greeks of the Bronze Age are generally
called. Wresting control of the waters from Crete, they lorded it over
the eastern Mediterranean from 1500 to 1200. Further south, Egypt
and the Levant carried on a thriving seaborne commerce. It is a great
misfortune that, for the ships involved in this widespread and in-
tense maritime activity, we have nothing better to go on than simplified clay models, tiny engravings on seals, crude graffiti, and a handful of vase-paintings.[1]

I THE AEGEAN

FROM the island of Syros comes a batch of terra-cotta mirrorlike
objects of the third millennium B.C. incised on which are the earliest
ship-pictures of the age (Fig. 22), sketches in profile of galleys propelled by multiple oars. The hull is slender, straight, and low; join-

[1] For the evidence, see App. and *Archaeological Reports* (1965-66) 44, fig. 1. The
lead models listed there are surely of dugouts.

ing it at a sharp, almost right, angle is a narrow and high-rising stem-post bearing at its top a fish-shaped device; the stern, finished off equally sharply with apparently nothing more than a vertical tran-som, has a needlelike projection at waterline level. The representa-tions are too sketchy to show steering oars, but such must have been the steering arrangement, since the steering oar appears on a con-temporary fragmentary painting (Fig. 46). And, though the draw-ing is too primitive to inspire faith in the exact number of oars shown, the clear implication is that there was a good number. We see, in effect, a sizable swift galley whose shape is particularly distinguished by the absence of curves. Its descent from a dugout seems beyond question, and this is just what we would expect in an area as well supplied with timber as the Aegean was during the Bronze Age.[2]

These craft reappear on a series of graffiti (Figs. 24, 25, 27) that span the second millennium B.C.: on a graffito found on Malta and dated ca. 1600, on a second of perhaps slightly later date from the east coast of Greece,[3] then on a third from Cyprus, dated between 1200 and 1100. The last, the graffito from Cyprus, reproduces every feature of the Syros ships down to the projection at the stern. It also includes a sail—presumably the mast was stepped and sail stowed away in all the other representations—which is shown bellying to-ward the high end. This settles once and for all a long-standing argu-ment about which end of these ships was the prow (see App.). A seal found on Crete, dating ca. 1400 B.C. (Fig. 41), i.e., only slightly later than the two earlier graffiti, pictures the same kind of hull and also provides some precious indication of size: ten oars are shown, the same number carried by the dispatch and transport galleys of the Homeric Age (44 below). Contemporary with the last graffito are two vase-paintings which make it appear that the Mycenaeans, lords of the Mediterranean for the previous three centuries, had by this time introduced some significant improvements. One (Fig. 28), a

[2] Loc. cit. previous note.

[3] For the date of ca. 1500 B.C. for this graffito, see Vermeule (op. cit. App.) 259 and 346, note 6.

painting on a Mycenaean clay box found at Pylos on the west coast
of Greece, shows the hull fitted with a short sternpost to add some
height aft and with bulwarks to screen the bows and quarters. The
other (Fig. 29), on a Mycenaean vase from Asine on the east coast
of Greece, gives the lofty stempost, which up to now had remained
stubbornly straight, a curve that seems a definite step along the road
to the bowed stemposts of the Greek galleys of the next age (Fig.
30; cf. 49 below). In a word, the glimpse we catch of the Aegean
galley through these crude representations raises the distinct possibility that, with its straight lines and angled ends, it was the prototype of the warships that would become standard in the fleets of
Greece and Rome.

The graffito at Malta mentioned above was found in company with
several dozen others, most of which are too sketchy to yield significant information.[4] One group of three (Fig. 31), however, share the
basic characteristics of the Aegean galley: they have a straight low
hull with a high stem rising from it at a distinct angle. Shorter and
higher than the galleys, they give more the impression of merchant
craft than of warships. A Greek vase-painting of several centuries
later,[5] the end of the second millennium B.C., shows very much the
same lines. And we shall see that the Greek merchantman of the
next period (Fig. 33; cf. 66 below) reflects their minimally curved
lines rather than those of the only other type of merchant vessel
known from this age, a bowl-shaped variety to be discussed shortly.

II CRETE

WE know Cretan ships almost wholly from tiny and often
stylized portrayals on seals. Despite the lack of detail, one point
seems fairly clear: the island's shipwrights went in chiefly for rounded hulls distinguishable at a glance from the straight-lined, angular-

---

[4] E.g., Woolner's "basic type," to which a majority of the identifiable sketches
belong, might be a careless drawing of either an angular or a curved hull.

[5] See Vermeule, cited in App.

ended Aegean versions.[6] In the earliest period, before 1600 B.C. or so, the prow alone was rounded (and finished off with a three-pronged or arrow-shaped device), and the stern was given an appendage or bifurcation (Figs. 34-36). This last feature is a puzzle, whose solution will have to wait until more evidence turns up. With the passage of time, as we shall see in a moment, both ends came to be rounded.

The Cretan vessel we can be surest of, since it is portrayed on a number of carefully engraved seals (Figs. 37-40), dates from shortly after the middle of the second millennium B.C. It has a slender rounded hull, with identical or nearly identical prow and stern, both devoid of any ornamental device. The rig, without exception, consists of a ponderous pole mast supported by stays fore and aft and carrying, relatively high, a single broad square sail. The latter appears to have a boom along the foot in the Egyptian fashion; this is perhaps deceptive, an effect of the stylization that characterizes these engravings.[7] Other carefully done seals of the same age show a hull so rounded that it seems almost crescent-shaped (Figs. 47, 48). Again, stem and stern are generally undecorated, although one end (Fig. 48) occasionally is finished off with an ornamental bifurcation. The ends can be of equal height (Fig. 47) or the stern slightly higher (Fig. 50).

One particular series of seals (Figs. 42-45), all of about the same age—ca. 1400 B.C.—and make, show a type of vessel very much like the above with two key differences. First, the prow consistently ends in a pronged ornament; this is a traditional Cretan device, for it appears on ships of 1600 B.C. and earlier (Figs. 34-35) as well as on the schematized ship-symbol that appears as a writing sign in tablets

---

[6] Some (e.g., Figs. 36 and 41) have angular lines, but they are very few compared with the rounded hulls.

[7] When Marinatos wrote, only one example was known, and this led him to suggest (216) that the ship portrayed was actually Egyptian. Cf. Xénaki-Sakellariou (*op. cit.* App.) 79-80, who is more cautious. The seal-pictures show no great affinity with Hatshepsut's vessels—and these are the sole undisputed representations of seagoing Egyptian ships extant (on the seagoing vessels pictured in certain Egyptian tombs, see 35-36 below).

of 1400 B.C. and later.[8] The stern is never shown in full so that, although it is clearly rounded like the prow, we know nothing about its decoration. Secondly, these vessels, instead of a mast and square sail, have some apparatus which the artist represents by two, occasionally three, poles linked by cross-hatching. Sometimes there is no cross-hatching, and the poles then seem to have whatever the cross-hatching indicates wrapped about them and lashed into place (Figs. 44-45). To interpret these ships as two- and three-masted, as is so often done,[9] cannot be right. Aside from the intrinsic unlikeliness of such rigs at this time—the rest of the evidence points to single masts only—other, unambiguous, seal-pictures show how the Cretan artist drew a mast and sail (Figs. 37-40). Perhaps the enigmatic uprights and what lies between them are a stylized representation of deck cargo; perhaps they are symbols of some sort, like the half moons that so often appear in connection with the uprights (Fig. 42).[10]

The only clue we have for the size of Cretan vessels is the number of oars shown on the few pictures done carefully enough to merit respect for what details are given. Two, for example, show 10 on a side (Figs. 40-41) and another 15 (Fig. 47); that the numbers are not mere artistic fancy is proved by the appearance of galleys with precisely this many rowers as standard types among the Greek craft of the next age. Ships so oared would have a length of ca. 50 and 75 feet respectively (55 below). Larger merchant vessels were decked (Figs. 52-53) and passenger vessels could carry a cabin or shelter upon the deck (Fig. 50).

The clay models that have been preserved, from Cyprus and Melos and other islands as well as Crete, seem to show small craft. The

[8] Evans-Myres, *Scripta Minoa* II (Oxford 1952) 22, no. AB 64 = Ventris' no. *86 appearing in, e.g., Document Dc 1117 (Bennett-Chadwick-Ventris, *The Knossus Tablets²*, University of London, Institute of Classical Studies, Bulletin Suppl. no. 7, 1959, p. 23).

[9] See App.

[10] Cf. Kenna, *Cretan Seals* (cf. App.) no. 241. If nonsymbolic, then L. Delaporte's suggestion for one such seal (*Musée du Louvre, Catalogue des cylindres* I, Paris 1920, 93, no. 17: "barque dans laquelle sont dressées deux nasses") is as good as any.

most significant are a few (Fig. 54; cf. 55) which have a distinct projection where the stempost joins keel (occasionally where stern-post joins keel as well; see Fig. 54 and cf. Fig. 23).[11] This projecting forefoot, evidenced as early as the third millennium B.C., is a feature that will be found during the whole of antiquity, on seagoing craft as well as small boats. Its reason for being is unclear. A bifid stem, which leaves a very similar projection at the waterline, is characteristic of many forms of primitive craft, skin boats, dugouts, and even planked boats.[12] Shipwrights, who are as conservative as seamen, may simply have perpetuated it as a traditional feature.[13]

III THE LEVANT

An invaluable wall-painting in an Egyptian tomb of about 1400 B.C.[14] shows a group of merchant vessels (Fig. 57) arriving, as the context makes clear, from Syria. They are manned by hands who, with their full beards and hooked profiles and gaily embroidered ankle-length robes, are unquestionably Syrian. The ships, however, have a disturbing Egyptian look about them: their well-rounded but spoon-shaped hulls, prominent overhangs fore and aft, straight stem-posts, deck beams brought through the sides, broad square sail with boom along the foot, are all features found in Hatshepsut's vessels (21 above). Yet, there is one key difference: Hatshepsut's were braced with a rope truss; these lack it, implying that the strength needed for sailing over open water was supplied structurally. Moreover, another tomb-painting (Fig. 58) of a very similar vessel, also Syrian, shows a sail that, with a downward curving yard, is distinctly unlike the Egyptian sails attested hitherto, which have either straight or upward curving yards.[15] These tomb-paintings portray either a

[11] For another example, see Marinatos no. 27.

[12] Hornell 170-71 (skin boats), 190 (dugouts), 202 and 210 (planked boats, including a pre-Viking Age example found in Schleswig).

[13] Cf. Hornell's explanation (210) of the ramlike projections on coasting vessels of Java and Madura as primitive survivals.

[14] Cf. N. Davies and R. Faulkner, "A Syrian Trading Venture to Egypt," *JEA* 33 (1947) 40-46 and pl. 8.

[15] W. Müller, "Neue Darstellungen mykenischer Gesandter und Phönicischer

new type of Egyptian ship manned for some reason by Syrian crews, or Syrian ships which, painted by an Egyptian artist, were given an Egyptian look. The second explanation seems more satisfactory.[16] Rounded hulls such as these vessels have, as we have seen, were common on Crete from 1600 B.C. on. An example appears among the Malta graffiti of 1600 (Fig. 56), and still another on a vase of ca. 1200-1100 found on Cyprus (Fig. 59); this last, like the Syrian craft, is clearly a large merchantman with a deck and roomy hold. Perhaps Cretans and not Syrians originated the design. In any event, it gained favor throughout the eastern waters, and—to judge from a reappearance there in subsequent centuries (Fig. 60; cf. 66 below)—held it. These eastern merchantmen could run to good size: a document of ca. 1200 B.C. refers to one of some 450 tons burden as if it was by no means unusual.[17]

For the warfleets of the Levant, the evidence goes back no farther than about 1200 B.C., the period when the eastern Mediterranean was inundated by the barbarian invaders we call, for lack of a better name, the Peoples of the Sea. A gigantic carving on the side of an Egyptian temple portrays a battle that took place in the Nile Delta between some of their contingents and Egypt's fleet (Fig. 61). The

---

Schiffe in altägyptischen Wandgemälden," *Mitteilungen der Vorderasiatischen Gesellschaft* 9 (1904) 113-80, pl. 3; cf. Säve-Söderbergh cited in the following note. The yard is exactly the type visible in both the Egyptian and foreign ships in a relief of ca. 1200 B.C. (Fig. 61). The painting is in too poor condition to tell if there was a boom along the foot; certainly none appears in the very similar sails in the relief. Müller's suggestion (136) that the vessel had a rope truss is most doubtful.

[16] Egypt, to be sure, had her own carriers that sailed to the Levant. The so-called Byblos boats, though by 1500 B.C. the term meant little more than seagoing vessel, originally must have been used on the run to Byblos; cf. Säve-Söderbergh, *op. cit.* (Two, note 37) 48-49. Säve-Söderbergh holds (52-60) that the ships pictured in the tombs, despite the Syrian crews, are Egyptian. However, in doing so, he discusses neither the Cretan evidence nor the downward curving yard, a distinctly un-Egyptian feature at this time.

[17] J. Nougayrol, "Nouveaux textes accadiens de Ras-Shamra," *CRAI* (1960) 163-71, esp. 165: mention of a shipment of 2,000 measures of grain across the Gulf of Alexandretta in a single bottom. Nougayrol points out that the measure involved can only be the *ḳor*, which at Ugarit at this time held 300 liters. A liter of grain weighs ca. 1.7 lbs.

few ships pictured for each side give no idea of how large the fleets were; they must have been impressive, for we know that even the king of so minor a power as Ugarit could dispose of 150 units at this time.[18] The Egyptian galleys clearly have departed from the traditional types portrayed in, e.g., Hatshepsut's reliefs. The hull, though rounded, no longer is distinctively spoon-shaped, and prow and stern are finished off in new ways, forward in a lion's head instead of the usual unadorned post and aft in a plain sternpost instead of one with the lotus bud. They are true fighting craft, apparently undecked, and with the rowers well protected behind a high bulwark. An important innovation is the absence of the truss, revealing that, like the merchantmen just discussed, these ships must have had some sort of inner structure; indeed, the whole general appearance gives the impression of a far sturdier hull than any attested from Egypt hitherto. The rig, too, has novelties: the boom with its web of lifts is gone, leaving the sail loose-footed; gone as well are the multiple lifts of the yard and the multiple stays of the mast; the yard is no longer straight but has a well-defined curve; and the mast has a top in which a lookout is stationed. As a matter of fact, the rig is identical with the enemy's. The most prominent lines in both seem to be brails (70 below) for shortening sail; if so, we have here the first appearance (its invention proper may have been much earlier) of an all-important feature, one that was to become standard throughout the rest of ancient history.[19] The new style of hull, the mast with simplified staying and a top, the loose-footed sail possibly fitted with

[18] C. Virolleaud, *Le palais royal d'Ugarit* v (Paris 1965) no. 62 (pp. 88-89).

[19] The yards curve downward as in Fig. 58 (cf. note 15 above) and on a scaraboid of the 13th B.C. found at Ugarit (see App.).

Two pictures on fragments of papyrus, dated loosely to the New Kingdom (ca. 1600-1100 B.C.), show Nile boats fitted with brails, like the ships on the relief, yet with two older features retained: (1) the yards curve upward in the earlier fashion (Bowen 123-24); (2) the sails are not loose-footed but have booms. See A. Fabretti, F. Rossi, and R. Lanzone, *Regio Museo di Torino: Antichità Egizie* i (Catalogo generale dei Musei di Antichità . . . del Regno, Serie i: Piemonte i, Turin 1882) 265, nos. 2032, 2033; the fragments, not illustrated in the catalogue, are on display in the Turin museum.

brails—these all seem features that Egypt adopted from elsewhere. Where they arose—in the Aegean, on Crete, or in the Levant—is anybody's guess.

The ships of the Peoples of the Sea are shown without oars and with sails furled. The text accompanying the relief makes it almost certain that the artist had in mind to portray a deadly surprise attack[20]—perhaps one so successful that it caught the enemy either at anchor or before he could run out his oars. Like the Egyptian galleys, these are undecked. Their straight profile and angular ends belong to the Aegean tradition, and the duck-head figurehead reappears on a Greek vase of slightly later date;[21] however, the Aegean tradition so far has produced no evidence for double-ended craft, such as these are. Perhaps this was a change that the Peoples of the Sea introduced into what was basically an Aegean design.

The relief, incidentally, confirms that the only specifically naval weapon known was the grappling iron; the ram is conspicuously absent.[22] A sea fight was still but a version of a land fight, one in which ships grappled one another to let their marines have it out on the decks with bow, sword, or pike.

IV SUMMARY

SCANT as it is, the evidence unmistakably reveals the second millennium B.C. as a crucial period for Mediterranean maritime history. It witnessed the development of the true seagoing ship, both galley and sailing craft, built with some system of internal bracing. It saw the introduction of the rig that was to become standard in the ancient world, the single broad loose-footed square sail, and perhaps it

[20] Cf. H. Nelson, "The Naval Battle Pictured at Medinet Habu," *Journal of Near Eastern Studies* 2 (1943) 40-55, esp. 46.

[21] See note 5 above. Basch (142-43) loosely lumps together all double-ended craft of this age, both those with an angular hull and others with a rounded hull (e.g., Fig. 60). As shown above (30-33), the two types are distinct. R. Barnett in *CAH* II² (1969), Chapter XXVIII, section IV, mentions a ship with duck's head figurehead on an unpublished vase of the 7th-6th century B.C. found on Cyprus.

[22] On Kirk's suggestion that the ships of the Peoples of the Sea had no rams because they were merchantmen, see App., note 4.

saw the introduction of brails, the standard ancient device for shortening sail. The Aegean produced a hull design distinguished by straight lines, angled ends, and a lofty prow; this, brought further along by the Bronze Age Greeks, served as prototype for the later Greek warship and very possibly the merchantman. Farther south a rounded form of hull came into use for both galleys and sailing craft, probably the product of Cretan or Levantine shipyards. The developments were such that, by the end of the period, they had weaned away even conservative Egypt from her traditional hull and rig.

APPENDIX
THE EVIDENCE

THE MATERIAL available up to 1933 was collected by Marinatos (172-80), 69 items of which nearly half were representations on seals. The following significant additions have appeared since:

SEALS

V. Kenna, *Corpus der minoischen und mykenischen Siegel* VIII (Berlin 1966) nos. 49 (= Fig. 45), 55, 106 (= Fig. 40); VII, II (Berlin 1967) nos. 85 (= Fig. 41), 227 (= Fig. 44)

V. Kenna, *Cretan Seals* (Oxford 1960) nos. 107, 188, 192 (= Evans, *Palace of Minos* IV 828, fig. 807), 241, 288

A. Xénaki-Sakellariou, *Les cachets minoens de la collection Giamalakis* (École française d'Athènes, Études crétoises X, Paris 1958) nos. 79, 81, 339, 341 (= Fig. 38). Cf. the author's résumé of new material on pp. 78-80

C. Schaeffer, *Ugaritica* IV (Paris 1962) 147, and figs. 114, 115 on p. 134; scaraboid of 13th B.C.

A number of seals in Marinatos' list have been published again with superior reproductions:

Marinatos no. 28 = N. Platon, *CMMS* II, 1 (Berlin 1969) no. 287b

no. 32 = P. Demargne, *Naissance de l'art grec* (Paris 1964) 111, ill. 144 (= Fig. 36)

nos. 34, 36, 40, 41 = Kenna, *Cretan Seals* nos. 64, 59, 106, 228

no. 42 = A. Sakellariou, *CMMS* I (Berlin 1964) no. 436 (= Fig. 43)

nos. 44, 50 = Kenna, *CMMS* VII, II nos. 104 (= Fig. 42), 72

no. 51 = Kenna, *CMMS* VIII, no. 139

GRAFFITI

D. Woolner, "Graffiti of Ships at Tarxien, Malta," *Antiquity* 31 (1957) 60-67 (= Figs. 24, 31, 56)

C. Schaeffer, *Enkomi-Alasia: Nouvelles missions en Chypre, 1946-1950* (Paris 1952) 102 and fig. 38 (= Fig. 27)

C. Blegen, "Hyria," *Hesperia*, Supplement VIII (Athens 1949) 39-42 and pl. 7.6 (= Figs. 25, 32)

VASE-PAINTINGS

O. Frödin and A. Persson, *Asine. Results of the Swedish Excavations 1922-1930* (Stockholm 1938) 300, no. 2 and fig. 207.2 (= Fig. 29)

40

Swedish Cyprus Expedition I (Stockholm 1934) 484, no. 262 (= Fig. 59)

E. Vermeule, *Greece in the Bronze Age* (Chicago 1964) 259, fig. 43f

D. Theochares, "Iolkos, Whence Sailed the Argonauts," *Archaeology* 11 (1958) 13-18, esp. p. 15

MODELS

C. Renfrew, "Cycladic Metallurgy and the Aegean Early Bronze Age," *AJA* 71 (1967) 1-20, esp. 5, 18. Lead models found on Naxos, the best preserved of which is 40.3 cm. long and 3.3 wide and the others correspondingly long and slender; they would seem to be, thus, models of dugouts. The low end, finished off in a sort of transom, is the stern, and the high end, coming to a point, is the prow.

The evidence, both old and new, has given rise to two major controversies, one concerning hull and the other rig.

The hull in a good many representations terminates at one end in a lofty vertical or nearly vertical post, while the other, with no upright fixture at all, trails off into a low horizontal extension at the waterline (e.g., Figs. 22, 23, 27, 34). Some take the high end for the prow, some the low.[1] To complicate matters, those who take the low end for the prow see in the horizontal extension the earliest form of that naval weapon par excellence of the ancient world, the ram.[2] Though there were enough clues to have settled the question long ago,[3] new evidence (the lead models and Fig. 27; cf. 31 above) provides an incontrovertible answer: it is the high end that is the prow. With this established, the argument for

[1] For a résumé of opinions, see Miltner 906, Marinatos 182-83, 125-27. All three view the low end as prow.

[2] This opinion, flourishing for a while (cf. Marinatos 183), then rejected (Marinatos 183, note 4; Miltner 906), has been revived by Kirk (125-27) and is back in the handbooks (cf., e.g., F. Matz, *Kreta, Mykene, Troja*, Stuttgart 1956, p. 77). See also note 4 below.

[3] For one, to take the low end as prow means that the fish emblem points backward (cf. Fig. 22), which goes not only contrary to sense but to the location of the emblem on the one example we have where prow and stern are indisputable (Fig. 28). For another, high prow and low stern are characteristic of early craft (see Evans, *Palace of Minos* II 240-41), which is why the specialists see no problem (cf., e.g., C. Hawkes, *The Prehistoric Foundations of Europe*, London 1940, p. 156; Behn in M. Ebert, *Reallexikon der Vorgeschichte* XI 240 [1927]; Woolner, *op. cit.* 63). Among those who take the low end as prow, only Marinatos has studied the question comprehensively, and behind his conclusion lie inconsistencies and inconclusive statistics; see L. Cohen, "Evidence for the Ram in the Minoan Period," *AJA* 42 (1938) 486-94, esp. 489, notes 7, 8. The most recent editor of Fig. 34 takes the high end as prow; see *CMMS* II, 1 no. 287b.

the ram at this period ceases to exist.[4] The horizontal extension still needs explaining, but, until conclusive evidence turns up, little is gained by guessing.

Turning to rig, we find that a series of seals[5] of the same age, type, and fashion of workmanship shows an enigmatic structure above the deck which has frequently been taken to represent two, even three, masts of equal height.[6] The seamen of this age were admittedly precocious, but hardly enough to invent a rig that was to drop out of sight until shortly before Columbus embarked for America; explanations more in line with the technological level of the age can be found (34 above).

[4] Kirk (125-27) simply accepts Marinatos' conclusion that the low end is the prow, making no effort to improve on the arguments, and holds that the projections were at first structural but then soon became a ram. That neither Homer (49 below) nor the Peoples of the Sea (38 above) knew anything of the ram he explains (126, note 37) by assuming that the ships involved were all merchantmen. This posits a distinction between oared merchant vessels and warships which is not only unproven but most unlikely at this age. And what of the Egyptian ships shown attacking the Peoples of the Sea (Fig. 61)—fighting craft pure and simple, yet with no ram? If the ram was known in the Bronze Age, the Egyptians would necessarily have adopted it, for it was a weapon like the naval gun—once one fleet had it, all had to have it.
  Two fragments of pottery from Thessaly of about 1600 B.C. have been restored by Theochares (op. cit. above) to show ships with rams. The restorations have been accepted uncritically; see W. Taylour, The Mycenaeans (London 1964) 162-63 and Williams, GOS 7 and pl. 1a. Yet they are almost entirely fancy—from a tiny fragment of what Theochares takes to show part of a prow (but which may not even be part of a boat at all), he has reconstructed a total ship. Even if a prow is intended, it is not at all one with a ram but only a traditional projecting cutwater (cf. THREE, notes 12 and 13).

[5] Marinatos nos. 42-51; Kenna, CMMS VIII nos. 49, 55; Kenna, Cretan Seals nos. 241, 288; Xénaki-Sakellariou, Coll. Giam. 339. Cf. Figs. 42-45.

[6] See Marinatos 200, note 4, for citation of literature. Marinatos (200-208), rightly aware that multimasted ships could not have existed at this date, sought to explain the rig as a square sail divided vertically down the middle into two panels—a blunder arising from his use of 18th and 19th century line-drawings of ship-pictures from Pompeii, which seemed to show such half-sails; cf., e.g., his fig. 6 with Fig. 151. Seals showing poles alone (e.g., Fig. 44) he sought to explain (207) as storm-tossed ships with their canvas torn and in disarray, a conclusion the evidence hardly warrants. Miltner, too, questioned (907) the characterization of these ships as multimasted. Recent editions of seals, however, unconcernedly continue to refer to two- and three-masted vessels; cf. Kenna, Cretan Seals nos. 241, 288; Xénaki-Sakellariou, Coll. Giam. 79, 81, 339.

CHAPTER FOUR

The Eastern Mediterranean:
1000-500 B.C.

BY 1000 B.C., after the last waves of barbarian invasion had washed ashore, a new cast of characters appears on the maritime stage in the eastern Mediterranean.

The Mycenaeans have left, replaced by the Phoenicians, who swiftly achieved the reputation as seamen and traders that they were to maintain for a millennium. The Phoenicians did not have matters to themselves for very long. In the ninth century B.C., Greeks were back on the scene—Greeks of the Iron Age this time, just starting on their celebrated career in history. By the middle of the next century they had developed the merchantmen and warcraft with which they planted and defended new coastal settlements from Marseilles to the shores of the Black Sea. For both Phoenician and Greek ships, we have the evidence of contemporary pictures. For the Greek we have, in addition, the numerous precious references in Homer's *Iliad* and *Odyssey*.

I THE HOMERIC GALLEY

A PREFATORY remark is in order. Homer composed probably in the eighth century or early seventh B.C.—but he purports to describe a war carried on by the Bronze Age Greeks, the Aegeans, of four to five hundred years earlier. What he says fits very well with what we know of eighth century galleys from other sources (49 below), but also fits the little we know of Aegean galleys, particularly since the poet seems to have been careful not to commit anachronisms.[1] It is safest to assume that he has in mind the vessels he himself sailed on, though certain statements may equally well be true of the previous age.

---

[1] E.g., he never mentions the ram, although it was a prominent feature of the warcraft of his own day; cf. 49 below.

43

HOMER constantly calls his heroes' ships "swift"—which is what we would expect of the low sleek hulls pictured in contemporary representations (e.g., Fig. 62).[2] The ships were "hollow,"[3] i.e., undecked. There was a scant deck forward for the lookout and a slightly larger one aft, where the captain or an occasional passenger could stretch out;[4] the crew did its dozing on the benches, which is why, unless utterly unavoidable, nights were spent ashore.[5] Gear and provisions were stowed under the decks and rowing thwarts, dry provisions such as grain in leathern sacks, and water or wine in clay jars or skin bottles.[6]

There were several standard sizes of galley. The 20-oared, the smallest mentioned, seems to have been used for ordinary dispatch and transport work;[7] the 50-oared—or penteconter, to give it the technical name it was known by later—seems to have been a common troop transport;[8] and there were large carriers which had as many as 100 rowers.[9] Homer fails to mention the triaconter, the 30-

---

[2] For statistics on Homer's use of epithets, see *GOS* 45-46.

[3] γλαφυρός (e.g., *Il.* 2.454, *Od.* 3.287) or κοῖλος (e.g., *Il.* 1.26, *Od.* 1.211).

[4] *Od.* 12.229-30: εἰς ἴκρια νηὸς ἔβαινον πρῴρης "I stepped on to the foredeck"; *Od.* 12.411-14: πρύμνῃ ἐνὶ νηὶ πλῆξε κυβερνήτεω κεφαλήν ... ὁ δ' ἄρ' ... κάππεσ' ἀπ' ἰκριόφιν "[The falling mast] struck the head of the helmsman in the stern of the ship ... and he ... fell from the deck." Vase-paintings frequently show the lookout on the foredeck; see Fig. 90. Odysseus slept on the afterdeck of the Phaeacian ship during the voyage to Ithaca (*Od.* 13.73-75), and Telemachus shared the afterdeck with Athena on the run to Pylos (*Od.* 2.416-18) and with Theoclymenus on the way back (*Od.* 15.285-86).

[5] E.g., *Il.* 1.476; *Od.* 9.150-51, 168-69.

[6] Grain in leather bags, *Od.* 2.354; water and wine in skins, *Od.* 5.265-66; wine in jars, *Od.* 9.164.

[7] Chryseis was sent back to her father in a 20-oared galley (*Il.* 1.309); Telemachus used one to go to Pylos (*Od.* 1.280); and the suitors planned to intercept him in one (*Od.* 4.669).

[8] Philoctetes' transports were 50-oared (*Il.* 2.718-20), as were Achilles' (*Il.* 16.169-70). The only other size of transport mentioned is even bigger; see the following note. Odysseus' transports also seem to have been at least 50-oared: on arrival at Circe's island, the total aboard was 46 (*Od.* 10.203-209) and this was after losing 6 to the Cicones (*Od.* 9.60) and 6 to Polyphemus (*Od.* 9.289, 311, 344). Homer has the Phaeacians use a 50-oared galley to take Odysseus to Ithaca (*Od.* 8.34-36, 48: the *kouroi* "youths," i.e. the total complement, numbered 52, probably 50 rowers and 2 officers). This may have been an honorific gesture (even as a state today will transport distinguished visitors in a battleship or cruiser rather than a destroyer).

[9] The Boeotian transports were manned by 120 *kouroi* (*Il.* 2.509-10), i.e., a total

oared galley, but it almost certainly existed in his day.[10] All types were long and narrow,[11] and so low that, when the beached Greek ships were under pressing Trojan attack, Hector could reach up and grab the ornament atop the sternpost, and Ajax could leap from the gunwale to the ground.[12] They were so light that Odysseus could get his vessel free of the Cyclops' island with one good shove on the boat pole,[13] and the crews could run them up on the beach at night.[14]

The hull was black[15]—either painted, or, more likely, smeared with pitch. Viewed in profile, stem and stern rose more or less straight for a distance and then finished off in a curve, for Homer compared the shape of his ships to the "straight horns"[16] of cattle as against, say, the curly horns of a ram; contemporary vase-paintings (Figs. 62, 65, 66, 72) show precisely what he means. He also calls them "rounded on both sides," which must mean that, viewed head on, they showed gracefully rounded bilges.[17] The bows were decorated with patches, red or purple or blue.[18] The hull was made up of keel, stempost, sternpost, frames, planks, gunwales, and beams—which may also have served as the rowers' benches—all put together with tree-

---

complement that must have included deck hands as well as officers and men. On these ships, see 59 below.

[10] Herodotus unquestionably considered it old: he mentions (4.148.3) the use of triaconters in the colonization of Thera, which he thought took place in very early times. For 30-oared galleys in the Bronze Age, see 34 above.

[11] For estimates of their size, see 54-55 below.

[12] Il. 15.704, 716-17 (Hector); 15.729 (Ajax).

[13] Od. 9.487-88.   [14] E.g., Od. 9.546.

[15] Ill. 1.141, 433, 524, etc.

[16] ὀρθόκραιρος "straight-horned" (e.g., Il. 18.3), but also κορωνίς "curved" (e.g., Od. 19.182).

[17] ἀμφιέλισσα, in the Iliad used exclusively and frequently of the Greek ships lined up along the beach (2.165 = 2.181, 9.683, 13.174 = 15.549, 17.612, 18.260) and hence as seen from front or back rather than from the side. Cf. Od. 6.263-64, cited in SIXTEEN, note 5.

[18] μιλτοπάρῃος "red-cheeked" Il. 2.637, Od. 9.125, etc.; φοινικοπάρῃος "purple-cheeked" Od. 11.124, etc.; κυανόπρῳρος "blue-prowed" Il. 15.693, Od. 9.482, etc. Bow-patches are a phenomenon common to craft of all ages; see, e.g., R. LeBaron Bowen, Jr., "Maritime Superstitions of the Arabs," The American Neptune 15 (1955) 5-48, esp. 5-24. Herodotus' statement (3.58) that "in olden days ships were all daubed with red" (τὸ δὲ παλαιὸν ἅπασαι αἱ νέες ἦσαν μιλτηλιφέες) would imply that red was commonest.

nails and joints.[19] There were two massive through-beams, one just behind the fore-decking, the other just forward of the after-decking.[20] Ship timber was oak, poplar, pine, fir; masts and oars were of fir.[21]

Oars were worked against tholepins;[22] a leather strap held them in place against the pin, and also kept an oar from going over the side if a rower lost his grip.[23] Homer mentions only the single steering oar,[24] a large-bladed one on the quarter with a handle socketed into the loom as tiller;[25] he may have in mind the Mycenaean Age in this regard, since eighth century pictures (Fig. 74) already show the double steering oars that were to be standard thenceforth.[26]

[19] τρόπις "keel" Od. 5.130, 7.252, 12.421, 422, 424, 438, 19.278; στεῖρα "cutwater" Il. 1.482 = Od. 2.428; ἄφλαστον either "sternpost" or more strictly "ornament atop the sternpost" as in Il. 15.716-17 (the ornament proper is called ἄκρα κόρυμβα "high peaks," Il. 9.241-42); σταμῖνες "frames" Od. 5.252; τοῖχοι, literally "walls," is used to mean "sides," "side-planking" Il. 15.382, Od. 12.420; ἐπηγκενίδες "gunwales" Od. 5.253 (cf. TEN, App. 2, note 5); ζυγά "beams," "thwarts" Od. 9.99, 13.21 (in Il. 2.293 a ship is called πολύζυγος "many-thwarted" and in 20.247, in a hyperbole, ἑκατόζυγος "hundred-thwarted"; in open boats such as Homer has in mind, the line of thwarts was a visible measure of size); γόμφοι "treenails" Od. 5.248; ἁρμονίαι "joints" Od. 5.248, 361. For the method of construction, see TEN, App. 2. Most of these terms continued in use; see TEN, App. 3.
The Iliad seems to contain a reference to a more primitive form of boatbuilding with sewn planks; see ONE, note 27.

[20] The after through-beam was the "7-foot threnos" onto which Ajax stepped when leaping down from the poop (Il. 15.728-29); the forward one was the ephol-kaion onto which Odysseus says he stepped when lowering himself off a beached ship into the water (Od. 14.350-52; cf. GOS 49). The latter means literally "tow-piece"; very likely it was so called because its protruding butts offered strong points for making fast tow lines. Apollonius Rhodius (4.1609), for example, has a Triton tow by "holding on to the olkaion of the hollow Argo" (ἐπισχόμενος γλαφυρῆς ὀλκήιον Ἀργοῦς); he must be referring to the same feature.

[21] Oak, poplar, pine, Il. 13.389-91 = 16.482-84; fir mast, Od. 2.424 = 15.289; fir oars, Il. 7.5, Od. 12.172.

[22] κληῖδες Il. 16.170; Od. 2.419, 4.579, 8.37, etc. The word means literally "keys," and they look very much like keys in contemporary representations (Figs. 67-69).

[23] τροπός Od. 4.782, 8.53. As the latter passage shows, strapping on the oars was one of the regular steps in the procedure of launching a galley. When rowers lost their grip, the oars, held by the straps, would clatter along the sides (Od. 12.203-204).

[24] πηδάλιον Od. 3.281; 5.255, 315; 8.558.

[25] οἰήιον Il. 19.43; Od. 9.483 = 9.540, 12.218.

[26] The poet of the Homeric Hymns (7th B.C.) knew the double steering oar; see Hymn to Apollo 418: ἀλλ' οὐ πηδαλίοισιν ἐπείθετο νηῦς "but the ship did not obey the steering oars."

Rig consisted of a single sail[27] placed amidships on a mast that could be raised and lowered. To set sail,[28] the crew raised the mast[29] from its crutch[30] aft by hauling on the forestays,[31] and set it in its tabernacle,[32] which probably projected a certain distance above the keel.[33] The sail was square, bent to a yard.[34] Standing rigging consisted of two forestays, one run to either bow,[35] and a single backstay.[36] Running rigging consisted of braces,[37] sheets,[38] and brails[39]

[27] Either ἱστίον (*Il.* 1.481, 15.627; *Od.* 2.427) or, more commonly, the plural ἱστία (*Il.* 1.433, 480; *Od.* 2.426, 3.10, etc.) because the sail was made up, as cloth sails always are, of a series of sewn strips.

[28] See, e.g., *Od.* 2.424-26 = 15.289-91 (and *Il.* 1.433-35 for taking in sail).

[29] ἱστός *Il.* 1.434, *Od.* 4.781, etc.

[30] ἱστοδόκη *Il.* 1.434. Since the mast was lowered by the forestays, it follows that it dropped sternward, and the crutch to receive it must have been aft. A lowered mast is so pictured on the François vase (*GOS*, Arch. 33 and pl. 11a; 600-550 B.C.). For later practice, see FOURTEEN, note 3.

[31] πρότονοι *Il.* 1.434, *Od.* 2.425 = 15.290 (cited in the following note).

[32] To get past the sirens, Odysseus reports (*Od.* 12.178-79; cf. 51-52, 161-62) that "they bound me hand and foot upright against the *histopede* and tied the ends of the rope to the [mast] itself" (μ' ἔδησαν ὁμοῦ χεῖράς τε πόδας τε / ὀρθὸν ἐν ἱστοπέδῃ, ἐκ δ' αὐτοῦ πείρατ' ἀνῆπτον. Presumably the mast had been left standing even though, with the coming of a calm [168-69], sail and yard had been stowed away [170-71].) Sailing craft of traditional make that are fitted with a retractable mast have a slender vertical casing into which the lowest portion of the mast fits; such a meaning for *histopede* here would suit the context very nicely. When Telemachus got under way (*Od.* 2.424-25 = 15.289-90), his sailors "raising the fir mast, set it inside the hollow *mesodme* and fastened it down with the forestays" (ἱστὸν δ' εἰλάτινον κοίλης ἔντοσθε μεσόδμης / στῆσαν ἀείραντες, κατὰ δὲ προτόνοισιν ἔδησαν). The *mesodmai* of a house are tie-beams running from rafter to rafter; so the "hollow *mesodme*" here could be a carling, running fore and aft between two thwarts amidships, that had a hole or notch in it which centered over the mast step. A very similar arrangement appears in a clay model of the 9th or 8th B.C. (Fig. 87).

[33] E.g., a line from Alcaeus (18.6 = D. Page, *Lyrica Graeca Selecta*, Oxford 1968, 148.6) implies that the *histopede* was a measure of the height of the water in the bilge: in describing a storm, he mentions that πὲρ μὲν γὰρ ἄντλος ἱστοπέδαν ἔχει "The bilge water surrounds the tabernacle."

[34] ἐπίκριον *Od.* 5.254, 318.

[35] Cf. notes 31 and 32 above. The forestays were run to the port and starboard bows respectively to provide some lateral bracing; Apollonius Rhodius, for example, writes (1.564): "They secured [the mast], stretching the forestays to either side" (δῆσαν δὲ προτόνοισι τανυσσάμενοι ἑκάτερθεν). When a sudden gust from ahead snapped the forestays on Odysseus' ship, the mast immediately fell sternward and brained the helmsman (*Od.* 12.409-12, cited in part in note 4 above).

[36] ἐπίτονος *Od.* 12.423; cf. note 39 below.

[37] ὑπέραι *Od.* 5.260.

[38] πόδες *Od.* 5.260. The helmsman held the lee sheet in his hand (*Od.* 10.32-33).

[39] κάλοι *Od.* 5.260; cf. 70 below. Paintings of the 6th B.C. indicate how they were

for shortening sail (70 below). Assigning brails to these ships need not be an anachronism for, as we have seen (37 above), there is evidence for them during the Bronze Age. Halyards are not mentioned by name but alluded to.[40] Sails were of linen, not one piece but, for added strength, patches sewn together.[41] Lines were of twisted papyrus fiber (and probably grass fibers) or of leather.[42]

The gear carried included mooring lines,[43] particularly stern lines[44] since these ships were generally beached stern to; stone anchors;[45] perhaps a drain plug for emptying the bilge when beached;[46] punting poles;[47] long pikes for fighting;[48] probably sidescreens for closing in the waist in heavy weather;[49] and the bags and jars mentioned above for holding provisions.

---

rigged (Figs. 89-91). They are always shown drawn taut, so they must have provided some bracing aft, which may explain why these ships carried but a single backstay.

[40] ἕλκον δ' ἱστία λευκὰ ἐϋστρέπτοισι βοεῦσιν "And they drew up the white sails with well-braided thongs" Od. 2.426 = 15.291. Miltner (910) groundlessly assumes that by the leather ropes here, the backstay is meant.

[41] They are of white (see previous note) linen cloth (φᾶρος Od. 5.258; σπεῖρον Od. 5.318, 6.269). Cotton was not used since, during the whole of the ancient period, it was an expensive import from remote India.

[42] Of papyrus fibers, Od. 21.390-91; of leather, Od. 2.426 (cited in note 40 above) and 12.423. For grass fiber used in light cordage, see Il. 2.135 (cited in ONE, note 27). Three specimens of rope were recovered from a Bronze Age shipwreck. At least one was two-strand. Two were of grass fiber, the third a mixture of grass and palm leaf fiber. See Bass, Gelidonya 160-61 and fig. 160.

[43] πείσματα Od. 6.269; 9.136; 10.96, 127; 22.465.

[44] πρυμνήσια Il. 1.436, 476; Od. 2.418; 9.137 and 178; 15.498.

[45] εὐναί (Il. 1.436, 14.77; Od. 9.137, 15.498) literally "beds," referring to the look of the flat stones used as they lay on the bottom; cf. ELEVEN, note 112.

[46] Hesiod, Op. 626: χείμαρον ἐξερύσας "by drawing the drain-plug."

[47] κοντοί Od. 9.487.

[48] ναύμαχα ξυστά "marine pikes" Il. 15.388-89, 676-78 (Ajax' pike was 33 feet long—but Ajax' strength was fabulous). These pikes appear in contemporary illustrations; see GOS, Geom. 6, 8.

[49] Alcaeus, fr. 6.7-8 in Lobel-Page, Poetarum Lesbiorum Fragmenta (Oxford 1955) = Page, op. cit. (note 33 above) 107.7-8: φαρξώμεθ' ὠς ὤκιστα [νᾶα(?)] / ἐς δ' ἔχυρον λιμένα δρό[μωμεν] "Let us fence our ship as quickly as possible and run for a safe haven"; Aeschylus, Suppl. 715: στολμοί τε λαίφους καὶ παραρρύσεις νεώς "the folds of the canvas and the vessel's sidescreens." On the meaning of φράττω in the passage from Alcaeus, see the note by J. Taillardat in RPh 39 (1965) 83-86.

II GALLEYS ON GEOMETRIC VASES

WHAT did Homer's galleys look like?

The answer is supplied in part by the ship-pictures on a series of vases that date roughly 850-700 B.C., the Geometric vases (Figs. 62-72, 74, 77), as they are called from the nature of their decoration.[50] These portray two classes of galley, one low and open, the other with an elaborate superstructure. To a certain extent both fit Homer's words—they are sleek and fast, have stem- and sternpost so curved as to give the effect of "horns," have round bow-devices that would correspond to Homer's bow-patches, mount a single square sail that can be raised or lowered, show a suitable number of rowers. It is only the open type, however, as we have seen, that he has his heroes use; he may have felt that this simpler form was the sole known in their day.

But both, without exception, show a feature that previous ages never knew and that Homer never mentions: the ram. Very likely it made its debut during the obscure period after 1000 B.C. that marked the transition from the Bronze to the Iron Age. Its introduction must have had as revolutionary an impact as, say, that of the naval gun twenty-five hundred years later. A warship was no longer merely a particularly fast transport to ferry troops or bring marines into fighting proximity with those of enemy ships; it had become an entirely new kind of craft, one that was, in effect, a man-driven torpedo armed with a pointed cutwater for puncturing an enemy hull.

In the wake of the new weapon came, inevitably, far-reaching changes in ship design and construction. From now on, in order to withstand the shock of ramming, all men-of-war had to be built more powerfully and of heavier materials; the bow area in particular had to be as massive as possible, for blows were felt here first and hardest. The changes are visible even in the simple profile drawings in silhouette which are all we have to rely on for this critical period:

[50] On the material available, see App. 1. For the dating of Geometric pottery, see J. Coldstream, *Greek Geometric Pottery* (London 1968) esp. 26, 30-31, 302-31 (esp. 321, 330-31).

the Bronze Age artist had rendered the bow area as open or with a latticed design; the Geometric artist shows it as a solid, ponderous mass.

Open galleys[51]—or aphracts, to adopt the technical term—no less than the galley with a superstructure now carry the ram. They are recognizable (Figs. 62-64) by a lighter hull with but a low rail along the sides, and by a bow area that is less massive, that often lacks room for the round bow-devices which almost invariably distinguish the other class of vessel. More stable because unencumbered by any superstructure, and more swift, they were probably used for carrying dispatches and transporting personnel, the roles Homer assigns them,[52] leaving their big sisters to serve as ships of the line. Conformably, they are never shown in combat: we see them in elaborate maneuvers (Figs. 62-63) or storm-tossed (*GOS* pl. 6e) or getting under way (*GOS* pl. 7d), but never fighting.

On the other hand, the galleys with a superstructure are almost always shown in combat. The Geometric artist especially fancied scenes of an attack on a beached ship, which enabled him to introduce elements of land-fighting as well. We are fortunate enough to have several pairs of pictures from two different areas of the same vase, and these reveal a deliberate effort to vary the scenes portrayed: in three instances,[53] one of the pair depicts a beached galley, its sails stowed out of sight, beset by an attacking force; while the other shows the galley, its sail set and pulling, drawing away from or out of danger (Figs. 65, 66).[54] All this serves to give us a welcome fund of information about the ship of the line.

---

[51] See App. 2.

[52] Cf. notes 7, 8 above.

[53] Figs. 65-66 (= *GOS*, Geom. 25). Fig. 67 (= *GOS*, Geom. 4) + *GOS*, Geom. 7; cf. E. Kunze, "Disiecta membra attischer Grabkratere," *Eph. Arch.* (1953-54, pt. 1) 162-71, esp. 166-67, for attribution to the same vase. *GOS*, Geom. 5(2) (= Kunze, pl. v.1) + *GOS*, Geom. 8 (= Kunze, pls. v.3/vi.1/vi.2; cf. Kunze 167 for attribution to the same vase).

[54] Kirk, who is quick to interpret in the light of artistic convention (see Apps. 1 and 2), characteristically views (103) the inclusion in the same scene of warriors on land and a ship under sail as one of "the unrealistic tendencies of the conventional Dipylon ship-style." Cf. also his remarks on his no. 28.

The superstructure consists primarily of a deck raised perhaps a good two feet above the line of rowers. In a certain number of paintings (Fig. 67), warriors appear on the deck and, at a distinctly lower level—the distance from a man's foot to his thigh, as will be clear in a moment—is the gunwale, sometimes shown with the key-shaped tholepins (Figs. 67-69) against which the oars work.[55] The oars themselves and the rowers do not appear, for these are all scenes depicting an assault on a beached ship; presumably the oarsmen are among those fighting to repel the attackers.

At bow and stern were short, rather high platforms (Fig. 72). The one forward—the most important fighting station—was surrounded by a protective palisade (Figs. 67-69). The deck proper, set at a slightly lower level (Fig. 67), ran the full length from the fore to the after platform (Fig. 72), but not the full width from gunwale to gunwale; along the sides the planking must have been left off. This can be deduced from the way the artists position the figures they include. In certain fighting scenes warriors are portrayed with their feet at the rowers' level, their thighs at the raised deck level, and their bodies towering above it (Figs. 68-69);[56] obviously they are standing on the rowing benches, or on some gunwale-level decking, at a point uncovered by the raised deck. Certain sailing scenes show rowers plying their oars from the same level as the raised deck (Fig. 72).[57] Obviously the part left uncovered must have been that along the side, where rowers would sit.[58]

Adding such a deck did vastly more than provide these galleys with a useful platform. It enabled the shipwrights to convert a light

---

[55] For other examples of the tholepins, see *GOS*, Geom. 5(1), illustrated in Kirk, pl. 39.1; *GOS*, Geom. 6, illustrated in *Ath. Mitt.* 17 (1892) 300, fig. 7.

[56] For another example, see *GOS*, Geom. 3, illustrated in *JHS* 78 (1958) pl. 13c.

[57] On the rowers in these scenes, see 55 below.

[58] Whether the raised deck ran along the center, leaving the sides open, or vice versa, has raised much discussion; cf. Kirk 128-29 and Williams, *GOS* 15-16. Since Kirk considered rowers shown on the raised deck as not really being there (see note 74 below), he was able to argue for side decks. As we shall see (Figs. 75-76), nearly contemporary Phoenician galleys had decking along the center only; it is hardly likely that Greek craft, so similar in most other fundamental respects, differed in this. Egyptian galleys were also decked this way; see 21 above.

hull, which was vulnerably low amidships and offered scant protection to the rowers, into a powerful one with high sides that offered a good deal of protection. Such a conversion had become a *sine qua non*, thanks to the ram: a light hull was fine for ships whose prime purpose was to catch up with and swing alongside an enemy, but not at all for those who must slam their beaks into his side or receive such an attack from him. In the earliest pictures (Figs. 30, 65-66), dated to the latter part of the ninth century B.C., the raised deck is represented as a thin line resting on a row of slender stanchions,[59] which were so placed as to frame the rowers.[60] The deck itself, bridging the two ends of the vessel, added welcome longitudinal stiffening; but the slender stanchions left the sides as frail as ever. By the middle of the eighth century B.C., the thin line is replaced by a complex of lines—a thick black band outlined by two thin lines (Figs. 67-69, 72, 77).[61] I take[62] the band to represent heavy planking, and the lines about it[63] to be wales that girdled the ship laterally for added strength (Fig. 69).[64] These additions not only reinforced the sides but, at the same time, by reducing the area between deck level and gunwale level to no more than a wide slot, gave a certain amount of screening to the rowers. By the end of the century the slot was further closed in when wide partitions replaced the narrow stanchions (Figs. 70-71). Thucydides in the fifth century remarked of the ships which went to Troy that they "were not fenced in (*cataphracta*) but were vessels fitted out rather pirate-style in the old-fashioned way." The Geometric vase-paintings reveal that the process of "fencing in" the

[59] For other examples, see *GOS*, Geom. 26 and pl. 6c, Geom. 27, illustrated in *AJA* 44 (1940) pl. 22.1. Williams (*GOS* 30) assigns a later date to Figs. 65-66 largely on the basis of its "more natural and developed perspective." For the date I have used, see Coldstream, *op. cit.* (note 50 above) 26, 321. On Williams' views concerning Geometric perspective, see App. 1.

[60] When the tholepins are included (Figs. 67-69), they are placed one each in the middle of the space bounded by a pair of stanchions.

[61] Cf. Williams, *GOS* 14-17.

[62] See App. 1.

[63] Kirk (100) dismisses these lines as decorative only.

[64] Wales are a standard feature in later pictures of both merchantmen and warships; cf., e.g., Figs. 82, 90, 94, 97.

52

"old-fashioned" open craft of the Homeric poems—or, to use the Greek terminology, converting them into cataphracts[65]—was well on its way by about 700 B.C.

III THE INVENTION OF THE TWO-BANKED GALLEY

THE introduction of the ram thus triggered the development of a powerful type of war galley with raised deck and screened sides. And the raised deck, in its turn, made possible a rearrangement of the oars that was to determine the entire future course of the ancient warship.

First, some introductory remarks on the oarage of galleys as revealed by the full span of their history. The earliest and simplest way to arrange rowers was to put one at each oar and to seat them all horizontally in a single line, as was done in Pharaonic Egypt (e.g., Fig. 18), very likely on Mediterranean craft of the Bronze Age (e.g., Fig. 22), and certainly on the ships Homer describes. As warcraft grew in size and needed more power, additional oarsmen joined the line—but only up to a certain point: a rower needs a minimum of three feet of room,[66] and putting too many of them in single file results in an overlong craft dangerously frail amidships. One solution to the problem, the best known to us since it was adopted in the notorious slave-driven galleys of the sixteenth to eighteenth centuries, was to replace the one-man oar with a big sweep pulled by a number of rowers; this provided the desired increase in muscle without any in the vessel's length (though necessarily in the beam, thereby producing a heavier ship). A variant arrangement, favored particularly by the Venetians in the fourteenth and fifteenth cen-

[65] For the passage from Thucydides, see FIVE, note 58. A cataphract ship was *ipso facto* a ship with a raised deck since the "fencing in" covered the space between the deck and the gunwale below; cf. 88 below. Aphracts, though they could have decks at either end, and some decking at gunwale level (see note 88 below), had no raised deck from which side screening could be hung and were hence "unfenced"; cf. Taillardat, *op. cit.* (note 49 above) 86-88.

[66] See Morrison, "Trireme" 21-22, *GOS* 155. Torr's statement (21, note 56) that the distance between tholepins was variable is a mistake.

turies, was to stick to individual oarsmen each pulling one oar, but to crowd three or four together in echelon on the same bench; this tripled or quadrupled the total number of rowers while adding much less proportionately to the vessel's length. Both multiple-rower sweeps and the variant just described required a stroke in which the oarsmen rose from the benches to a full standing position and then threw themselves back on the benches again.[67]

Still another way to gain more oar-power and yet avoid both an overlong hull and the fatiguing sit-stand-sit stroke was to put the oarsmen in superimposed lines instead of all in one. This was, for example, the method used in the Mediterranean from Byzantine times until the Venetians came up with their variant, in Indonesian waters until very recently, and, as we shall see in a moment, by the ancients.[68] Its great advantages were that it permitted a narrow, fast hull and allowed the rowers to remain seated during the whole of their stroke. That only expert crews could be used and that the hull, standing somewhat high above the water, lost in stability were the chief disadvantages.

To return to the Geometric vases. Among the ships pictured, those identifiable as aphracts were powered, as they are in Homer, by one line of oarsmen. The rowers indicated vary in number from 8 to 19,[69] but such figures must not be taken too strictly; vase-painters neither then nor later attempted photographic reproduction. When an artist shows 8 rowers on a side, he is perhaps portraying the 20-oared galley that we know from Homer was a standard size; when he shows 19, perhaps he is attempting a penteconter.[70] These ships would all be markedly long. A single-banked penteconter would run some 125

---

[67] *IH* 70-74, 117-25 and ills. 91, 92, 95, 147, 148.

[68] *IH* 42, 130 and ills. 51, 155. Tarn's doubts ("Warship" 206, note 85) of the validity of the evidence for Malay biremes and triremes are totally unwarranted; cf., e.g., C. Nooteboom, "Eastern Diremes," *MM* 35 (1949) 272-75.

[69] See App. 2. The second example listed there shows 13 rowers, and the eighth 19. If we assume that the space between each pair of stanchions accommodates 1 rower (cf. 52 above and note 60), the seventh in the list had 10 rowers, and the third 8.

[70] Cf. note 93 below.

feet in length: 75 to accommodate the rowers and 40 to 50 more for poop deck, foredeck (the larger of the two), and ram.[71] The beam would be about 13 feet.[72] A single-banked triaconter would run some 75 feet in length, and a single-banked 20-oared galley, 50.

So much for aphracts; now for ships furnished with a raised deck. The artists often paint in the oarsmen or—what is just as useful for our purpose—the tholepins about which the oars pivoted. Such pictures (Figs. 67-69, 77)[73] make it clear that the rowers sat at gunwale level, well below that of the deck, as we remarked earlier. But there are also pictures (Figs. 72, 74) in which the oars are indisputably worked from deck level.[74] In other words, the deck was not only a platform for personnel; it doubled as a platform for oars—these ships could be rowed *either* from the gunwale *or* from the deck. The artists furnish clues to what circumstances called for what level. In action, when the decks had to be kept clear and the oarsmen protected, rowing was done from below.[75] When there was no danger, rowing could be done from above where, in any kind of sea, oarsmen no doubt stayed a good deal drier (Figs. 72, 74).[76]

If the oars could be worked from either level, why not from both

[71] E.g., the artists of the black-figured vases generally give about 2/5 of the overall length to the poop deck and the complex of ram and fore deck (the latter gets the lion's share) and 3/5 to the rowing area. However, a carefully done relief (Fig. 73) which portrays a ship of the very same type as those on the black-figured vases, allots as much as half to ram and decks. So, too, apparently, does the best preserved of four sketchy seventh century wooden models of galleys found in the ruins of the Heraeum on Samos; see *Ath. Mitt.* 68 (1953) 111-18, fig. 27.

[72] The model mentioned in the previous note shows a length to beam ratio of 10 to 1, and the same was true of a 5th century Athenian trireme (82 below). The ship on whose stern Ajax took his stand measured 7 feet along the *threnos*, i.e., just forward of the poop (cf. note 20 above); presumably it was somewhat wider amidships.

[73] For additional examples, see notes 55 and 56 above; *GOS*, Geom. 10 and pl. 3c; cf. *GOS*, pl. 3d; *GOS*, Geom. 11 and pl. 4a; *GOS*, Geom. 15.

[74] See also *GOS*, Geom. 39 and pl. 7b, possibly Geom. 20 and 21. In these representations, the oars, plain as day, are worked from the raised deck. Kirk (130) would have it that this is just another instance of the Geometric artist's conventional treatment: the rowers, he claims, should be below, but the artist has put them above to avoid overcrowding and to improve the total composition.

[75] *GOS*, Geom. 11 and pl. 4a.

[76] In Fig. 72, a vessel is cruising in balmy weather; in Fig. 74 and *GOS*, Geom. 39 the ships are making a peaceful departure.

at once? A series of representations reveals that vessels were indeed so rowed from the latter half of the eighth century B.C. onward. A fragment (Fig. 77), dated ca. 750 B.C., shows the earlier type of galley, that with deck and open sides, driven by oarsmen along both the gunwale and the deck, while two others (Figs. 70-71), dated ca. 700 B.C., show the latter type, with deck and well-screened sides, also driven by superimposed oarsmen. Finally, Assyrian reliefs (Figs. 75-76, 78) from the palace of Sennacherib, who reigned between 705 and 681 B.C., show Phoenician vessels that, with two lines of rowers working their oars from inside the hull, are full-fledged two-banked craft.[77]

The germ of the two-banked galley, then, lay in adding a raised deck and, *ipso facto*, making available a second level at which oarsmen could be placed. As the Assyrian reliefs graphically reveal, the new ship in its fully developed form was no mere decked single-banker with the complement of rowers doubled. A shipwright could not help but realize that superimposing the rowers would enable him to take further strides in his favored direction, the building of stronger, more shock-resistant warcraft. A 50-oared galley with rowers in two banks, for example, would be perhaps half the length of its single-banked sister—some 65 as against 125 feet; a two-banked triaconter would run some 45 feet as against 75 for the single-banker.[78] The shorter craft would inevitably boast a hull that was more robust, more seaworthy—and that offered a good deal less of a target to an enemy ram.

The Assyrian reliefs are of key importance, for they reveal the final improvement in design that brought about a proper two-level galley. The eighth century prototype had been a top-heavy craft with one line of oarsmen at gunwale level and the other at the level of the

[77] The reliefs show the evacuating of Tyre by the king of Sidon, as R. Barnett has demonstrated; see *Archaeology* 9 (1956) 91-93, where he has published a newly discovered piece of the relief (ill. 9 on p. 93) which supplied the clue. That the ships are Phoenician is now a certainty. The originals have been lost except for one piece; for a photograph of this, see *Eretz-Israel: Archaeological, Historical and Geographical Studies* 9 (1969) 6 and pl. 1.

[78] Cf. 54-55 above and note 71.

raised deck. In the Phoenician version visible on the reliefs, we see an infinitely more compact ship, whose upper oars are worked from the gunwale itself, and the lower through ports in the hull. To fit everyone in with economic use of space, the oars of the two banks are staggered: each one of the upper is centered over the space between two of the lower. The first step toward the creation of the ship rowed in bireme fashion had been an increase in the side planking that left but a wide slot for the lower bank of oars (52 above); the last step, it appears, was the closing in of the slot, thereby making the hull solid right up to the level of the erstwhile raised deck.

The change inevitably brought in its wake a change in the position of the deck. Had this been left untouched, it would have been even with the rowers in the upper bank instead of being above them. It had to be raised—and, fortunately, one of the Assyrian reliefs (Figs. 75, 76) is detailed enough to reveal how this was done. We see what seems to be a line of stanchions forming squares that frame the rowers like those in the Geometric ship-pictures. But a glance at the position of the hands and arms shows the truth of the matter: the artist has carefully portrayed these as being *outside* the stanchions. The stanchions then, do not rest on the gunwale but are *inside* the hull, supporting a raised deck that runs over the centerline area only (Fig. 75).[79] This superstructure, offering no possibility for the seating of oarsmen, was a fighting deck pure and simple.

A galley with such a fighting deck must have proven particularly useful, because it was produced in a single-banked version as well. A relief (Fig. 79) from Asia Minor that dates between 725 and 680 B.C. and is clearly Phoenician in inspiration,[80] shows a vessel similar to the

[79] The plan, Fig. 75, is from Salonen, *Wasserfahrzeuge* 40, who was the first to point out this feature. Basch (148), in dismissing it as the mere result of artistic convention, is being unfair to a careful and consistent sculptor; see my note in *MM* 56 (1970) 31.

[80] Found at Karatepe in Cilicia; see R. O'Callaghan in *Orientalia* 18 (1949) 193-99 and pl. 27. For date and connection with Phoenicia, see H. Frankfort, *The Art and Architecture of the Ancient Orient* (Pelican History of Art 1954) 186-88. A horizontal supported by vertical stanchions runs from afterdeck to foredeck; two figures seated upon it prove that a deck, not a rail, is intended.

A Phoenician galley appears in a fragmentary Assyrian wall-painting of the 8th

Phoenician bireme in all key respects save one: it has rowers only along the gunwale (the hull, having to accommodate but one bank of oarsmen, is as a consequence somewhat shallower). The type, first attested here, apparently swiftly gained favor elsewhere (60-64 below).

Thus, the earliest evidence for important advances in naval construction comes from the east, predating that from Greece by at least a century (60 below). The credit, it would seem, ought to go to Phoenician shipwrights. Do they also deserve credit for the total development—for the introduction of the galley with a raised deck? Or, indeed, for the introduction of the ram itself? Both these features, to be sure, make their first appearance on Greek vases of the ninth century B.C.—but this may simply reflect the fact that Greeks liked to decorate ceramics with maritime scenes and Phoenicians did not. For the time being, at any rate, it is safest to leave such questions open.

What were the sizes of the earliest biremes? Among the Phoenician vessels pictured on Assyrian reliefs, there are examples of 8 oars per side, of 9, 11, 15, and one has at least 17 and very likely more.[81] Herodotus and Thucydides, in dealing with the warcraft of this age, speak of triaconters and penteconters; those in contexts implying dispatch or transport could be either single-banked or double, more likely single, but those described as serving as ships of the line are surely double-banked. When Homer specifies (note 8 above) that the ships in which Philoctetes or Achilles ferried their troops to Troy were penteconters, he probably has single-banked craft in mind—

B.C. (F. Thureau-Dangin and M. Dunand, *Til-Barsib*, Paris 1936, pp. 71-72, and frontispiece; for a far clearer reproduction see G. Garbini, *The Ancient World*, New York 1966, p. 86, fig. 50). Whether it is a combat craft or just a naval auxiliary is uncertain. The forefoot juts out in a long needlelike extension, but this may be only the structural projecting forefoot (cf. 35 above) and not a ram, since rams on Phoenician warships are distinctively cone-shaped (Figs. 75-76, 78-79).

[81] See Figs. 76, 78. The craft with 15 oars per side—a two-banked triaconter, with 8 oars in the upper bank and 7 in the lower—is illustrated in V. Place, *Ninive et l'Assyrie* III (Paris 1867) pl. 50 bis, no. 3 (= Köster pl. 16 and *Antiquity* 32, 1958, pl. 22a), a relief from the time of Assurbanipal III (668-after 633 B.C.).

but when Herodotus and Thucydides assume the pentaconter to have been the standard ship of the line until the arrival of the trireme, they surely have the two-banked version in mind.[82] And ships with rowing complements much larger than 50 necessarily were biremes. For example, Homer gives to the Boeotian contingent the distinction of using galleys that carried 120 men each, which must include the officers and deck personnel as well as rowers (cf. note 9 above). The figure cannot be dismissed as poetic fancy, for Thucydides, no armchair admiral, accepts it as sober fact.[83] Such ships had to be biremes: a galley with two banks per side of, say, 25 to 30 rowers is perfectly feasible (the two lower banks of a trireme, as we shall see, had 27), but a galley with 50 to 60 on each side—some 200 feet in length at the very least[84]—that would still be narrow enough to be a useful fighting ship is a structural impossibility.[85] The Boeotian galleys, as it happens, are the largest warships recorded until the invention of the trireme.

The rig carried by all the various war galleys of this age was

[82] Herodotus and Thucydides make it clear that, for many years before the widespread use of the trireme in the 5th b.c., the standard warship was the pentaconter; see Davison 19-22. On the other hand, they never mention two-banked craft, although, as we have seen, these had made their appearance before 700 b.c. This has been reckoned a problem (cf., e.g., Davison 23-24, Kirk 136-37), and various solutions have been offered. Kirk, for example, has suggested that the two-banked galley was but short-lived among the Greeks, lasting just long enough for them to discover that it was "unsuitable for the open sea"—a solution which, inter alia, overlooks the fact that two-banked galleys served the Roman and Byzantine fleets for nearly a millennium. Actually there is no problem. One arises because it has been presumed that, since a pentaconter was in origin single-banked, it always remained so. A pentaconter was a galley with 50 or so oars whether they were in one line or two (cf. AM 84-86, Anderson 4-5, GOS 155). Herodotus and Thucydides, to be sure, never characterize pentaconters as two-banked; neither does a modern historian bother to characterize, say, today's fighter aircraft as jet-propelled. As it happens, the one clear reference we have to a two-level pentaconter is from a tragic poet. Euripides, IT 407-408, describes Orestes' ship, which is a pentaconter (cf. line 1124), as moving "with a double-beat [dikrotos] surge of oars of fir" (ῥοθίοις εἰλατίνας δικρότοισι κώπας); cf. GOS 194-95.

[83] See note 103 below.

[84] Allowing only 30-50 feet for afterdeck, foredeck, and ram—no more than in a pentaconter; cf. note 71 above.

[85] Cf. Torr 3. Anderson 2 suggests that they were single-banked with two men to each oar. Perhaps—but the subsequent development of the galley that culminated in the trireme would seem to be against this.

standard: the single square sail set amidships on a retractable mast. On the Geometric vases the sails are filled in with cross-hatching (Fig. 66), probably to represent the sewn patches that made up the surface. The stays are best shown on the Phoenician ships (Fig. 78), a double forestay and single backstay, just as described in Homer (47 above). The mast-top, known from the Bronze Age (Fig. 61) though not mentioned by Homer, continued in existence, at least on some ships (Fig. 80).[86]

IV SIXTH-CENTURY WARCRAFT: THE BLACK-FIGURED VASES

For the seventh century B.C., representations of ships[87] are relatively few and portray only single-banked galleys, both aphracts[88] and the new version with fighting deck,[89] the first Greek evidence for the latter. For the next century, the period of the black-figured vases, the material is abundant.[90]

Though artists of this age have the same partiality for single-banked craft with their more slender, elongated lines (Figs. 88-90), they have left a good number of representations of two-banked galleys, which, as we would expect, are all the latest version with solid hull and the lower oars worked through ports (Figs. 81-83, 85).[91] These are, as it happens, of a lighter class, with no fighting deck.

[86] See also GOS, Arch. 8 and pl. 8c.

[87] A full catalogue of the evidence is presented by Williams in GOS 73-83.

[88] For examples of 7th century aphracts, see GOS, Arch. 2, 27, 28; cf. Kirk 119-23. GOS, Arch. 2 shows an interesting scene. The rowers, having dropped their oars to don battle dress, are standing at attention. Since their feet are on a level with the oars, they must be standing either on the rowing thwarts or, equally likely, on a line of gunwale-level decking that ran down the center. Williams (GOS 74) argues that, since the circular shield was in reality a small affair covering only the upper body, the men must be seated; other pictures, however (see Fig. 80), show the same kind of shield covering the entire body.

[89] GOS, Arch. 5 and pl. 9a (from the same source as Fig. 80, the famous vase by Aristonothos [= Torr, ill. 3.15]); GOS, Arch. 31 and pl. 10d, an ivory plaque from Sparta.

[90] Williams (GOS 84-117) furnishes a comprehensive catalogue of the evidence.

[91] Other examples are BM 60 (AM, pl. 6b; cf. GOS 112 and pl. 22b) and G. Micali, Storia degli antichi popoli italiani (Milan 1836²) pl. 103.3. Possibly the fragment by Exekias (Fig. 88) should be included; see 62-63 below.

On the other hand, single-banked galleys with a fighting deck over the centerline, more graceful versions of the type first attested in the previous century, are common (Fig. 84).[92]

The pictures are abundant and varied enough to illustrate galleys of all sizes:[93] single-banked[94] and double-banked penteconters,[95] sin-

[92] Other examples are *GOS*, Arch. 39, 53, 65, 67, 69, 74, 84, 90, 93 (pls. 12b, 14c and d, 17a and c-e, 18d, 21a and d).

Vessels with a fighting deck can generally be distinguished from those without. The latter have above the hull a line or set of lines, often very light, running past the rowers' faces to indicate a rail (cf. Fig. 89); the former invariably have a heavy line (cf. Fig. 84) supported by vertical stanchions (which may be omitted; cf. App. 1, note 6) running over the heads of the rowers. Now, a rhyton of ca. 600 B.C. in the form of a ship (*GOS*, Arch. 30, pl. 10c)—a poor object to use, but three-dimensional evidence is so scarce we cannot afford to be choosey—has the foredeck's palisade set well inboard of the sides. If we prolong the lines of the palisade, they follow exactly the course of a fighting deck covering only the area over the ship's centerline.

[93] Even in the case of these sophisticated artists, the number of rowers shown cannot be taken at face value precisely. This is obvious from the way the number of rowers' heads and the number of oars more often than not disagree. A very instructive example is the fragment of a mixing bowl done by the famous Exekias (*GOS*, Arch. 53; Fig. 88 shows two of the five ships figured), so noted for his skill and care (cf. Davison 24; Williams, *GOS* 91). Five ships are shown. Of the two best preserved, one has 23 heads and oars (Fig. 88, ship to left)—but the other has only 20 heads for 29 oars (Fig. 88, ship to right).

Another clear case in point is a bowl in Vienna with four ships painted about the rim (*GOS*, Arch. 51). Although the oars are very few—from 7 to 11—there are consistently two or three fewer or more heads than oars (*GOS* pl. 14a-b shows three of the ships, and the fourth has been reproduced in *Art in America* 29, 1941, p. 210, fig. 1). A bowl in Madrid (*GOS,* Arch. 65) has five ships painted about the rim (*CVA* III H *e* pl. 6.3 and 7.1-4 [Spain, pls. 24-25]). Of these, three are perfect triaconters, 15 heads and 15 oars. Yet, of the other two, one lacks an oar and a second was given 16 heads and oars. The vase in Vienna, as it happens, is connected with Exekias (*GOS* 91).

Often the artist paints in many more heads than oars, even double the number. The fact that he will double the number is significant: it seems likely that he was trying to indicate, sometimes summarily, the rowers on the farther side. E.g., a bowl in the Louvre, F 61 (*GOS*, Arch. 67), has one ship (*GOS* pl. 17c) with 23 heads to 13 oars and another (not illustrated) with exactly 20 heads to 10 oars; another on the same vase has 18 heads and 11 oars (*CVA* III H *e* pl. 2.4 [France 74]). *GOS*, Arch. 81 bis shows four ships, one of which has 9 oars and 16 heads; another 10 oars and 23 heads (or 24, allowing for one hidden behind the forward parapet); a third 11 oars and 23 (or 24) heads, and a fourth 12 oars and 18 heads. Paris Bibl. Nat. 322 (*GOS*, Arch. 81) has one ship (*GOS* pl. 18b) with 27 heads to 15 oars and another (*CVA* III H *e* pl. 53.5 [France 439]) with 28 to 16. London B 679 (*GOS*, Arch. 93) has one ship (*GOS* pl. 21d) with 26 heads and 16 oars.

All the above renders doubtful the value of putting much weight on the precise number of oarsmen depicted, as Williams (*GOS* 29, 39) occasionally does.

[94] *GOS,* Arch. 54 (pl. 14e) ca. 18 oars and ca. 28 heads; *GOS*, Arch. 55 (*CVA* III

gle-banked and double-banked (Fig. 85) triaconters,[96] single-banked 20-oared craft.[97] Vessels showing 18 to 22 oars a side (Fig. 89)[98] may stand either for 40-oared galleys or for penteconters,[99] even as those with 12 or 13 a side may stand for triaconters.[100] One ship, painted by Exekias, among the most skilled and careful of the black-figure artists, has 25 oars, and, in addition, a row of ports in the hull; it

---

H e pl. 2.1 [France 74], ship to left) 26 oars and 24 heads; Arch. 73, 25 oars; also *GOS*, Arch. 53, pl. 14d, if it is not two-banked (see 63 below).

[95] The galley in Figs. 81-82, though in effect a double-banked penteconter (12 oars in the lower bank, 10 visible in the upper) strictly speaking is a *hemiolia*; see 128 below.

[96] Single-banked: *GOS*, Arch. 63 = Louvre F 145 (*CVA* iii H e pl. 88.5 [France 629]), four ships, one with 16 oars, another with 17, a third with 14; *GOS*, Arch. 72 = Louvre Camp. 11247 (*CVA* iii H e pl. 155.8 [France 828]) 14 oars and 15 heads; Arch. 65 = Madrid 10902 (cf. note 93 above); Arch. 68 = Villa Giulia 959 (*CVA* iii H e pl. 56.2 [Italy 140]), 14 oars and heads; Arch. 81 = Paris Bibl. Nat. 322 (see note 93 above), one ship with 15 oars and another with 16; Arch. 93 = London B 679 (see note 93 above) one ship with 16 oars; Arch. 74 = Berlin 1800 (*GOS* pl. 17e) 15 oars and apparently the same number of heads.

Double-banked: Fig. 85, 8 oars in each bank; perhaps Fig. 83, 6 oars in each bank. Cf. the Assyrian relief of a galley (see note 81 above) with 8 oars in the upper bank and 7 in the lower.

[97] *GOS*, Arch. 61 and pl. 16c = Leningrad B 1525, one ship with 9 oars and ca. 16 heads; Arch. 67 = Louvre F 61 (see note 93 above), one ship with 11 oars and 18 heads and another (not illustrated) with 10 oars and 20 heads; Arch. 97 = Louvre Camp. 11248 (*CVA* iii H e pl. 155.10 [France 828]), 10 oars and 9 heads; Arch. 91 = Würzburg 527 (E. Langlotz, *Martin von Wagner-Museum der Universität Würzburg, Griechische Vasen*, Munich 1932, pl. 135) 10 oars and heads and (*GOS* pl. 21b) 11 oars and 10 heads; Arch. 35 and pl. 11d = Louvre E 735, 10 oars; Arch. 51 and pls. 14a and b = Vienna 3619, one ship with 10 oars and 9 heads, another with 11 and 8, a third with 9 and 7; Arch. 68 = Villa Giulia 959 (see note 96 above, pl. 56.1) 10 oars and 16 heads; Arch. 90 (= *AM* pl. 5a), 11 oars and 16 heads; Arch. 79 and pl. 18a = London E 2, 11 oars and 18 heads; probably Arch. 81 bis, which shows four vessels with oars ranging from 9 to 12 (see note 93 above).

[98] *GOS*, Arch. 66 and pl. 17b = Leningrad 86, one ship with 18 oars and 17 heads, another with 17 and 18; Arch. 56 and pl. 14g = Munich 781, 19 oars and 22 heads; Arch. 53 (see note 93 above; not illustrated), 20 oars; Arch. 55 = Louvre F 62 (my Fig. 89) 22 oars and heads; Arch. 70 = Louvre Camp. 11244 (*CVA* iii H e pl. 154.6 [France 827]), 20 oars and heads; Arch. 84 and pl. 18d = Salerno Museum *dinos*, 19 heads (no oars shown).

[99] Cf. note 93 above.

[100] *GOS*, Arch. 67 and pl. 17c = Louvre F 61, 13 oars and 23 heads and still another on the same vase (see note 93 above) has 11 and 18; Arch. 92 = Naples bowl (*Accademia dei Lincei, Monumenti Antichi* 22 [1913] pl. 60.2; cf. *GOS* pl. 21c), one ship with 13 oars and 18 heads, another with 12 oars; Arch. 68 = Villa Giulia 959 (see note 96 above, pl. 56.3), 13 oars and 16 heads.

looks for all the world like a two-banked 100-oared vessel being rowed from the upper bank alone.[101] Still another (Fig. 88) with a row of ports in the hull and, above, a line of 29 oars,[102] looks like the exact equivalent of the type that Homer ascribes to the Boeotians (note 9 above). As a matter of fact, numbers of galleys larger than the penteconter must have been common in this period, common enough to induce Thucydides to conclude that the penteconter was the smallest transport unit in the fleet dispatched against Troy.[103] Very likely it gained its favor after experience proved it to be a particularly efficient size. A special version of the two-banked galley was produced on the island of Samos, one somewhat wider and roomier than usual so it could serve as cargo-carrier as well as man-of-war.[104]

Sixth-century representations reveal a number of details that could not be discerned, or discerned as well, in the simpler drawings of previous ages. Horizontal lines indicate wales, stout timbers that

[101] GOS, Arch. 53, pl. 14d.

[102] Williams (GOS 97) interprets all the ships shown on this vase, with a row of ports below a line of oars, as two-banked.

[103] Thucydides 1.10.4: πεποίηκε γὰρ χιλίων καὶ διακοσίων νεῶν, τὰς μὲν Βοιωτῶν εἴκοσι καὶ ἑκατὸν ἀνδρῶν, τὰς δὲ Φιλοκτήτου πεντήκοντα, δηλῶν, ὡς ἐμοὶ δοκεῖ, τὰς μεγίστας καὶ ἐλαχίστας· ἄλλων γοῦν μεγέθους πέρι ἐν νεῶν καταλόγῳ οὐκ ἐμνήσθη. αὐτερέται δὲ ὅτι ἦσαν καὶ μάχιμοι πάντες, ἐν ταῖς Φιλοκτήτου ναυσὶ δεδήλωκεν· τοξόταs γὰρ πάντας πεποίηκε τοὺς προσκώπους· περίνεως δὲ οὐκ εἰκὸς πολλοὺς ξυμπλεῖν ἔξω τῶν βασιλέων "[Homer] has, with the 1,200 ships, made those of the Boeotians 120-man ships and those of Philoctetes 50-man ships, indicating, as I see it, the largest and smallest units. Of the size of the others, at any rate, the Catalogue of Ships makes no mention. That the men all were at the same time both rowers and fighters he has made clear for Philoctetes' ships, for he has included all the archers as oarsmen. Nor is it likely that many passengers were included in the voyage, aside from the chieftains." Thucydides thus makes it clear that, as he understands Homer's figures, they refer to the number of rower-fighters aboard each ship and do not include any significant number of nonrowing passengers. Morrison (GOS 46, 68) believes that Thucydides overrated the size of early galleys, but it is dangerous to doubt the word of a trustworthy historian who was an admiral in the bargain.

[104] The Samia or Samaina "Samos-craft." Plutarch, Per. 26.3-4: ἡ δὲ Σάμαινα ναῦς ἐστιν ὑόπρωρος μὲν τὸ σίμωμα, κοιλοτέρα δὲ καὶ γαστροειδής, ὥστε καὶ ποντοπορεῖν καὶ ταχυναυτεῖν. οὕτω δ' ὠνομάσθη διὰ τὸ πρῶτον ἐν Σάμῳ φανῆναι, Πολυκράτους τυράννου κατασκευάσαντος "The Samaina is a vessel boar-prowed at the nose and rather big-bellied and roomy so as to sail on open water as well as serve as a galley of the line [tachynautein; see FIVE, note 81]. It was so called because it made its first appearance at Samos, where Polycrates the Tyrant had some built"; cf. Herodotus 3.59. Polycrates reigned ca. 532 to 523 B.C. Photius and Suidas, s.v. Σαμίων ὁ δῆμος specify that the ship was two-banked (dikrotos; see 133 below).

girdled the ship horizontally (e.g., the twin lines just above the oar blades in Fig. 89 and below the oarports in Fig. 90). The stanchions that supported the cheeklike projections forward where the hull swelled out from its slender bow to form the rowing chamber are roughly indicated by one or more concave lines running from the upper wales vertically to the waterline (e.g., four lines on Fig. 85, two on Fig. 89, one on Fig. 90).[105] Even the bronze sheath that fitted over the wooden core of the ram is sketched in by one or more vertical lines to indicate where the metal coat ended (e.g., the chevron-like lines in Fig. 89),[106] while sometimes a long horizontal stroke indicates its seam (Fig. 88).[107]

The representations reveal as well some novelties. The oars occasionally nestle in ports or half-ports (Figs. 85, 90), and on aphracts a latticed bulwark fences in the amidships area (Fig. 90). The stempost is now a short upright instead of the long curved horn of the Geometric age. The sternpost ends in a graceful fanlike cluster made up of slender curving poles occasionally embellished with a turbanlike cap-piece (most elaborate version in Fig. 90); this fanlike form of stern-ornament will remain the hallmark of the warship. The earlier round geometric bow-device has become a true oculus, and the bow-patch tends to be combined with the decorative pattern of the palisade (Figs. 85, 89, 90).[108] And the ram, hitherto just a massive prong, is made to resemble a boar's snout.[109]

The rig, being the same as on merchantmen, is discussed below (68).

[105] Cf. Williams, GOS 96-97, who was the first to discuss these lines and point out their significance.

[106] The end of the bronze sheath is clearly indicated on the early 7th century Assyrian relief (Fig. 76); the complicated lines shown perhaps represent clamps and rivets.

[107] Cf. Williams, GOS 95-96, who was the first to discern the significance of these lines.

[108] On bow-patches, cf. Bowen, loc. cit. note 18 above.

[109] The boar-headed prow is mentioned in literature in connection with certain craft of the island of Samos. Cf. Plutarch, cited in note 104 above, and Herodotus 3:59: τῶν νεῶν καπρίους ἐχουσέων τὰς πρῴρας "the [Samian] ships with boar-shaped prows."

FROM the two-level man-of-war with both banks in the hull and with a fighting deck over the centerline area to the early trireme—the kind Thucydides described (see FIVE, note 54) as "not fully decked"—is but a step. And though it was taken before 500 B.C., it is best treated as part of the discussion of the trireme (FIVE).

V MERCHANTMEN

THE first merchant ships used in the Mediterranean must have been oar-driven. As the pace and volume of overseas commerce grew, large seagoing carriers came into being, and these were necessarily sailing ships. But the merchant galley stayed very much alive through the whole of ancient history (EIGHT) because it performed a particular function: in the Mediterranean, plagued by calms all summer when maritime activity was at its height, only an oared ship could offer speed and reliability.

At the outset, the merchant galley differed from the man-of-war solely in being somewhat roomier and heavier. Homer, when he likens the Cyclops' staff to "the mast of a broad-beamed, black-hulled, 20-oared merchantman that sails the great sea,"[110] neatly hits the key features: a sturdier mast, no doubt carrying a bigger sail and perhaps not retractable, and a wider hull. But all this had to change when the ram made its debut. From then on, war galleys, powerful vessels with reinforced hulls and special bow structure, were one thing, and their peaceable sisters another.[111] The distinction is clearly visible in one (Fig. 78) of the Assyrian reliefs referred to above. The Phoenicians are shown evacuating the city of Tyre in anything that would float,[112] and the bowl-shaped merchant galleys are totally unlike the beaked men-of-war. A clay model from Cyprus (Figs. 86-87) illustrates what a heavy merchant galley of this age looked like—and,

[110] *Od.* 9.322-23: ὅσσον θ' ἱστὸν νηὸς ἐεικοσόροιο μελαίνης / φορτίδος εὐρείης, ἥτ' ἐκπεράᾳ μέγα λαῖτμα. See 5.249-50 for another allusion to a merchantman's beaminess.

[111] Compare what happened when naval guns were introduced toward the end of the 14th century. Until then the sailing merchantman and warship had not been too different and were often used interchangeably. This ceased to be when the man-of-war was transformed into, in effect, a gun platform.

[112] See note 77 above.

except for a telltale series of ports pierced in the hull for oars, it could easily be mistaken for a sailing ship. The hull is deep and rounded, and it ends aft in a prominent sterncastle flanked by bulky towers to house the steering oars.[113] A Greek merchant galley of the sixth B.C. shows a similar heavy hull with rounded stern, but with a distinctive straight prow (Fig. 91); it belongs to a tradition observed in the Aegean as early as the second millennium B.C. (32 above).

The cargo carrier par excellence, however, was the sailing ship. The round hull, favored earlier by Crete and the Levant (32-35 above), is now attested from Phoenicia to Italy. The evidence, moreover, is abundant enough to give some idea of how widely, then as during all the rest of maritime history, sailing craft could differ from place to place. The Phoenicians, for example, went in for hulls that were particularly beamy and rounded; the best known type among their merchantmen, for example, was named *gaulos* "tub."[114] They also liked simple, upright stem- and sternposts of equal height. The class of ships called *hippoi* "horses" bore a horse's head as figurehead (Fig. 92), but this was the sole important distinction between the two ends—and sometimes both ends bore a figurehead.[115] Far-

[113] The housings for the stern oars have no ancient parallel, though there are parallels from later ages (cf. *IH* ill. 93).

[114] It occurs as a common name for Phoenician merchantmen from the 5th century B.C. to the 3rd: see Epicharmus (5th B.C.) and Antiphanes (4th B.C.) quoted by Athenaeus 7.320c and 11.500f; Aristophanes, *Birds* 598, and cf. the scholium citing Callimachus (3rd B.C.): Κυπρόθε Σιδόνιός με κατήγαγεν ἐνθάδε γαῦλος "A *gaulos* of Sidon brought me here from Cyprus"; Herodotus 3.136, 6.17, 8.97; Scylax, *Per.* 112, Müller, *GGM* I, p. 94 (4th B.C.): οἱ δὲ ἔμποροί εἰσι μὲν Φοίνικες· ἐπὰν δὲ ἀφίκωνται . . . , τοὺς μὲν γαύλους καθορμίζουσιν "The traders are Phoenician. When they arrive . . . they moor their *gauloi*." Few representations of Phoenician ships have been preserved, and none identifiable as *gauloi*, but those we have show well-rounded hulls, as, e.g., the transports in Fig. 78 and a clay model of what seems to be a coastal barge (J. Fevrier, "L'ancienne marine phénicienne et les découvertes récentes," *La Nouvelle Clio* 1-2, 1949-50, pp. 128-43, esp. 134-35, reproducing M. Dunand, *Fouilles de Byblos*, Paris 1937, no. 6681 = pl. 140 of Atlas).

[115] The distinctive figurehead decorated craft of all sizes. In the reliefs from Sargon's palace (Fig. 92), the *hippoi* are smallish oared boats engaged in hauling logs along the Syrian coast, whereas a story in Strabo (2.99) indicates that *hippoi* were used for circumnavigating Africa. For boats with the figurehead on both stem and sternpost, see King, *op. cit.* (Two, App. 2) pl. 13 = Salonen, *Wasserfahrzeuge* pl. 17.1 (9th B.C.).

The small boats pictured in Fig. 92 have a modern descendant that shows in-ther west, the Greeks, using the same sort of rounded hull (Fig. 94), decorated the ends far more ornately. Sometimes both curved inward;[116] sometimes they curved like those of Geometric Age warships: the sternpost described a great arc inward, while the stempost twisted outward (Figs. 93, 95-96).[117] In Italy there seems to have been developed a form of armed merchantman, one with a hull as round and roomy as elsewhere but with a bow that ends in a massive spur high above the waterline (Figs. 80, 93).[118] This cannot be a ram, which would be lower down at the waterline—and, in any event, would have no place aboard a sailing vessel. It may, however, be a

---

credibly little change, the *dghaisa*, a distinctive harbor craft of Malta, where the Phoenicians early established a colony; see A. Tilley, "A Phoenician Survival," *MM* 55 (1969) 467-69.

[116] See J. Brock, *Fortetsa* (*BSA*, Supplementary Paper, no. 2, Cambridge 1957) pl. 135 (= *GOS*, Geom. 1), 9th century vase from Cnossus. A sherd of the late 8th B.C. recently found in Elis shows what may be a round-hulled merchantman; see *Archaeological Reports for 1966-67*, p. 11, fig. 16. We cannot be certain since the prow is lost as well as part of the stern.

[117] Cf. also *AM* pl. 6a, 7th century vase from Cyprus, and Louvre E 751, 6th century vase from Cerveteri (illustrated in L. Stella, *Italia Antica sul mare*, Milan 1930, pl. XLIII opposite p. 160; the ship to the left). The vase in Fig. 93 has been reproduced by Torr (pl. 3.12) in a line-drawing that is most misleading, making the spur look far more like a ram than it should.

[118] Other examples are the ship on the vase from Cerveteri (see previous note) and a ship on a 7th B.C. ivory bucket found in Etruria (O. Montelius, *La civilisation primitive en Italie, Italie Centrale*, Stockholm 1895, pl. 225 [line-drawing]; *Enciclopedia dell'arte antica*, VII, Rome 1966, p. 358, fig. 448 [photograph]).

I believe a further example of the protective spur is to be found hidden under a fantastic animal shape on a 7th B.C. vase from Cerveteri in the Louvre (D 150; E. Pottier, *Vases antiques du Louvre* I, Paris 1897, pl. 34; Moll B VI a 8; Whibley 577, figs. 133, 134). As on the Aristonothos vase (Fig. 80), we see a pirate attack on a merchantman. As there, the attackers are using a decked single-banked galley, in this instance distinguished by a gigantic duck-head figurehead; they stand on the fighting deck, armed with sword and shield ready to board. Their prey is an oared merchant ship that has a prow ending in a massive point well above the waterline; this, which the artist has imaginatively rendered as a monster's snout, would be, stripped of its disguise, what is pictured elsewhere as a spur.

Some writers have argued that these round-hulled vessels are warships of Etruscan make, the spur being a ram; see Stella, *op. cit.* (note 117 above) 155-60, and S. Paglieri, "Origine e diffusione delle navi etrusco-italiche," *Studi Etruschi* 28 (1960) 209-31. Paglieri reproduces (221) a graffito of a round-hulled galley that has a sketchily indicated peculiar rectangular projection forward which he considers a ram; if so, it is one with a singular shape.

projection to keep would-be attackers at a distance.[119] One of the ships so armed (Fig. 80) also has a fighting deck and top, and we see warriors on both ready to stand off a pirate galley. Still farther westward, the Sardinians built their craft with no sternpost at all and, by contrast, a prominent stempost which projects stiffly upward and outward and ends in an animal head.[120]

The hull with straight lines and angled ends that was attested in the Aegean during the second millennium B.C. (32 above) reappears among Greek sailing ships as well as merchant galleys. We see it in two pictures of the sixth-fifth century B.C. which, because of the quality of their execution and the details shown, are particularly helpful, a vase-painting (Figs. 81-82) and a tomb-painting (Fig. 97). The two vessels are clearly close kin: both are fenced round with the same kind of latticed bulwark, both carry a landing ladder prominently displayed at the stern, and, most significant, both have a distinctive concave bow; that in the vase-painting is particularly striking, being as graceful as any clipper's bow.

VI RIG

THE ship-pictures of the sixth century B.C. are detailed enough to supply information about rig, which both confirms and expands Homer's evidence. It will be convenient to treat the rig of merchant craft and warcraft together.

The square sail continued to be the sole type used, a broad type bent by lacings[121] to a single long yard made of two spars fished together (Fig. 90). The yard was held to the mast by a parral, a collar of twisted cords (Fig. 91). Sails often had so deep a bunt that they ballooned out like spinnakers (Fig. 90). Apparently they were made of lightweight cloth, for they had to be reinforced with ropes running

---

[119] Cf. the beaks on galleys of the 15th century and later (*IH* ills. 88, 91, 142).

[120] The evidence consists of numerous bronze models. Though most are of small craft, some seem to represent seagoing vessels; see G. Lilliu, *Sculture della Sardegna Nuragica* (Cagliari 1966²) 387-411, esp. 409-11 (nos. 297, 298).

[121] Clearest in the famous cup by Exekias showing Dionysus reclining in a boat (*GOS*, Arch. 52 and pl. 13).

at fixed distances horizontally from edge to edge over the front surface; these, together with the brails (see below), which they crossed
at right angles, marked off the face of the sail like a checkerboard
(Fig. 91).[122] Masts were composite, secured by wooldings at evenly
spaced intervals,[123] and could be fitted with a top (Fig. 80).[124] Mast
step and partners are clearly indicated in a clay model of the ninth
or eighth B.C. (Figs. 86-87).

The standing rigging described by Homer, a double forestay and
single backstay (47 above), is attested both on Phoenician (Fig. 78)
and Greek ships (Fig. 98). More often a single forestay is shown
(Figs. 80, 92, 97). Frequently (e.g. Fig. 89) no stays of any kind appear, an omission which is surely the result of artistic convention.[125]
Shrouds, however, are so consistently left out that it seems purpose-
ful; the retractable masts of warships[126] probably had enough lateral
support from the tabernacle, and fixed masts must have been stout
enough not to need any.

Next, running rigging. To ease the work of raising the yard, there
were two separate halyards: the artists carefully record this detail
by showing lines descending along either side of the mast (Figs. 81-
82); in one representation we actually see two hands hauling away,
one on each.[127] Twin halyards go back to the Bronze Age, for the
ships of the Peoples of the Sea were so rigged (Fig. 61). A collar
set near the tip of the mast had earlike appendages to port and starboard to which the halyard blocks were attached (Fig. 91).[128] Braces
ran from the yardarms, sometimes (Fig. 90) with a purchase[129]

---

[122] For other examples, see GOS, Arch. 63, 79, 84, 91, 92 (pls. 16d, 18a and d,
21b and c).

[123] Clearest in the cup by Exekias (see note 121 above).

[124] For another example of a mast-top see JHS 74 (1954) 174, fig. 1 = AA 78
(1963) 561, fig. 38 (clay model from Cyprus, 7th-6th B.C.).

[125] E.g., Exekias, in his cup picture (note 121 above), leaves them off.

[126] For an excellent illustration of a retractable mast, see the famous François
vase (GOS, Arch. 33, pl. 11a).

[127] GOS, Arch. 31 and pl. 10d. For other examples, see Arch. 63, 65, 67, 69, 74,
79, 84 (pls. 16d, 17, 18a and d).

[128] For another clear example, see GOS, Arch. 52 and pl. 13.

[129] For other examples, see GOS, Arch. 51-56 (pls. 13, 14).

69

(i.e., from the deck, over a block at the yardarm, back to the deck). Sheets are always shown but never with any detail, save an occasional indication of the cringle to which they were made fast (Fig. 90).[130] The most prominent lines are the brails (Figs. 79, 82, 89, 90, 91, 97), the ancients' distinctive device for shortening or furling sail. These were a series of lines made fast at fixed intervals along the foot, whence they traveled up the front surface of the sail guided by fairleads, over the yard (Fig. 91), and then down to the deck aft. Sail could be taken in quickly and efficiently merely by hauling on the brails, which bunched the sail toward the yard much in the manner of a venetian blind. Since a number of brails could be bundled together, the shortening of sail required only two or so deckhands.[131] Save for emergencies, strictly speaking, men had to go aloft only when sail was furled, to secure it with gaskets once it had been brailed up.

The standard rig from earliest times down to the sixth century B.C. had been the single square sail. In that century, a dramatic advance took place—the invention of the foremast. There is but one piece of evidence for it—the tomb-painting mentioned above (Fig. 97)— but the evidence is conclusive: we see a second mast set, with a slight forward rake, midway between the mainmast (which stands amidships) and the prow, and carrying a sail that is a smaller-sized version of the mainsail. As will appear later (240 below), this new sail, often used to aid steering rather than as a driver, forever remained distinctly subordinate to the mainsail.

[130] For another example, see *GOS*, Arch. 63 and pl. 16d.
[131] See *GOS*, Arch. 63 and pl. 16d and cf. Fig. 151.

APPENDIX 1

THE SHIP-PICTURES ON GEOMETRIC VASES

GREEK NAVAL development before the sixth century B.C. was almost a complete mystery until, toward the end of the nineteenth century, archaeology finally produced the evidence of the paintings on Geometric vases, which span the period from, roughly, 850 to 700 B.C. These furnished the basis for some worthwhile pioneering studies, but then received no further serious treatment for better than half a century, until G. S. Kirk took them up in 1949. Kirk published a careful and comprehensive article which established a firm foundation for all future research by providing an accurate conspectus of the material then available.[1] However, his interpretations were marred, in some part by an over-readiness to dismiss certain significant features as artistic convention, and in greater part by a consistent distortion resulting from his conviction that the trireme, the end product of the development, had but a single line of rowers.[2] A decade later, R. T. Williams[3] made a new and up-to-date collection of evidence, re-examined it, and arrived at certain important and convincing conclusions, yet, at the same time, showed an even stronger regard than Kirk for the presumed effects of artistic convention.

Indeed, the question of artistic convention has proved a stumbling block in the interpretation of the Geometric ship-pictures ever since they were first studied. The earliest representations, dated ca. 850-750 B.C., show the deck as merely a thin horizontal line resting on slender vertical stanchions (Figs. 30, 65-66).[4] Those dating from 750-700, however, very commonly depict a type of vessel in which the single thin horizontal line is replaced by a thick band outlined by two thin lines (Figs. 67-69, 72, 77).[5] What does the artist mean to convey by this? Torr and Kirk hold that he intends a deck.[6] Pernice argues that, in a naive attempt at perspective, he has

---

[1] The pioneering studies were C. Torr, "Les navires sur les vases du Dipylon," *RA* 25 (1894) 14-27; E. Pernice, "Ueber die Schiffsbilder auf den Dipylonvasen," *Ath. Mitt.* 17 (1892) 285-306. For Kirk's article, see List of Abbreviations.

[2] Cf. FOUR, notes 54 and 74 and App. 2. See Kirk 137 for his view of the trireme.

[3] First in *JHS* 78 (1958) 121-30, with an important addendum in *JHS* 79 (1959) 159-60, then a final publication with a complete catalogue of all relevant representations in *GOS*, Chapter 2.

[4] See also FOUR, note 59.

[5] Cf. Williams, *GOS* 14-17.

[6] Kirk 127-28; Torr assumes it throughout his article (*op. cit.* note 1 above). In some instances (e.g., Fig. 77) no vertical stanchions are shown, omitted presumably

in mind to represent with the thick band the side of the ship away from the viewer, while Williams pushes the theory of primitive perspective still further to argue that the artist has combined both bird's-eye view and profile view: the thick band represents the far side, the space in between, crossed by apparent verticals, is a bird's-eye view of the benches, and below is the near side of the hull in profile.[7] Practically all are agreed that any rowers shown on the band[8] are to be considered not as an upper bank but as the naive result of an attempt to portray the oarsmen of the far side. The ships, in other words, may appear to be biremes, but, in point of fact, are all single-banked.

There is no question that the Geometric artist used certain conventions resulting in features that cannot be taken at face value. He did not, however, use them either exclusively or consistently. There is, for example, a vase in the Louvre[9] with a scene involving chariots that provides a neat case in point. In an upper band of pictures the artist has drawn a chariot with two wheels side by side. This very likely is a convention: we are to understand that one of the two belongs on the far side.[10] However, in a second band of pictures immediately below, he has drawn a chariot with but one wheel. Same artist, same vase, same set of scenes—and one part is probably not to be taken at face value, while the other indubitably is.[11] We are, therefore, perfectly entitled to take what we see in the ship-pictures at face value,[12] if necessary. And certain evidence does indeed make it necessary.

---

because the artist wished to avoid overcrowding his composition. Even more sophisticated artists were not averse to using the same convention; cf., among the ship-pictures on black-figured vases, *GOS* pls. 17c, 18d.

[7] Pernice in *JDI* 15 (1900) 93-95; Williams, *GOS* 12-17.

[8] As in Fig. 77 or *GOS*, Geom. 10; cf. Kirk 129-30, Williams, *GOS* 16, Miltner 911-12.

[9] Louvre A 552 (*CVA* III H b pl. 12.1 [France 788]).

[10] Cf. Williams, *GOS* 13.

[11] Williams argues (*GOS* 13, 17) that the Geometric artist shows a single wheel only when pressed for room—but the artist of the vase I have mentioned was definitely not pressed for room; to show one wheel was his deliberate choice. Williams admits (*GOS* 13, note) true perspective for "later phases of Geometric"; this vase pushes it back half a century.

[12] The naval experts have all along held that the superimposed lines of rowers should be taken at face value; cf. Assmann in *JDI* 1 (1886) 315-16, Cartault in *Monuments Grecs* 11-13 (1882-84) 52, Anderson 5. I am suggesting that everything we see—structure as well as rowers—be so taken. The much discussed bowl from Thebes in the British Museum (Fig. 74), showing two superimposed fighting decks each with a line of oarsmen, reveals how a Geometric artist went about portraying

On a sherd from a late Geometric vase, dated ca. 700 B.C., there is present a portion of a galley that—no question about it—is being rowed from two levels at once. It was Williams' notable contribution to have spotted the fragment and appreciated its importance.[13] And, as he goes on to point out, still another pair of sherds (Figs. 70-71), just slightly earlier in date, must also necessarily be interpreted as showing a two-banked galley.[14] We see a lower row of oarsmen set in squares much like those formed by the slender stanchions of the earlier Geometric representations. However, here the squares are separated by wide partitions and are bounded above by a broad band decorated with conventional motifs. Above the band appears another line of oarsmen. A band so decorated cannot possibly represent the far side of the hull, and, if so, the rowers on it cannot be interpreted as those of the farther side—they must belong to an upper bank.

The evidence of these three sherds leaves two possibilities. If we follow Williams and the others who dismiss all apparent upper-level rowers in the vases dated earlier than the sherds just discussed as the deceptive result of artistic convention, then we are willy-nilly driven to the conclusion that the two-banked galley suddenly sprang into existence shortly before 700 B.C. fully formed, like Athena out of Zeus' head.[15]

If, however, we take what we see on these vases at face value—which, as noted above, we are perfectly justified in doing—then a gradual technological development and a motivation for it can be demonstrated (cf. 49-56 above): the introduction of the ram had a fundamental effect on the design of war galleys, and the changes introduced led through various

---

the farther side: he made it exactly like the nearer and set it on top of the nearer. Williams (*GOS* 29) hesitates as to whether the artist of this picture intended a two-banked galley or not. The answer is no; in a two-banked galley, the lower oarsmen are never at the level of a raised deck but either along the gunwale or within the hull, rowing through ports. See 56-57 above.

[13] See *JHS* 79 (1959) 159-60 and cf. *GOS* 73. The fragment is *GOS*, Arch. 1.

[14] Figs. 70-71 = *GOS*, Geom. 43, 44; cf. also Geom. 40, 41. For Williams' discussion of these, see the following note.

[15] Miltner, otherwise an adherent of the primitive perspective school of thought, took (911-12) these sherds as evidence for upper side planking and an upper level of rowers, without appreciating the spurt in naval technology he was thereby assuming. Williams (*JHS* 78, 1958, pp. 125-26; cf. *GOS* 36-40), by a careful analysis of details and the garnering of fresh evidence (see note 13 above), provided the proof for Miltner's identifications and drew the conclusion that Miltner had failed to draw. Kirk (129-30), following Pernice (*op. cit.,* note 7 above, 95), interpreted even these sherds as showing nothing more than single-banked galleys in faulty perspective.

stages to the development of the two-banked ships attested by the sherds.[16] Even more, we can acknowledge the evident relationship between these sherds and the earlier ship-pictures: what was earlier represented as a wide black band now appears as a decorated one of the same width; what were earlier represented as slender stanchions have grown into wide partitions.

There is, too, a linguistic argument that can be adduced. The term aphract "unfenced" is not the antonym of cataphract, well known from its widespread use in subsequent naval history (88 below), but of $\pi\epsilon\phi\rho\alpha\gamma\mu\acute{\epsilon}\nu\eta$ $\nu\alpha\hat{\upsilon}\varsigma$ "fenced ship," aut sim.[17] When all boats were low and open, there was no need for such a word as "aphract"; it could only have been called into being by the entrance upon the scene of some type of high-sided vessel. If we take the Geometric paintings in question at their face value, they show precisely that type of vessel coming into being in the 8th B.C.

[16] Williams missed this technological development because he assumed the Mycenaean Age knew the ram (GOS 7), an assumption that arose from his failure, first to examine the evidence critically, particularly the identification of bow and stern (cf. THREE, App.), and second to distinguish between the massive bow projection that is a naval weapon and the projecting forefoot that is an age-old characteristic of craft of all size and uses (cf. 35 above).

[17] See Taillardat, op. cit. (FOUR, note 49) 86-89.

APPENDIX 2

APHRACTS IN GEOMETRIC AGE
REPRESENTATIONS

THE FOLLOWING, all of the 8th century B.C., show open galleys:

1. Part of an Athenian mixing bowl formerly in Königsberg = *GOS*, Geom. 16 and pl. 4c; ca. mid-8th B.C. Deckhands hold brails as a ship, after a fight on shore (there are telltale corpses aboard), prepares to sail off; cf. 50 above.
2. Part of an Athenian mixing bowl in the Louvre (Fig. 62) = *GOS*, Geom. 17; ca. mid-8th B.C. Ship at rest with 13 rowers in an unusual stance (see below).
3. Jug from Thebes = *GOS*, Geom. 33. For illustration, see *JDI* 3 (1888) 248, figs. a-b; late 8th B.C.
4. Fragments from the Agora of Athens (Fig. 64) = *GOS*, Geom. 35; late 8th B.C.
5. Fragment of a vase from the Agora of Athens = *GOS*, Geom. 37. For illustration, see *AA* (1963) 656, fig. 12; late 8th B.C.
6. Fragments from the Ceramicus at Athens = *GOS*, Geom. 36. For illustration, see *AA* (1963) 656, fig. 11; late 8th B.C.
7. Vase from Ischia = *GOS*, Geom. 32 and pl. 6e; late 8th B.C.
8. Bowl from Thebes = *GOS*, Geom. 42 and pl. 7d; late 8th B.C. Ship under way manned by nineteen rowers. The hull is marked by a series of prominent white circles bisected by vertical lines. Williams (*GOS* 36, 38), taking these for tholepins, has argued that the ship is a two-banked galley. Perhaps—but by no means certainly. Pictures from a later period (e.g., Fig. 85) show a similar series of circles below a line of undoubted oarports; the clear difference in treatment between the two indicates that the circles must stand for something other than oarports—perhaps nothing more than decoration.

The above examples show hulls that are distinct from those in all other representations of the period by virtue of their lightness and slenderness and, above all, by the absence of any superstructure. They portray open galleys, or aphracts to use the ancient terminology. That such craft were to be found among the preserved ship-pictures of the 8th B.C. has hitherto gone unrecognized for two reasons. For one, some of the evidence (nos.

75

4-7) was published only recently. For another, the representations in question have been the victims of much theorizing. No. 1, for example, shows a vessel with no upper structure and with some corpses on the hull line; Williams (*GOS* 16) argued that the lines which I interpret as representing an upper structure would have been there had not the artist deliberately omitted them in order to leave room for the corpses. Yet, elsewhere—e.g., *GOS*, Geom. 11 and pl. 4a—the artists were perfectly able to fit in both upper structure and corpses; in this case, I suggest that the artists showed no upper structure because the ship, in common with the others listed above, was an aphract and had none. No. 2 is another case in point. Kirk, convinced it was early in date, considered it a primitive make of craft and (99) dismissed the unusual stance of the oarsmen as a "singularly naive attempt simply to portray rowers in action." Williams, aware of the correct date, argued (*GOS* 17; cf. *JHS* 78, 1958, p. 124, note 13) that the artist intended to paint a ship with all the usual upper horizontal lines but, because he had set his picture right under the handles of the vase, found he had too little room and just left them out. Neither were aware that a ship with rowers in precisely the same posture appears on an Egyptian tomb-painting (Fig. 63) done two thousand years earlier.[1] Whatever the meaning of the oarsmen's unusual position,[2] it is clear that the vase-painter means what he shows. Kirk, by failing to recognize the continued existence of the open galley, was forced to posit (137) a naval development in which types and key features appeared and disappeared in a most unlikely fashion.

[1] R. Lepsius, *Denkmäler aus Ägypten und Äthiopien* (Berlin 1849-58) III, Abth. II, pl. 12 (Gizeh, Grab 86 [tomb of Nebemakhet]) = Boreux, fig. 109 on p. 324.

[2] Boreux (340, note 4) concludes that the Egyptian artist was merely indulging his fancy. Assmann suggests (*JDI* I, 1886, p. 315) that the artists were illustrating some ceremonial maneuver. It is a maneuver, but, in my opinion, not a ceremonial one—the rowers are moving the vessel broadside. They have lowered their oars in the water with the blades parallel to the hull. Working as a team (their arms would be thrust out in front of them rather than to the sides, but to portray this involves foreshortening, which was in neither artist's repertoire), they pull the looms inboard, levering against the gunwale, and thereby set the ship in motion, far side foremost.

CHAPTER FIVE

The Age of the Trireme:
500-323 B.C.

I THE "TRIREME QUESTION"

In the fifth century b.c., the ship of the line throughout the ancient world was the trireme,[1] and, except for a few centuries of experiment with larger types, it retained this distinction down to the days of the later Roman Empire.[2] "Trireme" is the English rendering of a word found only in Latin literature; the technical name for the ship, in the Roman navy[3] as well as the Greek, was *trieres* "three-fitted." Precisely what is meant by "three-fitted" has given rise to the famed "trireme question."[4]

Before the trireme made its appearance, the only ship-types mentioned are the triaconter and penteconter (58 above). These terms

---

[1] E.g., triremes were the capital ships in both the Persian and Peloponnesian Wars. Larger units did not appear in Athens' fleets until the second half of the 4th b.c., and then in only relatively small numbers. See *GOS* 223-27, 249.

Most of our detailed information concerns the Athenian navy in the 4th b.c., thanks to the preservation of extensive fragments of the navy yard records of the period. See *IG* ii² 1604-32, 377-322 b.c.; *SEG* xiii 48 (additions to *IG* ii² 1624); E. Schweigert in *Hesperia* 8 (1939) 17-25 (fragments of a second copy of *IG* ii² 1611); D. Laing in *Hesperia* 37 (1968) 244-54 (additions and corrections to *IG* ii² 1628); *SEG* x 355. Laing, who has conducted a thorough-going restudy of the originals, reports (p. 245, note 4) that *IG* ii² 1604 and 1605 probably come from the same stele; that 1613 and 1614 join and are respectively the upper and lower parts of the same stele, which is to be dated 353/2 (p. 253); that 1615 and 1617-19 are parts of the same document; that 1620 and 1621 are from the same stele and should probably be dated 348/7; and that 1628 and 1630 are parts of the same document.

There are also extant a few lines dating from the second half of the 5th b.c.; see *IG* ii² 1604a on p. 811.

[2] Starr 53.

[3] E.g., sailors of the Roman fleet use only *trieres*, the Greek term, on their tombstones, never the Latin *triremis*; see Seven, note 6, and Starr 52.

[4] Morrison, "Trireme" 17-24, gives a concise survey of the history of the question. The appendices in the second edition of Torr (154-214) contain a number of articles from the later stages of the debate. Tarn's footnote on p. 154 there (= *JHS* 25, 1905, p. 137) and Starr in *CPh* 35 (1940) 353 provide a good sampling of the earlier bibliography.

are clear enough: both refer to the total number of rowers in the crew. The *trieres* obviously was named on some different basis. And so, too, were a whole group of larger sisters, developed in the fourth and third centuries B.C., the *tetreres* or "four-fitted," *penteres* or "five-fitted," *hexeres* "six-fitted"—all the way up to a monster *tessarakonteres* "forty-fitted" (cf. 108 below).[5]

The traditional theory about the nature of these galleys goes back to the fifth century A.D., when they had already been out of use[6] so long that writers were reduced to guessing what they were like. This theory held that the names referred to superimposed banks of rowers in which each rower pulled his own oar; a *trieres* thus would be a galley with its oarsmen disposed in three such banks, a *tetreres* with its in four, a *penteres* with its in five, and so on.[7] So far as the trireme was concerned, the theory had the advantage of squaring with whatever scanty information was available. Of the larger types, practically nothing was known—a state of things that the passage of fifteen hundred years has improved but little.

[5] *Trieres, tetreres*, and the others in the *-eres* series refer to warships of more or less fixed specifications: fixed as to general length, breadth, number of rowers. The nomenclature is comparable, say, to the terms "first-rate," "second-rate," "third-rate," etc., of the sailing navies of the 17th and later centuries (cf. *IH* 105). There was never a *dieres* in Greek and Roman navies, since no "two-fitted" vessel of fixed specifications ever existed. Two-level penteconters, two-level triaconters, and other craft rowed in bireme fashion (cf. 61-62 above) were called *dikrotos* "two-banked," a general term that saw frequent use since there happened to be quite a number of different kinds of two-banked galleys (cf. Six, notes 94, 95, 124, 127-29 and App. I, note I). *Trikrotos*, on the other hand, never occurs as an official navy term since there was no need for it: there were no three-banked warships outside of the *trieres* or others of fixed specifications in the *-eres* series. *Monokrotos* is attested as the name of a light naval craft of some sort; e.g., in the Ptolemaic navy it is mentioned in connection with other small units (see Six, note 126), and Strabo (7.325) records that, after Actium, Augustus dedicated a squadron of ten ships "from *monokrotos* to 'ten' " (ἀπὸ μονοκρότου μέχρι δεκήρους), i.e., a light single-banked craft, a somewhat heavier double-banked craft, a trireme, and so on through the *-eres* series to a "ten." Had we more documents at our disposal, we very likely would have more instances of *monokrotos*.

Eventually *dieres* made the grade as a naval term. It appears in the 8th A.D., used of vessels in the Arab navy, where it is a synonym for *dikrotos* (Seven, note 62).

[6] Zosimus, writing in the 5th A.D., remarks (5.20.4) that "ships of the trireme type, . . . for a great many years there has been no construction of these" (τῶν τριηρικῶν . . . πλείστοις ἔτεσι τῆς τούτων ἐκλιπούσης δημιουργίας).

[7] Vegetius, *de re mil.* 4.37; cf. Morrison, "Trireme" 17.

But then, in the early sixteenth century, a rival theory arose.[8] The historian of naval antiquities of this time not unnaturally had trouble believing in galleys powered by 8, 9, 10 superimposed rows of oarsmen, to say nothing of 40. Furthermore, he was well acquainted with the standard warship of the Venetian navy, the so-called *a zenzile* galley. This, like the ancient trireme, used rowers each of whom pulled his own oar. The rowers, however, were not above one another in superimposed rows, but alongside each other in clusters of three to the same bench.[9] Such a craft obviously qualified to be described as "three-fitted"—why could it not be a descendant of the ancient trireme and reflect, at least in a general way, the same arrangement of rowers?

And this theory flourished[10] despite the fact that (a) the *a zenzile* galley was demonstrably *not* the product of a tradition reaching back unbroken to ancient times (indeed, it did not even go back as far as the Middle Ages),[11] (b) such a craft did not at all suit whatever information was available about triremes,[12] and (c) it did not at all square with new evidence that archaeology kept producing, notably the unquestioned examples of galleys with superimposed banks of oars on Greek vases (60 above). But it was defended stubbornly until a few decades ago, when finally a convincing demonstration, imaginatively utilizing all the pictorial evidence available, settled once and for all the key point at issue, if not all the details: the tri-

---

[8] Cf. Morrison, "Trireme" 18-20.　　[9] *IH* ills. 91-92.

[10] Its last great protagonist was W. W. Tarn. That the theory prevailed during the first half of this century owes more to his vigorous pen, positiveness, and appetite for polemic than to the cogency of his arguments. See the appendices in the second edition of Torr for the key articles he wrote on the subject.

[11] Many Byzantine galleys had two superimposed banks (149 below) and these and similar types remained in use until the 12th century (Anderson 36-41; cf. Morrison, "Trireme" 18). At a slightly later date there came into being a version in which one bank rowed from the gunwale and a second from an outrigger; see Anderson 53 and pl. 8a = *IH* ill. 88 (precisely the same arrangement appears on two galleys in a painting by the 14th century artist Gherardo Starnina [*La Tebiade*, in the Uffizi Gallery, Florence]).

[12] E.g., the oars of an *a zenzile* galley were over twice the length of those on a trireme (30 feet as against ca. 14; cf. Anderson 55), and only 24 benches could be fitted into a length of 170 feet whereas a trireme had at least 27 benches in a length of only 120 feet. Cf. Starr, *CPh* 35 (1940) 369; Morrison, "Nautical Terms" 131-32.

reme at least—the larger units are another matter—was rowed by three more or less superimposed banks of oarsmen (cf. Fig. 105).[13] That this arrangement produced a relatively unseaworthy hull was of secondary moment, since ancient warships rarely ventured far from land,[14] went to sea almost always during the mild summer months (270-72 below), and as much as possible avoided action under unfavorable weather conditions.[15] What counted was that, so arranged, the oarsmen could continue to row, as they had always rowed, from a seated position (cf. 54 above).

II INTRODUCTION OF THE TRIREME

THE trireme, a galley whose design was particularly suited for fighting with the ram,[16] was the culmination of an evolution sparked by the introduction of that weapon into naval warfare. The ram, as we have seen (49 above), was invented probably some time after 1000 B.C. It inevitably brought into being a more powerful vessel,

[13] Starr, "The Ancient Warship," *CPh* 35 (1940) 353-74, answered Tarn's arguments but, limiting himself solely to literary evidence, could not push any nearer to a solution. Morrison, in two fundamental articles ("Trireme" and "Nautical Terms") not only used all the literary material but by keen observation was able to add the evidence of certain important vase-paintings that had never been utilized before. This permitted him to offer a convincing and detailed interpretation of the famous Lenormant relief of ca. 400 B.C. which portrays part of the forward starboard side of an Athenian galley (*GOS* pl. 23a = *IH* ill. 43). Chapters 7 and 11 in *GOS* provide an exhaustive and up-to-date review of all the evidence and represent his last word on the subject; "Trireme," however, furnishes the best exposition of his argument.

The rowing arrangement that Morrison offers is not unlike those proposed by a number of others who have studied the problem (e.g., the four arrangements pictured in Anderson's fig. 6 on p. 18, Anderson's own in fig. 3, p. 15, Tursini's in Ucelli 386, fig. 354); however, Morrison has the unique distinction of (a) providing the evidence to justify it, (b) positing oars of all the same length as indicated by certain evidence (cf. note 23 below).

The artist of Fig. 105, though limited by the scant space at his disposal, has done his best to indicate three superimposed rows of oars. The same way of showing three banks appears on coins of Sidon minted ca. 380-374 B.C.; see Babelon 562-63, pl. 119.1-4 = Basch, pls. 7a, b. For a companion picture to Fig. 105, see Basch, pl. 10a.

[14] Cf. *AM* 102.

[15] Cf. Vegetius 4.43: *navalis pugna tranquillo committitur mari* "Sea battles are joined only when the sea is calm."

[16] Cf. *GOS* 313-20.

this inevitably led to attempts to improve speed and maneuverability, and the result was the two-banked warship. By 700 B.C., such craft, now fitted with raised decking over the centerline to carry a fighting contingent, were in use in Greek and Near Eastern navies (57-58 above).

And then, very likely during the next hundred years, the crucial step was taken of adding a third bank of rowers, and the *trieres* was born. Two-banked galleys were powered by one line of rowers working their oars through ports in the hull and a second working theirs on or just below the gunwale (62-64 above). Greek naval architects created the trireme by adding an outrigger above the gunwale and projecting laterally beyond it to accommodate a third line (Figs. 100, 102).

The new design, though it was eventually to dominate Greek naval architecture, was accepted only gradually. The shipwrights of Corinth seem to deserve the credit for launching the first Greek triremes;[17] when they did so is unsure, but sometime during the seventh century B.C. seems a safe guess.[18] By about 600 B.C., fleets outside of Greece had taken up the new craft.[19] Yet, during the next half century, the penteconter, presumably the two-banked version, continued to serve as the ship of the line.[20] Finally, toward the end of the sixth century B.C., the new warship, which far outclassed its predecessor, came into its own.[21]

[17] Thucydides 1.13.2: λέγονται . . . τριήρεις ἐν Κορίνθῳ πρῶτον τῆς Ἑλλάδος ναυπηγηθῆναι "Triremes . . . are said to have been built at Corinth earlier than anywhere else in Greece." Thucydides cautiously limits his statement to Greece; the only possible rival for the honor of the invention could be Phoenicia (see 94 below).

[18] Cf. GOS 129, 158-59.

[19] E.g., Necho, a ruler of Egypt at this time, built a number of triremes (Herodotus 2.159.1). Basch (231-32) suggests, probably rightly, that these were modeled on the Phoenician type of trireme.

[20] Davison (18-24) argues from this that the trireme could not have existed in Greek navies until after this time, and that its introduction must be dated as late as the third quarter of the 6th B.C. He overlooks the shellback's time-tested ability to resist change. Iron merchantmen had been in use in England for three decades before the Admiralty launched its first iron warship in 1860. The U.S. Navy in that year was still blithely depending on its magnificent wooden frigates—and kept doing so until the celebrated moment in 1862 when the *Merrimack* brought it abruptly to its senses. See *IH* 230-33.

[21] GOS 160-61.

III THE ROWING ARRANGEMENTS

FROM a statement in Thucydides, it is clear that the rowers of a trireme each pulled one oar.[22] From entries in preserved fragments of the Athenian naval records of the fourth century B.C., it is clear that the oars were short and all practically of the same size, either 9 cubits (13'6" = 4.18 m.) or just half a cubit longer;[23] remarks by ancient writers indicate that the slightly longer were used amidships.[24] The dimensions of the remains of certain ship-sheds of the fourth century B.C. (Fig. 197) supply some approximate figures for the trireme's size: length overall ca. 115 to 120 feet; breadth, including outrigger, ca. 16; width across the bottom ca. 10, and from gunwale to gunwale probably ca. 12.[25] Thus the ratio of the width of the hull proper to the length was 1:10.

The remaining dimensions, e.g., a freeboard of ca. 4½ feet and an overall height of ca. 8½ feet above the waterline (Fig. 100), as well as the arrangement of the rowers, can be deduced from a relief of 400 B.C., done to scale, that shows the side of a trireme[26] and from several other representations of about the same date or a little later (Fig. 106).[27] As in the earlier two-banked ships, the lowest line of rowers,

[22] Thucydides, describing the transfer of some trireme crews, remarks (2.93.2): "It was decided that each sailor, taking his oar and cushion and oarstrap, should go on foot, etc." (ἐδόκει δὲ λαβόντα τῶν ναυτῶν ἕκαστον τὴν κώπην καὶ τὸ ὑπηρέσιον καὶ τὸν τροπωτῆρα πεζῇ ἰέναι κτλ.). The pictures of two-banked galleys (Figs. 81-83, 85, 88) and the Lenormant relief (see note 13 above) all show one man to an oar.

[23] The spare oars (perineo; for the meaning, see GOS 289) carried by a trireme came in only two sizes, indicating that these were the only sizes used. For the lengths, cf. IG II² 1606.43, 1607.9, 22, 23, 55, 98: ἐννεαπήχεις "9-cubit"; 1607.14, 51: ἐννεαπήχεις καὶ σπιθαμή (aut sim.) "9½-cubit."

[24] The evidence is cited and discussed by Morrison, GOS 290.

[25] Cf. GOS 285.

[26] The Lenormant relief, mentioned in note 13 above; cf. also Morrison, "Trireme" 35-38.

[27] See also GOS, Clas. 2 (the Talos vase) 3.

The scene in Fig. 106 (end of 4th B.C.) and that on the Talos vase are from the story of the Argonauts. Both picture the after portion of the Argo and render the key features in the same way: some distance above the keel a pair of wales brackets the ports for the thalamite oars (two are visible); then, just under the gunwale, come the ports for the zygite oars (two are complete, and there are indications of two others); then from the gunwale spring struts to support the raised deck (those shown curving aft support the deck on the side near the viewer, those curving for-the thalamites as they are generally called,[28] worked their oars through ports (*thalamiai*).[29] Since these were only some 18 inches or so above the waterline, a leather bag (*askoma*) fitted snugly about the oar and its opening to keep out the sea (cf. Figs. 130, 132);[30] in any sort of chop, the thalamite oars were no doubt secured and the bags sealed. The Athenian naval records show that there were 27 rowers in this bank on each side.[31] The next higher row, the zygites, had the same number.[32] Each sat above and slightly forward of the corresponding thalamite and worked his oar through a port just below the gunwale (Figs. 102, 106). On special benches built over the gunwale (Fig. 100) sat the highest row, the thranites, each slightly forward of and slightly higher than the corresponding zygite, and outboard of him (Fig. 101);[33] the tholepins for their oars were set in an outrigger that projected about two feet or so from the side

ward support it on the side away from the viewer). The single horizontal line midway between gunwale and deck stands for the gunwale on the side away from the viewer. As on other contemporary representations (*GOS*, Clas. 10, the *Argo*; Clas. 4-5, Theseus' ship), the outrigger is not indicated; the artist, limited to incised lines, may well have felt that its inclusion would have overcomplicated his picture.

[28] In the 5th and 4th centuries b.c., the word for this oarsman is θαλαμιός; Aristophanes' θάλαμαξ (*Frogs* 1074) is a comic term, and θαλαμίτης, whence the English "thalamite," a creation of the Byzantine period. See Morrison, "Nautical Terms" 128-29 and *GOS* 269.

[29] Morrison, "Nautical Terms" 125-26, *GOS* 269-70.

[30] Morrison, "Nautical Terms" 126-27, *GOS* 283-84. The thalamites, deep in the hold, had the least chance of escaping if their vessel was struck a mortal blow; cf. Appian, *Bell. Civ.* 5.107: in the battle off Mylae, 36 b.c., Agrippa rammed an enemy ship and "it at once took in water; of the rowers, the thalamites were all cut off, but the others, breaking through the deck [*katastroma*], swam free" (τὴν θάλασσαν ἀθρόως ἐδέχετο, καὶ τῶν ἐρετῶν οἱ μὲν θαλαμίαι πάντες ἀπελήφθησαν, οἱ δ' ἕτεροι τὸ κατάστρωμα ἀναρρήξαντες ἐξενήχοντο).

[31] See *GOS* 270 for the figures for each of the three banks.

[32] The 5th and 4th century word was ζύγιος (Morrison, "Nautical Terms" 128), derived from *zygon* "thwart" (see Ten, App. 3, s.v. "beams"); like the rowers of the original single-banked galleys, the zygites sat on or at the level of the vessel's thwarts and worked their oars from the gunwale or near it.

[33] The model in Fig. 101 differs in a few details from the plan in Fig. 100. The latter has been improved to show the thranite's bench anchored at its outboard end in the outrigger.

Thranite derives from θρῆνυς "stool," a reference perhaps to the little auxiliary bench on which these rowers sat; cf. Morrison, "Nautical Terms" 129-30.

of the ship (*parexeiresia* "by-rowing apparatus").[34] Their stroke was the most wearing, for their oars, pivoting so high up, struck the water at a relatively sharp angle (Fig. 102).[35] The hull curved aft into a graceful run, which squeezed out the lower banks there but left room for the highest; since there were 31 thranites to a side as against 27 in the two other banks, the 4 extra must have been located here (Fig. 99),[36] the sternmost of them being the stroke oar.

If one looked at the vessel's side, the oars seemed to make a quincunx pattern · · · · · . The important cluster, however, was the group of three in an oblique line ·\·\·\·\ ; the thranite, zygite, and thalamite oars in such a segment was the unit that counted and gave the vessel its name, *trieres* "three-fitted." There were 27 such units on each side, which, plus the 4 thranites rowing alone aft, made a total of 170 oarsmen.[37] The longer oars were used throughout the midships section, the slightly shorter ones fore and aft where the hull narrowed toward prow and stern.[38] The use of oars of just about the same length in all banks not only simplified the problem of spares, but enabled oarsmen to be shifted easily from bank to bank.[39]

[34] The meaning of παρεξειρεσία, long misunderstood (cf. Torr 62, note 141), was first conjectured by Assmann (1608-1609) and confirmed by Tarn ("Warship" 141, 219-20). Cf. Morrison, "Nautical Terms" 127-28, *GOS* 281-83. The term occurs as late as the 6th A.D., in a passage in Agathias (5.21). Describing a fleet of improvised reed boats, he notes that the builders "fashioned a plank to hold the tholepins [*kopeter*; see note 51 below] on each side and, as a natural consequence, a sort of outrigger [*parexeiresia*]" (κωπητῆρας ἐφ' ἑκατέρᾳ πλευρᾷ καὶ οἷον παρεξειρεσίας αὐτομάτους ἐμηχανήσαντο).

[35] Thus, at the outset of Athens' great expedition to Syracuse in 415 B.C., they— but not the other two banks—received extra pay (Thucydides 6.31.3).

[36] Cf. Morrison, "Nautical Terms" 129; *GOS* 173, 283. The triremes depicted on Phoenician coins of the 4th B.C. in some cases show exactly 31 oars in the topmost bank; see 95 below.

[37] A galley's set of oars was called a *tarrhos*; cf., e.g., *IG* ii² 1629.684-86 (325/4 B.C.): παρὰ Νεοπτολέμου . . . ταρροῦ τετρηριτικοῦ ἀπελάβομεν Ⱶ Η Ⱶ Δ Γ "From Neoptolemos . . . for the oars [*tarrhos*] of a quadrireme, we collected 665 [drachmas]."

[38] Unlike galleys of the 15th century and later, whose outriggers followed a straight line (Morrison, "Nautical Terms" fig. 1; *IH* ill. 148), that on a trireme followed the curve of the hull. See *GOS* 290 and cf. P. Gille's reconstruction in *Journal des Savants* (1965) 65 (on the dubiousness of the other features in this reconstruction, see *MM* 54, 1968, pp. 21-22, 279).

[39] See TWELVE, note 44.

IV HULL OF THE GREEK TRIREME

FORWARD, the hull at the waterline ended in the ram (*embolos*),[40] which now, instead of having a single point as heretofore, was at least two-pronged.[41] To form the ram, the cutwater was reinforced by massive horizontal timbers (cf. Fig. 107) prolonged into a core over which fitted a bronze casing; this, an expensive item of equipment,[42] could be removed and salvaged.[43] Half-way up the stempost, the point where the waling pieces on port and starboard came together was capped (cf. Fig. 107) by a subsidiary spur (*proembolion* "fore-ram").[44] Shortly abaft the ram there sprang out from the hull at right angles the forward faces or cheeks of the outrigger (*epotides*; e.g. Fig. 107 shows the port *epotis*), which were made particularly strong since either was likely to be struck when ships collided in com-

[40] ἔμβολος is the form found in the Athenian naval records; cf. the passage cited in note 43 below. In literature, the neuter also occurs (e.g., Athenaeus 5.204a, cited in SIX, note 47; Plutarch, *Ant.* 66.2). Pliny, *NH* 32.3, speaks of "rams . . . [armed] with . . . bronze and iron" (*rostra . . aere ferroque . . . armata*) but in a passage that, dealing with the miraculous powers of the sucking fish (one on the rudder can stop a 400-rower quinquereme in its tracks), does not bother overly with facts. When Vitruvius (10.15.6) describes a battering timber as having a "hard iron ram like warships" (*de ferro duro rostrum, ita ut naves longae solent habere*) he means only that it has a reinforced tip like the prow of a war galley; in the case of battering rams this happens to be of iron.

[41] *GOS*, Clas. 16, pl. 27a (coin of 4th B.C.) and a gravestone relief of ca. 400 B.C. (A. Conze, *Die attischen Grabreliefs*, Berlin 1893-1906, pl. 122) show what seems to be a two-pronged ram. Thereafter it is consistently pictured as three-pronged, from ca. 300 B.C. (Fig. 107) on (e.g., *Fitzwilliam Coll.* 7427-29, 7458, 7460 [Sinope and Cius, ca. 300 B.C.]; 3654 [Philip V, 221-179 B.C.]; 4637-38 [Magnetes, 197-146 B.C.]).

[42] In 324 B.C., a ram alone cost ca. 131 drachmae (*IG* ii² 1629.1144-47), whereas a whole set of oars for a quadrireme could be bought for 665 (1629.684-85, cited in note 37 above).

[43] When old ships were broken up, the rams were taken off and stockpiled; see, e.g., *IG* ii² 1623.121-23 (334/3 B.C.): τὴν δὲ παλαιὰν διαλύσειν καὶ τὸν ἔμβολον ἀποδώσειν εἰς τὰ νεώρια "to break up the old ship and restore the ram to the shipyard." The bosses just aft of the trident in Fig. 107 may be part of the fitting that held the casing in place. A number of entries (1606.27, 32, 89 [374/3]) report that certain captured ships came stripped, they had "not even the upper bronze piece" (οὐδὲ τὸ χάλκωμα τὸ ἄνω). Since the word translated here "bronze piece" (*chalkoma*) is used by writers to mean "ram" (Diodorus 20.9.2; Plutarch, *Ant.* 67.3), the entry may indicate that the casing was in two pieces, an upper and a lower; after a jarring blow, the upper might well stay in place while the lower fell off.

[44] *IG* ii² 1614.27-30 (353/2 B.C.): Εὐετηρία . . . προεμβόλιον οὐκ ἔχει "The Eueteria . . . has no fore-ram."

bat.[45] The lower part of the hull was girdled with horizontal waling pieces;[46] the Lenormant relief shows two, one above and one below the line of ports for the thalamite oars (Figs. 99-100, 102).[47]

The stempost, though still, as on sixth century galleys, an unadorned slender upright, was now somewhat higher and was given a pronounced outward curve.[48] The sternpost (*aphlaston, akrostolion, akroterion*),[49] again like those of the sixth century, was made up of a number of timbers that arched fanlike inboard (cf. Fig. 108), but the cap-pieces are gone, and the *stylis* (see 346 below) makes its debut.

Just below the gunwale were the oarports for the zygite rowers. Outboard of the gunwale was the outrigger; it ran, curving with the line of the ship, from the *epotides* forward to just before the steering oars aft (cf. Fig. 108), and it accommodated the tholepins (*skalmoi*)[50] for the thranite oars.[51] Each thole had a strap (*tropos, tropo-*

[45] In describing the Battle of Naupactus (413 B.C.), Thucydides relates (7.34.5) that seven Athenian ships were put out of action because, rammed prow to prow, their outriggers were smashed by the Corinthian ships which "had reinforced outrigger cheeks for this very purpose" (ἐπ' αὐτὸ τοῦτο παχυτέρας τὰς ἐπωτίδας ἐχουσῶν). Theophrastus, *Hist. Plant.* 5.7.3, reports that "the cutwater, to which the false keel is fixed, and the outrigger cheeks are made of manna-ash, mulberry, or elm, for these elements have to be strong" (τὸ δὲ στερέωμα, πρὸς ᾧ τὸ χέλυσμα, καὶ τὰς ἐπωτίδας, μελίας καὶ συκαμίνου καὶ πτελέας· ἰσχυρὰ γὰρ δεῖ ταῦτ' εἶναι). The outrigger cheeks served as catheads for the anchors (Euripides, *IT* 1350-51: οἱ δ' ἐπωτίδων / ἀγκύρας ἐξανῆπτον "And they made fast the anchors from the outrigger cheeks [*epotides*]").

[46] Called, at least in later centuries, *zosteres* "girdles"; see TEN, App. 3, s.v. "wales."

[47] Two also in *GOS* pl. 27a.

[48] For the clearest representation of a stempost of the period, see the relief cited in note 41 above.

[49] The technical term was *aphlaston* (derivation uncertain); it is found as early as Homer (*Il.* 15.716-17: Hector seizes a ship by the stern and does not let go, "holding on to the *aphlaston* with both hands" [ἄφλαστον μετὰ χερσὶν ἔχων]). The other two terms are more general in signification, being used in contexts other than naval. *Akroteria* is preferred by earlier writers (Herodotus 3.59, 8.121; Xenophon, *Hell.* 2.3.8, 6.2.36) or those who quote from them (Polyaenus 5.41, Athenaeus 12.535c); *akrostolion* is preferred by the later (Diodorus 18.75.1, 20.87.4; Strabo 3.157; Plutarch, *Alc.* 32.1; Appian, *Mithr.* 25). Both of these terms, though usually used of the stern-ornament, could on occasion refer to a bow-ornament as well (e.g., Herodotus 3.59; Athenaeus 5.203f, cited in SIX, note 47). H. Wade-Gery in "Note on *Akroteria*," *JHS* 53 (1933) 99-101, lists numerous examples of vase-paintings showing figures, often of Nike the goddess of victory, holding stern-ornaments as trophies, and literary references indicate that these were called *akroteria*.

[50] Vitruvius 10.3.6: *remi circa scalmos strophis religati* "oars, made fast about

ter)[52] that looped about it and was made fast to the oar; this held the oar in place against the pin. At an appropriate distance below the line of zygite ports were the ports for the thalamite rowers, each fitted with an *askoma* (83 above).

About three feet above the thranite rowers was a deck (*katastroma*) extending from prow to stern and from gunwale to gunwale (Fig. 105; cf. Figs. 99-100, 103-104).[53] The one- and two-banked predecessors of the trireme had a raised deck over the centerline only (57-58 above), and the triremes built by the Athenians before the Second Persian War still clung to this curtailed type.[54] About 467 B.C., Cimon, favoring boarding tactics over ramming, extended the deck to the gunwales in order to make room for more marines.[55] Though Athens quickly returned to her traditional preference for the ram, she maintained the continuous raised deck: it was essential for the next important step to be taken in naval architecture—the

tholes [*scalmi*] with straps." Thus Aeschylus can describe (*Persians* 677-79) the triremes in the Battle of Salamis as *triskalmoi* "with three tholepins" (cf. *GOS* 154-55).

[51] The plank along the outrigger for the thranites, into which the tholes were set, was perhaps called *kopeter*; see L. Bergson, "Was bedeutet ΚΩΠΗΤΗΡ?" *Eranos* 55 (1957) 120-26.

[52] The layman's term was *tropos* "loop" (cf. Aeschylus, *Persians* 375-76 and FOUR, note 23); the technical term was *tropoter* (see Thucydides, cited in note 22 above). θαλαμιῶν τροπουμένων in Aristophanes, *Ach.* 553, may mean not "thalamite oars fitted with straps" but "oarports fitted with leather," i.e., *askomata*; see Morrison, "Nautical Terms" 126.

[53] The fighting deck, with its parapet screened by a line of shields, is clearly indicated on coins of Sidon of 400-384 B.C.; see G. Hill, *A Catalogue of the Greek Coins in the British Museum, Phoenicia* (London 1910) p. 139 and pl. 18.3 (for a particularly good reproduction of one of these coins, see *Archeologia, Tresors des âges* no. 21, March-April 1968, p. 52). See also Basch, figs. 5-9, 12-15.

[54] Thucydides 1.14.3: αὗται οὔπω εἶχον διὰ πάσης καταστρώματα "These ships did not yet have overall decking."

[55] Plutarch, *Cimon* 12.2: Cimon took 200 triremes that "Themistocles had originally built particularly for speed and maneuverability and made them broader and bridged in their decks so that, carrying many soldiers, they were the better armed for attacking the enemy" (πρὸς μὲν τάχος ἀπ' ἀρχῆς καὶ περιαγωγὴν ὑπὸ Θεμιστοκλέους ἄριστα κατεσκευασμέναις, ἐκεῖνος δὲ τότε καὶ πλατυτέρας ἐποίησεν αὐτὰς καὶ διάβασιν τοῖς καταστρώμασιν ἔδωκεν, ὡς ἂν ὑπὸ πολλῶν ὁπλιτῶν μαχιμώτεραι προσφέροιντο τοῖς πολεμίοις). In other words, he filled in the area along each side that had hitherto been left uncovered, thereby creating a complete bridge between foredeck and poop and, in the process, making a ship of distinctly broader aspect; cf. *GOS* 162-63.

creation of the armored trireme, or cataphract as the Greeks called it. Cataphract means "completely fenced in";[56] with the raised deck providing a point of support, a screen, most likely of leather,[57] could be fitted from it down to the line of thranite tholes thereby "fencing in" this area. Thus the whole crew was protected against enemy missiles: overhead by the deck and, along the sides, by the outrigger screens and the walls of the hull. The step was taken sometime in the fifth century B.C., very likely not long after Cimon's modification.[58]

[56] Cf. Taillardat cited in FOUR, note 49, who aptly cites (88) Sophocles, *Ant.* 958, in which the poet describes Lycurgus, immured like Antigone in a cavern to die, as "completely fenced in [kataphraktos] in a bond of rock" (πετρώδει κατάφρακτος ἐν δεσμῷ).

[57] The "hairy *pararrhymata*" referred to in the naval records; see ELEVEN, note 99. Each trireme was issued two, one for the port and one for the starboard outrigger. They must have been either one long strip of leather or a series of strips, kept rolled up at the ready along the edge of the raised deck. The form in the singular, *pararrhyma* (or *parablema*), is collective, referring to the whole screen along one side whether made of one or several pieces.

[58] Cataphracts are mentioned by Thucydides, 1.10.4: in the early days the Greeks "did not have cataphract vessels but vessels fitted out rather pirate-style in the old-fashioned way" (οὐδ' αὖ τὰ πλοῖα κατάφαρκτα ἔχοντας, ἀλλὰ τῷ παλαιῷ τρόπῳ λῃστι-κώτερον παρεσκευασμένα). The last words must refer to fast, undecked craft. Polyaenus (3.11.13) reports of Chabrias (1st half of 4th b.c.) that "to counter the battering of the waves, under the outrigger on either side he ran skins and, making fast [the text is corrupt here] . . . to the deck ran them upward as a screen for the outriggers. This prevented the ship from being swamped and the sailors from being drenched by the waves. Moreover, the sailors, not seeing the waves coming at them because of the addition of this screen, did not jump up in fright and rock the boat" (πρὸς τὰς ἐπιβολὰς τῶν κυμάτων ὑπὸ τὴν παρεξειρεσίαν ἐκατέρου τοίχου δέρρεις κατελάμβανεν καὶ †κατὰ μόνας ἀρτίους† τῷ καταστρώματι κατὰ τὸ ὕψος φράγμα κατε-λάμβανεν αὐτὸ πρὸς τὰς παρεξειρεσίας. τοῦτο δὲ ἐκώλυε τὴν ναῦν ὑποβρύχιον φέρεσθαι καὶ τοὺς ναύτας ὑπὸ τῶν κυμάτων βρέχεσθαι· καὶ τὰ ἐπιφερόμενα κύματα οὐχ ὁρῶντες διὰ τὴν τοῦ φράγματος πρόσθεσιν οὐκ ἐξανίσταντο διὰ τὸν φόβον οὐδὲ τὴν ναῦν ἔσφαλλον). Since spray-shields are mentioned by Aeschylus in a play written before 470 B.C. (see FOUR, note 49) and cataphracts were known to Thucydides, this passage can hardly be taken to prove that Chabrias invented either. His innovation consisted in reversing the usual procedure: instead of running the leather screens from the deck down over the sides of each outrigger, he ran them from the gunwale out and past the under surface of the outrigger and then up the side to the deck (cf. Figs. 100, 102). This not only prevented water from welling in there but also cut off the sea from the thranites' view as they bent down to begin their stroke. This passage has long been misunderstood because of editorial tinkering with the text; Morrison, *GOS* 288, note 21, gives the proper readings.

On the aphract trireme of later times, see 123-24 below.

At the prow the foredeck was fenced about by a solid parapet,[59] while the poop had just a railing set on stanchions.[60] There does not seem to have been a deckhouse; for some reason, this useful structure is not attested until Roman times.[61]

Very little is known about the shape and construction of the hull. We can conclude from the flat stone slips found in the shipsheds (Fig. 197) that it was flat-bottomed, and, from the need to accommodate three banks of at least 27 oarsmen in a row, that it was relatively straight-sided longitudinally and vertically. Every effort was made to keep the hull as light as it could possibly be; as has been aptly said, an ancient warship was much like an overgrown racing shell. Not only triremes but the larger types as well were light enough to be drawn up by their crews on the beach at night, to be portaged over considerable distances on rollers,[62] even to be divided up into sections that could be transported long distances overland for quick assembly in a new location.[63] Precautions were constantly taken to see that waterlogging did not add unwanted weight.[64] This

---

[59] Cf. *GOS* pl. 27a. Possibly the term for it was *entorneia*. A papyrus letter dated 250 B.C. (*Sammelb.* 9215 = P. Fraser and C. Roberts, "A New Letter of Apollonius," *Chr. d'Eg.* 24, 1949, pp. 289-90) mentions that the king has ordered "the cutting of [native] timber for the *entorneia* of the warships" (πρὸς τὴν ἐντορνείαν τῶν μακρῶν νηῶν κόψαι ξύλα), namely acacia, tamarisk, and willow. The editors, citing the only other instance of the word (Hero, *Belopoiea* 97.5), where it means a raised rim or flange, suggest that "here it would refer to the defensive breastwork of ships." Egyptian native timber, available only in short lengths (cf. 11 above), could well be used for such a purpose.

[60] E.g., *GOS* pl. 26a, and the coins of Phaselis of the 3rd B.C. (well described and illustrated in *British Museum Quarterly* 27, 1963, 24-25, pl. 3.4-6).

[61] See the coin-pictures cited in SEVEN, App., note 1.

[62] Cf. Thucydides 3.15.1 (the Spartans set up windlasses to haul warships across the Isthmus of Corinth); Polyaenus 5.2.6 (Dionysius I had his men haul 80 triremes a distance of 20 stades, or 2 1/3 miles, in a day).

[63] See SIX, App. 1.

[64] See *GOS* 280. When Xerxes reached a convenient spot on the shores of Thrace, at his orders the commanders "beaching the ships, dried them out" (τὰς νέας ἀνέψυχον ἀνελκύσαντες Herodotus 7.59.3). Lysander did the same when he took over the Spartan fleet in 407 B.C. (Xenophon, *Hell.* 1.5.10). Thucydides (7.12.3) has Nicias report from Syracuse to the Athenian Assembly a year and a half after the fleet had been commissioned that at first the fleet was in fine condition "because of the dryness of our ships (τῶν νεῶν τῇ ξηρότητι). . . . But now our ships are waterlogged" (νῦν δὲ αἱ τε νῆες διάβροχοι). He then goes on to point out that he has no chance

is why warships, whether Odysseus' undecked craft or ponderous quinqueremes, were not moored offshore but, just like racing shells, were pulled up on dry land, and, when back in the navy yard, were stored in covered sheds (363 below).[65] Some ballast, probably in the form of sand,[66] was carried at the bottom of the bilge. There was, understandably, storage space for little else beyond essential gear, missiles, and a minimal supply of drinking water.[67] Stocking provisions for 200 men was utterly out of the question; a fleet was either accompanied by supply ships, or it arranged to put in at shore each night to allow the crews to forage for their dinner.

We have no information whatsoever about warship construction. The naval records reveal that Athenian triremes, despite their fragile nature, lasted 20 years on the average, and some considerably longer.[68] They must have been built with great care,[69] which makes

---

to "beach his galleys for drying out" (ἀνελκύσαντας διαψῦξαι), whereas the enemy "has much more chance to dry out his" (ἀποξηρᾶναι τὰς σφετέρας μᾶλλον ἐξουσία).

[65] For beaching in general, see GOS 311. For quinqueremes run up on shore, see Polybius 1.51.12, 3.96.5. For heavy transport triremes, see Diodorus 20.47.1-2 (Demetrius had drawn up on shore and fenced about with a palisade and ditch not only "more than 110 fast triremes but also 53 of the heavier troop-transport type" [ταχυναυτούσας μὲν τριήρεις πλείους τῶν ἑκατὸν δέκα, τῶν δὲ βαρυτέρων στρατιωτίδων πεντήκοντα καὶ τρεῖς]).

[66] See the following note.

[67] Thus, when Belisarius' fleet, beset by calms, took 16 days to go from Zacynthus to Sicily, "it so happened that everyone's drinking water went bad" (ξυνέπεσεν ἅπασι διαφθαρῆναι τὰ ὕδατα Procopius, Bell. Vand. 1.13.23). The sole exceptions were Belisarius and his staff, thanks to the foresight of his wife who "had glass jars made, filled them with water, made a cubicle of planks in the bilge where the sun could not enter, and buried the jars in sand there" (ἀμφορέας ἐξ ὑάλου πεποιημένους ὕδατος ἐμπλησαμένη οἰκίσκον τε ἐκ σανίδων ποιήσασα ἐν κοίλῃ νηὶ ἔνθα δὴ τῷ ἡλίῳ ἐσιέναι ἀδύνατα ἦν, ἐνταῦθα ἐς ψάμμον τοὺς ἀμφορέας κατέχωσε 1.13.24). The water for all the others must have been in clay jars carried on deck in the full glare of the sun. That these were ships of a much later age (cf. 148 below) means little; a millennium later Venetian galleys were still using sand as ballast, and pilgrims to the Holy Land, like Belisarius' wife, kept jars of wine cool in it (Lane 21).

[68] W. Kolbe, "Zur athenischen Marineverwaltung," Ath. Mitt. 26 (1901) 377-418, esp. 386-97. Kolbe (397) puts the average life of a trireme at 20 years. The longest-lived one known was in use 26 (388-89). These figures agree with those we can work out for warcraft of the subsequent centuries; see 120 below.

[69] The Venetians considered 18 to 20 years of life the mark of a well-built galley. Normally their ships lasted 13 years, and poorly built ones less than half that. See Lane 263.

it a reasonable assumption that they were put together the way merchantmen were, with planks edge-joined by means of close-spaced mortises and tenons (202 below). Another argument for this style of construction is that it would better enable so light a hull to take the strains of combat: planks made fast not only to frames but also to each other would be that much less likely to start under shock.

Still another device to help the hull stand up to the punishing demands upon it were the powerful cables called *hypozomata* "undergirds." All warships carried a fixed number of these—a fourth century trireme had four—and could be issued extras on occasion.[70] The technique of frapping a ship,[71] i.e., passing heavy cables under the keel during bad weather in order to reinforce provisionally a weakened hull, is known from at least the first century A.D. and lasted as long as the wooden sailing ship;[72] the most celebrated example is the time when St. Paul's vessel was struck by a gale and the sailors "used helps to undergird the ship."[73] Did the *hypozomata* of a warship, like the "helps" used for undergirding a merchantman, pass vertically under the hull? Or did they pass horizontally from stem to stern? A comprehensive review of the evidence[74] ends a controversy that has raged for over a century:[75] they ran from stem to stern; they "girded" the ship horizontally "under" the line of the gunwale (cf. Figs. 119, 125).[76] A short but powerful loop passed vertically about the stern (cf. Fig. 108)[77] to furnish a point for anchoring

[70] *GOS* 295.

[71] Falconer, in his Marine Dictionary (revised by Burney, 1815), s.v. *frap*, states: "To frap a ship (*ceintrer un vaisseau*) is to pass four or five turns of a large cable-laid rope round the hull or frame of a ship, in the middle, to support her in a great storm or otherwise, when it is apprehended that she is not strong enough to resist the violent efforts of the sea."

[72] Smith (66) reports an instance that took place in 1837.

[73] Acts 27.17: βοηθείαις ἐχρῶντο, ὑποζωννύντες τὸ πλοῖον.

[74] *GOS* 294-98.

[75] See Smith 172-77 for references to some of the older writers and Bursian's *Jahresberichte* 73 (1892) 103-104 for the following generation. See, for more recent studies, E. Schauroth in *Harvard Studies in Classical Philology* 22 (1911) 173-79; R. Hartmann in *RE*, Supplbd. IV, s.v. *Hypozoma* (1924).

[76] See *GOS*, Clas. 20 and pl. 27b for still another example.

[77] Other examples: Fragment of the Telephus frieze from Pergamon, 2nd B.C. (*JDI* 15, 1900, pl. I, no. 33); relief on an Etruscan urn of the 2nd-1st B.C. showing

the girding cables at this awkward spot. There seems to have been some sort of device (*tonos* "stretcher") to keep them at the proper tension.[78]

On the trireme's rig, see 235-37 below.

V TYPES OF GREEK TRIREMES

THOUGH all Greek triremes of this age were more or less alike in size, overall structure, and rowing arrangements, there was variation in emphasis from navy to navy. The Athenians, using few marines and trusting to their dexterity in ramming, armored their ships as lightly as possible. Others, notably the Corinthians and Syracusans, preferring to ram prow to prow and decide the issue by boarding, beefed up the forward structure—ram, bows, outrigger cheeks—and reinforced the deck to carry a sizable contingent of fighting men.[79]

Triremes also varied measurably in quality. The earlier naval records (377-369 B.C.) reveal that the Athenian admiralty distinguished between "new" and "old" ships. Later (357 B.C.) ever finer distinctions were introduced: "selects" (*exairetoi*), which probably corresponded to the previous category of "new" ships, "firsts," "seconds," and "thirds."[80]

Up to this point we have limited the discussion to combat ships only, what the Greeks called "fast triremes" or "fast sailing triremes."[81] These were not ships of a special build but simply ships in

---

Helen being led aboard ship by Paris (L. B. Ghali-Kahil, *Les enlèvements et le retour d'Hélène*, Paris 1955, pl. 99.1); relief of Jason building the Argo, first half of 1st A.D. (H. von Rohden and H. Winnefeld, *Die antiken Terrakotten* IV, Berlin and Stuttgart 1911, pl. 32 and ill. 14 on p. 13).

[78] *GOS* 296. [79] *GOS* 281-82, 313, 317-20.
[80] *GOS* 248.

[81] ταχεῖαι (e.g., Thucydides 6.31.3, 43.1), ταχυνναυτοῦσαι (e.g., Diodorus 20.47.2, cited in note 65 above; Arrian, *Anab.* 2.21.1, cited in note 82 below; *IG* II² 1623.284; Polyaenus 1.48.4, 3.10.6). Aeschylus, *Persians* 341-43, reports that, of Xerxes' 1,000 ships, 207 were "arrogant for their speed" (ὑπέρκοποι τάχει). H. Wallinga has made the attractive suggestion (by letter) that this is the poet's paraphrase of ταχεῖαι, that these made up the fully manned striking force of the Persian fleet, the other 800-odd being in good part reserves with skeleton crews. Thus, in estimating the

first-rate shape and with full crews, ready to go into action; they were "fast" because the hulls were dry and light and every rowing bench was manned.[82] Not as fast as the full-fledged combat type was the *stratiotis* "soldier ship," a form of trireme which served for such noncombat functions as guard duty and doubled as transport.[83] There was also the troop transport proper (*hoplitagogos*).[84] How these craft were equipped and rowed, or what their capacity was, is uncertain; a passage in Thucydides indicates that a *stratiotis* could carry at least 85 or so troops.[85] Since speed was no object, there is every likelihood that as a rule they operated with far fewer rowers than a "fast" trireme (cf. 94 below).

Then there was the trireme horse-transport (*hippagogos, hippegos*). Until 430 B.C., horses were moved overseas in any sort of vessel that suited the purpose.[86] In that year the practice was adopted of converting old triremes to do this job. The naval records show that, in the fourth century at least, these were rowed by 60 oarsmen and held 30 horses.[87] Apparently, all the thalamite and zygite rowers

---

[82] total Persian personnel, instead of calculating 200 men per ship for these 800, one need calculate but 60 or so (this was, e.g., the number of rowers aboard a horse-transport trireme), thereby arriving at a more realistic figure.

[82] Thus, when Alexander at the siege of Tyre mounted catapults upon "whatever triremes he had that did not rate as fast" (ἐπὶ τῶν τριηρῶν ὅσαι αὐτῶν οὐ ταχυναυτοῦσαι ἦσαν Arrian, *Anab.* 2.21.1), he must have selected ships that were waterlogged from overlong use or were shorthanded or both.

[83] *GOS* 247-48. Cf. the passage from Diodorus cited in note 65 above.

[84] *GOS* 247-48.

[85] Thucydides 6.43: a fleet of 60 Athenian "fast" triremes, 40 *stratiotides*, and 34 non-Athenian triremes transport a land force of 6,400 men. The 60 "fast" ships, being stripped for action, would have no passengers aboard. Assuming that the non-Athenian ships were all *stratiotides*, the average number of men carried per ship was 87 (6,400 ÷ 40 + 34); if some or all were not, the figure would be correspondingly higher.

[86] A Boeotian fibula of the 7th B.C. shows a horse being transported (*GOS*, Arch. 9; illustrated in *JDI* 31, 1916, pl. 18.1, Moll B 1 44); the vessel is identical with the fighting ships shown on other fibulae (e.g. *GOS*, pl. 8c).

[87] Cf. *GOS* 248-49. The distance from thole to thole, three feet (cf. Four, note 66), would take a horse nicely. E.g., Jal, *Gloss. naut.* s.v. *platea*, cites a 13th century statute of Marseilles that allows 73 cm. (2' 4¾") and s.v. *écurie*, p. 617, a vessel plan of 1811 that allows 76 cm. (2' 5⅞"). Jal adds (s.v. *platea*) that government vessels in his own day allow from 89 to 97 cm. (2' 11" to 3' 2¼") in width and 2.59 m. (8' 6") in length.

were removed and, most likely, the sternmost thranite oar, leaving 30 thranites on either side to drive the ship. Twenty-nine horses standing athwartship would be fitted in under the rowers, and a thirtieth just forward of the poop; the shallowness of the hold caused by the vessel's run there would leave no room for a rower above.

VI THE PHOENICIAN TRIREME

GREECE's one naval rival was Phoenicia. The maritime cities of Byblos and Aradus and Sidon and Tyre, all boasting a long tradition on the sea, furnished many units that faced the Athenians at the Battle of Salamis,[88] the ships that Alexander sent overland to Babylon,[89] and entire fleets for the Seleucids, including the one that Hannibal led in his last combat against the Romans.[90] The Phoenicians had two-banked galleys as early as the Greeks (Figs. 75-76, 78); when the Greeks switched to the trireme as capital ship, they followed suit;[91] and they turned to the building of still larger units as soon as, or earlier than, the Greeks did.[92]

Phoenician architects were by no means slavish imitators of the Greeks. They had a centuries-old tradition to guide them and, in designing a trireme (Figs. 103-105), they followed a path all their own.[93] In order to accommodate a third level of rowers, the Greeks had added an outrigger; the Phoenicians accomplished the same end

---

[88] Herodotus 8.85.   [89] See SIX, App. 1.   [90] Livy 37.8.

[91] Basch (231-32), adopting Davison's view of a late date for the Greek trireme, argues that the trireme was a Phoenician invention, but this view is by no means as sure as he believes; cf. note 20 above and my note in *MM* 56 (1970) 340.

[92] E.g., they were the first to build "sevens" and perhaps "fours"; see 97 below and SIX, App. 2, note 2.

[93] The evidence is chiefly the coins minted by the Phoenician cities of Aradus, Byblos, and Sidon between, roughly, 450 and 350 B.C.; see Hill, *op. cit.* (note 53 above) pls. 1, 2, 11, 17-21; Babelon, pls. 116-21. By analyzing the representations on these coins and noting their resemblance to a clay model found in Egypt (Figs. 103-104), Basch (152-62) was able to identify the Phoenician type of trireme. The vessel pictured in Fig. 105, presumably a unit of the Persian navy, is a trireme of this Phoenician type. Persia, which had no navy of its own, pressed into service contingents of its maritime subjects, including strong squadrons from the Phoenician cities.

by increasing the vessel's height. The lowermost rowers, the thalamites in Greek terminology, worked their oars through ports in the hull and the zygites over the gunwale, more or less as on Greek triremes. The thranites, however, worked theirs on a railing that ran, at an appropriate height, directly over the gunwale (Fig. 103)—rather than outboard of it, as an outrigger would (cf. Fig. 100)—and rested on a series of stanchions planted in the gunwale. A second set of stanchions, planted in the railing itself, supported the raised fighting deck, which extended from gunwale to gunwale (Fig. 104); along this there was usually a line of shields to form a pavesade.[94] The probability is that Phoenician galleys had always been beamier than Greek. In adapting them to three-level rowing, the architects very likely retained the beaminess, and the extra room allowed them to fit all the oarsmen into the span of the hull.[95] When it came to length, however, Phoenician and Greek were of one mind. Several representations of Phoenician triremes that are manifestly done with care show 31 oars in the topmost bank—exactly the same number as the Greek.[96] It follows that the two types of trireme were more or less equally long.

The absence of an outrigger also affected the design of the prow. Greek ships were distinguished by a stubby two- or three-pronged ram, with subsidiary spurs above, flanked by the massive projections of the outrigger cheeks. Phoenician ships had no such projections for one, and, for another, retained the style of ram used during the seventh and sixth centuries B.C., a long one that tapered to a single point and had no subsidiary spurs, since these would have served no purpose on such a ram.

Lastly, whereas the Greeks left the prow relatively unadorned, Phoenician triremes had either a figurehead or a tutelary statue car-

[94] The clearest example is the clay model, Figs. 103-104; see Basch 157-59 for the identification as a Phoenician trireme (an identification which confirms the Hellenistic date originally assigned to the piece; cf. GOS 180). For other examples, see Basch, pls. 7c, 8.

[95] Cf. Basch 157.

[96] Basch, pl. 7a, b. Hill, op. cit. (note 53 above) pl. 40.13 shows 31 stanchions; see Basch 155. Some representations show a few less, 28 to 30; see Basch 154, note 3.

ried somewhere in the bows.[97] Furthermore, projecting at an angle from the tip of the prow was a spar, resembling a bowsprit, to which lines such as braces and brails could be made fast (Fig. 105).[98] The sternpost ends in the traditional Phoenician horse's head (cf. 66 above) instead of the Greek fanlike *aphlaston* (64 above).[99]

[97] Herodotus remarks (3.37) that the Phoenicians "carry around on the prows of their triremes" (ἐν τῇσι πρῴρῃσι τῶν τριηρέων περιάγουσι) likenesses of their dwarf-sized divinities called *Pataiḳoi.* This has generally been taken to refer to figureheads but could equally well apply to tutelary statues carried in the bows. It has been suggested (G. Perrot and C. Chipiez, *History of Art in Phoenicia,* translated by W. Armstrong, London 1885, p. 18) that such figureheads are depicted on Phoenician coins, and this has been generally accepted; see, e.g., Babelon 523, 527 and pls. 116.23-24, 117.4 (coins of Aradus, 350-322 B.C.).

[98] Cf. Basch 228. The fastening of the lines is clearest in his pl. 10a.

[99] Clearest in Basch pl. 10a.

CHAPTER SIX

The Warships of the Hellenistic Age: 323-31 B.C.

FOR OUR PURPOSES, the Hellenistic Age embraces the three centuries opened by the death of Alexander in 323 and closed by the Battle of Actium in 31 B.C. Its naval hallmark was the adoption of larger units as the ship of the line, galleys big enough to relegate triremes to the light craft in a fleet.

I TETREREIS, PENTEREIS, POLYEREIS

IN 399 B.C., Dionysius, ruler of Syracuse, began to build *tetrereis* "fours" and *pentereis* "fives." The second was his own invention;[1] Carthage perhaps gets the credit for the first.[2]

The new ships took some time to gain acceptance but, within half a century, they were in all navies, both Greek and Phoenician. By 330, Athens had 18 "fours" (as against 392 triremes).[3] Six years later the number had gone up to 43, and 7 "fives" had been added.[4] By 351 B.C., there were "fives" in the fleet of the city of Sidon.[5] In 332 B.C., during Alexander's siege of Tyre, "fours" and "fives" were in ac-

[1] Diodorus 14.42.2: ἤρξατο δὲ ναυπηγεῖσθαι τετρήρεις καὶ πεντηρικὰ σκάφη, πρῶτος ταύτην τὴν κατασκευὴν τῶν νεῶν ἐπινοήσας "[Dionysius] began to build 'fours' and vessels fitted as 'fives,' the first to design the construction of such ships." Cf. 14.41.3: διενοεῖτο γὰρ . . . κατασκευάσαι . . . ναῦς τετρήρεις καὶ πεντήρεις, οὐδέπω κατ' ἐκείνους τοὺς χρόνους σκάφους πεντηρικοῦ νεναυπηγημένου: "And he designed and constructed . . . 'fours' and 'fives' at a time when no vessel fitted as a 'five' had as yet been built"; 14.44.7: ἀπέστειλεν . . . πεντήρη πρῶτον νεναυπηγημένην "He dispatched a 'five,' such being built then for the first time."
[2] Pliny, *NH* 7.207: *quadriremum Aristoteles* [*auctor est fecisse*] *Carthaginiensis* "Aristotle [states] that Carthage invented the 'four' "; Clemens Alexandrinus, *Strom.* 1.75.10: Καρχηδόνιοι γὰρ πρῶτοι τετρήρη κατεσκεύασαν "Carthage was the first to build a 'four.' "
[3] *IG* II² 1627.275-78; cf. *GOS* 249.
[4] *IG* II² 1629.808-11.
[5] Diodorus 16.44.6: τριήρεις καὶ πεντήρεις εἶχε πλείους τῶν ἑκατόν "[Sidon] had over 100 triremes and 'fives.' "

tion,[6] and Alexander had both types in the contingents he collected at Babylon.[7] Dionysius II (367-344 B.C.), going his father one better, introduced "sixes" into the Syracusan navy.[8]

When Alexander died, the fight for the spoils of his empire raged on the water as well as on land. Ptolemy I of Egypt inherited the lion's share of Alexander's fleet. Antigonus the One-Eyed, who had a keen appreciation of sea power, together with his son Demetrius, a brilliant tactician and bold designer of ships, set out to match it, and this touched off the greatest naval arms race in ancient history, as each side launched bigger and bigger *polyereis* or polyremes, as we call these oversize galleys which required veritable regiments of rowers to drive them.[9]

In 315, Antigonus set about constructing a naval aggregation that included up to "sevens" at least. By 301 Demetrius had "eights," "nines," "tens," "elevens," even a "thirteen." A dozen years later he added a "fifteen" and a "sixteen"; Lysimachus, his enemy, and Antigonus Gonatas, his son, both launched ships that were equal to or more powerful than these; and Ptolemy II topped them all with first a "twenty" and then two "thirties." Toward the end of the third century B.C., Ptolemy IV built a brobdingnagian "forty," but this was intended for display not action.

The superdreadnoughts were short-lived, having all but run their course by the middle of the third century. At the Battle of Chios in 201 B.C., Philip V, though he had Demetrius' "sixteen" at his disposal,[10] did not bother to use it and put his flag aboard a "ten";[11]

---

[6] Arrian, *Anab.* 2.22.3-5.   [7] See App. 1.

[8] Aelian, *Var. Hist.* 6.12: ναῦς μὲν ἐκέκτητο οὐκ ἐλάττους τῶν τετρακοσίων, ἐξήρεις καὶ πεντήρεις "[Dionysius II] had no less than 400 ships, 'sixes' and 'fives.'" Conformably, Pliny, *NH* 7.207, gives the credit for the "six" to Syracuse.

[9] See App. 2.

[10] It was in his navy in 197 B.C., for the terms of the peace treaty Rome made him sign at that time included the provision "to hand over . . . to Rome . . . the cataphract vessels except . . . the 'sixteen'" (ἀποκαταστῆσαι . . . Ῥωμαίοις . . . τὰς καταφράκτους ναῦς πλὴν . . . τῆς ἐκκαιδεκήρους Polybius 18.44[27].6). Livy 33.30.5 renders this *Romanis . . . naves omnes tectas tradere praeter . . . regiam unam inhabilis prope magnitudinis, quam sedecim versus remorum agebant* "hand over . . . to Rome . . . all the cataphracts . . . except the king's one personal ship of almost unmanageable size, which was powered by 16 files of oars." In the loose usage of

moreover, he inaugurated a new era of naval tactics by emphasizing the use of destroyers (*lemboi*; see 125 below).[12] At Actium in 31 B.C., Antony had nothing bigger than "tens," and Augustus nothing bigger than "sixes."[13] The heaviest unit in the Roman navy thereafter was the Imperial flagship, a "six."[14]

II THE OARAGE OF THE *POLYEREIS*

How galleys larger than a trireme were oared is a puzzle on which as much ink has been spilled as on the trireme problem. For long it was held that they were all extensions of the trireme, i.e., powered by superimposed banks of rowers each of whom pulled but one oar; this system might work for "fours," "fives,"[15] even "sixes," but thereafter it produced vessels that could stay afloat only on a drawing board.[16] Naval historians frequently pointed out the practical shortcomings,[17] and the classical scholar, W. W. Tarn, gave it the coup de grâce by demonstrating that it could not be reconciled with the ancient evidence.[18] A later theory held that the ships were rowed with but a single bank of multiple-rower sweeps and were named according to the number of men assigned to pulling each sweep.[19] This was a considerable improvement, since it had in its

---

Roman writers, *remi* "oars" and *remiges* "oarsmen" were interchangeable; see Tarn, "Warship" 205. On the meaning of "file of oarsmen," see 113-14 below.

[11] Polybius 16.3.3, cited in note 118 below.

[12] Polybius 16.2.9, 4.8-12 (cited in notes 109, 111 below).

[13] Plutarch, *Ant.* 64.1: τὰς δὲ ἀρίστας καὶ μεγίστας ἀπὸ τριήρους μέχρι δεκήρους ἐπλήρου: "[Antony] manned his best and biggest ships, from trireme to 'ten'"; cf. Dio Cassius 50.23.2. Augustus had at Actium the fleet with which he had crushed Sextus (Dio Cassius 50.19.3), and this included up to "sixes" (Florus 4.11.6). Sextus, too, had been content with nothing greater than a "six," for he came to the meeting with the triumvirs in a "six" (Appian, *Bell. Civ.* 5.71, 73), presumably his most impressive unit.

[14] Starr 53; cf. SEVEN, note 1.

[15] Cf. Smith 196-98.

[16] For a résumé see Tarn, "Warship" 137, note 1; Cook in Torr² 197-201 = *CR* 19 (1905) 371-73.

[17] Cf. Jal, *Flotte* 104; Anderson 22.

[18] "The Greek Warship," *JHS* 25 (1905) 137-56, 204-218 (= Torr² 154-89).

[19] This was, e.g., L. Weber's theory (*Die Lösung des Trierenrätsels*, Danzig 1896): cf. Tarn, "Warship" 137. It is followed, with variations, by Tarn himself (*Hell. Dev.* 136-38).

favor the example of the seventeenth and eighteenth century galleys, some of which used as many as—but not more than—eight men to an oar; yet this very same example set the limit beyond which the system could not in all reasonableness be carried.[20] The solution lies in combining the two systems, i.e., reconstructing these oversize ships with superimposed banks of multiple-rower sweeps.[21] Except for the monster galleys bigger than a "sixteen," which are discussed separately below, this solution produces results that are both feasible and in accord with whatever evidence is available.

Polyremes—to use the convenient term for all sizes larger than the trireme—seem to have developed in three stages: first, from the trireme to the "six," by the first half of the fourth B.C.; second, from the "six" to the "sixteen," between 315 and 288 B.C.; third, certain still more powerful types culminating in Ptolemy II's "thirty," between 288 and 246 B.C.[22]

Treating these stages individually will help clarify the line of development.

From the Trireme to the "Six"

The sole facts available to go on are these: that the Rhodian navy used "fours" the way the Athenians used the trireme, as a swift and highly maneuverable vehicle for attacks with the ram; that the

[20] In order to help explain the very large warships of the Hellenistic Age, the upholders of this theory casually assume the common use in the 17th and 18th centuries of as many as 10 men to an oar (e.g., Rodgers 254, 261; Tarn, *Hell. Dev.* 135-36). Actually the standard number was 5 or 6 (cf. Guglielmotti, *Vocab.*, s.v. *galera*); flagships went up to 7 (Masson 201); galeasses 8 (Pantera 152: *galeazze . . . non si deveno armare a meno d'otto huomini per remo* "Galeasses ought to be manned by no less than 8 men to the oar"). To my knowledge there are no examples of units commonly used in action that put more men than this to an oar; cf. A. Jal. *Archéologie navale* (Paris 1840) 1 392: "deux, trois, quatre, et même huit hommes . . . s'asseyaient sur le banc." On the French *La Royale* with its reputed 19 to the oar, see now Anderson 79-81; it may possibly have had nine, but Anderson considers even that excessive.

[21] Both naval experts and scholars are agreed on this. Cf., e.g., Bauer 462-64; Rodgers 256; Morrison, *GOS* 291 and "Trireme" 43; Anderson 24-25, 28-29. Jal, *Flotte* 204-11, offers a variant: multiple-rower sweeps in superimposed banks but set on occasion in clusters of two sweeps.

[22] See App. 2.

Carthaginians in the Punic Wars used "fives" in the same way[23] (on the Roman "five," a heavy ship designed for boarding, see note 41 below); that an Athenian "four" of the fourth century B.C. very likely had fewer oars than a contemporary trireme.[24]

In sum, certain "fours" and "fives" were very like triremes save that they had a reduced number of oars. Could not such ships have been created simply by "double-banking," to use the modern term, i.e., assigning two men to an oar instead of one?[25] In this fashion we can explain the development from a trireme right up to a "six." The reduced number of oars would be the inevitable consequence of using longer oars to accommodate two men: the inboard part, describing a greater arc, would need more room. From the ancient point of view, increasing the oar-power in this particular way had one great advantage: with only two men to an oar, the traditional system of rowing from a seated posture could be maintained. Still another tradition that it would help maintain would be the use of oars of the same length, with the convenience for manufacture and stocking of spares that this afforded. A "five" with two thranites, two zygites, and one thalamite, would necessarily have to break with tradition and use shorter oars in the lowest bank (although even in this arrangement, two of the banks would have oars of the same length), but a "six" with three banks of two-man oars, or a "four" with two banks of two-man oars, would very nicely carry on the tradition.

The argument thus far has been based solely on development, on the consideration that the system described above is the easiest way

[23] Cf. *AM* 162-63, 165, 168-72.

[24] See Morrison, *GOS* 290-91. He points out that a set of oars for a trireme cost at least 1,000 drachmas in the 5th B.C., whereas a set for a "four" in 325/4, when the purchasing power of the drachma was assuredly less, cost but 665 (*IG* ii² 1629.684-85, cited in FIVE, note 37). That a "four" had fewer oars is the sole possible explanation.

[25] This is what Anderson (23) suggests. The evidence that one man used one oar (FIVE, note 22) applies only to the trireme and smaller units; there is no justification whatsoever for applying it to any of the larger galleys. The statement, so often repeated (Bauer 461; Tarn, *Hell. Dev.* 131; Anderson 22) that the oars of a trireme were interchangeable with those of a "four" or "five" is based on a misunderstanding; see Morrison, "Nautical Terms" 132-35.

a "four" or "five" could have evolved from the trireme. But there are two scraps of evidence, admittedly far from conclusive, that can be adduced in support.

The first concerns the "five." It is very likely that a "five" of the fourth century B.C. measured only about 10 feet or so across the outrigger at its forward end.[26] Now a trireme of that period was 16 feet amidships (82 above). It would appear, then, that the two types ran about the same width—as we could have gathered from the fact that both were stored in the same boathouses.[27] A single line of five-man sweeps could not possibly fit into so narrow a hull, nor could an upper line of three-man sweeps. Only by assuming the three levels of a trireme with the thranite and zygite oars double-banked can we arrive at an arrangement of five rowers that would fit.

The second concerns the "four." The famed statue of the Winged Victory of Samothrace stands on a ship's prow. This monument has often been connected with the naval victories of Demetrius or his son Gonatas and interpreted as showing the prow of some oversize galley.[28] We now know that it was a dedicatory monument erected by Rhodes between 200 and 180 B.C.[29] Since the ship par excellence of the Rhodian navy at the time was the "four,"[30] this is what should be portrayed. The outrigger shows two oarports, one staggered slight-

[26] Diodorus 17.115.1-2: for Hephaestion's funeral Alexander "built a square pyre, a stade long on each side. . . . He adorned the entire periphery, where gilded quinquereme prows, 240 in all, made up the lowest layer. Each had upon its outrigger cheeks two kneeling archers 6 feet high and armed male figures 7 1/2 feet high" (ᾠκοδόμησε τετράπλευρον πυράν, σταδιαίας οὔσης ἑκάστης πλευρᾶς. . . . περιετίθει τῷ περιβόλῳ παντὶ κόσμον, οὗ τὴν μὲν κρηπῖδα χρυσαῖ πεντηρικαὶ πρῷραι συνεπλήρουν, οὖσαι τὸν ἀριθμὸν διακόσιαι τεσσαράκοντα, ἐπὶ δὲ τῶν ἐπωτίδων ἔχουσαι δύο μὲν τοξότας εἰς γόνυ κεκαθικότας τετραπήχεις, ἀνδριάντας δὲ πενταπήχεις καθωπλισμένους). The total of 240 means that there were 60 such prows to a side, and a stade being roughly 600 feet, this would make each prow 10 feet wide. It was Morrison (GOS 285-86) who drew attention to this significant passage.

[27] Cf. Morrison, GOS 286 and 364 below.

[28] Notably by Tarn, "Warship" 208, MM 19 (1933) 70. Others have, more properly, taken it as a much smaller craft; see the résumé of reconstructions in Anderson 25-27.

[29] M. Bieber, The Sculpture of the Hellenistic Age (New York 1961²) 125-26.

[30] AM 152, 168-71. At the Battle of Side in 190 B.C., "The Rhodians had 32 quadriremes and 4 triremes" (Livy 37.23.4: Rhodiorum duae et triginta quadriremes et quattuor triremes fuere).

ly higher than the other. If we assume that the higher port carried a two-man thranite oar and the lower a similar zygite oar, and that there were no thalamites,[31] we reach a solution that fits both the monument's dimensions and its reason for being. This solution presupposes that naval architects had made a significant innovation in rowing arrangements, that they had taken the zygite oars, which in a fifth century trireme worked through ports, and repositioned them to work through an oar-box at a level just below the thranite oars. Representations of a somewhat later date (Figs. 114, 116) reveal an important change in the nature of the outrigger that supports this presupposition. In them, instead of an outrigger which follows the curve of the hull and serves only one bank of oars, there is an oar-box which makes a straight line from prow to stern and accommodates all the oars on a side (118-19 below).

FROM THE "SIX" TO THE "SIXTEEN"

One feature above all others governs this stage of development: boarding now became an important naval tactic, and galleys more and more ceased being man-propelled missiles to become carrying platforms for fighting men and—a new naval weapon—catapults.[32] Seaworthiness, never very important in a warship, counted for less than ever; the new floating platforms rarely ventured out of sight of land—and certainly never voluntarily spent a night on the water.

The moving spirit in the great advance in naval architecture and a pioneer in the use of catapults on ships was Demetrius, Besieger of Cities. Demetrius' starting point was the trireme with its superimposed three banks—we have sure evidence that no ship ever went beyond that number.[33] His revolutionary move was to substitute the

[31] Reconstructing it thus puts it well within the possibilities suggested by those who have most recently studied the monument. E.g., Anderson (27) concludes that its oars were "worked by one man each or by several," while Morrison (GOS 286) suggests it was either a "four" or a "five."

[32] See note 88 below. Tarn has no grounds for arguing (Hell. Dev. 120-22) that catapults were not used in Hellenistic naval warfare.

[33] Over a century ago, Jal (Flotte 109) had pointed this out. The greatest galley ever launched, the "forty," had only three banks; see 108-109 below, and cf. Aristides Or. 25(43).4, cited in note 55 below.

103

multiple-rower sweep for the one-man or two-man oar used hitherto. This involved one clean break with tradition: to operate such sweeps, rowing in the traditional seated fashion necessarily gave way to a stroke in which the oarsmen rose to their feet to dip the blade and fell back on the bench for the pull.[34]

One particular advantage of the multiple-rower sweep is that it reduces drastically the number of trained men needed, for only the rower at the tip of the loom has to be skilled;[35] the others supply just muscle. This factor was unquestionably of importance to Demetrius, who had to fill the benches of a vastly expanded fleet at a time when experienced oarsmen were in short supply.[36] A parallel can be drawn with the seventeenth and eighteenth centuries; at that time, when most of the rowers were totally untrained, largely poor unfortunates who had fallen afoul of the law, the preferred arrangement for galleys was a single bank of oars with five men to each oar, while heavy

[34] Cf. FOUR, note 67. The multiple-rower sweep is generally referred to by scholars as a *remo scaloccio* "big-ladder oar," the technical term for it in the Italian fleets of the 17th century. To operate a *remo scaloccio*, the rowers climbed, ladderlike, on to the *pedagna*, then higher to the *banchetta*, and then fell back on the *banco*, highest of all; see Guglielmotti, *Vocab.*, s.v. *remo* and *scaloccio*. Lucan seems to be referring to this kind of stroke when he writes (3.542-43): *tunc caerula verrunt / atque in transtra cadunt et remis pectora pulsant* "[The rowers] sweep the blue sea and fall back on the benches and bring the oar[looms] against their breast." Tarn, "Warships" 150-52, makes much of Appian, *Bell. Civ.* 4.85, as evidence for a standing stroke, but with scant justification. The passage describes the difficulties inexperienced crews had when caught amid the rips and currents of the Strait of Messina: "Salvidienus' crews were thrown into confusion, neither standing securely because of their inexperience, nor any longer able to pull the oars, nor keeping the steering oars under control" (οἱ δ' ἀμφὶ τὸν Σαλουιδιηνόν, οὔτε ἑστῶτες βεβαίως ὑπὸ ἀηθείας οὔτε τὰς κώπας ἔτι ἀναφέρειν δυνάμενοι οὔτε τὰ πηδάλια ἔχοντες εὐπειθῆ, συνεταράσσοντο). Tarn misinterprets the passage, for it says nothing about oarsmen operating from a standing position. Appian simply points out, in highly condensed language, that each element in the crew had a special difficulty—the steersmen in keeping control of the tiller, the oarsmen in handling the oars, the marines (not the rowers) in keeping their footing.

[35] The *vogue-avant* of the French galleys, *vogavante* of the Italian. Cf. Pantera 133: *questi [vogavanti] bisogna che siano de i migliori vogatori della galea, perche guidano il remo, et fanno la maggior fatica, però commandano à tutti gl'altri dal banco i servitii della galea* "These [vogavanti] have to be from the best oarsmen aboard the galley because they guide the oar and do the most work; for that reason they are in charge of the ship's services for all the others on the bench."

[36] On the shortage of rowers from the mid-4th B.C. on, cf. *AM* 124-25.

galeasses had as many as eight.[37] Obviously ancient "fives" to "eights" could have been rowed in the same way.[38] In the fourth century B.C., a "five," as suggested above, may have had two banks of two-man oars and a third bank of one-man oars; however, once the ancients had turned to the multiple-rower sweep at the end of the century, there is no reason for their not designing a "five" which simply put five men on each oar—or even a "four" with four men on each. We have almost certain evidence that "fours" and "fives" were oared in this fashion from 100 B.C. on,[39] while the Roman quinqueremes of the First Punic War (264-241 B.C.), which were thrown together in a frantic hurry and shoved into the water with green crews,[40] very likely had their oars arranged in the same fashion (they had ca. 270 rowers,[41] which would come out to 27 five-man oars a side).[42]

[37] Standard French galleys of the 16th century used 24 sweeps per side with three men on each (Masson 58-59); of the first half of the 17th century, 24 sweeps with five men on each (Masson 141); of the last half, 26 sweeps with five men on each (Masson 202). Galeasses, heavy craft which were a mixture of galley and sailing ship, had up to eight men to an oar (Pantera, cited in note 20 above).

[38] But it is a mistake to assume, as, e.g., Tarn does (*Hell. Dev.* 136), that "nines" and "tens" were; cf. note 20 above.

[39] See SEVEN, App. Basch (239-40) claims, perhaps rightly, that the Phoenicians from the outset, the beginning of the 4th B.C. on, used a single bank of multiple-rower sweeps on their "fours" and "fives," basing his argument on a coin of 373 B.C. (cf. his pl. 11 and pp. 234, 239) which shows a single-banked galley that is most likely a quinquereme.

[40] See note 82 below and TWELVE, note 32.

[41] Polybius 1.26.7: τῆς ναυτικῆς δυνάμεως περὶ τέτταρας καὶ δέκα μυριάδας ὡς ἂν ἑκάστης νεὼς λαμβανούσης ἐρέτας μὲν τριακοσίους, ἐπιβάτας δ' ἑκατὸν εἴκοσιν "The total naval force [was] about 140,000, so that each ship took 300 crewmen and 120 marines." Though Polybius literally says "300 rowers," his method of calculation shows that this is a loose expression for the entire crew. A trireme carried 16 officers, ratings, deck hands, etc. above the 170 oarsmen (305 below). Since a Roman quinquereme very likely had the same rig (235-38 below), it should have had roughly the same number of nonrowing personnel.

This fleet of "fives" was patterned upon a captured Carthaginian cataphract (Polybius 1.20.15). Since the Roman ships were specially designed to use as their principal weapon an oversize complement of marines and were consequently markedly clumsy and slow whereas the Carthaginian ships, specially designed to use the ram as their principal weapon, were consequently markedly maneuverable and swift, the conclusion is inescapable that Rome's shipwrights simply followed the basic lines and otherwise freely adapted; cf. Thiel, *History* 171-77, esp. 176-77. One way in which they adapted, I would suggest, was in foregoing the two or three banks of their model in favor of a single bank of five-man sweeps, a change that would in great part explain why the Roman version was so much slower. Toward the end

On the other hand, there is no reason for assuming that the ancients would give up their predilection for superimposing rowers. Indeed, since the experience of the seventeenth and eighteenth centuries indicates that 8 men to an oar was just about the practicable maximum, anything larger than an "eight" would necessarily have more than one level of rowers.[43] However, for anything smaller, there very likely were single-banked and double-banked models in use, depending upon a navy's preference. Alongside the single-banked "five" just mentioned, there could well have been a two-banked type with 3 men on the upper oars and 2 on the lower.[44] "Sixes" could be single-banked, or they could be two-banked, either with 4 men on each upper oar and 2 on each lower or with three on all oars. Similar single-banked and two-banked versions may have existed in the case of "sevens" and "eights" as well. But a "nine" certainly had at least two banks with, say, 5 men to each upper oar and 4 to each lower. And Demetrius' great "sixteen" either had two banks, each using elongated 8-man sweeps, or three, with 16 men distributed over a thranite-zygite-thalamite set (cf. 84 above). Each increase in oar-power resulted, to be sure, in an increase in a vessel's

---

of the war, Rome again built a fleet after the model of a Carthaginian prize, this time of crack quinqueremes patterned exactly upon the blockade-runner of Hanno the Rhodian (Polybius 1.59.8).

[42] This would make a Roman quinquereme a close parallel to a French galley of the late 17th century, which carried 26 five-man sweeps on each side. The dimensions of such galleys were: overall length including ram and stern overhang 55 m. (180½'), along the keel 47 m. (154 1/6'), of the rowing space ca. 33 m. (108¼'), width from gunwale to gunwale 6 m. (19¾'), from outrigger to outrigger 8 m. (26¼'). See Masson 202-203; G. la Roërie and J. Vivielle, *Navires et marins, de la rame à l'hélice* 1 (Paris 1930) 103; Anderson 68.

[43] See notes 20 and 37 above. Some writers (e.g., Rodgers 254, 258), to fit in more rowers to a sweep, assume it was worked on a push-pull system, i.e., with oarsmen on both sides of the loom, those on the fore side pulling, those on the other pushing. Anderson (79-81) has demonstrated that the *La Royale*, contrary to what is often asserted, was not rowed in this way, and, indeed, that there is not a single well-substantiated instance from the 17th or 18th centuries. His own reconstruction of the largest Hellenistic galleys with oars so manned, as he himself admits (81, note), was a solution born of desperation.

[44] Cf. SEVEN, App. This is the arrangement preferred by Anderson (24) and Morrison (*GOS* 291).

beam[45]—but this was all to the good: a wider ship meant more stability for firing catapults and broader decks to accommodate more marines.

POLYREMES LARGER THAN A "SIXTEEN"

With Demetrius' last effort, we reach a point in the development that gives every indication of marking a radically new departure in naval architecture. For one, the next galley in the progression, the challenger of the "sixteen," is called an "eight." For another, the facts reported about it and the still larger types that followed it, make abundantly clear that the oarage systems suggested above, which explained all types up to the "sixteen," will not do for its big sisters.[46]

[45] By using data from the multirower galleys of the 17th and 18th centuries, we can deduce in a general way, the impressive dimensions these oversize warships must have reached. Let us take, for example, any powered by eight-man sweeps, whether a single-banked "eight" or a two-banked "sixteen." In 17th century galleys using multiple-rower sweeps, 4 feet of space was left between the tholepin and the nearest rower, and each rower was allotted just under 2 feet (Anderson 69). Thus, an eight-man sweep had an inboard length of just under 20 feet (8 x 2' + 4'). Double this for port and starboard and add 4 feet or so for a catwalk down the center, and we arrive at a beam of ca. 44 feet from outrigger to outrigger. The length would be at least five times the beam (cf. the table of dimensions of 17th and 18th century galleys in Anderson 68), or a minimum of 220 feet.

[46] Tarn (*Hell. Dev.* 136-38) suggested that these big ships were rowed by up to 10-man sweeps gathered in clusters, e.g., of two 10-man sweeps in the "twenty," of three in the "thirty," and of four in the "forty." Since this suggestion seemed to make some sense out of the numbers involved, it has received general, though half-hearted, acceptance; cf. Starr in *CPh* 35 (1940) 372, Anderson 30. But there are three unanswerable objections to Tarn's arrangement.

1. It flies in the face of the evidence. It takes no cognizance whatsoever of the fact that the thranite oars of the "forty" were singled out for mention because they were longer than the others—yet, in Tarn's arrangement, all oars are the same length. It takes no cognizance of the fact that there *were* thranite oars—which would imply the existence of zygite and thalamite, but not of a fourth group. Lastly, it takes no cognizance of the fact that the "forty's" thranite oars, the longest known, would just about take 10 men, whereas all the other oars, presumably smaller, could not.

2. There are no parallels for a 10-man sweep; see notes 20 and 37 above.

3. It has no purpose. Tarn posited sweeps in clusters by analogy with his conception of a trireme as having, like a Venetian *a zenzile* galley (cf. 53-54 above), its oarsmen in clusters of three. Clusters of 1-man oars make excellent sense—the men can thereby be seated all along the same bench. Clusters of multiple-rower sweeps make no sense whatsoever, and this is why there is no demonstrable instance of them in the history of oared warships. There is nothing to be gained by putting

In tackling the problem, the best starting point is the "forty," the showpiece launched by Ptolemy IV, because its specifications, being so utterly extraordinary, became public knowledge and were recorded and have come down to us:[47]

	cubits	feet
length	280	420
beam	38	57
height from waterline to tip of stern	53	79½
height from waterline to tip of prow	48	72
draft (when empty)[48]	under 4	6

two or three great sweeps close together; indeed, they are far more effective spread evenly apart (cf., e.g., Anderson 30).

The only other solution offered has involved the push-pull system (e.g., Anderson 29; cf. Morrison, "Trireme" 43). On such a system, see note 43 above.

Bauer, who realized that Greek galleys never had more than three superimposed banks (462), and who was aware that the largest-sized oar one should reckon on is an 8-man oar (464), could only offer the desperate solution of assuming four superimposed banks for the "thirty" and five for the "forty" (464).

[47] Athenaeus 5.203e-204b: τὴν τεσσαρακοντήρη ναῦν κατεσκεύασεν ὁ Φιλοπάτωρ τὸ μῆκος ἔχουσαν διακοσίων ὀγδοήκοντα πηχῶν, ὀκτὼ δὲ καὶ τριάκοντα ἀπὸ παρόδου ἐπὶ πάροδον, ὕψος δὲ ἕως ἀκροστολίου τεσσαράκοντα ὀκτὼ πηχῶν. ἀπὸ δὲ τῶν πρυμνητικῶν ἀφλάστων ἐπὶ τὸ τῇ θαλάσσῃ μέρος αὐτῆς τρεῖς πρὸς τοῖς πεντήκοντα πήχεις. πηδάλια δ᾽ εἶχε τέτταρα τριακονταπήχη, κώπας δὲ θρανιτικὰς ὀκτὼ καὶ τριάκοντα πηχῶν τὰς μεγίστας, αἳ διὰ τὸ μόλυβδον ἔχειν ἐν τοῖς ἐγχειριδίοις καὶ γεγονέναι λίαν εἴσω βαρεῖαι κατὰ τὴν ζύγωσιν εὐήρεις ὑπῆρχον ἐπὶ τῆς χρείας. δίπρωρος δ᾽ ἐγεγόνει καὶ δίπρυμνος καὶ ἔμβολα εἶχεν ἑπτά· τούτων ἓν μὲν ἡγούμενον, τὰ δ᾽ ὑποστέλλοντα, τινὰ δὲ κατὰ τὰς ἐπωτίδας· ὑποζώματα δὲ ἐλάμβανε δώδεκα· ἑξακοσίων δ᾽ ἦν ἕκαστον πηχῶν . . . γενομένης δὲ ἀναπείρας ἐδέξατο ἐρέτας πλείους τῶν τετρακισχιλίων, εἰς δὲ τὰς ὑπηρεσίας τετρακοσίους· εἰς δὲ τὸ κατάστρωμα ἐπιβάτας τρισχιλίους ἀποδέοντας ἑκατὸν καὶ πεντήκοντα "The 'forty' was built by Philopator [Ptolemy IV, 221-203 B.C.]. It was 280 cubits long, 38 from gangway to gangway, and 48 high to the prow-ornament [akrostolion]. From the stern-ornament [aphlaston] to the part where the ship entered the water was 53. It had four steering oars that were 30 long, and thranite oars—the longest aboard—that were 38; these, by virtue of having lead in the handles and being heavily weighted inboard, because of their balance were very easy to use. It was double-prowed and double-sterned, and had seven rams. Of these, one was the chief ram and the others subordinate, and, [of the latter], certain were on the outrigger cheeks. It had 12 horizontal undergirds, each 600 cubits. . . . During a trial run it took aboard over 4,000 oarsmen and 400 other crewmen and, on the deck, 2,850 marines."

[48] Athenaeus 5.204c: τοὺς θεμελίους κατῳκοδόμησε λίθῳ στερεῷ πρὸς πέντε πήχεις τὸ

steering oars (4)	30	45 long
thranite oars (the long-		
est aboard)	38	57 long

personnel

oarsmen	4,000
officers, ratings,	
deckhands	400
marines	2,850

To the above must be added the vital piece of information that the ship was "double-prowed" and "double-sterned."

The mention of thranite oars implies the existence of zygite and thalamite; the ship, in other words, was an overblown trireme. However, if we attempt to distribute 40 rowers over a thranite-zygite-thalamite set, we immediately run into an impasse. We are specifically told that the thranite oars were the longest; each, therefore, would have to accommodate at least 14 of the 40. We are further specifically told what the length was, namely 57 feet, and the writer clearly conveys the impression that he felt this to be a truly remarkable size. Yet a 57-foot oar, which would permit an inboard length of no more than 19 feet, will barely accommodate 10 rowers, to say nothing of 14. As a matter of fact, it is not too much longer than the oars that were handled by 6 or 7 men in the galleys of the seventeenth and eighteenth centuries.[49]

βάθος, καὶ διὰ τούτων φάλαγγας ἐπικαρσίας κατὰ πλάτος τῆς τάφρου διώσας συνεχεῖς τετράπηχυν εἰς βάθος τόπον ἀπολειπούσας "He set up a bottom stratum of hard rock [along the launching canal] at a depth of 5 cubits [7½']. Transversely across the whole length of this [bottom stratum], running the width of the canal, he put in a line of rollers, leaving a depth of 4 cubits [6']." The rollers, therefore, were 1 cubit in diameter, and the vessel's draft when empty—presumably when launched the hull was but a shell—was about 6 feet. Its loaded draft, of course, must have been a good deal more. The steering oars were 45 feet long; assuming they were carried at a 45° angle and had one-third their length immersed, the draft when loaded would be in the neighborhood of 12 feet.

[49] The five-man oar on a French galley was ca. 11.83 m. long (38¾'), of which ca. 8 m. (26¼') was outboard (la Roërie, op. cit. note 42 above, I, 112); the six- or seven-man oar was 2 m. (6½') longer, or 13.83 m. in all (Jurien de la

There is yet another impasse. We are told that the ship had 2,850 marines and 400 deckhands, officers, ratings and the like. The fighting deck, which covered more or less the part of the hull given over to the rowers, would provide a surface of, say, 350′ x 70′ (overall length of 420, less 70 to allow for the projections at bow and stern; breadth of 57, plus the lateral projection of the outrigger on either side). Such a surface would accommodate 3,250 persons only if they were lined up as if on parade—30 men abreast in lines three feet behind each other—with no room at all for maneuvering the men and certainly none for the operation of catapults. Yet the whole point of this ship, as of any supergalley, was to plow into clusters of smaller craft with its catapults volleying stones and darts, its archers firing arrows, and boarding parties readying at given points to hurl grapnels and eventually leap.

All who have studied this vessel have, to my mind, overlooked a clue of the highest importance: that it was "double-prowed" and "double-sterned." This is sometimes interpreted to mean that the ship was double-ended like a canoe, so it could be rowed in either direction without having to have its huge bulk turned all around.[50] However, such an interpretation not only mistakenly takes "double-prowed and double-sterned" to mean "double-ended," but completely overlooks the features that very definitely distinguish the two ends: the prow was marked by the *akrostolion*, and the stern by the *aphlaston*; the latter was, as traditional on Greek ships, somewhat the higher; and there was but a single principal ram at the prow, whereas a double-ended ship would have one at each end. In a word, we are dealing with a mammoth version of what today is called a catamaran

---

Gravière, *Les derniers jours de la marine à rames*, Paris 1885, p. 190). Smith (201) mentions that a 57-foot oar is not much longer than those used on the galleys of Malta.

[50] Smith 200, Torr 74. The author of the description of the "forty" goes on to tell of Philopator's vast river barge (cf. FOURTEEN, note 67). This too he characterizes as "double-prowed and double-sterned" (Athenaeus 5.204e), and the details that follow show clearly that the two ends were distinct. Tarn was aware of the special nature of the "forty's" hull ("Warship" 143, note; *Hell. Dev.* 141-42), but failed to appreciate its significance for the rowing arrangements.

(Figs. 112-113). There were two hulls, each 53 cubits wide—the specifications include four steering oars, so each hull had the customary pair, one to port and one to starboard. A vast platform—sufficient to accommodate the mass of marines and hands—must have yoked the two hulls. Now, *if we assume that both sides of both hulls were supplied with oars*—and the two hulls need be only 30-odd feet apart (Fig. 112) to enable the inner oars to work freely—a feasible system of oarage emerges. There was a total of 4,000 oarsmen. This would mean 2,000 in each hull, or 1,000 a side. The vessel was 420 feet long, which would allow ample space for 50 multiple-rower sweeps (Fig. 113).[51] And, with not 40 men but only 20 to distribute over a thranite-zygite-thalamite set (50 x 20 = 1,000), there are a number of workable alternatives. The best would be to assign 8 men apiece to the thranite oars; this would observe the limit of men per oar attested in the seventeenth and eighteenth centuries, and is just about the right number for a 57-foot oar. Then 7 could be put on the zygite, and 5 on the thalamite. Other arrangements are equally possible. The essential point is to recognize that we are dealing with a double-hulled vessel. Once this is done, everything falls into place. We are enabled to reconstruct a fighting craft which provides the proper deck space for the huge number of marines reported as well as a stable platform for them and the catapult battery to operate from; and, by assuming there were rowers on both sides of each hull, we arrive at a system of oarage that is well within the bounds of credibility and squares in every respect with the evidence.

It may be objected that such a vessel is structurally impossible or that oarsmen could never be taught to row with the precision that those on the interior benches would have to have. For any con-

[51] On galeasses of the 17th century, which could run to eight men per oar (see note 20 above), the distance from thole to thole was 1.75 m. (ca. 5' 9"); see Jal, *Flotte* 197.

Figs. 112 and 113 reconstruct this overblown trireme with an outrigger for the thranite oars. A reconstruction with oar-box for all banks (see 118 below) could probably be worked out as well.

111

vinced by such arguments, I have an alternative theory to offer: that the "forty" was a catamaran with the hulls side by side, like two ships lashed together, and that the interior benches were fully manned by rowers who had no oars but were stationed there to serve as a spare crew. The exterior oarsmen would drive the vessel up to the field of action, and the fresh crew would take over as soon as battle was to be joined; in this way, a lumbering catamaran had a chance of keeping up with its more agile adversaries. The deck of such a craft would still be ample enough to accommodate the 3,250 marines and nonrowing personnel.

Since Ptolemy IV was no innovator but merely a builder of show-pieces, there is no reason to credit him with the design of the cata-maran supergalley. It is far easier to assume that examples were right under his eyes—in Ptolemy II's "twenty" and two "thirties." If we try to explain these as single-hulled ships, we run into the same impasse we did with the "forty": a three-banked "thirty" would have to have at least 11 men on its thranite oars—yet these oars were presumably even shorter than the 57-foot sweeps singled out for mention in the description of its much larger successor. Assume, however, that it was double-hulled with rowers to port and starboard of each hull, and all becomes easy: we need distribute only 15 men over a thranite-zygite-thalamite set; it might even have been two-banked, with 8 men on each upper oar and 7 on each lower. Similarly with the "twenty"; that might well have been a two-banked catamaran with 5 men to each oar.

THERE remain but two more of these largest of all ancient galleys to discuss.

The first is the *Leontophoros,* the ship "remarkable for its size and beauty" that Lysimachus built to meet the challenge of De-metrius' fleet led by a "fifteen" and a "sixteen."[52] The only detail we

[52] Memnon 13 = Jacoby, *FGH* no. 434, 8.5, vol. III B, p. 344: ὀκτήρης μία ἡ Λεοντοφόρος καλουμένη, μεγέθους ἔνεκα καὶ κάλλους ἤκουσα εἰς θαῦμα· ἐν ταύτῃ γὰρ ἑκατὸν
μὲν ἄνδρες ἕκαστον στοῖχον ἤρεττον, ὡς ὀκτακοσίους ἐκ θατέρου μέρους γενέσθαι, ἐξ

have is the size of the crew—but it is enough to put us on the right track: "In each file, 100 men rowed so that there were 800 in each part [of the two parts], 1,600 in both; those assigned to fight from the decks [totalled] 1,200; and [there were] 2 helmsmen." Nobody has so far made any sense of any of the figures reported for this ship. It was an "eight," yet it could lock horns with a "sixteen." A contemporary "five" carried a maximum of 300 rowers, any ordinary "eight" would presumably carry only proportionately more, say 500 or so—yet this "eight" had 1,600, fully two-fifths of the number carried by the mammoth "forty." A contemporary "five" carried 120 marines,[53] yet this "eight" had 1,200, fully two-fifths of the number carried by the "forty." And the arrangement of the rowers as described seems incomprehensible.[54]

The significant hint lies in the way the totals are far more closely related to a "forty" than to any ordinary "eight." If we assume that this vessel, too, was built as a catamaran, everything falls perfectly into place. There were two helmsmen—naturally, one for each hull. There were 1,200 marines—far too many for the fighting deck of even the biggest ordinary polyreme, but easily accommodated on a spacious platform yoking two hulls. And, if we assume that there were two banks of 4-man oars—a legitimate arrangement for an "eight"—and 50 oars in each bank, the arrangements cease being incomprehensible and work out exactly:

a) "In each file 100 men rowed"—i.e., taking a bird's-eye view of

---

ἑκατέρων δὲ χιλίους καὶ ἑξακοσίους· οἱ δὲ ἀπὸ τῶν καταστρωμάτων μαχησόμενοι χίλιοι καὶ διακόσιοι· καὶ κυβερνῆται δύο.

[53] For the crew and marines of a quinquereme, see Polybius 1.26.7, cited in note 41 above.

[54] Jal (*Flotte* 195-96) decided that either Memnon had made a mistake or the Byzantine encyclopedist, who has preserved his words for us, somehow made alterations as he copied. Tarn originally did not attempt to extract sense from the passage: "whatever Memnon's description exactly means" was his earliest comment ("Warship" 208, note 93; cf. "Ded. Ship" 211). Ultimately he decided it was a "sixteen" which Memnon had misreported as an "eight" (*Hell. Dev.* 136-37, 141). This hardly helps, since it explains neither the rowing arrangements as described nor the extraordinarily high number of marines. The same objections can be made to Anderson's suggestion (29) of an "eight" in three levels.

each hull, in each bank there were 50 oarsmen in the file nearest the port tholes, and another 50 nearest the starboard, for a
total of 100; similarly for the file of rowers one inboard from
the tholes, and so on. If we translate this Greek phrase into
the terminology of seventeenth and eighteenth century galleys, we would say that each bank of each hull had 100 *vogavanti*, 100 *posticci*, 100 *terzaroli*, 100 *quartaroli*.[55]

b) "so that there were 800 in each part (of the two parts)"—i.e.,
in each of the two hulls (the Greek word used is *meros* "part,"
and not *toichos* "ship's side"). Reverting to the terminology of
the seventeenth and eighteenth centuries, there were, including
both walls of a hull, 200 *vogavanti*, 200 *posticci*, 200 *terzaroli*,
and 200 *quartaroli* for a total of 800.

c) "1,600 in both"—i.e., the two hulls together give a grand total
of 1,600.

Many years ago, Tarn characterized the *Leontophoros* as a "new
or abnormal development of some sort,"[56] a judgment I repeated
at the beginning of this section when I pointed out that this ship
gave every indication of signalizing some radical new turn in naval
architecture. The new turn, I suggest, was the debut of the catamaran supergalley.

It is not hard to imagine where the idea originated. From the
fifth century B.C. on, and no doubt even earlier, commanders often,
for one reason or another, lashed two vessels together.[57] It simply

---

[55] The Greek word for "file" is *stoichos* and the Latin *versus* (see Livy, cited in
note 10 above and in App. 2, note 12). Thus, Aristides (*Or.* 25 [43].4), in recalling
the naval glories of Rhodes, says that in her harbor "it was possible . . . to see warships [*triereis*] of two banks [*dikrotoi*] and of three banks [*trikrotoi*] and up to
seven and nine files [*stoichoi*]" (τριήρεις δὲ . . . ὑπῆρχεν ἰδεῖν δικρότους καὶ τρικρότους
καὶ εἰς ἑπτὰ καὶ εἰς ἐννέα στοίχους). In a ship with, say, seven files, the seven men
in each rank would be distributed over one, two, or three benches, depending upon
whether the ship was single-, double-, or triple-banked.

[56] "Ded. Ship" 211.

[57] Polyaenus gives three instances in which commanders lashed galleys in pairs
and raised but one sail on each pair to trick the enemy into mistaking the size of
the force by half: Thrasylus at the end of the 5th B.C. (1.47), Chabrias in the first
half of the 4th (3.11.3), Diotimus in the 4th (5.22.2). Marcellus used quinqueremes
yoked in pairs at the siege of Syracuse (Polybius 8.4[6].2 = Livy 24.34.6-7).

remained for some imaginative officer or shipwright to conceive of yoking a pair permanently, either side against side or far enough apart so that the inner oars could still be used. The credit, it would seem, does not go to the great innovator, Demetrius; though we cannot be sure, one gets the feeling that his mightiest effort, the "sixteen," was single-hulled. The *Leontophoros* was built in a ship-yard of Heraclea on the Black Sea, probably at the orders of Demetrius' bitter enemy, Lysimachus; perhaps one of his admirals deserves the credit or some anonymous naval architect of Heraclea.

If the first great catamaran galley was launched ca. 280 B.C., then such ships stayed in use at least down to the time of Ptolemy IV (221-203 B.C.), the builder of the "forty." In other words, they should not have been unique rarities, and there is some evidence to confirm this.[58]

The last ship we have to deal with is the galley designed by Antigonus Gonatas, Demetrius' son, that led his fleet to victory over Ptolemy II, whose aggregation included units up to a "fifteen" and very likely the *Leontophoros* as well. Gonatas' vessel seems to be identical with one described by Pausanias as having "as many as nine rowers from the decks"[59] and by Pollux as being *triarmenos*,[60] which should mean "three-masted" but in this case may mean "three-leveled."[61] It somehow must have gone the father's "sixteen" one better. If we assume that the latter had two banks of eight-man oars, then the son's improvement may have been the addition of some provisional nine-man sweeps worked from the fighting deck.

---

[58] Homer attributes to Nestor a very special bowl for mixing wine, one with a set of two handles parallel to each other, on either side; see *Il.* 11.632-37 and the explanation, derived from Aristarchus, in the scholium to 11.632. Now, Athenaeus describes it (11.489b) as a bowl "with handles set alongside each other, just like the ships that have double prows" (παρακειμένως ἔχοντα τὰ ὦτα, καθάπερ αἱ δίπρῳροι τῶν νεῶν). Obviously such ships must have been familiar enough to his readers to enable him to use them in a simile.

[59] See App. 2, notes 14 and 15.

[60] Pollux 1.82: Ἀντιγόνου τριάρμενος. Tarn argues convincingly ("Ded. Ship" 209-210) that this Antigonus can only be Gonatas.

[61] For *triarmenos* "three-masted," see ELEVEN, notes 74, 75. Anderson (28) has suggested that it be taken to mean "three-leveled" here.

115

I am not happy with this explanation, but I cannot think of a better and it at least satisfies the requirements of both descriptions.[62]

III HULL, RIGGING, ARMAMENT

WE know even less about the hulls of polyremes than about their oarage. Most apparently, were cataphract,[63] that is, closed in by a fighting deck overhead and screened in along the sides (88 above). A coin minted by Demetrius the Besieger of Cities to celebrate his naval triumph at Salamis in 306 B.C. shows a prow with the goddess of victory on it (Fig. 107); the vessel portrayed should be his flagship, a "seven" (see App. 2). It indicates that the fighting deck followed the line of the gunwale rather than of the oar-box— which, as we shall see in a moment, came to replace the outrigger— leaving the corners of the latter projecting.[64] Probably this area was given over to massive bracing rather than oar space, since it was in constant danger of collision during combat. The coin also indicates that, at least at this date, the sidescreens were removable, for it pictures the prow with the space between deck and outrigger open.[65] The poop was finished off, as in the fifth century (86 above), with the traditional *aphlaston* and *stylis* (Figs. 109-110) and no cabin (Figs. 116, 120).[66] The stempost from the end of the fourth century

[62] There are no grounds whatsoever for suggesting, as Tarn does (*BCH* 46, 1922, pp. 473-75), that this was the dedicated ship placed in the so-called building "des Taureaux" excavated on Delos. Gonatas won the Battle of Cos sometime around the middle of the 3rd century B.C., whereas the building dates toward the end of the 4th; see J. Marcadé in *BCH* 75 (1951) 88-89 (he suggests it may have housed a dedication by Demetrius for his victory off Salamis in Cyprus in 306).

[63] Thus Polybius uses cataphract as an inclusive term for all warships larger than a trireme; see 16.2.9-10, 5.62.3 (cited in note 94 below).

[64] Clearest in the relief of a galley prow on a cippus of the mid-3rd B.C. in the Tunis museum; see *Antiquités africaines* 1 (1967) 29, pl. 10.1.

[65] In a coin that is later (118-105 B.C.) but obviously modeled on Demetrius', oars are shown extending from this open space; see P. Naster, *La collection Lucien de Hirsch* (Bibliothèque royale de Belgique, Cabinet des médailles, Brussels 1959) no. 1877.

[66] For other examples of the *aphlaston* and *stylis* and absence of a cabin, see the coins cited in SEVEN, App., note 1. Conformably, when artists of this age illustrate Odysseus' story, they show his ships with lofty rounded stern and no cabin: see, e.g., P. von Blanckenhagen "The Odyssey Frieze," *Röm. Mitt.* 70 (1963) 127 and

B.C. on is an unadorned upright as before, but now it is massive, no longer curves outward, and ends in a rounded or conical head (Fig. 107).[67] By the middle of the third century B.C. the rounded head has become a volute in shape. The stempost ending in a volute perpetuated itself thereafter for many centuries.[68] The ram is three-pronged (FIVE, note 41).

Polyremes lay low in the water, even the biggest of them. The steering oars of Ptolemy's "forty" were only 45 feet long (note 47 above); thus the helmsman on the poop deck could not have been more than 20 to 25 feet above the water (Fig. 112).[69] Antony's "tens" at Actium measured only 10 feet from water level to gunwale,[70] a picture of a Roman "four" (Fig. 126) shows a craft that is as low in the water as any seventeenth century galley, and the light destroyerlike vessels called *lemboi* (125 below) could trade ram or boarding attacks with "fours" and "fives."[71] Possibly the greatest single advantage of the supergalleys, the "elevens" and larger, was that they stood high enough to make boarding very difficult.

The introduction of the multiple-rower sweep brought in its wake a significant change in the nature of the outrigger. In the trireme of the previous century it had been a sort of continuous bracket fol-

---

pl. 46; G. Iacopi, *L'antro di Tiberio a Sperlonga* (Istituto di Studi Romani, I Monumenti Romani IV, Rome 1963) figs. 48-50.

[67] Fig. 107, minted in 300 B.C. or so, is the earliest example; cf. R. Thomsen, *Early Roman Coinage* (Copenhagen 1957-1961) III, p. 148. It also appears on coins of Demetrias minted ca. 290 B.C. (*Fitzwilliam Coll.* 4567) and on certain Roman coins of the mid-3rd B.C. (Thomsen I, p. 83, no. 99).

[68] The earliest example is a coin struck by Antigonus Gonatas to celebrate a naval victory in 258 B.C. (*Hunterian Coll.* I, pl. 23.18; *IH* ill. 46). It is common on Roman coins of the mid-3rd B.C. (Thomsen, *op. cit.* [previous note] I, ills. on pp. 79, 81, 84-87), and Greek coins of the 2nd B.C. (*Fitzwilliam Coll.* 4642, 5362). For the stem with volute on ships of the Roman Imperial Navy, see 146 below.

[69] Cf. note 48 above.

[70] A deduction from Orosius 6.19.9 where, in describing the size of Antony's ships, he says "They stood 10 feet in height above the sea" (*decem pedum altitudine a mari aberant*). This would *a fortiori* apply to Antony's largest, which were "tens" (see note 13 above).

[71] Polybius 2.10.3-5: *lemboi* lashed in fours let the enemy ram, caught him and held him fast, and then boarded, in this manner capturing four "fours" and sinking a "five"; 16.4.8-12 (cited in note 109 below): Philip's *lemboi* gave the Rhodian contingent, which included "fives," a hard time.

lowing the curve of the hull, which was added to provide room for
a third tier of oarsmen in the traditional hull built to accommodate
two levels. However, as the galleys of the seventeenth and eighteenth
centuries plainly show,[72] multiple-rower sweeps were best accommodated in a *telaro*, as it was called in the Italian navies, a rowing
frame or oar-box that followed a straight line and not the curve
of the sides; the ship in bird's-eye view looks like a long narrow
rectangle with the point of the prow emerging at one end and the
bulge of the stern at the other.[73] We see the forward end of such
a straight-sided oar-box in the Victory of Samothrace, the after end
in a relief of the 2nd century B.C. (Fig. 108; cf. Fig. 114),[74] and the
whole in a mosaic that seems to portray a galley of the Hellenistic
period (Fig. 116). The frame accommodates all the levels of oars,
each oar of an upper level being set in echelon slightly above the
corresponding oar of the lower level (Figs. 114, 116).[75] The best-

[72] See *IH* ills. 86, 88, 91, 142, 147-49; Ucelli fig. 291.

[73] A similar rearrangement can be discerned in the pictures of Phoenician ships
on the coins of the Phoenician cities. Down until the end of the 4th B.C., they depict the standard Phoenician galley, one without an outrigger (94-95 above). In the
next century there is a distinct change. As on other coins of the age, only prows are
now shown, and these consistently display the bulging cheeks of a massive *telaro*.
Compare, e.g., in the coinage of Aradus, *SNG, Danish National Museum, Phoenicia*
nos. 1-5, 10-23 (5th B.C. to mid-4th) with nos. 25-29 (3rd B.C.) and 59-69 (2nd-1st
B.C.); in the coinage of Sidon, nos. 204-208 (354-333 B.C.) with no. 219 (1st B.C.).
Basch (234-37) connects the change, perhaps rightly, with Alexander's conquest
of the Phoenician cities.

[74] Fig. 114 is one of a series of reliefs in the Palazzo Spada, Rome. They belong
to the Roman period, being copies done ca. mid-2nd A.D. from originals of a few
decades earlier; see P. Zanker in Helbig-Speier, *Führer durch die öffentlichen
Sammlungen klassischer Altertümer in Rom.* II, *Die städtischen Sammlungen*
(Tübingen 1966⁴) 761-62 and 766-68. The reliefs reveal a mélange of elements from
all different periods: Greek Archaic, Greek Classical, Hellenistic—even some that
belong to the artist's own day (Zanker 767). Thus the arrangement of the oars
could very well have been inspired by Greek pictures of the Hellenistic Age. The
helmsman's shelter, on the other hand, looks like something the sculptor added on
his own, using a shape that earlier Roman artists had chosen for mythological ships
(see SEVEN, note 28).

[75] The prow as pictured on Demetrius' coin of ca. 300 B.C. (Fig. 107) reappears on
some pieces of Calenian pottery signed by the potter L. Canoleios (*JDI*, Suppl. VIII,
1909, no. 128a = *CVA* Naples IV e pl. 2 [Italy 1024]; Assmann 1606, fig. 1675),
who worked ca. 250-180 B.C. (cf. *Enciclopedia dell'arte antica* II 272). Canoleios
shows three levels of oars working through the outrigger. The value of his evidence, however, is doubtful. One of the scenes portrayed (Assmann 1606, fig. 1675 =

known example of this arrangement is the base on which the Victory of Samothrace stands; as suggested above, it may represent the prow of a two-level "four." It goes without saying that the seating of the oarsmen in these or any ships where all levels were worked through a straight-sided oar-box was considerably different from that in a trireme of the fifth century B.C. Any vessel with three-man sweeps or larger would necessarily be fitted for a standing stroke, perhaps with the steplike arrangements we know from French or Italian galleys of the sixteenth century and later.[76]

The base on which the Victory of Samothrace stands also provides some welcome detail on how oars were fitted into the rowing frame—at least on some vessels. Oblong oarports were cut into the frame, one for each oar (Fig. 118). A tholepin bisected the port; the oar fitted into the space abaft the pin and pivoted against it, probably held in place by an oar-strap.[77] The after end of the port was carefully planed away to a tapered edge—a feature whose purpose we can only guess at. It would enable the oars to be swung closer alongside the hull, and this would keep them out of the way of an enemy attempt to shear them off, which was a common form of attack.[78]

Most polyremes were built to last, like the triremes of the previous century (90 above). Demetrius' "sixteen," launched ca. 288 B.C., was the same ship that Philip had been allowed to keep in 197 and that had carried Paulus up the Tiber in 167—in other words, a Methuselah of almost 130.[79] The quinqueremes that were the backbone of the Roman fleets from the mid-third to the mid-second century B.C.

Moll B vi d 181) is of Odysseus and the sirens; whereas Homer was content to give Odysseus a light single-banked 20-oared ship, Canoleios puts him aboard a triple-banked affair with even more oars than an ordinary trireme.

[76] Cf. note 34 above, and *IH* ill. 92.

[77] Compare the very similar arrangement on Byzantine galleys (SEVEN, note 49).

[78] Cf. Polybius 16.4.8-12, cited in note 109 below.

[79] See App. 2, note 12. Livy 35.26.5-9 describes a quadrireme that was over 80 years old and still afloat; it was leaky and rotted, but that was in good part due to neglect.

seem to have remained in service about 25 years on the average,[80] bettering the figure indicated for Athenian triremes. A squadron of ships Caesar refitted for use in 48 B.C., probably triremes but perhaps larger, had been laid up since Pompey's war against the pirates 19 years earlier.[81] On the other hand, in emergencies vessels were thrown together with green timber in a matter of months.[82]

Of rig we know absolutely nothing. In the absence of any positive evidence, we can only conclude that these large ships were rigged in the same way as triremes.[83]

With the increased emphasis on boarding over ramming, armament was the area in which the polyremes varied most from their

[80] At the outbreak of the second Punic War, Rome had 200 quinqueremes which almost certainly were the same as those built in 242 B.C. or captured from Carthage in that year (Thiel, *Studies* 35, note 11). They were, then, 24 years old.

For the Syrian War of 192 B.C., Rome had at least 115 ships, most of which no doubt dated from the building programs of 217 and 214 B.C. (Thiel, *Studies* 264). They were, conformably, 22 to 25 years old.

[81] *Bell. Civ.* 2.23: *x longis navibus . . . quas naves Uticae ex praedonum bello subductas P. Attius reficiendas huius belli causa curaverat* "Ten warships which, laid up at Utica after the war against the pirates, P. Attius had refitted for this war."

[82] Pliny, *NH* 16.192: *apud antiquos primo Punico bello classem Duilli imperatoris ab arbore LX die navigavisse, contra vero Hieronem regem ccxx naves effectas diebus XLV tradit L. Piso. secundo quoque Punico Scipionis classis XL die a securi navigavit* "The ancients in the First Punic War put the fleet of Admiral Duilius in the water on the 60th day from the tree [i.e., the hewing of timber]. L. Piso reports that in the war against King Hiero 220 ships were built in 45 days. In the Second Punic War, Scipio's fleet was in the water on the 40th day from the ax." These were fleets of quinqueremes. Smaller craft could take even less time. E.g., in 49 B.C., for operations around Marseilles, "Caesar . . . had 12 warships built at Arles. These, completed and equipped within 30 days from the day the timber was cut, etc." (*Bell. Civ.* 1.36: *Caesar . . . naves longas Arelate numero xii facere instituit. quibus effectis armatisque diebus xxx, a qua die materia caesa est, etc.*). The ships could not have been larger than triremes and perhaps were smaller.

In the 16th century, the Arsenal at Venice, with a full stock on hand of seasoned timber of all required shapes, averaged two years for turning out a galley (Lane 135).

Galleys constructed of green timber were inferior to properly made craft since the latter, being dry, were lighter and therefore faster (cf. 89 above); e.g., the vessels referred to above that Caesar had built in a month "having been made in a hurry out of green timber did not have the same degree of speed [as the enemy's]" (*Bell. Civ.* 1.58: *factae enim subito ex humida materia non eundem usum celeritatis habebant*).

[83] Antigonus Gonatas' great ship is called a *triarmenos*, which should mean "with three sails" (cf. ELEVEN, note 74, 75) but in this case seems to mean "with three levels of oars"; see 115 above.

predecessors. Athens' light triremes had carried only 14 or so marines, and even the heavier triremes limited themselves to 30 or 40 (304-305 below). A Roman "five" of the mid-3rd B.C. carried all of 120, while the supergalleys could accommodate whole regiments.[84] In the first Punic War, the Romans, totally new to naval warfare and up against the crack Carthaginian fleet, were saved by a special kind of boarding device,[85] the *corvus* "raven" as it was called. This was probably the sailors' slang term for it; we would call it a "crane" rather than a "raven" (Fig. 111). It consisted of a gangplank, 36 feet long and 4 wide, with a heavy spike at the outboard end and, at the other, a long slot which fitted around a pole set up like a mast in the bow of the ship; when raised, the plank stood upright snugly against the pole and, when lowered, projected far over the bow. One tackle between the farther end and the head of the pole controlled the raising and lowering, and two, made fast to the deck on either side, swiveled it from side to side. A vessel so equipped would warily keep its prow headed toward the enemy and, as soon he closed in to ram, drop the "raven"; the spike would embed in his deck, and a boarding force could rush over the plank. The *corvus* did not stay in use very long, for Carthage must have soon come up with a defense, but it served its purpose: by that time Carthage had lost greatly in ships and men, and Rome had gained greatly in experience.[86]

Though ramming remained a standard fighting tactic, it was now generally subordinate to action by the marines—the firing of missiles and heaving of grapnels.[87] The missiles with the longest range were

[84] See notes 41, 47, 52 above.

[85] H. Wallinga, *The Boarding-Bridge of the Romans* (Groningen 1956) is an exhaustive study of the *corvus* with a minute analysis of the *locus classicus*, Polybius 1.22.

The "ravens" that, according to Appian, *Bell. Civ.* 5.106, were used at the Battle of Mylae in 36 B.C. cannot be the same as the boarding bridge of over two centuries earlier. They very likely were some form of grapnel.

[86] Thus Rome won the Battle of the Aegates Islands by superior ships and seamanship; cf. Thiel, *History* 311-14.

[87] Called "iron hands" ($\chi\epsilon\hat{\iota}\rho\epsilon\varsigma$ $\sigma\iota\delta\eta\rho\alpha\hat{\iota}$). They were in use at least as early as the 5th B.C. (Polyaenus 1.40.9 and Diodorus 13.50.5: Alcibiades attacks with the ram the enemy ships under way and hauls off with grapnels those moored).

121

darts shot from arrow-shooting catapults.[88] Grapnels were also fired
from catapults;[89] in 36 B.C. Agrippa, Augustus' gifted commander,
introduced a specially sophisticated type called the *harpax* "gripper."[90] Once the range had shortened, archers were able to go into
action, and there were javelins and stones for close work.[91] Since the
ships were so low, to give the marines height enough to sweep an
enemy's deck, collapsible towers, which could be swiftly set up or
dismantled, were carried fore and aft (cf. Figs. 130, 132).[92] About

[88] Arrow-shooting catapults are listed among the gear stored in the naval warehouse at the Peiraeus between 330 and 322 B.C. (*IG* ii² 1627.328-41, 1628.510-21,
1629.985-97, 1631.220-29). Alexander mounted catapults on triremes at the siege of
Tyre (Arrian, *Anab.* 2.21.1-2; cf. Five, note 82) and Demetrius on his most powerful lighter units at the siege of Rhodes (Diodorus 20.85.3). For the rank of "cata-
pultist" in the Rhodian navy, see 309 below. At Actium rock-throwing catapults
were also used, loaded on occasion with pots of hot coals and pitch (Dio Cassius
50.32.8, 34.2).

[89] Athen. 5.208d, cited in Nine, App. pt. 4. The grapnels are here called *korakes*
"ravens"; on this term, cf. note 85 above.

[90] Appian, *Bell. Civ.* 5.118: "Agrippa devised the so-called gripper, a timber 7 1/2
feet long bound with iron and having ring-bolts at each end. To one of these ring-
bolts was fixed the 'gripper' proper, an iron hook, and, to the other, multiple lines;
these, drawn by winches, pulled in the 'gripper' once it had been fired by catapult
into an enemy craft" (ἐπενόει δὲ καὶ τὸν καλούμενον ἄρπαγα ὁ Ἀγρίππας, ξύλον πεντά-
πηχυ σιδήρῳ περιβεβλημένον, κρίκους ἔχον περὶ κεραίας ἑκατέρας· τῶν δὲ κρίκων εἴχετο
τοῦ μὲν ὁ ἄρπαξ, σιδήριον καμπύλον, τοῦ δὲ καλῴδια πολλά, μηχαναῖς ἐπισπώμενα τὸν
ἄρπαγα, ὅτε τῆς πολεμίας νεὼς ἐκ καταπέλτου λάβοιτο).

[91] Cf. Dio Cassius 50.32.5 (Actium).

[92] Towers erected on merchantmen are reported as early as the 5th B.C. (Thucydides 7.25.6: during the operations in Syracuse Harbor). The first recorded use in
a strictly naval action was at the Battle of Chios in 201 B.C. (Polybius 16.3.12). Eu-
damus, admiral of the Rhodian contingent that fought Hannibal's powerful fleet in
190 B.C., had turrets on his flagship, which was a "four" (Livy 37.23.4, 24.6). When
Crassus besieged Rhodes in 43 B.C., he equipped his ships with "collapsible towers,
which were then set up" (πύργους ἐπτυγμένους, οἳ τότε ἀνίσταντο Appian, *Bell. Civ.*
4.72). Marcellus put *turres contabulatas* "towers of several stories" as well as siege
engines on the yoked quinqueremes he was using to besiege Syracuse in 214 B.C.
(Livy 24.34.6). For towers fore and aft, see Appian, *Bell. Civ.* 5.106; for dousing
them in a hurry, see 5.121 (Sextus' crews, as they fled at Naulochus in 36 B.C., τούς
τε πύργους κατέρριψαν "threw down the towers") and Dio Cassius 50.33.4 (when
Antony's men saw their leader flee, they raised sail, τούς τε πύργους καὶ τὰ ἔπιπλα
ἐς τὴν θάλασσαν ἐρρίπτουν "and kept tossing towers and gear into the sea").
A coin of Augustus (*Numismatische Zeitschrift* 34, 1902, pp. 117-18 and pl. 6.9-11;
Schweizer Münzblätter 19, 1969, p. 33, figs. 1-3) shows a prow that seems to have
a queer structure on it which some have taken to be a fighting tower (e.g., H. Seyrig
in *Revue Numismatique* sér. 6, 6, 1964, p. 19). Professor B. L. Trell, whose specialty
is the depiction of architecture on coins, assures me that it represents a building in
the background.

190 B.C., the Rhodians introduced the use of the fire pot—the last naval weapon to be invented until the very end of the ancient world. Containers of blazing fire were slung from the tips of long poles projecting over the bows (Fig. 115). If an enemy neared for an attack, the pots were dropped on his deck; if he flinched, he laid himself open to a stroke from the ram.[93]

IV LIGHT CRAFT

In the Hellenistic Age, when the ship of the line was generally at least a "four" and more often larger, the trireme could be specifically excluded from the ranks of the cataphracts and classed among the lighter units.[94] In other words, once the heavier galleys had appeared on the scene, in certain navies these were the ships that were

[93] Though the use of fire in the form of brands and the like is attested much earlier (e.g., Thucydides 7.53.4 describes an attack with a fire ship at the siege of Syracuse), the fire pot was introduced as an emergency measure by a Rhodian admiral in 190 B.C., as we learn from a fragment of Polybius preserved in Suidas, s.v. πυρφόρος (= Polybius 21.7[5]): "The 'fire-bearer,' which Pausistratus the Rhodian admiral used, was funnel-shaped. Along the inner surface of the gunwale on either side of the prow were rigged two looped lines. By means of these, poles were made fast with their ends sticking out over the water. To the tip of each pole the funnel-shaped container full of fire was attached with a piece of iron chain in such a way that, in attacking an enemy frontally or broadside, the fire could be dumped on his ship while, at the same time, it was a safe distance away from its own ship, thanks to the downward tilt [sc. of the pole]" (πυρφόρος, ᾧ ἐχρήσατο Παυσίστρατος ὁ τῶν Ῥοδίων ναύαρχος. ἦν δὲ κημός. ἐξ ἑκατέρου δὲ τοῦ μέρους τῆς πρώρας ἀγκύλαι δύο παρέκειντο παρὰ τὴν ἐντὸς ἐπιφάνειαν τῶν τοίχων, εἰς ἃς ἐνηρμόζοντο κοντοὶ προτείνοντες τοῖς κέρασιν εἰς τὴν θάλατταν. ἐπὶ δὲ τὸ τούτων ἄκρον ὁ κημὸς ἁλύσει σιδηρᾷ προσήρτητο πλήρης πυρός, ὥστε κατὰ τὰς ἐμβολὰς καὶ παραβολὰς εἰς τὴν πολεμίαν ναῦν ἐκτινάττεσθαι (ms. ἐκταράττεσθαι) πῦρ, ἀπὸ δὲ τῆς οἰκείας πολὺν ἀφεστάναι τόπον διὰ τὴν ἔγκλισιν). Livy (37.11.13) and Appian (Syr. 24) add the information that the containers were of iron. Appian further reports (Syr. 27) that the fire pots were used later in the same year as regular fleet equipment.

[94] The most revealing passage is Polybius 5.62.3: Antiochus III in 219 B.C. acquired 40 ships in the harbor of Tyre, of which "20 were cataphract, varying in build though none was smaller than a 'four,' while the rest were triremes, biremes [dikrota], and keletes" (κατάφρακτα μὲν εἴκοσι, διαφέροντα ταῖς κατασκευαῖς, ἐν οἷς οὐδὲν ἔλαττον ἦν τετρήρους, τὰ δὲ λοιπὰ τριήρεις καὶ δίκροτα καὶ κέλητες). The passage does not rule out that triremes could be cataphract (as they had been earlier; cf. 88 above). However, at this particular time, the cataphract trireme seems to have been the exception in major fleets; cf., e.g., Polybius' listing (16.2.10) of the forces opposed to Philip at the Battle of Chios: "65 . . . cataphracts . . . 9 triemioliai, 3 triremes" (κατάφρακτα . . . ἑξήκοντα καὶ πέντε . . . ἐννέα τριημιολίαι καὶ τριήρεις τρεῖς).

closed in with a fighting deck above and screens along the sides for use in hard combat, while the trireme was relieved of any ponderous superstructure in order to perform more efficiently in its new role as a member of the light contingents. Thus it reverted to the form it had had before the Second Persian War (87 above) and, as will be shown later (143 below), was perpetuated in this form by the Roman navy.

Smaller naval powers, which still relied on the trireme as their ship of the line, continued to use the cataphract version.[95] While the trireme thus maintained itself, and indeed in two forms, the pente-conter vanished from the major fleets: there were but nine at Arte-misium and four at Salamis.[96] Lesser nations, on the other hand, still found it useful. Rhodes sent a pair along with the Athenian aggrega-tion that sailed to Syracuse in 415 B.C. and a trio from Etruria joined it on arrival; Trapezuntum was still using penteconters at the be-ginning of the fourth B.C.; and so were many Greek city-states of South Italy in the mid-third B.C.[97] We have no way of telling whether these were the original single-banked or the later two-level version (cf. 61 above). Thereafter the penteconter disappears from naval his-tory—at least under that name.[98] As we shall see shortly, some craft

[95] Appian, *Mithr.* 17 describes Mithridates' fleet as consisting of "300 cataphracts and 100 biremes [*dikrota*]" (νῆες κατάφρακτοι τριακόσιαι, δίκροτα δὲ ἑκατόν); in *Mithr.* 92 he relates that the pirates who supplied the king with the bulk of his navy had beefed up their contingents to include "biremes [*dikrota*] and triremes" (cited in note 124 below). The cataphracts of the first passage must be identical with the triremes of the second. The city of Heracleia Pontica in 100 B.C. or so supplied Rome with "two cataphract triremes" (δυσί τε τριήρεσι καταφράκτοις Mem-non 29 = Jacoby, *FGH* no. 434, 21, vol. III B, p. 351). Caesar, describing the events of 48 B.C., refers (*Bell. Civ.* 2.23) specifically to a *triremis constrata* "decked tri-reme."

[96] Herodotus 8.1 (Artemisium), 8.48 (Salamis).

[97] Thucydides 6.43.1 (Rhodes), 6.103.2 (Etruria); Xenophon, *Anab.* 5.1.15 (Trape-zuntum); Polybius 1.20.14 (South Italy).

[98] E.g., Strabo 2.99 describes the ship Eudoxus built to continue his attempt to circumnavigate Africa (end of 2nd B.C.) as "a *lembos* . . . the equivalent of a pente-conter" (λέμβον . . . πεντηκοντόρῳ πάρισον).

Memnon (fr. 37 = Jacoby, *FGH* no. 434, 27.2, III B, p. 355) ascribes penteconters to Mithridates' fleet in 74 B.C. (πεντηκοντέρων τε καὶ κερκούρων ἀριθμὸς ἦν οὐκ ὀλίγος "[Mithridates] had no small number of penteconters and *kerkouroi*"), but he very likely is committing an anachronism; he wrote in the 1st or 2nd century A.D.

of the second and first B.C. are characterized vaguely as "two-level"; these may be penteconters or some descendant.

The triaconter can be traced down to perhaps the second century B.C. In the first half of the fourth, Athens' naval commanders sometimes operated in triaconters the way Homer's heroes had in their light galleys half a millennium earlier: they manned them with rower-fighters and used them for plundering coastal settlements.[99] The naval records indicate that Athens still had triaconters in active service in the second half of the century.[100] Alexander had a good number, including some with oars in two levels, in the fleet he assembled at Babylon and on the Indus.[101] Thereafter the triaconter must have gradually passed out of use, for, by about 200 B.C., newly introduced types of light galley had taken its place.[102]

The first of these was the *lembos*. Though the word often means no more than skiff (162 below), it has a very special naval sense. One of the peoples living along the coast of Illyria who made a profession of piracy developed a craft ideally suited to their needs called a *lembos*.[103] Shortly before 200 B.C., Philip of Macedon, finding it

---

[99] Polyaenus 3.9.63: Iphicrates, cruising off Phenice in 100 triaconters, observing the populace along the beach, ordered "his captains to cast anchor from the stern and moor in a row, and his men to arm themselves and each to lower himself in the sea alongside his own oar and to keep in line" (τοῖς μὲν κυβερνήταις ἄγκυραν ἀφιέναι κατὰ πρύμναν καὶ τὴν καταγωγὴν ἐν τάξει ποιεῖσθαι, τοῖς δὲ στρατιώταις ὁπλισαμένους ἕκαστον κατὰ τὴν αὑτοῦ κώπην εἰς τὴν θάλασσαν αὐτὸν καθιέναι καὶ τὴν αὑτοῦ τάξιν διαφυλάττειν). The townspeople were dumbstruck by the boldness and efficiency of the maneuver, the landing party stormed ashore with no opposition, and the raid netted captives and booty.

Clearly Iphicrates was using single-banked triaconters.

[100] E.g., *IG* II² 1627.410-14 (330/29); 1629.91-110, 128-44 (325/4 B.C.).

[101] See App. I. For the two-banked triaconter, see App. I, note I and 62 above.

[102] E.g., the 30-oared craft, which was the maximum size Antiochus III was permitted to retain under the terms of the Treaty of Apamea in 188 B.C., are classed as aphracts by Polybius (21.43.13: ἀφράκτων ⟨ὦν⟩ μηδὲ⟨ν πλείοσι⟩ τριάκοντα κωπῶν . . . ἐλαυνόμενον "aphracts, none driven . . . by more than 30 oars") and as *naves actuariae* by Livy (37.38.8: *naves actuarias, quarum nulla plus quam triginta remis agatur* "naves actuariae, none driven by more than 30 oars"). On these passages, see A. McDonald and F. Walbank, "The Treaty of Apamea (188 B.C.): The Naval Clauses," *JRS* 69 (1969) 30-39.

[103] *Lemboi* are frequently mentioned in connection with the squabbles between Rome and the Illyrians during the last half of the 3rd B.C. and the first half of the 2nd. All the Illyrian leaders had sizable aggregations of them, Teuta (Polybius

equally suited for certain of his needs, promoted it into a full-fledged fleet unit.[104] What distinguished all *lemboi* was the special build that gave them their celebrated speed and maneuverability; in other respects there was considerable variation. Some versions had 50 rowers, some as few as 16;[105] some had their oars all in one level, some in two;[106] most had a ram,[107] while some, which were used for express transport or carrying dispatches, perhaps did not.[108] The com-

---

2.9.1, 2.11.14, 2.12.3), Demetrius of Pharos (4.19.7-8), Scerdilaidas (4.29.7, 5.95.1, 5.101.1), Gentius (Livy 44.30.13-14, 45.43.10).

[104] Polybius 5.109.1-3: "Philip . . . reckoned that, for these enterprises of his, he needed ships and a naval service. Not for naval warfare—he had no hopes of being able to match Rome in a formal sea battle—but rather to transport troops, to move swiftly wherever he wanted, to make sudden unexpected appearances before the enemy. Since he considered the ships built by the Illyrians best for the purpose, he decided upon the construction of 100 *lemboi*" (Φίλιππος . . . ἀναλογιζόμενος ὅτι πρὸς τὰς ἐπιβολὰς αὐτοῦ χρεία πλοίων ἐστὶ καὶ τῆς κατὰ θάλατταν ὑπηρεσίας, καὶ ταύτης οὐχ ὡς πρὸς ναυμαχίαν—τοῦτο μὲν γὰρ οὐδ' ἂν ἤλπισε δυνατὸς εἶναι, Ῥωμαίοις διαναυμαχεῖν—ἀλλὰ μᾶλλον ἕως τοῦ παρακομίζειν στρατιώτας, καὶ θᾶττον διαίρειν οὗ πρόθοιτο, καὶ παραδόξως ἐπιφαίνεσθαι τοῖς πολεμίοις· διόπερ, ὑπολαβὼν ἀρίστην εἶναι πρὸς ταῦτα τὴν τῶν Ἰλλυριῶν ναυπηγίαν, ἑκατὸν ἐπεβάλετο λέμβους κατασκευάζειν).

Conway-Whatmough-Johnson, *The Prae-Italic Dialects of Italy* ii (Cambridge, Mass. 1933) 64, suggest that the word *lembos* may be Illyrian in origin, and H. Krahe, *Gymnasium* 59 (1952) 79 offers some linguistic evidence in support.

In addition to Macedon's navy, *lemboi* are reported in Syria's (Antiochus III had 200 in 197; see Livy 33.19.10) and Sparta's (in 192 B.C., Nabis had collected a flotilla of "3 cataphracts plus *lemboi* and *pristeis*" [*tres tectas naves et lembos pristesque*], Livy 35.26.1), while Rome frequently pressed detachments of Illyrian *lemboi* into fleet service (from Issa, Livy 31.45.10 and 32.21.27; from Dyrrachium, Issa, and King Gentius, Livy 42.48.8).

[105] See Strabo 2.99 (cited in note 98 above), Polybius 2.3.1 (5,000 Illyrians aboard 100 *lemboi*), Livy 34.35.5 (in the peace treaty of 195 B.C., Rome made Sparta give up all her navy "save two *lembi* powered by no more than 16 oars" [*praeter duos lembos, qui non plus quam sexdecim remis agerentur*]). On Mommsen's restoration of "30-oared *lemboi*" in Polybius 21.43.13, see note 102 above and the article by McDonald and Walbank cited there.

[106] Livy 24.40.2 (Philip attacked Apollonia in 214 B.C. with 120 *lembis biremibus* "two-banked *lembi*."

[107] Though a ram is never specifically mentioned, no ship without it could go in for the tactics described in Polybius 2.10.3-5 (see note 71 above) and 16.4.8-12 (cited in note 109 below).

[108] Polybius (18.1) reports that Philip sailed to a parley in 197 B.C. "with five *lemboi* and one *pristis* (πέντε λέμβους ἔχων καὶ μίαν πρίστιν), which Livy renders (32.32.9) "with five *lembi* and one ship with a ram" (*cum quinque lembis et una nave rostrata*). Either Livy, writing two centuries later, had no great knowledge of these ships, or noncombat *lemboi* dispensed with the cost (cf. Five, note 42) of a ram. Two types of *lembos*, one with ram and one without, would involve no great varia-

bat *lembos* would dart in among the enemy's heavier units to break up their formation, interrupt their tactics, even do damage to their oars.[109] The mark of its success is that one variety eventually served as model for a standard unit in the Roman Imperial Navy (141-42 below).

A galley called the *pristis* "shark" must have been similar to the *lembos*, for mention of it occurs always in connection with *lemboi*. There were contingents of the two types in the Spartan fleet at the time of Nabis[110] and in the Macedonian at the time of both Philip[111] and Perseus.[112]

IN THE navy of Philip's great rival, Rhodes, it was not the *lembos* that replaced the triaconter but a special light craft of Rhodian in-

---

tion in design. Both could have been made with the projecting cutwater, and the combat *lembos* would have this shod in bronze (cf. EIGHT, notes 9, 12).

[109] Polybius 16.4.8-12: "[At the Battle of Chios] if it had not been for the *lemboi* that the Macedonians introduced into the line of cataphracts, the fight would have been short and easy. But these hindered the Rhodian action in a great many ways. For, once the original line had been disturbed as the result of the initial clash, all got thrown together in a melee, so that the Rhodians could neither carry out the maneuver of breaking through the enemy line nor turn their ships around nor execute any of their favorite tactics. One moment the *lemboi* were attacking the oars to hinder the crews, the next they were back heading for the prows, the next they were heading for the sterns to get in the way of the helmsmen and their oars. Against prow-to-prow attacks the Rhodians did work out a technique: they depressed the bows of their own ships; thus they received the enemy's blows above the waterline and rendered their own blows fatal by inflicting their wounds below the waterline."

(εἰ μὲν οὖν μὴ μεταξὺ τῶν καταφράκτων νεῶν ἔταξαν οἱ Μακεδόνες τοὺς λέμβους, ῥᾳδίαν ἂν καὶ σύντομον ἔλαβε κρίσιν ἡ ναυμαχία· νῦν δὲ ταῦτ' ἐμπόδια πρὸς τὴν χρείαν τοῖς Ῥοδίοις ἐγίνετο κατὰ πολλοὺς τρόπους. μετὰ γὰρ τὸ κινηθῆναι τὴν ἐξ ἀρχῆς τάξιν ἐκ τῆς πρώτης συμβολῆς πάντες ἦσαν ἀναμὶξ ἀλλήλοις, ὅθεν οὔτε διεκπλεῖν εὐχερῶς οὔτε στρέφειν ἐδύναντο τὰς ναῦς οὔτε καθόλου χρῆσθαι τοῖς ἰδίοις προτερήμασιν, ἐμπιπτόντων αὐτοῖς τῶν λέμβων ποτὲ μὲν εἰς τοὺς ταρσούς, ὥστε δυσχρηστεῖν ταῖς εἰρεσίαις, ποτὲ δὲ πάλιν εἰς τὰς πρώρρας, ἔστι δ' ὅτε κατὰ πρύμναν, ὥστε παραποδίζεσθαι καὶ τὴν τῶν κυβερνητῶν καὶ τὴν τῶν ἐρετῶν χρείαν. κατὰ δὲ τὰς ἀντιπρώρους συμπτώσεις ἐποίουν τι τεχνικόν· αὐτοὶ μὲν γὰρ ἔμπρωρα τὰ σκάφη ποιοῦντες ἐξάλους ἐλάμβανον τὰς πληγάς, τοῖς δὲ πολεμίοις ὕφαλα τὰ τραύματα διδόντες ἀβοηθήτους ἐσκεύαζον τὰς πληγάς).

[110] Livy 35.26.1 (cited in note 104 above).

[111] At the Battle of Chios he had "150 *lemboi* and *pristeis* together" (λέμβοι δὲ σὺν ταῖς πρίστεσιν ἑκατὸν καὶ πεντήκοντα Polybius 16.2.9); cf. Polybius 18.1, cited in note 108 above.

[112] Livy 44.28.1.

vention. This was the *triemiolia*, offspring of a somewhat disreputable parent, the *hemiolia*.

The *hemiolia* shared with the *myoparo*—to which we will come in a moment—the distinction of being the ancient pirate's favored craft.[113] The name, meaning "one-and-a-half," derives from the vessel's peculiar oarage, specifically designed to meet the special requirements of a pirate ship. The sails of a war galley were solely for cruising; when going into action, a commander ordered them stripped off and either stowed away or even left ashore, since he had no use for them during a fight (235-36 below). The skipper of a pirate craft faced a totally different situation. His job was not to attack an enemy, waiting like a fighting cock to receive the onslaught, but to overhaul and capture a merchantman that was straining every inch of canvas to get away. In a good wind a sailing vessel could make better than 5 knots (288 below), while the best a galley could do was only about 7 (TWELVE, note 37), and that for no very long time. After a pull under such odds, rowers would be drooping on the benches, in no condition to muster for a boarding party. In short, a pirate had to give chase under sail as well as oars. But this raised a problem: when he had caught up with his quarry and was ready to grapple and make the capture, he had to do what the warships did, somehow get the sailing rigging out of the way and clear for action, a complicated process that demanded men and room. The solution, it appears, was the *hemiolia*. Basically it was a light, fast two-banked galley (Fig. 117).[114] But it was so constructed that

---

[113] Thus Theophrastus, *Char.* 25.2, can characterize a coward as someone who "when aboard ship, keeps saying that every headland is a *hemiolia*" (πλέων τὰς ἄκρας φάσκειν ἡμιολίας εἶναι).

[114] The vase-painting pictured in Figs. 81-82 shows two successive scenes. In the first, a pirate is overhauling his prey. There is a good wind blowing: the skipper of the merchantman, unaware of danger, has taken up on the brails and is traveling under shortened sail; the pursuer, in hard chase, has let his canvas fly so that every inch is drawing and he has every available oar manned. In the second, he is readying for the kill, on the point of carrying out the complicated maneuver of taking down sail. The oarsmen abaft the mast in the upper bank, having left their rowing stations and secured their oars, are some of them on their feet, one handling the windward sheet, another the halyards, a third the brails.

the rowers and oars in the top bank abaft the mast could be swiftly removed leaving, first, a large clear area in the after part of the ship in which to douse and secure sail, and, second, a dozen or so hands available to do it. The skipper of a *hemiolia* could race after a prize under canvas as well as oars (Fig. 81) and, at precisely the proper moment, put his vessel into the wind and give the command to secure the rigging (Fig. 82); seconds later sail and yard would be stowed out of the way, the mast unstepped and lowered into the crutch aft, and a boarding party stationed along the gunwale ready for the signal to jump.

The *hemiolia* eventually gained enough respectability to qualify for entrance into certain fleets that had use for a vessel of this kind.[115] More important, it produced an offspring so eminently respectable that it became a standard unit in the finest navies of the Hellenistic world—the *triemiolia*. It is first found in Rhodes' fleet[116] and was particularly favored by the island.[117] If we consider the name of the vessel with its obvious connection with the *hemiolia*, and at the same

[115] E.g., Alexander had *hemioliai* on the Hydaspes (Arrian, *Anab.* 6.1.1) and the Indus (*Anab.* 6.18.3). The Ptolemaic navy included all units "from the *hemiolia* to the 'five'" (ἀπὸ ἡμιολίας μέχρι πεντήρους Appian, *Praef.* 10). Philip V used *hemioliai* in the Battle of Chios (Polybius 16.6.4: παραλαβὼν τέτταρας πεντήρεις καὶ τρεῖς ἡμιολίας "taking with him 4 'fives' and 3 *hemioliai*") as well as in the naval actions against Scerdilaidas a few decades earlier (Polybius 5.101.2: he collected 12 cataphracts, 8 aphracts, 30 *hemioliai*). The Rhodian navy included them (*IG* XII Suppl. p. 139, no. 317, 2nd-1st B.C., with mention of "commodore of the *hemioliai*" [ἀγεμόνος τᾶν ἡμιολιᾶν]; Blinkenberg 15, for no good reason, suggests that the latter is a mistake for *triemioliai*), and the navy of the little island of Astypalaea (*IG* XII. 3.201, 2nd-1st B.C., with mention of "commander of the *hemioliai*" [ἄρχοντος τᾶν ἡμιολιᾶν]). The Romans used them for fast transport of troops from Sicily to Utica in 149 B.C. (Appian, *Pun.* 75).

[116] In 304 B.C., Menedemus with three *triemioliai* became a sort of Rhodian John Paul Jones: he burned freighters, captured ships full of provisions, even hijacked a quadrireme that was carrying the wardrobe of Demetrius' queen (Diodorus 20.93.3-4, the earliest mention of the *triemiolia*).

[117] The inscriptional evidence has been collected by Blinkenberg 13-17. Even earlier, F. Bechtel (*Die griechischen Dialekte.* II, *Die westgriechischen Dialekte*, Berlin 1923, p. 634) had pointed out that the *triemiolia* seemed intimately connected with the island of Rhodes.

Blinkenberg's long argument (21-44) that the hull profile one sees on the Victory of Samothrace and on certain other monuments made by Rhodians (e.g., Fig. 108) is the distinguishing mark of the *triemiolia* is utterly pointless. Any galley of whatever size that had an oar-box had that sort of profile.

time keep in mind the special nature of the Rhodian fleet, we can make a reasonable guess as to what kind of ship it was and how it came into being.[118]

Rhodes had taken upon itself the thankless job of sweeping the seas clean of pirates, one they carried out successfully until, about the middle of the second century B.C., Rome's inane foreign policy made it impossible for them to continue. The traditional units of a Greek navy were unfit for this purpose: a swift *hemiolia* could show its heels to a penteconter or trireme, whose rig was designed primarily for cruising. What was needed was a vessel that could not only give chase but have a clean advantage in the fight to follow. The simplest and most logical explanation of the *triemiolia* is that it was a design worked out by the Rhodians as the answer to this problem. Pirates had taken the two-banked galley, rearranged the oars in the after part of the upper bank, and created the *hemiolia* to chase merchantmen; Rhodes' naval architects, fighting the devil with fire, took one of the faster aphract models of the trireme, adapted it in the same way, and created the *triemiolia* to run down *hemioliai*.[119] A three-banked ship was more than a match for any pirate craft:

[118] Most writers assume that there is a connection between *triemiolia* and *trieres*, that the *triemiolia* was some sort of trireme: cf. Wilcken, *UPZ* 151, note to lines 2-4; Miltner, *RE*, s.v. *trihemiolia* 143; Starr in *CPh* (1940) 366; Tarn, "Warship" 141, note 11. Photius, for whatever his evidence is worth, calls the *triemiolia* a trireme (s.v. ἡμιολία) and a passage in Polybius (16.3.3-4 refers to its thranite oars: "[At the battle of Chios] Philip's flagship, a 'ten,' . . . when a *triemiolia* came into its path, gave it so hard a blow in the middle of the hull under the thranite thole that [the big ship] got caught fast" (ἡ δὲ τοῦ Φιλίππου δεκήρης, ναυαρχὶς οὖσα . . . ὑποπεσούσης γὰρ αὐτῇ τριημιολίας, ταύτῃ δοῦσα πληγὴν βιαίαν κατὰ μέσον τὸ κύτος ὑπὸ τὸν θρανίτην σκαλμόν, ἐδέθη). The question that has never been answered is: what kind of trireme was it?

[119] Torr, since he had at his disposal only three instances of the use of *triemiolia* (one of which he considered a mistake for *hemiolia*; cf. note 121 below), argued (15, note 41) that *hemiolia* and *triemiolia* were just two names for the same thing, the latter formed by false analogy with such words as τριημιπόδιον. Since the time he wrote, over two dozen examples in inscriptions and papyri have turned up, which show a clean distinction between the two words: in all its occurrences, *triemiolia* refers exclusively to a standard fleet unit, never a pirate craft. Blinkenberg, although he gives the texts of all the passages containing the word (save several in inscriptions and papyri which he missed; see note 121 below), failed to see this essential distinction and (p. 6) followed Torr, thereby vitiating much of his subsequent discussion.

For the *triemiolia* being aphract, see note 128 below.

it was heavier and larger and had sufficient height to enable archers to shoot down on the enemy, and even the lightest types had some decking to protect the crew and to accommodate marines; but the standard models were made, like all war galleys, to go into action without sails aboard. By designing a trireme whose upper bank, the thranite oars, was like that on a *hemiolia*, this disadvantage was obviated. And what could be more natural than to name such a craft *"triemiolia"*?[120] Once it had proven its worth, it was adopted by the Ptolemies, who had sea lanes to protect and were in friendly contact with Rhodes, by Athens, and presumably other navies.[121]

The *triemiolia* was not a permanent contribution to the ship-types of the ancient world. It shared the fortunes of its inventors: it came into being some time before 300 B.C., when Rhodes' naval power was approaching its zenith, and it passed away when Cassius in 42 B.C. stripped the island of its fleet. It needed as large and as well-trained a crew as any trireme; the Roman Imperial Navy passed it up in favor of the more easily manned liburnian (cf. 141 below). So attached were the Rhodians to these ships that for a century they kept a few alive—carefully giving them Roman-style names—to bring out on ceremonial occasions.[122]

[120] *Triemiolia* is a loose compound of *trieres* and *hemiolia*. The proper form would be theoretically *trieremiolia*—which actually does occur (Athenaeus 5.203d)—but this is a mouthful difficult to pronounce. Bechtel (*loc. cit.* note 117 above) cites other examples of such haplology.

[121] For *triemioliai* in the Ptolemaic navy, see Wilcken, *UPZ* 151.1-4 (259 B.C., cited in THIRTEEN, App., note 8) and *P. Hamb.* 57 (160 B.C. The sailors manning the ship in this document were not from the League of the Islands, but from the islands off the Egyptian coast west of Alexandria; cf. W. Schubart in *Gnomon* 2, 1926, p. 745). A passage in Athenaeus (5.203d) lists *triemioliai* among the ships in Egypt's navy but Torr (15, note 41), apparently followed by Blinkenberg (8, 11), argued, citing Appian, *Praef.* 10, that Athenaeus really meant *hemioliai*. The occurrences in the papyri prove conclusively that there were *triemioliai* in the Ptolemaic fleet; Athenaeus means what he says.

For *triemioliai* in the Athenian navy, see L. Robert, "Trihémiolies athéniennes," *RPh* 18 (1944) 11-17 (summarized in *REG* 56, 1943, p. 336) and also *Hellenica* 2 (1946) 123-26. These articles present four Athenian inscriptions that include the expression οἱ πλέοντες ἐν ταῖς τριημιολίαις "those sailing in the *triemioliai*": *IG* ii² 3218 (1st B.C.), 3494; *Inscr. de Délos* 1508; *Hesperia* 11 (1942) 292, no. 57 (208/7 B.C.).

[122] An inscription of A.D. 23 (Blinkenberg no. 39 = Lindos ii, no. 420.11-14) mentions a τριημιολίᾳ ᾇ ὄνομα Εὐανδρία Δ . . . καὶ . . . τριημιολίᾳ ᾇ ὄνομα Εἰρήνα Δ i.e., the fourth *triemiolia* to have the name *Euandria* (= *Virtus*), and the fourth

The *myoparo,* mentioned a moment ago, was favored by pirates over even the *hemiolia.*[123] This was because it was smaller and less complicated, merely an extraordinarily swift yet seaworthy open galley, hence cheaper to build and maintain and easier to man; pirates who were successful probably graduated from the *myoparo* to the *hemiolia.*[124] A number of navies used the ship,[125] but whether the pirate version unaltered or a considerably different one, we cannot say; we have practically no idea of either its size or appearance. The single representation preserved (Fig. 137), a crude and sketchy affair, shows a beaked prow, which is what we would expect, and a heavy hull with high freeboard, which seems most unlikely.

The *hemiolia* and *myoparo* were promoted to fleet service from the pirate packs. Others, such as the *phaselos,* made it from the ranks of the oared merchant marine (167-68 below).

Among the naval craft of the Hellenistic Age are galleys that are

the name *Eirena* (= *Pax*). A *triemiolia* named *Euandria* occurs in A.D. 10 (Lindos II, no. 392 b 7) and in two inscriptions of ca. A.D. 45 (*Annuario* 2, 1916, p. 142, no. 11 and *Eph. Arch.*, 1913, p. 10, no. 9 = Blinkenberg nos. 36-37). A number of inscriptions of Flavian date mention the *triemiolia Euandria Sebasta* (= *Virtus Augusta*): *IG* XII.1.58, XII.3.104, *Annuario* 2, 1916, p. 146, no. 18 (these three = Blinkenberg nos. 41-43); Blinkenberg no. 40 = Lindos II, no. 445. Several inscriptions mention the *Eirena Sebasta* (= *Pax Augusta*): *IG* XII Suppl. p. 108, no. 210 = *SEG* I 345; *Clara Rhodos* 6-7 (1932-33) 433, no. 53 (= Blinkenberg no. 44).

[123] E.g., Plutarch, *Luc.* 13.3 (ληστρικὸν μυοπάρωνα); Cicero, *Verres* 2.5.73 (*myoparone piratico*; cf. 2.3.186 and 2.5.89, 97, 100); Jerome, *Vita Hil.* 41 = Migne 23, 51c (pirates made their appearance in two *haud parvis myoparonibus* "by no means small *myoparones*").

[124] Cf. Appian, *Mithr.* 92: the pirates who aided Mithridates used "first *myoparones* and *hemioliai,* then biremes [*dikrotoi*] and triremes" (μυοπάρωσι πρῶτον καὶ ἡμιολίαις, εἶτα δικρότοις καὶ τριήρεσι).

[125] The Carthaginians included *myoparones* in the fleet they readied in 147 B.C. to defend their city against Rome (Appian, *Pun.* 121). Of the ten ships the Romans in 83 B.C. ordered Miletus to build for their use, at least one was a *myoparo* (Cicero, *Verres* 2.1.86-90). In the fighting off Illyricum in 47 B.C., a Pompeian admiral whose flagship, a "four," had been sunk, escaped in the ship's boat, and, when this too was sunk, "he swims to his *myoparo*" (*adnatat ad suum myoparonem* Caesar, *Bell. Alex.* 46); it sounds as if the flagship had a *myoparo* assigned to it as escort. In 37 B.C. Antony sent Octavian a contingent that Plutarch (*Ant.* 35.4) describes as "20 *myoparones*" but Appian (*Bell. Civ.* 5.95) as "10 *phaseloi*"; cf. EIGHT, note 58. Lucullus, in crossing from Greece to Egypt in 87-86 B.C., had a contingent of Rhodian aphracts and "3 Greek *myoparones*" (τρισὶν Ἑλληνικοῖς μυοπάρωσι Plutarch, *Luc.* 2.3).

referred to merely as "two-banked" (*dikrotos* in Greek, *biremis*[126] in Latin) with no further qualification. Theoretically this term could include any vessel with its oars in two levels up to a ponderous polyreme powered by great sweeps. The contexts show, however, that these craft are smaller than a trireme and thus have two tiers of one-man oars.[127] Since details are never given, all that can be done is to list the possibilities. Any ship so described may be (1) a two-level penteconter or triaconter, both of which were known from the sixth century B.C. on; (2) a two-level *lembos*, which came into use at the latest in the third B.C.; (3) a liburnian, a standard member of Roman fleets from at least the middle of the first B.C. and perhaps a few dec-

---

[126] The Greek term occurs in both literature and documents (see notes 127-29 below), the Latin only in literature, save for a single occurrence in the inscription recording Augustus' will (see the following note), which is at least semiliterary. The form *bicrota* occurs in one Latin inscription (*CIL* v 1956), but otherwise the only two-level warship known to Latin epigraphy, aside from possibly two-level polyremes, is the liburnian (cf. 141-42 below).

The Greek term *monokrotos* "single-banked," used of units in the Ptolemaic fleet, occurs in *BGU* VIII 1744.16. They must be small since they are mentioned in connection with *dikrotoi* (see following note) and aphracts (*BGU* VIII 1744.11, 1745.15).

[127] E.g., Augustus in a celebrated *naumachia* used (*Mon. Anc.* 23.4.46) *triginta rostratae naves, triremes aut biremes* "30 war galleys, triremes or biremes"; the Greek translation renders the last two terms τριήρεις ἢ δίκροτοι "triereis or *dikrotoi*." The Achaeans in 207 B.C. gave Philip V (Livy 28.8.7) *tres quadriremes et biremes totidem* "3 quadriremes and as many biremes." Biremes were chosen to be used in an attack that involved carrying them over a natural breakwater, no doubt because they were light (Caesar, *Bell. Civ.* 3.40). The big two-banked polyremes, since in their case the number of banks was of minor significance, were referred to by the appropriate name in the *-eres* series; see FIVE, note 5.

Dikrotoi were standard units in the navies of the Ptolemies (*BGU* VIII 1744.12, 1745.3, 18; mid-1st B.C.), Rhodes (Lindos II, no. 707, cited in the following note), Miletus or Smyrna (*Inscr. de Délos* nos. 1855-1857; 1st B.C. [see FIFTEEN, note 53]), Mitylene (Cicero, *ad Att.* 5.11.4, cited in note 129 below), Syria (Polybius 5.62.3, cited in note 94 above), Mithridates (see note 95 above). The *dikrotoi* used by Rome before the creation of the Imperial fleets very likely were commandeered from any of the above navies, such as the units that fought in the actions off Illyricum in 47 B.C. (*Bell. Alex.* 47.2), those collected by Brutus and Domitius in 44 for the showdown against Antony and Octavian (Cicero, *ad Att.* 16.4.4), those collected by Gallus as part of the fleet for the Red Sea expedition (Strabo 16.780; cf. EIGHT, note 58). A Rhodian commander, serving under Marcus Antonius Creticus in his antipirate campaign of 74 B.C., records on a monument that he was commander of "an oared vessel, two-level" (ἐπικώπου πλοίου δικρότου *Annuario* 2, 1916, p. 143, no. 12; cf. *RE* Supplbd. 5, 804). The way he refers to the craft rules out any standard fleet unit. Perhaps he captained a naval auxiliary, a merchant galley beefed up with a second bank of oars.

133

ades earlier (142 below); (4) a totally new type of warship that arose in this age. Later on, in the Roman period, the mention of biremes in naval contexts can refer only to liburnians (142 below).

Another pair of terms of general import used in this age is cataphract and aphract "fenced" and "unfenced" (*kataphraktos* and *aphraktos* in Greek; *tecta* or *constrata* [*navis*] and *aperta* in Latin). The first refers, as we have seen, to armored "fours" and larger craft, and sometimes includes and sometimes excludes triremes. The second takes in all the smaller units of a fleet—at least when used in documents or by ancient historians;[128] elsewhere the term can be loosely applied to any small galley with one level of oars.[129]

To SUMMARIZE: the light craft of the period included aphracts of various types and sizes. The largest was the trireme, followed by its close relative, the *triemiolia*. Two-level galleys included certain

---

[128] Thus the career of a Rhodian officer could proceed as follows: captain of a *dikrotos*; service in aphracts and cataphracts; captain of an aphract (*Lindos* II, no. 707, 40-30 B.C.: τριηραρχήσαντα δικρότου . . . καὶ στρατευσάμενον ἔν τε τοῖς ἀφράκτοις καὶ ταῖς καταφράκτοις ναυσὶ καὶ . . . τριηραρχήσαντα ἀφράκτου). He was promoted obviously to increasingly larger ships, and an aphract bigger than a *dikrotos* can only be a trireme, since anything still bigger was necessarily cataphract in build (see note 94 above). Similarly, in Segrè, "Dedica" 228, lines 4-6, the officer honored had "served in *triemioliai* and cataphracts" (cited in THIRTEEN, note 39), from which it follows that the *triemiolia*, basically a form of trireme, was aphract.

In inscriptions it is not at all uncommon for crews or officers to use such general terms as "aphract" or "cataphract" instead of identifying the vessel by oarage. Cf. *REG* 46 (1933) 442, note 1, a dedication (2nd-1st B.C.) to their commander on the part of a Samian crew who identify themselves as οἱ στρατευσάμενοι ἐν τῇ καταφράκτῳ νηὶ τῇ ἀποσταλείσῃ κτλ. "those serving in the cataphract that was dispatched etc.," or *SEG* xv 112 (225 B.C.), a dedication to their trierarch by a group of Athenians "who sailed with him in the aphract" (lines 22-23: τοῖς συνπλεύσασιν ἐν τῷ ἀφράκτῳ. The ship had earlier [line 4] been referred to as a *ploion*). Similarly, a sailor today might very well refer to his ship as, say, a cruiser without specifying medium, heavy, or light.

For citations from the historians, see notes 94 and 95 above.

[129] Thus Cicero, when he had available "aphracts from Rhodes and biremes (*dikrotoi*) of Mitylene" (*aphracta Rhodiorum et dicrota Mytilenaeorum, ad Att.* 5.11.4), complained of "the fragility of the Rhodian aphracts" (*aphractorum Rhodiorum imbecillitatem, ad Att.* 5.13.1), that "they cannot take any kind of sea" (*nihil . . . fluctum ferre possit, ad Att.* 5.12.1). That these aphracts were naval units and not just local craft is clear from *ad. Att.* 6.8.4, where he refers to the "Rhodian aphracts and other warships" (*Rhodiorum aphractis ceterisque longis navibus*).

penteconters and triaconters in the early part of the age, certain *lemboi* in the later, and those vessels vaguely denominated *dikrotoi* (or *dikrota*); on the fringe of this category is the two-level galley's close relative, the *hemiolia*. Lastly there were the single-banked penteconters and triaconters in the early part of the age, the single-banked *lemboi* in the later, the *myoparo*, and the various craft that laymen vaguely referred to as aphracts. Save for the trireme, none of these galleys had outriggers; the oars were worked over the gunwale, or, in the two-leveled galleys, as they had been centuries ago (cf. 62-64 above), the upper on or near the gunwale and the lower through ports in the hull (Fig. 119).

Navies as a rule maintained a good number of what in Greek is called *kataskopos* (*naus*) and in Latin *speculatoria navis* "reconnaissance craft."[130] These terms do not designate a special type of vessel but rather any of the smaller galleys discussed above when employed for such duty. Similarly, *tesserarios* (*tesserarius*) is used of a galley employed as a dispatch boat.[131]

---

[130] Plutarch, Pomp. 64.1: λιβυρνίδων δὲ καὶ κατασκόπων ὑπερβάλλων ἀριθμός "an immense number of liburnians and reconnaissance craft"; Caesar, *Bell. Afr.* 26.3: *litteris . . . per catascopum missis* "letters sent via a reconnaissance craft"; Livy 36.42.8: *una et octaginta constratis navibus, multis praeterea minoribus, quae aut apertae rostratae aut sine rostris speculatoriae erant* "with 81 cataphracts and, in addition, many smaller vessels, either aphracts with rams or ramless reconnaissance craft." Cf. 22.19.5, 30.10.14, 35.26.9; Caesar, *Bell. Gall.* 4.26.

[131] *IG* XII.5.941 (ca. 20 B.C.) refers to a C. Iulius Naso, "commander of the dispatch boats in Asia" (*praefectus tesserariarum in Asia navium . . . ὁ ἐπὶ τῶν τεσσαραρίων ἐν ᾿Ασίᾳ πλοίων*). Since the *tesserarius* was the officer in the Roman army who passed along the watchword, a ship so called would seem to be a dispatch boat. On the Althiburus mosaic (Fig. 137) two types are shown, one with a rounded bow and the other with projecting cutwater.

APPENDIX 1

SHIPS IN SECTIONS FOR TRANSPORT

ARRIAN reports (*Anab.* 7.19.3) that, of the fleet Alexander collected at Babylon, "a part had been brought from Phoenicia, namely 2 quinqueremes from the Phoenician navy, 3 quadriremes, 12 triremes, and up to 30 triaconters; these, cut up into sections, were transported from Phoenicia to Thapsacus on the Euphrates and there reassembled" (τὸ δὲ ἐκ Φοινίκης ἀνακεκομισμένον, πεντήρεις μὲν δύο τῶν ἐκ Φοινίκων, τετρήρεις δὲ τρεῖς, τριήρεις δὲ δώδεκα, τριακοντόρους δὲ ἐς τριάκοντα· ταύτας ξυντμηθείσας κομισθῆναι ἐπὶ τὸν Εὐφράτην ποταμὸν ἐκ Φοινίκης ἐς Θάψακον πόλιν, ἐκεῖ δὲ ξυνπηχθείσας αὖθις). The story is repeated by Strabo (16.741) and by Quintus Curtius (10.1.19), who typically inflates the contingent into a fleet of 700 "sevens" (cf. Tarn, "Warship" 150).

According to Arrian (*Anab.* 5.8.5), the fleet of small craft and triaconters Alexander had on the Hydaspes had also been brought overland, from the Indus: "The ships were cut up into sections and transported to him, the shorter divided into two sections, the triaconters into three; the sections, loaded on yoked teams, were carried up to the bank of the Hydaspes. There the craft were reassembled" (καὶ ξυνετμήθη τε τὰ πλοῖα καὶ ἐκομίσθη αὐτῷ, ὅσα μὲν βραχύτερα διχῇ διατμηθέντα, αἱ τριακόντοροι δὲ τριχῇ ἐτμήθησαν, καὶ τὰ τμήματα ἐπὶ ζευγῶν διεκομίσθη ἔστε ἐπὶ τὴν ὄχθην τοῦ Ὑδάσπου· κἀκεῖ ξυμπηχθὲν τὸ ναυτικὸν αὖθις).[1]

The technique of transporting ships in sections was long known in the East. Ramses III in the 12th B.C. had ships so moved from Coptos to the Red Sea.[2] The legendary Semiramis, according to Diodorus (2.16.6, 17.2), summoned shipwrights from Syria and Cyprus and other maritime areas, had them build collapsible river craft, and furnished camels to transport 2,000 of these. And, just this past century, Muhammed Ali had a small fleet built at Alexandria and conveyed in sections down the Nile to Cairo and from there on camels to Suez.[3]

---

[1] It is hard to see how triaconters could be divided into just three sections, even though Alexander's, being two-banked (*dikrotoi*, Arrian, *Anab.* 6.5.2; for the demonstration that this term can apply only to the triaconters, see Tarn, "Warship" 144-45), were short, compact vessels (perhaps 45' long; see 56 above). Arrian, however, is a trustworthy historian, and in this case he is following the account of Aristobulus, an eye witness. Larger craft, of course, must have been divided up into a great many sections.

[2] R. Faulkner in *CAH* II² (1966), Chapter XXIII, section xiii.

[3] See P. Newberry, "Notes on Seagoing Ships," *JEA* 28 (1942) 64-66.

APPENDIX 2

THE NAVAL ARMS RACE,
315–ca. 250 B.C.

W. W. Tarn devoted a number of studies[1] to the history of naval construction in this period, exercising much ingenuity in an attempt to extract a coherent story from the details reported chiefly by Diodorus and Plutarch.

It had taken perhaps half a century to go from a "four" to a "six," and well over a quarter of a century from a "six" to a "seven." With the introduction of the "seven," the pace of development quickened dramatically. In but 25 years or so, the gamut was run from a "seven" to a "sixteen," thanks to the drive, inventiveness, and daring of Demetrius the Besieger of Cities. The first mention of "sevens" occurs in connection with the ambitious naval program that he and his father Antigonus inaugurated in 315. When Demetrius fought the Battle of Salamis (the Salamis on Cyprus) in 306 with the fleet so created, he had 10 "sixes" and 7 "sevens,"[2] while his opponent, Ptolemy I, had nothing larger than "fives";[3] Demetrius' flagship was a "seven," since this type was the biggest under his command.[4] There is one problem, however. Diodorus, who reports the above, in listing the ships Antigonus had put on the stocks in 315 includes 3 "nines" and 10 "tens."[5] Where were these in 306 if the largest Demetrius used then were "sevens"? Tarn suggests that Diodorus' mention of "nines" and "tens" at this early date is a mistake.[6] If not, we must assume that Demetrius for some reason did not take them into action; possibly, since the enemy had nothing larger than "fives," he felt the "sixes" and "sevens"

---

[1] "The Greek Warship," *JHS* 25 (1905) 137-56, 204-18; "The Dedicated Ship of Antigonus Gonatas," *JHS* 30 (1910) 209-22; "Alexander's Plans," *JHS* 59 (1939) 124-35; *Hellenistic Military and Naval Developments* (Cambridge 1930) 129-41.

[2] Diodorus 20.50.3: ἐπτήρεις ἐπτὰ Φοινίκων . . . ἐξήρεις δέκα "7 Phoenician 'sevens' . . . 10 'sixes.' "

[3] Diodorus 20.49.2: of Ptolemy's ships, "the biggest was a 'five,' the smallest a 'four' " (ἦν ἡ μεγίστη πεντήρης, ἡ δ᾽ ἐλαχίστη τετρήρης).

[4] Diodorus 20.50.2: τούτων δ᾽ ἦσαν αἱ μέγισται μὲν ἑπτήρεις "Of these [ships in his fleet], the biggest were 'sevens' "; 20.52.1: Δημήτριος ἠγωνίσατο τῆς ἑπτήρους ἐπιβεβηκὼς ἐπὶ τῇ πρύμνῃ "Demetrius fought, taking his stand on the poop of his 'seven.' "

[5] Diodorus 19.62.8: the total agglomeration included "90 'fours,' 10 'fives,' 3 'nines,' 10 'tens,' 30 aphracts" (τετρήρεις μὲν ἐννενήκοντα, πεντήρεις δὲ δέκα, ἐννήρεις δὲ τρεῖς, δεκήρεις δὲ δέκα, ἄφρακτοι δὲ τριάκοντα).

[6] "Alexander's Plans" 127.

were all he needed and preferred to mass his manpower on them rather than spread it thin on the bigger units.

The "nines" and "tens" that Diodorus mentions certainly came into being before 301 B.C., because, by that time, Demetrius' fleet included an "eleven"[7] and a "thirteen."[8] In 288 he raised the ante even further and launched the greatest ships he was to design, a "fifteen" and a "sixteen," both being vessels that were remarkable, considering their size, for their beauty and efficiency.[9] His bitter enemy, Lysimachus, had earlier requested and obtained permission to attend a naval review of Demetrius' growing aggregation,[10] and, about the time that Demetrius had added his two mightiest, Lysimachus was ready with an answer—the *Leontophoros*, a ship, equally remarkable for beauty and size; it is called an "eight,"[11] but it obviously must have been an "eight" of a type that was a match for

---

[7] Theophrastus, *Hist. Plant.* 5.8.1: μῆκος μὲν ἦν τῶν εἰς τὴν ἐνδεκήρη τὴν Δημητρίου τμηθέντων τρισκαιδεκαόργυιον "The timber cut for Demetrius' 'eleven' [*hendekeres*] was in 13-*orguia* [= about 75′] lengths." The building of an "eleven" would naturally have preceded a "thirteen."

[8] Plutarch, *Dem.* 31.1: after the defeat at Ipsus (301 B.C.), Demetrius demanded from Athens, and got back, the ships he had left there "which included the 'thirteen'" (ἐν αἷς ἦν καὶ ἡ τρισκαιδεκήρης). It was his flagship at the time; e.g., he once "entertained Seleucus on the 'thirteen'" (ἐκεῖνον [Seleucus] ἐν τῇ τρισκαιδεκήρει δεξάμενος Plutarch, *Dem.* 32.2).

[9] Plutarch, *Dem.* 43.4-5: οὐδεὶς γὰρ εἶδεν ἀνθρώπων οὔτε πεντεκαιδεκήρη ναῦν πρότερον οὔτε ἐκκαιδεκήρη "No mortal had ever before seen either a 'fifteen' or a 'sixteen,'" . . . τῶν . . . νεῶν οὐκ ἦν τὸ καλὸν ἀναγώνιστον, οὐδὲ τῷ περιττῷ τῆς κατασκευῆς ἀπεστεροῦντο τὴν χρείαν, ἀλλὰ τὸ τάχος καὶ τὸ ἔργον ἀξιοθεατότερον τοῦ μεγέθους παρεῖχον "The beauty of the ships was by no means neglected, nor did they lose in effectiveness because of the vast scale of their construction. As a matter of fact, they had a speed and maneuverability more remarkable than their size." Cf. *Dem.* 20, cited in the following note.

[10] Plutarch, *Dem.* 20.4: Λυσίμαχος μὲν . . . ἔπεμψε παρακαλῶν ἐπιδεῖξαι . . . τὰς ναῦς πλεούσας · ἐπιδείξαντος δὲ θαυμάσας ἀπῆλθε "Lysimachus . . . wrote to him, requesting that he show him . . . his ships under way. Demetrius did, and Lysimachus went off in amazement." Tarn ("Ded. Ship" 211, note 15) suggests this took place ca. 300 B.C., when Demetrius' biggest was a "thirteen."

[11] See Six, note 52. There is no certainty about the origin of this ship. All we know for sure is that the city of Heracleia (on the Black Sea) supplied it for Ptolemy Keraunos' fleet in 280 B.C. (cf. note 13 below). Tarn attributed its building to Lysimachus ("Warship" 208, note 93) on the basis of its name, the lion being Lysimachus' special symbol (cf. Tarn, *Antigonos Gonatas*, Oxford 1913, p. 131). The suggestion has much to be said for it, since a ship of such size has no place in a city-state fleet, nor could it be paid for out of a city-state treasury. Lysimachus might very well have had it built in one of Heracleia's shipyards; the city was within easy reach of good shipbuilding timber (E. Semple, *The Geography of the Mediterranean Region*, New York 1931, pp. 273-74).

a "sixteen."[12] This superdreadnought led the Ptolemaic navy to victory over Antigonus Gonatas in 280 B.C.[13] Gonatas took almost two dozen years to prepare for revenge and achieved it at the Battle of Cos, probably ca. 258 B.C., in which he utterly routed Ptolemy II, largely through the help of a superdreadnought of his design which he then dedicated to Apollo,[14] almost certainly at the sanctuary on Delos.[15] This ship must have been at least as powerful as, and very likely more powerful than, the *Leontophoros*. Ptolemy, with Egypt's vast wealth behind him, put an end to the

[12] I take this vessel to be a double-hulled "eight" and, in that way, the equal of a "sixteen"; see 112-14 above. Tarn's first idea was that it was "a new or abnormal development of some sort" ("Ded. Ship" 211), but afterward he changed his mind (*Hell. Dev.* 133, 135) and argued that it was really a "sixteen"—and indeed the very same one as that built by Demetrius, the fortunes of war having put it in Lysimachus' hands. This is impossible: Demetrius' ship was built either at Athens, Corinth, Chalcis, or near Pella (Plutarch, *Dem.* 43.3: [of the fleet that included the "sixteen"] τὰς μὲν ἐν Πειραιεῖ τρόπεις ἔθετο, τὰς δὲ ἐν Κορίνθῳ, τὰς δὲ ἐν Χαλκίδι, τὰς δὲ περὶ Πέλλαν "Some keels he had laid at Athens, some at Corinth, some at Chalcis, and some near Pella"), whereas one of the known facts about Lysimachus' ship is that it came from, or was built at, Heracleia (see previous note), which Lysimachus then ruled. For ordinary "eights," see, e.g., Polybius 16.3.2.

We can trace some of the subsequent history of Demetrius' last two great ships. After Demetrius' downfall in 285 B.C., Ptolemy I presumably got the "fifteen" and "sixteen" (cf. Tarn in *CAH* VII 92) and may have dedicated the "fifteen" to Apollo of Delos (Tarn, *BCH* 46, 1922, pp. 473-75). The "sixteen" probably was recaptured by Antigonus Gonatas at his victory over Ptolemy off Cos in 258 (Tarn's suggestions in *Hell. Dev.* 133, note 5, are pure guesswork), since it was still in Macedon's navy in 197 B.C.; by then it was so old and antiquated that Rome, after the victory over Philip V, had no hesitation in letting him keep it (see SIX, note 10). Eventually, after 167 B.C., Aemilius Paulus took the old battlewagon to Rome and sailed up the Tiber in triumph on it (Livy 45.35.3: *Paulus . . . regia nave ingentis magnitudinis, quam sedecim versus remorum agebant . . . adverso Tiberi ad urbem est subvectus* "Paulus . . . was carried to Rome right up the Tiber . . . on the king's personal ship, an enormous vessel powered by 16 files of oarsmen").

[13] Memnon, frg. 13 = Jacoby, *FGH* no. 434, 8.6: αὐτῶν δὲ τῶν Ἡρακλεωτίδων τὸ ἐξαίρετον ἔφερεν ἡ Λεοντοφόρος ὀκτήρης "And of these ships from Heracleia, it was the 'eight,' the *Leontophoros*, that distinguished itself."

[14] Athenaeus 5.209e: Ἀντιγόνου ἱερὰν τριήρη, ᾗ ἐνίκησε τοὺς Πτολεμαίου στρατηγοὺς περὶ Λεύκολλαν τῆς Κῴας ἐπειδὴ καὶ τῷ Ἀπόλλωνι αὐτὴν ἀνέθηκεν "the sacred trireme of Antigonus with which he defeated Ptolemy's admirals off Leucolla in Cos, when he vowed the ship to Apollo." Tarn is certainly right in arguing ("Ded. Ship" 212-15) that the sanctuary which received the offering was the one on Delos, the island in the very center of the waters where the great naval struggles were taking place.

[15] Tarn is probably right in identifying ("Ded. Ship" 216-18) this "sacred trireme" of Antigonus with the one at Delos reported by Pausanias 1.29.1: τὸ δὲ ἐν Δήλῳ πλοῖον οὐδένα πω νικήσαντα οἶδα, καθῆκον ἐς ἐννέα ἐρέτας ἀπὸ τῶν καταστρωμάτων "I know that nobody ever conquered the ship in Delos, having as many as nine rowers from the decks." On the meaning of the last phrase and the size of this ship, see 115 above.

race by outclassing all rivals once and for all with the launching of a "twenty" and two "thirties." His fleet at its most powerful, the mightiest the ancient world was to know, included

17	"fives"	30	"nines"	4	"thirteens"
5	"sixes"	14	"elevens"	1	"twenty"
37	"sevens"	2	"twelves"	2	"thirties"

Although the list is given in a secondary source,[16] there is no reason to question its trustworthiness, particularly since the existence of the "twenty" and "thirty" and other units have been confirmed by contemporary documents.[17] It must represent the navy toward the end of Ptolemy II's reign (he died in 246). Earlier the fleet included at least one "ten,"[18] Demetrius' "fifteen," which may have been taken out of service to be dedicated to Apollo of Delos, and Demetrius' "sixteen," which was probably lost to Antigonus Gonatas at the Battle of Cos.[19] Why "tens" were abandoned as a class we cannot say. The "fifteen" and the "sixteen" needed no replacing now that the fleet boasted a "twenty" and two "thirties."

When Ptolemy IV built a "forty," the race had already run its course; the ship was intended only as a showpiece.[20]

[16] Athenaeus 5.203d.

[17] *OGIS* 39: a base of a statue found in the sanctuary of Aphrodite at Paphos inscribed [B]ασιλεὺς Πτολεμαῖος [Πυργ]οτέλην Ζώητος, ἀρχιτεκτονήσ[αντα] τὴν τριακοντήρη καὶ εἰκ[οσήρη] "King Ptolemy for Pyrgoteles [or Ergoteles], son of Zoës, builder of the 'thirty' and 'twenty.'" Two of Ptolemy's supergalleys are mentioned in papyri. The first occurrence, of a "nine" in the abbreviated form τὴν θ' (*P. Cairo Zen.* 59036 = *Select Papyri* II 410.21, 257 B.C., cited in THIRTEEN, note 29), was for so long the only one that the editor was not fully convinced he had read the abbreviation correctly, and it is still sometimes queried (e.g., W. Peremans and E. Van't Dack, *Prosopographia Ptolemaica* v, Louvain 1963, no. 13800). But now we can add two more: *Sammelb.* 9780, mid-3rd B.C., is a letter from someone angling for the job of *skeuophylax*, guard of gear and stores, aboard a "nine" (this time spelled out in full), and *P. Col. Zen.* 63, 257 B.C., an account of miscellaneous expenditures, lists (recto, col. II, lines 2-3) a loan to the captain of a "ten" (also spelled out in full).

[18] See the previous note.

[19] See note 12 above.

[20] Plutarch, *Dem.* 43.5: θέαν μόνην ἐκείνη παρέσχε. καὶ μικρὸν ὅσον διαφέρουσα τῶν μονίμων οἰκοδομημάτων, φανῆναι πρὸς ἐπίδειξιν, οὐ χρείαν, ἐπισφαλῶς καὶ δυσέργως ἐκινήθη "This ship [the "forty"] was only for show. Hardly differing from buildings that are fixed on the ground, it moved unsteadily and with difficulty, to make appearance for display, not use."

The Roman Imperial and Byzantine Navies

I THE ROMAN IMPERIAL NAVY

ACTIUM wrote finis to the formal sea battle for over 300 years. Octavian, who had matched "sixes" against Antony's "tens" and won, not only had firsthand experience of the effectiveness of light units, but also the historical vision to discern that, with no rival naval power on the horizon, Rome's chief task on the water was anti-piracy control, communications, transport. So, when he founded the navy that was to serve the empire for the next three centuries, he gave priority to speed and maneuverability over weight and, in the interests of efficiency, limited and standardized the units. There was a single "six," the flagship of the main squadron based at Misenum. There were perhaps two "fives" and a fair number of "fours." But the standard units were the trireme for the major Italian fleets[1] and the destroyerlike liburnian for the provincial fleets.[2]

The liburnian[3] was a fast, two-banked galley[4] adapted from a craft developed among the Liburnians, piratical-minded dwellers of the

---

[1] We know the names of 1 "six," 1 "five," 9 "fours," 52 triremes, and 13 liburnians belonging to the fleet at Misenum, and 2 "fives," 7 "fours," 22 triremes, and 2 liburnians belonging to the fleet at Ravenna; cf. Kienast 120, note 153a. These figures, it must be remembered, refer to names that happen to occur on tombstones and are in no sense an indication of the strength of the fleets at any given moment. Thus I suspect that the Ravenna fleet had only one "five" at a time to serve as flagship, even though two differently named ships of that size are recorded.

[2] E.g., only liburnians have been attested in the Alexandrian flotilla; see FIFTEEN, note 57.

[3] On the liburnian, see the exhaustive article by S. Panciera, "Liburna," *Epigraphica* 18 (1956) 130-56, and his entry *Liburna* in *Dizionario epigrafico di antichità romane*, s.v. (1958).

[4] Lucan 3.534: *ordine contentae gemino . . . liburnae* "the liburnians, content with twin banks"; Appian, *Ill.* 3: Ῥωμαῖοι τὰ κοῦφα καὶ ὀξέα δίκροτα Λιβυρνίδας προσαγορεύουσιν "The Romans call their fast open double-banked ships [*dikrota*] 'liburnians.'"

141

Dalmatian coast and its offshore islands. We know that in the third and second centuries B.C. these people were using *lemboi*, that some of these were pressed into Rome's service, and that at least one model of *lembos* had two levels of oarsmen (126 above). All this makes a strong case for there being a connection between the *lembos* and *liburnian*. There were, as noted above, many kinds of *lemboi*; no doubt the liburnian was the one that the Romans found particularly useful for their purposes. The earliest certain mention of liburnians is at the Battle of Naulochus in 36 B.C.,[5] but there is no reason why they could not have been in use long before.

As just mentioned, the Imperial navy included only liburnians, triremes, quadriremes, and quinqueremes and one "six." The light units that Roman writers refer to as biremes must therefore be liburnians.[6] On Trajan's column are pictured numerous galleys with two banks (upper ship in Fig. 127, lower in 128). These cannot be two-level "fours," "fives," or "sixes" because, for one, the artist specifically shows but one man to each oar, and, for another, in several scenes the vessels are portrayed on the Danube,[7] a locale where such heavy units have no place. They therefore can only be liburnians. As shown in these reliefs, the liburnian is a small, aphract galley with no outrigger; the upper oars are worked through a latticed bulwark and the lower through ports just below the gunwale.[8]

---

[5] Augustus put his flag aboard a liburnian (Appian, *Bell. Civ.* 5.111).

[6] Roman writers consistently use layman's language when referring to warships instead of naval jargon, *biremis* for *liburna* being but one example. They almost always speak of triremes, quadriremes, and quinqueremes, whereas the official terms were *trieres* (*CIL* VI 1063.17, 3095, 32771; IX 41, 43), *quadrieres* (*CIL* VI 1063.15, *AE* 1927.3), and *penteres* (*CPL* 193). Cf. the use of *tetreris* in the caption in Fig. 126.

[7] E.g., K. Lehmann-Hartleben, *Die Trajanssäule* (Berlin 1926) pl. 19, no. 34, and pl. 24, nos. 46/47.

[8] A glass vessel, found in Afghanistan, is adorned with a picture of the lighthouse of Alexandria and a two-banked galley; see J. Hackin, *Recherches archéologiques à Begram* (*Mémoires de la délégation archéologique française en Afghanistan* IX, Paris 1939) 43 and fig. 39. The lower level of oars are a second bank, not the oars of the other side of the ship as the author suggests; the artist is following normal procedures of representation, as the canoe in the picture (fig. 38), with an oar on only one side, reveals. Quite possibly the vessel was a "souvenir of

142

The Roman navy used both models of trireme, the cataphract (Fig. 125)[9] and the lighter aphract (lower ship in Fig. 127, upper in 128). The quadriremes and quinqueremes, as we can see from representations on Roman coins (Figs. 120-123),[10] were either single-banked (i.e., driven by four-man and five-man sweeps respectively) or two-banked (i.e., in the case of quadriremes, two levels of two-man oars, and, in the case of quinqueremes, presumably an upper level of three-man oars and a lower of two-man oars; cf. Fig. 126). All these types were inherited by the Romans from the previous age. There is, however, one novel feature: very very few now have a rowing frame, an oar-box.

The outrigger had been an essential element in the Greek triremes of the fifth and fourth centuries B.C. In the form of the straight-sided rowing frame, it continued on in Hellenistic galleys (117-19 above), was still in use in the mid-first century A.D.,[11] and may have supplied the inspiration for the *telaro* that was a standard feature in the larger galleys of a later age.[12] However, alongside the Greek trireme with its outrigger, there was always in existence the Phoenician type without one (94-95 above). From the late first century B.C. on,[13] Rome's naval architects, in designing not only triremes but bigger units as well, like the Phoenicians, dispensed with the rowing frame and placed the oars in ports pierced in the hull. Ships in many cases were still fitted with projections that run along either side from the bow to the steering oar and look very much like a rowing frame, but these house no oars—the oars are consistently shown emerging from

---

Alexandria" brought back by someone who had traveled to the great port. Liburnians were common in the fleet Rome maintained at Alexandria; see note 2 above.

[9] On the arrangement of the oars in Fig. 125, see Anderson 33-34 and fig. 9.

[10] See App.

[11] One of the Nemi barges was fitted with a straight-sided rowing frame; see Ucelli 169-74, 256-57, fig. 184 and pl. 7.

[12] For examples, see Six, note 72. Cf. Ucelli 257.

[13] Hellenistic coins (e.g., Fig. 107 and those cited in App., note 1) consistently show the rowing frame, as do the earliest Roman issues (App., note 2). Thereafter it appears but rarely (e.g., Fig. 129). Oars emerging from the hull, with the frame-like projection above, can be seen on coins of 38-36 B.C. (*BM Republic* II, p. 518, no. 155 = *Numismatische Zeitschrift*, 1905, pl. 2.16a; Moll E IV f 7).

the side of the hull *below* the projections. A comparison of Fig. 116
with Fig. 124 or Fig. 133, and Fig. 129 with Fig. 131 points up the
difference graphically. In Fig. 116 and Fig. 129 the ships are rowed
in the older fashion, with oars working through the rowing frame;
in the other representations the oars come through the hull just below the projection. If the two arrangements were for a time in use
simultaneously, the latter soon became standard. We see it in pictures and models of triremes, both aphract (lower ship in Fig. 127)
and cataphract (Fig. 125), of two-banked polyremes either with the
projection (Figs. 122, 130) or without it.[14] These vessels, accommodating all the oars within the span of the hull, would be of necessity beamier than the type with a rowing frame.

The projection just described seems to appear for the most part
in the bigger units, quadriremes or larger, which were all, of course,
cataphract. A series of paintings from Pompeii (e.g., Fig. 133) shows
single-banked ships which, to judge from their size and the impressive number of marines on the fighting decks, are quadriremes or
quinqueremes.[15] All have the projection. It can be seen on certain
coin pictures (Figs. 122-123) that portray galleys of at least the same
size.[16] It is present, though less prominent, in the well-known relief

[14] See also Anderson pl. 7a = Moll B IV 42a (a galley with two lines of ports in
the hull, each port of the upper immediately above one of the lower, as in Fig. 125);
L. Basch, "Un modèle de navire romain au Musée de Sparte," *AC* 37 (1968) 136-71,
esp. fig. 1 and pls. 1, 2 (clay model of a cataphract of indeterminate size with a
projection. Basch argues [147-48] that the projection is a rowing frame, but a wall-
painting from Pompeii that he himself cites as a parallel [fig. 5 on p. 143] shows
the truth of the matter—the oars emerge from *below* the projection); Basch, fig. 8
(clay model without rowing frame. As Basch points out, this piece, dated much
earlier hitherto, is probably Roman).

[15] Fig. 133 has been selected from no fewer than eleven similar scenes, four in
the House of the Vettii and seven originally in the Temple of Isis and now in the
Naples National Museum. Cf. *IH* ill. 50, Moll B XI a 5. All consistently show the
projection with oars below. And so does a very similar vessel—a heavy single-banked
cataphract with marines on the fighting deck—from a tomb of Augustan date near
Rome; see C. Pietrangeli, "Frammento di trabeazione romana del cemeterio dei
Giordani," *Bullettino della Commissione Archeologica del Governatorato di Roma* 67
(1939) 31-36.

[16] See App. Gems picturing Roman warships have the same feature; see Moll
E III 6693, 6694.

from Praeneste (Figs. 130, 132)[17] of what is at least a quadrireme[18] and very possibly still bigger. We can only guess at its purpose. Perhaps it served as a massive bumper to shield the oarsmen. Possibly, too, it helped protect the oars. One standard method of attack was to go after an opponent's oars, to try to shear them off and thereby render him a standing target for a mortal stroke of the ram;[19] on these Roman cataphracts, at the critical moment the rowers could swing the oars as close alongside the hull as possible, where, under the shelf of the projection, they could not easily be reached. Lastly, as the Praeneste relief (Figs. 130, 132) shows, it was a convenient jump-off point for marines preparing to board.

From the relatively large number of representations of warships available, we can gain a fair idea of what cataphract galleys of this age looked like. The walls of the hull rise to the level of the projection just described (Figs. 130, 132-133) or, where there is none, to the gunwale (Fig. 125), completely enclosing the rowers and eliminating the need for the removable screens of an earlier age (88 above). The oars work through ports in the hull, and—at least on some ships—a series of louvers furnish ventilation (Figs. 130, 132). The projection itself rests on a series of brackets that spring out from the hull just above the oarports (Fig. 124). Exactly when this mode of construction was introduced, whether in Roman times or earlier, during the Hellenistic age, is anybody's guess. Over the rowers is the deck (katastroma) on which the fighting personnel are stationed. It is guarded by a low bulwark (Figs. 125, 130, 133) to

[17] The Praeneste relief (Figs. 130, 132), on the basis of the crocodile shown on the prow as well as on stylistic grounds, is usually dated around the time of the Battle of Actium and often connected with it; for discussion of the piece and bibliography, see E. Simon in Helbig-Speier, *Führer durch die öffentlichen Sammlungen klassischer Altertümer in Rom. I, Die päpstlichen Sammlungen im Vatikan und Lateran* (Rome 1963⁴) no. 489. (Simon's comments on the ship must be treated with caution: she describes the prow as "Hellenistic," though prows of identical shape are common on galleys of the Roman Imperial Navy, and the oblique fittings in the projection, which I have called louvers, she identifies as the blades of lances; whatever they are, they cannot be that.)

[18] This is Anderson's (33, fig. 8) estimate of its size.

[19] Cf. Polybius 16.4.8-12, cited in Six, note 109.

which shields may be attached as added protection. Just forward of the bow oars the bulwark breaks to leave a narrow entry for coming aboard (Figs. 127, 130, 133).[20] At about the same point may be affixed a rectangular frame with a carving of the ship's patron deity or of a figure connected with its name (Figs. 125, 127, 129-131).

At both prow and stern the architects of the Imperial navy introduced significant changes. When the Roman Republic first launched a fleet in the middle of the third century B.C., it adopted from the Hellenistic galleys the stempost that ended in a volute and the three-pronged ram,[21] and both features lasted until at least the end of the Republic (Figs. 120-121, 124-126, 129-133).[22] Shortly after the middle of the first century A.D. a single-pointed ram makes its appearance,[23] and thereafter the older form is no longer seen (Figs. 122-123, 127).[24] The stempost ending in a volute was retained (Fig. 122),[25] although, from the beginning of the second century A.D. on, some galleys were given a massive forecastle that took up the whole area in the bows

[20] See Lehmann-Hartleben, *op. cit.* (note 7 above) pl. 38 for another example of the entry.

[21] See the coins cited in Six, note 68.

[22] See also the coins cited in App., notes 3-5 and *BM Empire* I, Augustus 670 and pl. 16.13 (29-27 B.C.). Cf. Vergil, *Aen.* 5.143: *rostrisque tridentibus* "three-pronged rams," Valerius Flaccus 1.688: *aere tridenti* "three-pronged bronze."

[23] *BM Empire* I, p. 285, nos. 1, 2 and pl. 49.1, 2 (A.D. 68). Mattingly says (p. clxxxviii) that "the galley is borrowed from the famous legionary coin of Mark Antony" (cf. App., note 4), but the vessel portrayed on those consistently is given a three-pronged ram. The single-pointed ram also appears on coins of Galba (A. Robertson, *Roman Imperial Coins in the Hunter Coin Cabinet* I, London 1962, no. 46 and pl. 27), Vespasian (*BM Empire* II, p. 74 and pl. 12.8), Titus (Robertson, no. 64 and pl. 46).

The ships in the Pompeian wall-paintings cited in note 15 above, dated shortly after A.D. 63, all have the three-pronged ram. Possibly it was still in use, but, equally possibly, the painter chose to represent old-fashioned types. In a house at Herculaneum which was painted shortly before the eruption (Casa dell'Atrio Corinzio), there are two spirited pictures (unpublished) of sea battles, the galleys in which all have the new single-pointed ram; see A. Maiuri, *Ercolano: I nuovi scavi (1927-1958)* I (Rome 1958) 261-65 (the pictures are in the room numbered 3 in his plan on p. 261).

[24] See the coins cited in App., notes 6-9.

[25] See the coins cited in note 23 above. Also *BM Empire* III, Hadrian 247, 509, 1028, 1394, 1403 and pls. 51.10, 56.20, 68.19, 85.1, 85.5; V, Severus 847 and pl. 51.4, Caracalla 267-68, 859 and pls. 33.14, 33.15, 52.6.

and incorporated the stempost into its structure, leaving just a re-
sidual volute or none at all (Fig. 127).[26]

Aft, the traditional fan-shaped *aphlaston* continues in favor (Figs.
120-123, 128-129, 131, 133), but the *stylis* (346 below) is gone;
ships of the Republic carry nothing (Figs. 120-121),[27] those of the
Imperial navy replace it with legionary standards (Figs. 122-123,
128). A cabin for the commander, perhaps a Roman invention,[28] is
now a regular feature, taking the form of an arched doghouse under
the curve of the *aphlaston* (Figs. 128, 133). Representations show
the massive and complicated construction of the prow about the ram
(Figs. 130, 132),[29] the fighting towers in position (Figs. 130, 132),
and what seems to be a *hypozoma* (Figs. 119, 125).

A number of galleys are shown carrying sail (Fig. 120), a particularly welcome detail since it comes after a hiatus of nearly half
a millennium. The main driver is still a square sail stepped amidships, but it is now abetted by an *artemon* in the bows (Fig. 127),
an addition that made its debut at the earliest during Hellenistic
times (238 below).

[26] See also *BM Empire* iii, Hadrian 543, 1391, 1398 and pls. 57.20, 84.13, 85.3;
the relief cited in note 29 below.

[27] See the coins cited in App., notes 2-5.

[28] See Six, note 66. Galleys pictured on Roman coins right down to the end of
the 1st b.c. continue to show no cabin; see Fig. 120 and *BM Republic* ii, pp. 526-30,
564-65 and pls. 116, 120.16. The cabin may have made its appearance in the time
of Augustus. There is a wall-painting in Pompeii depicting Polyphemus attempting
to smash Odysseus' ship as it pulls out of range; see P. von Blanckenhagen and
C. Alexander, *The Paintings from Boscotrecase, Röm. Mitt.*, sechstes Ergänzungsheft
(Heidelberg 1962) pls. 40, 43. The painting is an original work, not an adaptation
of some earlier picture, and dates to the last decade of the first century b.c. (von
Blanckenhagen-Alexander 10-11, 48-51). The ship is shown with a cabin, and it is
quite possible that the painter had noted such a feature on contemporary galleys.
The shape he gave it, with a peaked instead of arched roof, may reflect his imaginative idea of what suited a mythological vessel, cf. Six, note 74.

Cabins in the form of arched doghouses appear frequently in later wall-paintings
(e.g., Fig. 133) and consistently in the coin-pictures (e.g., those cited in App., notes
6-9).

[29] See also *Huitième congrès international d'archéologie classique, Paris 1963*
(Paris 1965) pl. 17.2 (relief of a galley prow on the gravestone of a Roman sailor).

II THE BYZANTINE NAVY

In A.D. 324, Constantine and Licinius prepared for a final show-down on the sea. Licinius collected 350 triremes. Constantine chose to go into action with only 30-oared and 50-oared galleys. He won, and the trireme, for so long queen of the Mediterranean, was forced to yield the throne.[30]

Thus, by the fourth century A.D., naval architecture, going full cycle, returned to the ships that had served the Greek fleets of a thousand years before, galleys of one bank with each rower working his own oar. The reversion was brought about no doubt by chronic shortages of money and men. Two centuries later we find the same kinds of ships in Justinian's navy. By this time they have acquired a new name, *dromon* "racer."[31] It was to have a long career—though applied to a somewhat different kind of craft.

The one-level galleys of Constantine's and Justinian's fleets lived on to become the *moneres* and *galea* of the navy of the Byzantine Empire during its flourishing days in the tenth century, the time we know it best.[32] The name *dromon* was now given to the ships of

---

[30] For the makeup of Constantine's and Licinius' fleets, see Zosimus 2.22.1-2, 2.24.1; cf. Kienast 138-39.

[31] Procopius describes (*Bell. Vand.* 1.11.15-16) the vessels used in the expedition to North Africa in A.D. 533 as "single-banked and having decks overhead to reduce to a minimum the chance of the rowers being hit by enemy missiles. Men today call these ships 'dromons' because they are able to sail very swiftly" (μονήρη . . . καὶ ὀροφὰς ὕπερθεν ἔχοντα, ὅπως οἱ ταῦτα ἐρέσσοντες πρὸς τῶν πολεμίων ἥκιστα βάλλοιντο. δρόμωνας καλοῦσι τὰ πλοῖα ταῦτα οἱ νῦν ἄνθρωποι· πλεῖν γὰρ κατὰ τάχος δύνανται μάλιστα). These dromons, in other words, were single-banked cataphracts.

A passage from a dispatch of Theodoric (A.D. 525/6) is often taken (Torr 17, Kienast 156) as evidence that triremes were still being built at that time. In the dispatch Theodoric, lamenting that Italy no longer has a fleet, orders the construction of 1,000 dromons (Cassiodorus, *Varia* 5.16: *decrevimus mille . . . dromones fabricandos*). When the fleet was completed he describes it in purple prose as a "fleet-forest, floating homes, the army's feet . . . a trireme-conveyance revealing so great a number of oars, yet carefully hiding the men's faces" (*classeam silvam, domos aquatiles, exercituales pedes . . . trireme vehiculum rcmorum tantum numerum prodens, sed hominum facies diligenter abscondens* Cassiodorus, *Varia* 5.17). In a passage of this nature, the words *trireme vehiculum* need not be taken to the letter. Cf. Paulinus of Nola, *Poemata* 24.73, where an ordinary ship that carried a casual traveler who brought Paulinus some letters is referred to as a quadrireme.

[32] For the Byzantine navy, see the fundamental articles by R. Dolley, "The Warships of the Later Roman Empire," *JRS* 38 (1948) 47-53 and "Naval Tactics in the

the line of this fleet, two-banked galleys with a minimum of 25 oars in each bank, or 100 in all.[33] There were three sizes. The smallest was called *ousiakos* because it was manned by one *ousia* or company, numbering 100 or slightly more;[34] the men were charged with pulling the oars and also taking care of the fighting,[35] which, in effect, would mean those in the upper bank. The largest, the *dromon* proper, had a minimum crew of 200. Of these, 50 were permanently assigned to the lower bank, and the remaining 150 were stationed above.[36] Possibly 100 of them manned the oars, 2 to each, and 50 were marines;[37] when the occasion warranted it, the second oarsman

---

Heyday of the Byzantine Thalassocracy," *Atti dell' VIII Congresso di Studi Bizantini* 1 (Rome 1953) 324-39. The texts of the relevant authors have been re-edited by A. Dain in his *Naumachica* (Paris 1943).

For the *moneres* and *galea*, see Leo, *peri thalassomachias* 10 (Dain, p. 21): "smaller and very swift dromons, such as those called *galeai* ['galleys'] and *monereis* ['single-bankers'], quick and agile, which you use for patrol and other duties requiring speed" (δρόμωνας μικροτέρους γοργοτάτους, οἰονεὶ γαλέας καὶ μονήρεις λεγομένους, ταχινοὺς καὶ ἐλαφρούς, οἷσπερ χρήσῃ ἐν ταῖς βίγλαις καὶ ταῖς ἄλλαις σπουδαίαις χρείαις).

[33] Leo, *peri thal.* 8 (Dain, p. 20): "Let each [dromon] have at least 25 thwarts on which the rowers will be placed, so that all the lower thwarts will total 25, the upper similarly 25, for a grand total of 50. Let the rowers sit 2 on each thwart, one to starboard and the other to port, so that all the rowers—who will also double as marines—both those above together with those below, will total 100 men" (ἑκάστη δὲ ἐχέτω ζυγοὺς τὸ ἐλάχιστον κέ ἐν οἷς οἱ κωπηλάται καταστήσονται, ὡς εἶναι ζυγοὺς τοὺς ἅπαντας κάτω μὲν κέ, ἄνω δὲ ὁμοίως κέ, ὁμοῦ ν'. καθ' ἕνα δὲ αὐτῶν δύο καθεζέσθωσαν οἱ κωπηλατοῦντες, εἰς μὲν δεξιά, εἰς δὲ ἀριστερά, ὡς εἶναι τοὺς ἅπαντας κωπηλάτας ὁμοῦ καὶ τοὺς αὐτοὺς καὶ στρατιώτας τούς τε ἄνω καὶ τοὺς κάτω ἄνδρας ρ').

[34] Constantinus Porphyrogenitus, *de caerimoniis*, 2.45, p. 384A: "Four *chelandia ousiaka* with 108 men to each" (χελανδίων οὐσιακῶν δ' ἀνὰ ἀνδρῶν ρή). This concerns the preparations for an attack on Crete in A.D. 949. For *chelandia* = *dromones*, see Dolley, "Warships" 48.

[35] See Leo, *peri thal.* 8, cited in note 33 above.

[36] Leo, *peri thal.* 9 (Dain, p. 21): "And have other larger dromons than these [i.e., than those manned by a crew of 100] built, holding 200 men, or more or less than this figure depending on the need of the moment created by the enemy. Of these, 50 will work the lower bank of oars, and the 150 will all take their station above, armed, and fight the enemy" (καὶ ἕτεροι δὲ δρόμωνες κατασκευαζέσθωσάν σοι τούτων μείζονες, ἀπὸ διακοσίων χωροῦντες ἀνδρῶν ἢ πλέον τούτων ἢ ἔλαττον κατὰ τὴν χρείαν τὴν δέουσαν ἐπὶ καιροῦ κατὰ τῶν ἐναντίων· ὧν οἱ μὲν ν' εἰς τὴν κάτω ἐλασίαν ὑπουργήσουσιν, οἱ δὲ ρ' καὶ ν' ἄνω ἑστῶτες ἅπαντες ἔνοπλοι μαχέσονται τοῖς πολεμίοις).

[37] Dolley, "Warships" 48, puts three men on each upper oar. This would surely necessitate longer oars in this bank than in the lower, and we have no evidence for that.

could be called upon to leave the bench and take up weapons.[38] In between the *ousiakos* and *dromon* was the *pamphylos*, whose crew varied from 120 to 150 or 160.[39] Exactly how these were divided between rowers and marines, we do not know.

The word *dromon*, thus, was used in two senses: generically to refer to any of these three types of two-banked ships of the line, and specifically to refer to the largest of the three.[40] Within the class of *dromons* in its specific sense, the very biggest could have as many as 120 oars[41]—i.e., two banks of 30 a side—with a rowing complement of 220.[42] The biggest crew we hear of is 300, made up of 230 oarsmen and 70 marines.[43] Here again we do not know precisely how the oarsmen were distributed; perhaps the upper bank had 2 to each oar, which would put 180 men permanently assigned to the benches and leave 40 to 50 who could serve as rowers or marines as occasion called.

All three models seem to have been similar in build. The *dromon* proper must have been somewhat broader and also somewhat longer to permit more spacious decking fore and aft for accommodating its larger fighting contingent.

All three models were aphract. There were gangways along each

---

[38] For upper-bank oarsmen sharing in the fighting, see Dolley, "Naval Tactics" 332.

[39] Constantinus Porphyrogenitus, *de caer.* 2.45, p. 384A-B: "six *chelandia pamphyla* with 120 men each . . . six *chelandia pamphyla* with 150 men each" (χελανδίων παμφύλων ς′ ἀνὰ ἀνδρῶν ρκ′ . . . χελανδίων παμφύλων ς′ ἀνὰ ἀνδρῶν ρν′; this was in A.D. 949); 2.44, p. 377C: "seven *pamphyloi*, three with 160 men each and the other four with 130 men each, total 1,000" (πάμφυλοι ζ′ ἔχοντες οἱ μὲν γ′ ἀνὰ ἀνδρῶν ρξ′, οἱ δὲ ἕτεροι δ′ ἀνὰ ἀνδρῶν ρλ′, ὁμοῦ ‚α; this was in A.D. 906).

[40] Cf. Dolley, "Warships" 48.

[41] Constantinus Porphyrogenitus, *de caer.* 2.45, p. 388B: "for equipping the 20 dromons . . . 120 oars each, total 2,400" (εἰς ἐξόπλισιν τῶν κ′ δρομονίων . . . κωπία ἀνὰ ρκ′· ὁμοῦ ‚βυ′; this was in A.D. 949).

[42] Constantinus Porphyrogenitus, *de caer.* 2.45, p. 384B: "4 dromons, with 220 men each" (δρόμονες δ′ ἀνὰ ἀνδρῶν σκ′). A company (*ousia*) consisted of 108 (cf. note 34 above) or 110 men, and these dromons were manned by 2 companies. Cf. 2.45, p. 384A: "20 dromons with 2 companies each, [total] 40 companies" (δρόμονες κ′ ἀνὰ οὐσιῶν β′· οὐσίαι μ′).

[43] Constantinus Porphyrogenitus, *de caer.* 2.44, p. 377C: "seven dromons with 230 oarsmen and 70 marines, total 2,100" (δρόμονες ζ′ ἔχοντες ἀνὰ ἀνδρῶν κωπηλατῶν σλ′ καὶ ἀνὰ πολεμιστῶν ό· ὁμοῦ ‚βρ′).

side,[44] and a "stiffener," i.e. a girder that ran down the center and held the yard crutches,[45] very likely doubled as a catwalk. To give the rowers some protection, a light frame was rigged along the gangways on which shields could be hung.[46] Aft the ships seem to have been constructed like their Roman predecessors with a poop deck and, upon this, the commander's cabin;[47] on either side was the housing for the steering oars.[48] The ships followed Roman practice as well in dispensing with the rowing frame, for both banks of oars were worked through ports in the hull.[49] There was a stout raised

[44] Dain, *Naumachica* 5.2.6-7 (p. 65): "The center of the poop and of the vessel up to the prow is called the 'undecked' [*asanidon*]. The covered areas along either side of the hull are called the 'deck' [*katastroma*] or the 'platform' [*thranos*] or the 'planking' [*sanidomata*]. Above are the first bank of oars, the heavy armed soldiers, the bowmen, and the light armed soldiers; below the 'planking' is the second bank of oarsmen, who do nothing but row since the fighting personnel is on the 'deck' above" (τὸ δὲ μέσον τῆς πρύμνης καὶ νεὼς μέχρι τῆς πρῴρας ἀσάνιδον. τὰ δὲ ἑκατέρωθεν τῶν τοίχων κατάστεγα, κατάστρωμα λέγεται καὶ θρᾶνος καὶ σανιδώματα, ὧν ἄνωθεν ἡ πρώτη εἰρεσία καὶ οἱ ὁπλῖται καὶ τοξόται καὶ πελτασταί, κάτωθεν δὲ τοῦ σανιδώματος ἡ δευτέρα ἥτις δι' ὅλου ἐρέττει, τυχόντων ἐπὶ τοῦ καταστρώματος ἄνωθεν τῶν πολεμούντων).

[45] Dain, *Naumachica* 5.2.10 (p. 65): "What are called 'rests' are on the stiffener-keel, three of them made fast in a line, and the yard, when lowered, rests on them" (καὶ οἱ λεγόμενοι καθορμεῖς ἐπὶ τῆς τρόπιος στερεᾶς, προσηλοῦνται κατὰ στοῖχον τρεῖς ὄντες, ἐφ' ὧν ἡ κεραία καταγομένη ἐπίκειται). Since the yard would not be lowered down into the bottom of the hold, the "stiffener-keel" (*tropis sterea*) must have run above the keel at or near deck level. Cf. the upper ship in Fig. 127.

[46] Dain, *Naumachica* 5.2.13 (p. 65): "And there [i.e., by the gangways] is the *castellum*, where the marines hang their shields" (ἐκεῖσέ που τὸ καστέλλωμα γίνεται, ἔνθα τὰς ἀσπίδας οἱ στρατιῶται κρεμῶσι).

[47] Dain, *Naumachica* 5.2.5 (p. 64): "the platform aft, where there is also a cabin, or at any rate a berth [*krabatos*], set up for the admiral or the captain" (τὸ δὲ πρὸς τὴν πρύμναν ποδόστημα, ἔνθα δὴ καὶ σκηνὴ πήγνυται τῷ στρατηγῷ ἢ τριηράρχῳ, ἤγουν κράβατος); Leo, *peri thal.* 8 (Dain, p. 20): "And on the poop should be the berth for the captain . . . both to distinguish and set apart the commanding officer as well as to protect him at times of combat from missiles hurled by the enemy" (καὶ ὁ τοῦ ναυάρχου δέ . . . κράβατος ἐπὶ τῆς πρύμνης γινέσθω, ὁμοῦ μὲν ἀφωρισμένον δεικνύων τὸν ἄρχοντα, ὁμοῦ δὲ καὶ φυλάττων ἐν καιρῷ συμβολῆς ἀπὸ τῶν ῥιπτομένων βελῶν παρὰ τῶν ἐναντίων).

[48] Dain, *Naumachica* 5.2.6 (p. 64): "The parts on either side of the stern, in which the steering oars rest, are called the 'spreaders' [*petasoi*] or the 'splits' [*schista*] or the 'cheek-pieces' [*epotides*]" (τῆς δὲ πρύμνης τὰ μέρη πάλιν ἑκάτερα πέτασοι καὶ σχιστὰ καὶ ἐπωτίδες λέγονται, ἐν οἷς ἐπίκεινται τὰ πηδάλια).

[49] Dain, *Naumachica* 5.2.12-13 (p. 65): "The strake from which the [lower] oars project is called the 'shield strake' [*thyreon*]; that to which they are bound, the thole [*skalmos*]; that by which they are bound, the oar-strap [*tropoter*]. The strake over the tholes is the 'over-thole strake' [*episkalmis*]. The oars are fitted through what are called the 'borings' [*tremata*] . . . Above this bank of oars is a wale, then

platform forward for the fighting personnel[50] and, on big units, a smaller platform—a feature first attested in this age—amidships.[51]

Although *dromons* continued for a while to be equipped with the ram and a ram attack was still an element in battle,[52] they were primarily intended for fighting at close quarters. Their weapon par excellence was one that had played but a small role in the ancient world—fire. In the eye of the ship, nestling under the forward platform, was the cannonlike *siphon*, in effect a flamethrower and a formidable one.[53] It consisted of a long tube of wood lined with bronze, to the inboard end of which was coupled an air pump; some of the inflammable stuff known as "Greek fire" was poured in, ignited, the pump was activated, and a sheet of flame belched forth from the muzzle (Fig. 134).[54] The largest ships had flamethrowers

---

another 'shield strake' for the upper bank of oars. On top of the whole is the gunwale [*epenkenis*]" (ἡ δὲ σανὶς δι' ἧς αἱ κῶπαι ἐξέρχονται θυρεόν, καὶ ὅθεν μὲν ἐκδέδενται σκαλμός, ᾧ δὲ ἐνδέδενται τροπωτήρ. τὸ δὲ ἐπὶ τῶν σκαλμῶν ἐπισκαλμίς. δι' ὧν δὲ ἤρτηται ἡ κώπη τρήματα . . . ταύτης δὲ ἄνωθεν τῆς εἰρεσίας περίτονον, εἶτα πάλιν θυρεόν, ἔνθα ἡ ἄνωθεν εἰρεσία. ἄνωθεν δὲ πάντων ἡ ἐπηγκενίς).

[50] Leo, *peri thal.* 6 (Dain, p. 20): "[The dromon] should by all means have forward in the bows the flamethrower [*siphon*], girdled with bronze in the usual fashion, through which the prepared fire mixture is shot at the enemy. Topping a flamethrower of this sort there should be a false walk of planks, also fenced about with planks, on which the fighting personnel will take its stand" (ἐχέτω δὲ πάντως τὸν σίφωνα κατὰ τὴν πρώραν ἔμπροσθεν χαλκῷ ἠμφιεσμένον, ὡς ἔθος, δι' οὗ τὸ ἐσκευασμένον πῦρ κατὰ τῶν ἐναντίων ἀκοντίσει. καὶ ἄνωθεν δὲ τοῦ τοιούτου σίφωνος ψευδοπάτιον ἀπὸ σανίδων, καὶ αὐτὸ περιτετειχισμένον σανίσιν, ἐν ᾧ στήσονται ἄνδρες πολεμισταὶ κτλ.). These fighting platforms were called *xylokastra* "wood-castles"; see the following note.

[51] Leo, *peri thal.* 7 (Dain, p. 20): "And the so-called wood-castles can be set up around the middle of the mast on the largest dromons, fenced with planks, from which a number of men can fire into the middle of the enemy vessel either massive stones or iron weights by means of which an [enemy] ship can either be holed, etc." (ἀλλὰ καὶ τὰ λεγόμενα ξυλόκαστρα περὶ τὸ μέσον τοῦ καταρτίου ἐν τοῖς μεγίστοις δρόμωσιν ἐπιστήσουσι περιτετειχισμένα σανίσιν, ἐξ ὧν ἄνδρες τινὲς εἰς τὸ μέσον τῆς πολεμίας νηὸς ἀκοντίσουσιν ἢ λίθους μυλικοὺς ἢ σίδηρα βαρέα . . . δι' ὧν ἢ τὴν ναῦν διαθρύψουσιν κτλ.). I take this amidships-castle to be a platform with parapet set halfway up the mainmast. Dolley ("Warships" 51) took it to be a platform similar to the one at the prow; as he points out, this raises problems.

[52] Cf. Leo, *peri thal.* 69 (Dain, p. 31); paraphrased by Dolley, "Naval Tactics" 331.
[53] See the passage cited in note 50 above.
[54] Cf. J. Partington, *A History of Greek Fire and Gunpowder* (Cambridge 1960) 15-17; M. Mercier, *Le feu grégeois* (Paris 1952) 24-40.

amidships and aft as well as forward,[55] and sometimes even the marines were equipped with miniature hand models.[56] A second important arm were the catapults, which shot not only missiles but grenades, pots of "Greek fire" that would explode on impact.[57] The fighting platform amidships must have had overhangs projecting laterally past the gunwales; on these heavy weights were suspended like the "dolphins" of ancient times (239 below), and when an enemy came so close that any were poised over his rowers, the lashings that held them were loosed or cut away.[58]

There are no contemporary representations of these vessels, though we have a few dating from several centuries later (Fig. 134).[59] These reveal that the ram had by this time disappeared, replaced by a long projecting spur at deck level that served as boarding bridge. They also show the ships rigged with a lateen sail; very likely the same rig had been carried earlier. There was a mainmast amidships, set up permanently, as was later practice, rather than retractable as on ancient galleys,[60] certainly another forward, and a mizzen on the largest ships.[61]

THE details given above, as mentioned before, all date from the tenth century. The previous centuries are almost a blank. Some scraps

[55] Constantinus Porphyrogenitus, de caer. 2.48, p. 388B: "for the equipment of the 20 dromons: three flamethrowers each, total 60" (εἰς ἐξόπλισιν τῶν κ′ δρομονίων—σιφώνια ἀνὰ γ′, ὁμοῦ ξ′).

[56] Leo, peri thal. 65 (Dain, p. 30): "by means of little flamethrowers [siphones] fired by hand ... called a hand-flamethrower [cheirosiphon]" (τῶν διὰ χειρὸς βαλλομένων μικρῶν σιφώνων ... ἅπερ χειροσίφωνα λέγεται).

[57] Leo, peri thal. 60 (Dain, p. 30): "catapults on the poop, prow, and both sides of the dromon" (καὶ τοξοβαλίστραι δὲ ἔν τε ταῖς πρύμναις καὶ ταῖς πρώραις καὶ κατὰ τῶν δύο πλευρῶν τοῦ δρόμωνος). For pots of "Greek fire," see peri thal. 64. In 61, Leo recommends pots full of vipers, scorpions, and the like.

[58] See the passage cited in note 51 above.

[59] Anderson, fig. 11 and pl. 7b (= IH ill. 51).

[60] Dain 5.2.9 (p. 65): "In the middle of the ship, over the keel is fitted the mast-step, in which the mast stands. . . . The lower part of the mast, that which is nailed fast to the step, is called the 'leg'" (τῆς δὲ πλεούσης μέσον ἐπὶ τῆς τρόπιος προσαρμόζεται ἡ τράπεζα, ἧς ἐντὸς ὁ ἱστὸς ἵσταται ... τοῦ δὲ καταρτίου τὸ μὲν προσηλωμένον τῇ τραπέζῃ κατώτερον μέρος πτέρνα καλεῖται).

[61] See Dolley, "Warships" 52, who cites examples of ships with at least two masts, and dromons (in the specific sense) with three.

of information have been preserved about the navy of the Arab conquerors of Egypt around the year A.D. 700, which unquestionably consisted of units taken over from their Byzantine predecessors or modeled upon such units. The ship of the line, though here called a *karabos*, is a two-banked galley; furthermore, it is described as *kastellatos*, which may mean that, like the *dromon*, it was fitted with the *castellum*, the light frame along the sides for hanging shields.[62] Another unit attested is the *dromonarion*; it must be a smaller and lighter craft since at times, along with *akatia* (159 below), it was used to patrol the mouth of the Nile whereas the two-banked craft were never assigned such duty. It may be the same as the *dromonion*, or small *dromon*, that occurs in the tenth century Byzantine fleet.[63]

[62] See *P. Lond.* IV 1433.64, 129, 179, 227, 319 for mention of διήρεις κάραβοι "two-banked *karaboi*," and 1434.35, 1435.98 and 103, 1441.102, 1464 for κάραβοι καστελλᾶτοι "*karaboi* with *castella*" (specified as two-banked in 1449.94). On *castellum* "shield-frame," see note 46 above.

[63] For *dromonarion*, see *P. Lond.* IV 1435.95, the only instance in which it is spelled out in full, being abbreviated elsewhere. For only *dromonaria* and *akatia* in the squadron guarding the mouth of the Nile, see 1434.22, 135; 1435.10, 95 (for translations of these documents, see *Der Islam* 4, 1913, pp. 87-96). *Dromonaria* also served with the regular fleet units; see 1337.2-3, 1464; *P. Ross.-Georg.* IV 6.4. For the *dromonion* of the 10th century Byzantine navy, see Dolley, "Warships" 53.

In *P. Cairo Masp.* 67359, cols. II.2 and VI.9, δρο(μοναρίων) is probably to be read instead of δρο(μόνων).

APPENDIX

COIN EVIDENCE FOR SINGLE-BANKED "FOURS," "FIVES," AND LARGER UNITS

COINS of the Hellenistic Age for some reason tend to picture only the prows or sterns of galleys.[1] On Roman issues, however, we often see a whole ship, complete with oars.

The earliest of these date from the beginning of the first century B.C., though the craft portrayed memorialize actions that took place one to two centuries before.[2] Next comes a number of issues connected with the naval wars of 38-36 B.C. between Octavian and Sextus Pompey (Fig. 120).[3] Then follows a famous series that Antony struck in 32-31 for the use of the armed forces he had gathered in anticipation of the showdown at Actium (Fig. 121).[4] Of the coins in all three groups, none shows a galley with three banks of rowers; one, dated 38-36 B.C., shows a galley with two banks;[5] the rest show galleys that have but a single bank.

Throughout the following three centuries the coins, now picturing units of the Roman Imperial Navy, continue to show in the great majority of cases single-banked galleys (Fig. 123).[6] Some of Hadrian's coins bear two-banked (Fig. 122)[7] and three-banked[8] craft, and the

[1] E.g., *Fitzwilliam Coll.* 3536, 3574-76, 3584-89, 3654, 4260, 4567, 4637-38, etc.; Moll E IV a 2, 3, 5, 6, 11-15, 25, 26, 31-36, 45, 55, 61, 62, 64, 67, 74-76 (prows). *Hunterian Coll.* II, p. 504.2, 4; III, p. 34.34, p. 38.11; *Fitzwilliam Coll.* 9296, 9297; *BM Central Greece* pls. 24.6, 7, 9-14; Moll E IV a 22, 23, 71-72, III, II2 (sterns).

[2] *BM Republic* I.1204-30 = pl. 30.16-18 and II.597-616 = pl. 94.12-14 (Moll E IV c 54, 58, 109, 110) probably recalling (cf. I, p. 193) the fleet at Sardinia in 169 B.C.; II.636-42 = pl. 95.7-9 (Moll E IV c 114) probably recalling (cf. II, p. 298) a naval victory of 241 B.C.

[3] *BM Republic* II, pp. 511-16, 518-20, 564-65 (Moll E IV c 125, 126, 129, 131; E IV f 3-8).

[4] *BM Republic* II.183-226 = pl. 116 (Moll E IV c 132, 144).

[5] *BM Republic* II, p. 564, no. 21, pl. 120.16; clearer photograph in Sydenham-Haines-Forrer, *The Coinage of the Roman Republic* (London 1952) pl. 30.1350; also in Moll (E IV c 153).

[6] Augustus: Cohen 94; Hadrian: *BM Empire* III.243-47 and pls. 51.9-10, 508-509 and pls. 56.19-20, 543-46 and pls. 57.20 and 58.1-2, 621-26 and pl. 59.8, 1028 and pl. 68.19; Moll E IV f 17-19, 21-23, 25. Aurelius: *BM Empire* IV.500 and pl. 62.17, IV.1047-52 and pl. 75.5. Aurelius and Commodus: *BM Empire* IV.1618-30 and pls. 87.11, 13. Caracalla: *BM Empire* V.267-70 and pls. 33.14-15.

[7] *BM Empire* III.1394 and pl. 85.1; Moll E IV f 20. A variant of *BM Empire* III.1393 is two-banked; for a good illustration, see P. Strack, *Untersuchungen zur römischen Reichsprägung des zweiten Jahrhunderts* II (Stuttgart 1933) pl. 16.840.

[8] *BM Empire* III.1391 and pl. 84.13 = Moll E IV f 15, 1462A and pl. 86.8.

155

three-banked galley recurs on a few issues of subsequent emperors.[9]

Now we know (141-42 above) that the Roman Imperial Navy limited itself to four standard types of fighting ship: the light two-banked affair called a liburnian, the trireme, the quadrireme, and the quinquereme; there was also one "six," flagship of the fleet based at Misenum.[10] The three-banked galleys pictured on imperial coins are surely triremes. The two-banked could be liburnians, although we will have a further word to say on this score in a moment. But what about the single-banked, which, after all, form the great majority of the coin pictures? Unquestionably numbers of single-banked auxiliary craft of various kinds were attached to the naval bases and fleets, but there is no reason whatsoever why any of these should be granted the distinction of being commemorated on a coin. What vessels, then, should? The coins of ca. 100 B.C., which memorialize actions of the third and second centuries B.C., we would expect to portray quinqueremes, the standard Roman ship of the line at that time. Those dating from the late Republic should show quadriremes or quinqueremes or one of the still heavier units that spearheaded the fleets of the age. The issues of the Empire should show the quadriremes or quinqueremes or the "six" of the Imperial navy, the units, in other words, that best symbolized the navy's power and that served as its flagships. All this leads inevitably to one conclusion: the single-banked ships on these coins represent galleys larger than the trireme. A corollary conclusion is that the two-banked craft occasionally portrayed on Imperial coins, though they may possibly be liburnians, are much more likely two-banked quadriremes or quinqueremes.

The tiny representations cannot furnish the detail to enable us to distinguish the various sizes of galley, to identify one as a quadrireme, say, and another as a quinquereme. But they do supply conclusive evidence that many of the galleys bigger than a trireme were powered by a single line of multiple-rower sweeps.

[9] Severus: *BM Empire* v.847 = pl. 51.4; Caracalla: *BM Empire* v.859 = pl. 52.6; Gordian III: F. Gnecchi, *I medaglioni romani* (Milan 1912) II, p. 91, no. 39 and pl. 105.8 = Moll E IV e 178.

[10] The *Ops* (*ILS* 2835); cf. SEVEN, note I.

Merchant Galleys

MERCHANT GALLEYS[1] were used to carry dispatches (e.g., the *keles*), passengers (e.g., the *phaselus*), or cargo (e.g., the *kerkouros*), particularly cargo that required rapid transport.[2] The smaller types depended almost wholly on their oars (e.g., some *phaseli*), with sail being distinctly auxiliary (Figs. 135, 139). The larger, like those of Genoa or Venice in a later age, depended almost wholly on sail, and ran out the oars only when approaching or leaving harbor, rounding a point in the teeth of the wind, or the like; on such craft the mast perhaps was not even retractable but permanently fixed (Fig. 140).[3] The standard rig was a mainmast with one broad square sail (Figs. 138, 141); heavier types carried an artemon as well (Fig. 140). Like some war galleys, oared merchant ships often had a small rounded shelter aft (Figs. 139-140) with insignia-poles nearby (Fig. 139); they could even have a sternpost ending in

---

[1] The term *ploion* "ship" is frequently used of galleys as well as sailing craft, either merchant galleys or light noncombatant naval auxiliaries. E.g., Thucydides (7.7.3) reports that Gylippus sent envoys to Sparta and Corinth with a plea to transport troops to Syracuse's aid "in sailing vessels [*holkades*], galleys [*ploia*], or any other conveyance" (ἐν ὁλκάσιν ἢ πλοίοις ἢ ἄλλως ὁπωσοῦν). See *GOS* 244 for further examples, and *SEG* xv 112, discussed in Six, note 128.

[2] E.g., the transport of animals for circus games; see Fig. 141 and G. Gatti, "Di un musaico figurato scoperto a Veii," *Bull. Com.* (1900) 117-23 (mosaic picturing the loading of an elephant on to a galley). Not always, however, for a sarcophagus in the Villa Medici at Rome (3rd A.D.) shows caged lions carried in a sailing vessel; see *l'Urbe* 11 (1948), fasc. 2, 3-6 and pl. 1.

[3] The generic term for a merchant galley was *histiokopos* "sail-oar-er" in Greek, and *actuaria* in Latin; see Gellius 10.25.5: *actuariae, quas Graeci* ἱστιοκώπους *vocant vel* ἐπακτρίδας *"actuariae*, the ships the Greeks call *histiokopoi* or *epaktrides.*" The Greek term reflects the importance of the sailing rig for these galleys as against warships. The term *epaktris*, which Gellius offers as a synonym of *histiokopos*, probably originally meant "fishing boat"; cf. *epakter* "fisherman" and *epaktron* "fishing boat" (Nicander, *Ther.* 823-24: μογερούς ἁλιῆας . . . κατεπρήνιξεν ἐπάκτρου εἰς ἅλα "caused poor fishermen . . . to dive headlong out of their boats [*epaktra*] into the sea").

157

an *aphlaston* (Fig. 140), usually the mark of a man-of-war, though the goose-head, distinctive sign of the peacetime carrier (347-48 below), was probably more common (Fig. 139).

The hulls, as might be expected, were beamier than those of men-of-war, with a length to width ratio of 5½ or 6½ to 1 (e.g., the *kybaia*, the *kerkouros*) instead of 10 to 1. Despite the numerous varieties of craft attested by name, preserved representations permit us to distinguish only two basic hull-forms: one with rounded or straight prow (Figs. 135, 138-140), the other with a concave prow ending in a cutwater that jutted forward into a ramlike point (Fig. 141).

The ancients frequently give the size of a merchant galley by citing the number of tholes to a side. E.g., a craft described as *triskalmos* "three-tholed" would be six-oared.[4]

Many more types of merchant galleys are known than of sailing ships, even though there must have been more of the latter in service at any given time. This is because the galleys, being useful as naval auxiliaries, come in for frequent mention in ancient accounts of sea warfare. (Some were once thought to be naval craft pure and simple, but recent evidence, from sources other than literature, makes it clear that all had peacetime uses.) The following types can be identified by name—though it must be remembered that such names were used with as little precision in the ancient world as in the modern.[5]

[4] Plutarch, *Aem. Paulus* 6.3 mentions a πλοῖον τρίσκαλμον "boat with 3 tholes" and Ephippus (cited in note 16 below) a *keles* with 5 tholes. The odd numbers show that the term necessarily refers to the oars on one side, not both. Thus, when Cicero (*ad Att.* 16.3.6) describes an *actuariola* as being *decemscalmus* "10-tholed," he is talking of a 20-oared craft and not a 10-oared one as many have assumed (see Torr 106; Winstedt in the Loeb Classical Library translation, 1918; D. Shackleton Bailey, *Cicero's Letters to Atticus* VI, Cambridge 1967, p. 169). Other examples: Velleius 2.43.1 (*quattuor scalmorum navem* "4-tholed boat"); Plutarch, *Caesar* 38.1 (πλοῖον . . . τὸ μέγεθος δωδεκάσκαλμον "a boat . . . 12-tholed in size").

[5] Cf. such English terms as *ship* (either any rather large-sized craft or, very specifically, a certain type of three-masted square-rigged sailing vessel), *yacht* (any pleasure craft, whether driven by sail or power, from the poor man's outboard motorboat to the millionaire's miniature liner), *sloop* (used of craft ranging from a tiny pleasure boat to a three-masted sloop-of-war).

akatoi, keletes, lemboi

ἄκατος (ἀκάτιον) *actuaria* (*actuariola*). The Greek *akatos* could be used vaguely to mean no more than "boat" and the Latin to mean "merchant galley,"[6] but both also had a specialized sense as the name of a certain kind of oared galley.[7] That it was always oar-driven is attested not only by an identified representation (Fig. 137) but by references to its oars[8] and, most significant, by its use in emergencies as man-of-war.[9] Though definitely smaller than, e.g., a trireme,[10] it apparently was of fair size, using 30 or even 50 rowers, all unquestionably in one bank; smaller versions were referred to by the diminutive *akation* (*actuariola* in Latin).[11] Rig consisted of a

[6] Cf. e.g., Theognis 457-58: οὗτοι σύμφορόν ἐστι γυνὴ νέα ἀνδρὶ γέροντι · / οὐ γὰρ πηδαλίῳ πείθεται ὡς ἄκατος "A young wife is not suited to an old husband, for she won't obey the rudder like an *akatos*." Athenaeus (1.28c) quotes Critias, Plato's relative, to the effect that the Carians invented φορτηγοὺς ἀκάτους "cargo-bearing *akatoi*," by which he can only mean cargo vessels as against warships. Charon's ferry is often called an *akatos*: cf. *Anth. Pal.* 7.464.1; 9.242.8, 279.1. A ship's boat can also be called an *akatos*; see Nicander, cited in Twelve, note 19. For *actuaria* "merchant galley," see note 3 above.

[7] One of the vessels in the Althiburus mosaic (Fig. 137) is identified as an *actuaria*, and ancient historians frequently list *akatoi* as a class opposed to men-of-war on the one hand and pure sailing vessels on the other; see Thucydides 7.59.3 (the Syracusans in 413 B.C. closed the harbor mouth by anchoring a line of triremes, *ploia* "sailing ships," and *akatoi*) and Plutarch, *Mor.* 466b-c (the fearful and seasick will try to switch from an *akatos* to a *gaulos* [p. 66 above] to a trireme). A papyrus document concerning a maritime loan (*Sammelb.* 9571.6, 2nd A.D.) styles the carrier involved a πλοῖον ἄκατον "a vessel, *akatos*-type"; it plied between Ascalon and Alexandria. For the Latin, cf. *Dig.* 49.15.2, where *actuariae* are distinguished from warships (*naves longae*) and sailing ships (*naves onerariae*).

[8] See note 11 below.

[9] Caesar, *Bell. Alex.* 44.3: "[Vatinius, one of Caesar's commanders, finding himself short of men-of-war] fitted ships of the *actuaria* type with rams, even though their size was not at all right for naval combat" (*navibus actuariis . . . magnitudo nequaquam satis iusta ad proeliandum, rostra imposuit*).

[10] The smaller, emergency rig with which triremes were fitted included an "*akatos*-mast" and "*akatos*-yard"; see 236-37 above.

[11] Lucian, *Vera Hist.* 1.5 (a crew of 50 is collected to man an *akatos*). Livy 38.38.8 refers to *actuarias, quarum nulla plus quam triginta remis agatur* "actuariae, none driven by more than 30 oars," and this squares with Cicero, *ad Att.* 16.3.6, where he uses the diminutive *actuariola* of a 20-oared vessel (cf. note 4 above). Similarly, Polybius (1.73.2) uses the diminutive *akation* in referring to the types with less than 50 oars: τριήρεις καὶ πεντηκοντόρους καὶ τὰ μέγιστα τῶν ἀκατίων "triremes, penteconters,

159

single square sail (Fig. 137). The hull was of the pointed cutwater type.[12]

Akatoi were used on rivers as well as open water,[13] and had a long history, being attested from the beginning of the fifth B.C. to the eighth A.D.[14]

κέλης (κελήτιον), *celox*. The *keles* (diminutive *keletion*), as the name indicates, was built particularly for speed. It was single-banked[15] and, apparently, rather small, having few rowers[16] and carrying only modest amounts of cargo.[17] An identified representa-

---

and the largest of the *akatia*." The smallest *akatia* were so small they could be handled by one person (Polyaenus 8.46).

[12] It is so pictured in the Althiburus mosaic (Fig. 137). Moreover, *actuariae* could be quickly fitted out with rams to serve as men-of-war (see note 9 above); had they had a rounded or straight prow, lengthy rebuilding would have been involved, whereas the other type would need little more than a bronze sheath with barb over the pointed cutwater.

[13] For *akatoi* on the Nile, see *PSI* 558.4 and *P. Cairo Zen.* 59430.12-13, both 3rd B.C. (the restoration of ἀκα]τίου in *P. Athen.* 63.18 [2nd A.D.] is highly questionable). For use on the open water, see the citations in notes 7, 9, 11 above.

[14] They are mentioned by Pindar (*Nem.* 5.2) and are listed among the units in the Arab fleets after the conquest of Egypt (*Sammelb.* 5639.10, 5640.14).

[15] Polybius (5.62.3, cited in SIX, note 94) lists among the naval units Antiochus III seized at Tyre in 219 B.C. triremes, biremes (*dikrota*), and *keletes*.

[16] Ephippus, the 4th century B.C. writer of comedy, mentions 10-oared *keletes* (5.18 [Edmonds II, p. 158]: κέλητας πεντασκάλμους; cf. *GOS* 245 and note 4 above). Caesar attempted to cross the Adriatic in midwinter in a light-oared vessel identified by Appian (*Bell. Civ.* 2.56) as a *keletion*. Synesius (*Epist.* 4.165) mentions a 4-oared *keletion* (κελήτιον δισκάλμου). Polyaenus (4.7.4) mentions an ἐπίκωπος κέλης "oared *keles*," which was simply a harbor skiff that carried passengers from the dock at Ephesus to ships at anchor; cf. note 56 below.

Thucydides (4.9.1) describes the capture of a triaconter and a *keles* which together had 40 heavy-armed soldiers; Torr (109) is possibly correct in taking the figure to represent the total complement of both craft, which would leave 10 rowers for the *keles*, but it more likely refers just to the men-at-arms aboard.

[17] *P. Cairo Zen.* 59015 (259/8 B.C.) mentions a *keles* which carried from Asia Minor to Alexandria a cargo consisting of 145 18-*chous* jars of oil and 34 half-jars. Including the weight of the jars, this works out to about 14 tons burden (18 *chous* = ca. 60 liters, and a liter of oil weighs .9 kg.; a 25-liter jar weighs in the neighborhood of 17 kg. [Lane, "Tonnages" 218], and a 50-liter jar something under twice as much). Estimating the jars at their gross content (170-71 below and cf. Wallinga, "Nautika" 18-19), in cubic capacity the cargo would require some 31-35 cubic meters, which, appropriately stowed, would reduce to some 23½-27½ cubic meters (cf. Wallinga, "Nautika" 34-35) and fit into a hold 12 m. x 2 x 1 or slightly larger.

tion shows them as straight-prowed (Fig. 137; possibly the vessel in Fig. 139 is a *keles*).[18]

Because of its speed, the *keles* was a valued naval auxiliary[19] as well as a favorite of pirates.[20] It served for carrying dispatches[21] or passengers,[22] particularly on occasions when time counted.[23] The Greek papyri from Egypt make it clear that *keletes* also carried cargo, presumably when quick transport was desired.[24] A variant of the *keles* called the *epaktrokeles* probably was designed with greater carrying capacity.[25] Nothing is known of the rig of a *keles*, though it surely carried at least a mainsail.

The type is mentioned in literature and documents no later than the first century B.C. It is included (Fig. 137) in the Althiburus

[18] The relief portrays the fashioning of the tunnel at the time of Claudius to drain Lake Fucinus. The vessels shown on the lake would be either for passengers or cargo. If the former, then they are perhaps *celoces*, and their similarity in general to the *celox* pictured on the Althiburus mosaic (Fig. 137) would favor the identification.

[19] E.g., Livy (21.17.3) reports that Rome's fleet in 218 B.C. consisted of 220 quinqueremes and 20 *celoces*. Plautus refers (*Capt.* 874) to a "public *celox*" at Athens; presumably it belonged to the navy.

[20] Thucydides 4.9.1 (see note 16 above; the ships belonged to Messenian pirates); Livy 37.27.4 (*piraticos celoces et lembos*).

[21] Xenophon, *Hell.* 1.6.36 (news of the Spartan defeat at Arginusae carried to Eteonicus by a *keles*); Herodotus 8.94 (the mysterious craft that brought news of Salamis to the fleeing Corinthian squadron was a *keles*).

[22] Thucydides 1.53.1 (heralds sent to the Athenian squadron by the Corinthians aboard a *keletion*), 8.38.1 (Spartan admiral sails from Miletus aboard a *keles*).

[23] Brasidas, crossing from Torone to Scione by night, sent a trireme ahead and followed in a *keletion*, since enemy triremes would automatically go for the larger ship (Thucydides 4.120.2); clearly his craft could keep up with a trireme—or escape from one in case the decoy was attacked. Lucullus, out to sneak across from Athens to Alexandria in 87-86 B.C., started off in a *keletion* (Appian, *Mithr.* 33).

[24] Cf. note 17 above. A *keles* was used for carrying delicacies from Syria for the table of Egypt's finance minister in the mid-3rd B.C.; cf. P. *Cairo Zen.* 59002 (260 B.C.), 59672 (mid-3rd). In several other instances they may either have been employed on this run or on the Nile; see P. *Cairo Zen.* 59110 (257 B.C.), P. *Mich. Zen.* 22.2, and P. *Cairo Zen.* 59548, all three referring to the same vessel. In *PSI* 613.5 (ca. 257/6 B.C.), the restoration κέλητος (*Berichtigungsliste* III) seems certain, which means that *keletes* hauled grain on the Nile. See also *Sammelb.* 9367, no. 9.3 (163 B.C.), where a κελ( ) αβυ( ) is listed as a Nile transport; possibly the abbreviation is to be expanded to "a *keles* of the Abydos-type."

[25] Aeschines, *in Timarchum* 191: ταῦτα πληροῖ τὰ λῃστήρια, ταῦτ᾽ εἰς τὸν ἐπακτρο-κέλητα ἐμβιβάζει "These are the factors that fill the pirate ranks, that put men aboard the pirate galley [*epaktrokeles*]." For *epaktris* as a term for "merchant galley," see note 3 above.

161

mosaic, which dates to the third-fourth A.D., but this may be merely artistic whim.[26]

λέμβος, lembus. This term, it seems, originally meant little more than "skiff," being applied indiscriminately to ship's boats,[27] harbor craft,[28] fishing boats,[29] rivercraft.[30] From at least the third century B.C. on, there was an oar-powered vessel called a *lembos* serving as an auxiliary in war fleets.[31] One type, a naval auxiliary like the *akatos* and *keles*, perhaps had no ram; another, used for combat, did.[32] It was considerably bigger than a *keles* and apparently bigger than naval *akatoi,* having at times as many as 50 rowers[33] and sometimes two superimposed banks.[34] Of its shape we know only that its prow, as we would expect of a vessel that could double as a combat unit, was sharply pointed.[35]

In addition to service as a naval unit, the *lembos* was used for carrying cargo both across open water and on rivers.[36]

[26] Under the vessel the mosaicist has included (*ILS* 9456) a line from Ennius (*labitur uncta carina per aequora cana celocis* "The pitch-smeared hull of the *celox* slips through the grey waters" = *ROL* I, p. 164), which may mean that the picture was included for its literary associations.

[27] Demosthenes 34.10, 32.6. Athenaeus (6.242f) cites Anaxandrides (4th-3rd B.C.) as authority for the use of *lembos* as a nickname for a hanger-on.

[28] Plautus, *Merc.* 193, 259 (*lembus* used to carry people to a ship at anchor); Lycurgus, *in Leoc.* 17 (*lembos* used to carry belongings to a ship at anchor).

[29] Theocritus 21.12; Accius quoted by Nonius 534.1 = *ROL* II, p. 410.

[30] Polybius (3.42.2; 3.43.2, 3, 4 etc.) refers to vessels on the Rhone as *lemboi,* and Livy (21.26-28) translates the term not *lembi*—which he reserves for the strictly naval craft—but *naves.*

[31] See Six, note 104.

[32] See Six, notes 107, 108.

[33] See Six, note 105. In Livy 44.28.14-15, each of 10 *lembi* carried, in addition to its crew, 20 captured Goths and two captured horses from Chios to Thessalonica. Cf. note 36 below for the capacity of a river *lembos.*

[34] See Six, note 106.

[35] Aristotle remarks (*de animalium incessu* 710a) of the breasts of birds with curved talons that they are "sharp for efficiency in flight, just like the prow of a ship of the *lembos*-type" (ὀξὺ μὲν πρὸς τὸ εὔπορον [var. εὔτονον] εἶναι, καθάπερ ἂν εἰ πλοίου πρῷρα λεμβώδους).

[36] *P. Cairo Zen.* 59015 (259/8 B.C.): one *lembos* carried 258 18-*chous* jars (= ca. 12,900 liters) and 102 half-jars (= ca. 2,550 liters) of oil from Samos and Miletus to Alexandria, while another carried 122 18-*chous* jars and 140 half-jars. The cargo of the bigger, in the neighborhood of 25 tons (for the weight of oil and the jars, cf. note 17 above), is almost twice that carried by the *keles* which appears in the same

IN SUM, these three types of galley, serving basically the same functions, differed somewhat in shape (the *akatos* and probably the *lembos* had a concave prow with pointed cutwater, the *keles* a straight prow) but more importantly in size: the *lembos* was the largest and roomiest, the *akatos* somewhat smaller, and the *keles* the smallest and swiftest.

kerkouroi, kybaiai, phaseloi

THESE three types of merchant galley were cargo and passenger carriers, generally of much greater size than the three discussed above.

κέρκουρος (κερκουροσκάφη), *cercurus*. This vessel was for long attested as a naval auxiliary, but new information confirms it as an oared cargo carrier of substantial size.

The name *kerkouros* is the Greek rendition of the Assyrian word *qurqurru*, a type of Mesopotamian riverboat.[37] The Persian fleet in 480 B.C. included a number to transport cargo, and so did the aggregation Alexander collected on the Hydaspes in 327 B.C.;[38] in both cases, they could very well have been local working craft impressed into service. However, by the third century B.C. at least, they were common in Mediterranean fleets.[39]

The Greek papyri from Egypt have made the nature of the *ker-kouros* much clearer. They reveal that, during the last three pre-

---

document (see note 17 above). Thus the evidence is consistent: as both naval auxiliary and cargo galley, the *lembos* was distinctly larger than the *keles*. Cf. *P. Petrie* II 20 IV 4, 14 as emended (*Berichtigungsliste* III): a *lembos* of 900 artabs burden (= ca. 22½ tons) carried grain to Memphis.

[37] Cf. Salonen, *Wasserfahrzeuge* 51.

[38] The Persian fleet included (Herodotus 7.89, 97) triremes, triaconters, pente-conters, *kerkouroi*, and *hippagoga*. Alexander's fleet included (Arrian, *Anab.* 6.2.4) triaconters, *hippagoga, kerkouroi*, and miscellaneous rivercraft (ἄλλα ποτάμια); since the *hippagoga* carried the horses, and the various rivercraft the light cargo, the *kerkouroi* must have been for the heavier cargo.

[39] They are reported in the Roman fleets during the First Punic War (Diodorus 24.1), in both Roman and Carthaginian fleets in the Third Punic War (Appian, *Pun.* 75, 121), in the Syrian fleet in 197 B.C. (Livy 33.19.10), and in Mithridates' fleet in 74 B.C. (see Memnon fr. 37, cited in Six, note 98).

163

Christian centuries, it was the standard large carrier of grain on the Nile. One particularly illuminating document furnishes a list of the *kerkouroi* mobilized at Ptolemais to transport the annual grain revenues (the year in question is ca. 171 B.C.) downriver to Alexandria. Of 22 ships about which details are preserved, the smallest carried 225 tons, most carried between 250 and 275, and the largest as much as 450.[40] From another set of documents we learn the specific dimensions of one *kerkouros* (cf. Fig. 136): about 45 cubits (67½' or 20.6 m.) long, 7 cubits (10½' or 3.2 m.) wide, and with 10 rowers per side.[41] This *kerkouros* obviously was of modest size;[42] the big

[40] *P. Teb.* 856. The document is imperfect, so the figures are not complete. The fleet consisted of at least:

| | *Burden* | | |
Kerkouroi	(artabs)	= (tons)	(lines)
2	9,000	225	(129, 188)
9	10,000	250	(97, 116, 124, 126, 186, 191, 202, 205, 206)
5	11,000	275	(103, 107, 109, 114, 125)
4	12,000	300	(6, 99, 118, 187)
1	16,000	400	(127)
1	18,000	450	(112)

There are two other examples of *kerkouroi* of 10,000 artabs burden: *W. Chrest.* 442 (late 3rd B.C.) and *P. Teb.* 825 (perhaps 176 B.C.). The smallest *kerkouros* recorded is 3,000 artabs burden: *P. Teb.* 824 (perhaps 171 B.C.). The ship's boat of Hiero's monster grain carrier was a *kerkouros* of 3,000 talents burden, which is about the same as 3,000 artabs (see NINE, App., pt. 4).

The rating of the capacity of these boats was conservative, to judge by a document (*P. Oxy.* 2415) which, though of considerably later date (3rd A.D.), concerns the same general type of vessel, Nile grain transports. It is a list of the various craft in a fleet which gives both the rated capacity and the amount actually loaded aboard. In most cases about 10 per cent more was put aboard (e.g., lines 25-26: rated capacity 2,600 artabs, actually loaded 2,860; line 30: rated capacity 3,300, actually loaded 3,630), and one boat carried as much as 15 per cent more (line 28: rated capacity 2,700, actually loaded 3,100).

[41] In *P. Cairo Zen.* 59053, 257 B.C., Amyntas writes to Zenon asking him to secure coverings and other equipment for certain boats and encloses a list of desiderata. The list is preserved in *P. Cairo Zen.* 59054:

Memorandum to Zenon of items to be furnished for the *kerkouros* (lines 1-3: ὑπόμνημα Ζήνωνι ὧν δεῖ κατασκευασθῆναι εἰς τὸν κέρκουρον):

stern-awning 11 cubits long and 6½ wide, tapering for 3 cubits to reduce to a width at the sternward tip of 2½ cubits (lines 4-8: πρυμνητικὴ μῆκος πηχῶν ια', πλάτος πηχῶν ϛL', ἔχουσα συναγωγὴν εἰς πήχεις γ' τοῦ ἄκρου συνοξῦναι τοῦ περὶ τὴν πρύμναν πλάτος πήχεις βL')

another awning 10 cubits long and 6½ wide (9-10: ἄλλη μῆκος πηχῶν ι', πλάτος πηχῶν ϛL') another awning 11 cubits long and 6½ wide (11-12: ἄλλη μῆκος πηχῶν ια', πλάτος πηχῶν ϛL')

[these two were surely to cover the amidships section of the craft, one on either side of the mast]

another, bow-awning, 8 cubits long to reduce along the bows for 6 cubits to a width at its tip of 2½ cubits (13-17: ἄλλη πρῳρατικὴ μῆκος πηχῶν η' συνοξῦναι περὶ τὴν πρῷραν ἐπὶ πήχεις ϛ' τοῦ ἄκρου πλάτος πήχεις βL')

For the *kybaia* (18: κυβαίας):

stern-awning 11 cubits long and 7½ wide, tapering for 3 cubits for a width across the taper of 2½ cubits (19-22: πρυμνητικὴ μῆκος πήχεις ια', πλάτος πήχεις ϛL', συναγωγὴν ἔχουσα ἐπὶ πήχεις γ' πλάτος τῆς συναγωγῆς πήχεις βL')

another awning 10 cubits long and 7½ wide (23-24: ἄλλη μῆκος πηχῶν ι', πλάτος πήχεις ϛL')

another awning 11 cubits long and 7½ wide (25-26: ἄλλη μῆκος πηχῶν ια', πλάτος πηχῶν ϛL')

bow-awning 8 cubits long and 7½ wide (27-28: πρῳρατικὴ μῆκος πηχῶν η', πλάτος πηχῶν ϛL')

A linen curtain, maximum length 60 cubits, and, if not that much, whatever length there is (29-31: αὐλαίαν λινὴν μάλιστα μὲν οὖσαν εἰς ἐξήκοντα πήχεις, εἰ δὲ μή, ὁπόσων ἂν ὦσιν)

a shelter-cabin [*skene*], maximum size five-bed, but, if it is a little less, it will make no difference (34-36: σκηνὴν μάλιστα μὲν πεντακλινικήν, ἐὰν δὲ μικρῷ ἐλάσσω ᾖ, μηθέν σοι διαφερέτω)

a woolen curtain for the shelter-cabin [*tholos*], 26 cubits long and 3½ wide (37-38: αὐλαίαν ἐρεᾶν θόλῳ μῆκος πηχῶν κϛ', πλάτος πήχεις γL')

20 rowers' cushions for the *kerkouros* (44: ὑπηρέσια κερκούρου κ')

PSI 533 as emended (*Berichtigungsliste* 1), another memorandum concerning the above items, adds a bit more information.

"Memorandum to Zenon from Amyntas. If you can, bring downriver to us a four-bed or five-bed shelter-cabin [*skene*], and a woolen curtain to go around the shelter-cabin [*skene*—obviously the equivalent of *tholos*, the expression used in the previous document], and another curtain, of linen, 60 cubits long. If no wooden parts [sc. for the roof and uprights] are available, they will be available in the city. . . . About the bow-awning and stern-awning [literally "bow-shade," "stern-shade"] in Naucratis, remember to bring them downstream when you sail" (1-7: ὑπόμνημα Ζήνωνι παρὰ 'Αμύντου. ἐὰν ἐκποιῇ σοι, σκηνὴν κατάγαγε ἡμῖν τετράκλινον ἢ πεντάκλινον, καὶ αὐλαίαν περὶ μὲν αὐτὴν τὴν σκηνὴν ἐρεᾶν, ἄλλην δὲ λινὴν πηχῶν ἐξήκοντα · τὰ δὲ ξύλινα ἐὰμ μὴ ᾖ, ἐν πόλει ἔσται . . . 15-17: καὶ περὶ τοῦ σκιοπρῴρου καὶ σκιοπρύμνου τῶν ἐν Ναυκράτει ὅπως μνησθῇς καταγαγεῖν ὡς ἂν καταπλῇς).

A shelter-cabin embracing five beds, each, say, 5½' x 2½' (1.68 x .76 m.), would run about 13½' along the fore and aft axis and 6' athwartship (4.1 x 1.8 m.). The woolen curtain, 26 cubits long = 39', would be just enough to wrap around all four sides (13½ + 6 + 13½ + 6 = 39). Since the piece was 3½ cubits wide = 5' 3", the roof of the shelter-cabin must have stood that high above the deck. The 60-cubit awning may have been for fencing in the sides of the ship.

A reconstruction of the *kerkouros* on the basis of the above data is given in Fig. 136. The length to beam ratio works out to 6½:1.

[42] If we assume that the ship carried cargo wherever the awnings ran full width, and if we further assume, *exempli gratia*, a depth in the hold of 2 cubits, there would be a block of space for cargo 31 cubits x 6½ x 2 = 403 cubic cubits. Since a cubic cubit held 3 3/8 artabs (cf. E. Husselman in *TAPA* 83, 1952, p. 73), some 1,360

fellows that held 450 tons may have run as much as 50 m. long.[43] Seagoing *kerkouroi*, it is clear, were also big.[44] Smaller versions known as κερκουροσκάφαι "*kerkouros*-skiffs" were in use on the Nile.[45]

Rig consisted of a single sail.[46] The hull had rather sharp bows and a full stern (Fig. 136).

The type is not attested after the first century B.C.[47]

κυβαία (κυβαίδιον), *cybaea*. The *kybaia* has up to now been thought to be a large-sized sailing ship. Papyrological evidence reveals that, like the *kerkouros*, it was a cargo galley.[48] The papyrus

artabs could be carried in the hold. But we must surely reckon on a sizable deck load, additional cargo space fore and aft of the rowing area, etc.

[43] Taking the attested length to beam ratio of 6½:1 (note 41 above), a vessel 50 m. long would be ca. 7.7 m. wide. Of the overall length, only about 70 per cent would be fully available for cargo (cf. the two previous notes). If we assume a depth in the hold of 2 m., the cargo area would be ca. 35 m. x 7.7 x 2 = 539 cubic meters. Since a cubic meter is equivalent to 10 cubic cubits (see Husselman, cited in the previous note) and a cubic cubit held 3 3/8 artabs, 539 cubic meters of space would accommodate just about 18,000 artabs = 450 tons. However, the dimensions very likely should be reduced all around to allow for deck cargo, stowage in the forepeak, etc.

[44] Plautus, *Stichus* 367-69: *conspicatus sum interim / cercurum, quo ego me maiorem non vidisse censeo. / in portum vento secundo, velo passo pervenit* "In the meantime I spied a *cercurus*, than which I don't think I've ever seen a bigger. With its sail spread before a following wind, it entered the harbor."

Livy reports (23.34.4) an instance in which a fleet commander dispatched some *cercuri* to overtake a suspicious vessel, and Torr (110) has concluded from this that the *cercurus* was notable for its speed. This does not follow since, for all we know, the overtaken vessel may have been a ponderous sailing ship. All other evidence seems to show the *cercurus* as roomy rather than fast.

[45] *P. Lille* 22.5, 23.5 = *W. Chrest.* 189 (both 221 B.C.); *P. Teb.* 1035 (2nd B.C.); *P. Ryl.* IV 576 (3rd B.C.). The last had a capacity of 200 artabs (= 5 tons).

[46] See the passage from Plautus cited in note 44 above.

[47] The latest occurrence is in *BGU* 1303.31 (end of Ptolemaic period). In *P. Mich. Inv.* 4607.2 (*TAPA* 83, 1952, p. 78), second half of 4th A.D., the editor reads κερκ(ούρων) which is very doubtful; πλοί(ων) *aut sim.* would be far more usual at this time. In *P. Gron.* 6.3 (5th A.D.), the editor reads Βίκτορος κερκου(ρίτου) "Victor, the *kerkouros*-rower"; I suggest Κερκου( ), i.e., a patronymic or designation of geographical origin.

[48] The length to beam ratio is that of a galley (see the following note). The earlier identification as a big sailing ship (cf. Torr 111) was a wrong inference from Cicero's references to a *cybaea* Verres had built for his personal use at public expense (*in Verrem* 2.4.17, 2.5.44, 50, 59). Cicero refers to the ship as *onerariam navem maximam* "big as the biggest merchantman" (2.4.19, 150) and a *navem vero cybaeam maximam, triremis instar* "a ship truly big as the biggest *cybaea*, the likes

document that provides the dimensions of a *kerkouros* also provides those of a *kybaia*: the two ships, as it turns out, were of the same length, i.e., somewhat over 40 cubits, but the *kybaia* was definitely beamier (7½ cubits instead of 6½) and with squarish bows, as its name "cubic," "boxlike" would imply.[49] It was used on both open water and the river to carry grain, wine, or other cargo.[50] Apparently it was constructed in a variety of sizes: Cicero attests to the largest size, those in the papyri seem relatively small,[51] and there was even a smaller version called a *kybaidion*.[52]

There are no references to the type after the first century B.C.

φάσηλος (φασήλιον), *phaselus*. The word literally means a kind of bean; presumably the original craft of this type were slender and low, reminiscent of a bean pod. *Phaseloi* ran the gamut of size from very modest affairs that were little more than skiffs[53] to craft capable of traveling throughout the Mediterranean[54] and of carrying hundreds of men.[55] They surely sailed most of the time, using the

---

of a trireme" (2.5.44). The comparison with a trireme is understandable, now that we realize the *kybaia* was an oared ship—Verres had himself built a merchant galley as big as a large warship, a good 120 feet long (82 above).

[49] See note 41 for the dimensions of a set of awnings for a *kybaia*. It was identical with the set ordered for a *kerkouros* save that the *kybaia*'s was one cubit wider, and the bow-awning was a full rectangle with no taper. The length to beam ratio, therefore, was about 5½:1 instead of 6½:1, and the vessel must have been boxlike forward, like the modern Nile ghayassah (*IH* ill. 234).

[50] *P. Cairo Zen.* 59012 (259 B.C.) is the cargo manifest of two *kybaiai* that carried a load of wine, oil, and miscellaneous table delicacies from Syria to Pelusium and thence to Alexandria. *P. Cairo Zen.* 59320 (249 B.C.) mentions a *kybaia* that hauled grain down the Nile to Alexandria.

[51] For the citation from Cicero, see note 48. For small *kybaiai*, cf. the two mentioned in the previous note, which were carrying about 160 full-sized jars and 50 half-jars, and small amounts of miscellaneous cargo.

[52] *PSI* 594.3 (3rd B.C.).

[53] Juvenal refers to the pot-boats of the Nile (ONE, note 16) as *phaseli*. Lucan (5.518) uses the term of a skiff that a fisherman stood on its side against the seaward wall of his cottage as protection. Roman poets refer to the fragility of the *phaselus* (Horace, *Odes* 3.2.28-29; Ovid, *Epist. ex Ponto* 1.10.39) or express fear of its getting too far from shore (Seneca, *Herc. Oet.* 695-96).

[54] The famous *phaselus* described by Catullus (*Carmina* 4), he conceives of as having traveled from the Black Sea to Italy. Propertius imagines (3.21.17-20) a trip from Rome around Italy, across the lower Adriatic, and through the length of the Gulf of Corinth in a *phaselus*.

[55] Sallust (*Hist.* 3.8, Maurenbrecher) mentions a *cohors una grandi phaselo vecta*

167

oars only when unavoidable.[56] They transported passengers rather than cargo.[57] The larger versions could be pressed into service as men-of-war.[58] A smaller version, called a *phaselion*, appears as a cargo-carrier on the Nile.[59]

Nothing definite is known about their shape. Their use as passenger vessels and references to their speed[60] imply graceful lines, while their use as men-of-war may imply a prow with pointed cutwater.[61] They probably carried but one mast.[62]

References to the *phaselos* are limited to the first century B.C. and the first A.D.

"a cohort carried in a large *phaselus*" (a cohort at full strength was 600 men).

[56] The presence of oars is specifically mentioned by Catullus (4.4-5, 17) and Propertius (3.21.11-12, cited in note 62 below) and distinctly implied by the use of *phaseli* as men-of-war. Torr (120) takes Cicero's remark (*ad Att.* 14.16.1: *conscendens . . . in phaselum epicopum has dedi litteras* "I am sending off this letter as I get aboard . . . an oared phaselus") to imply that some were powered by sail alone. The remark implies rather that some *phaseli*, like the one Cicero took to go the short distance from Puteoli to Pompeii and back, were dories that normally relied on oars alone instead of on oars and sail, as a proper galley would; cf. the *epikopos keles* in note 16 above.

[57] Cicero used one (see previous note) for getting about the Bay of Naples; Martial mentions (10.30.12-13) yachting off Formiae in a gaily painted *phaselus*; Atticus crossed the Adriatic from Brundisium to Epirus in a *phaselus* (Cicero, *ad Att.* 1.13.1). The passages cited in note 54 above both refer to the carrying of passengers.

[58] When Aelius Gallus was planning his Red Sea expedition in 25 B.C., mistakenly anticipating a sea action, he prepared a fleet of 80 biremes (δίκροτα), triremes, and *phaseli*; then, realizing his mistake, he had 130 cargo carriers (σκευαγωγά) built instead (Strabo 16.780). In 37 B.C., Octavia gave her brother the gift of a squadron of what Appian (*Bell. Civ.* 5.95) describes as δέκα φασήλοις τριηριτικοῖς, ἐπιμίκτοις ἔκ τε φορτίδων νεῶν καὶ μακρῶν "10 trireme-equipped *phaseloi*, a cross between cargo ships and men-of-war" (Plutarch, *Ant.* 35.4 describes the gift as εἴκοσι μυοπάρωνας "20 myoparones," which can perhaps be explained by assuming that *phaseloi* often looked like *myoparones* [cf. 132 above] but could run much bigger). "Trireme-equipped" would mean that the vessels were given rams, screens, turrets, and the like, but not that they were given three banks of rowers; Appian (cf. *Praefatio* 10) uses the word "trireme-equipped" generically, of gear belonging to war galleys of all sizes.

[59] *P. Ryl.* IV 576 (3rd B.C.). The boat in question was 200 artabs (= 5 tons) burden.

[60] Catullus 4.1-5.

[61] See the passage from Appian cited in note 58 and cf. note 9 above.

[62] Cf. Propertius 3.21.11-13: *nunc agite, O socii, propellite in aequore navem / remorumque pares ducite sorte vices / iungiteque extremo felicia lintea malo* "Come now, shipmates—drive our ship into the sea. Draw lots for equal turns at the oars, and raise our lucky canvas to the tip of the mast." The ship is identified as a *phaselus* in line 20.

CHAPTER NINE

Sailing Ships

IN GREEK, sailing ships were known as *strongyla ploia* "round ships" (as against "long ships," the men-of-war)[1] or *holkades*,[2] or just loosely *ploia* "ships."[3] Sometimes they are further identified by the cargo carried.[4] Seagoing sailing ships were often called "twenty-ers," a term derived perhaps from a phrase in Homer.[5] The only seagoing type identifiable is the Phoenician *gaulos* (66 above).

The Latin term for sailing ship was *navis oneraria* "ship of burden,"[6] and the only identifiable types are the *ponto* and the *corbita*. Of these, the first was strictly local, native to the southern coast of France.[7] The other was in general use and noted particularly for its great size;[8] its name means "basket," and an identified representa-

---

[1] E.g., Herodotus 1.163; Theophrastus, *Hist. Plant.* 5.7.1 (cited in TEN, note 51); Polyaenus 5.16.3, 6.16.4.

[2] E.g., Herodotus 3.135; Thucydides 7.7.3; Euripides, *Cyclops* 505, cited in TEN, App. 3, s.v. *beams*; Strabo 11.496; Dio Cassius 42.2.3. The name comes from a root meaning "tow"; the *holkas*, powered solely by sail, had to be towed in and out of harbor or in emergencies (cf. *GOS* 244-45).

[3] A *holkas* is always a sailing ship but *ploion* can on occasion refer to a galley; cf. EIGHT, note 1.

[4] E.g., "grain-carriers" (ὁλκάδες σιταγωγοί Thucydides 6.44.1, πλοῖα σιτηγά Demosthenes 50.20); "wine-carriers" (ὁλκάδας οἰναγωγούς Pherecrates 143.4-5 [Edmonds I, p. 260]); "stone-carriers" (λιθηγὸς [ναῦς] *P. Cairo Zen.* 59172.6, 3rd B.C.).

[5] The word means literally "with 20 oars." It is first used of a seagoing merchant ship by Homer (*Od.* 9.322-23, cited in FOUR, note 110) who may very well have had in mind a merchant galley powered by that many rowers. But thereafter it was applied indiscriminately to sailing ships of all sizes including Hiero's superfreighter (Demosthenes 35.18 [the ship carrying 3,000 jars discussed in App., pt. 1]; Nicostratus, cited in FOURTEEN, note 76; Theodoridas in *Anth. Pal.* 6.222 [βουφόρτων εἰκοσόρων "cattle-carrying 'twenty-ers'"]; Athenaeus 5.207c, cited in App., pt. 4). Compare our term "full-rigged ship," which can be used of any vessel with a certain rig whether modest or huge. Morrison (*GOS* 245) thinks that the "twenty-er" mentioned by Demosthenes was actually a merchant galley, but its size (see App., pt. 1) and the usages listed above make this doubtful.

[6] See the passage from the Digest cited at the end of EIGHT, note 7.

[7] Caesar, *Bell. Civ.* 3.29: *pontones, quod est genus navium Gallicarum* "pontones, which are a type of Gallic ship." In *Bell. Civ.* 3.40, Caesar refers to them as *naves onerariae*, which reveals that they were merchantmen.

[8] Lucilius (quoted by Nonius 533 = *ROL* III, p. 166) says of Polyphemus' walking

169

tion (Fig. 137) shows a vessel that, with its rounded and big-bellied hull, is distinctly basket-shaped.[9]

I SIZE

First, some remarks on units of measurement.[10] The ancients had not one universal unit.[11] As in later times, varying purpose brought about the use of varying systems for indicating how much a vessel could carry.[12] The size of grain ships was expressed in the different measures used for grain: the artab in Egypt;[13] probably the *medimnus* at Athens;[14] the *modius* at Rome.[15] For wine and oil carriers, both Greeks and Romans reckoned size by the number of amphorae, or shipping jars, able to be loaded aboard;[16] when doing so, they had in mind not so much the net as the gross volume of each

stick that it was bigger "than the biggest mast in any *corbita*-ship" (*quam malus navi in corbita maximus ulla*). Cf. Plautus' joke (*Casina* 778) about two trencher-women who could eat "a *corbita*-cargo of food" (*corbitam cibi*). As big ships, they were dependent on sail alone and slow, so Plautus describes (*Poenulus* 507) a pair of slowpokes as "slower than *corbitae* in a calm" (*tardiores quam corbitae sunt in tranquillo mari*). Cicero, heading for Greece, had his choice (*ad Att.* 16.6.1) of traveling from Messina or Syracuse directly to Patras in a *corbita* or traveling in short hops in *actuariolae*.

[9] A vessel identically shaped appears in a 5th century A.D. mosaic found in Lebanon; see M. Chéhab, *Mosaiques du Liban* (Bulletin du Musée de Beyrouth, 14 and 15, Paris 1957, 1959) pl. 67.3. The author suggests (110) that it is a *corbita*, and he may very well be right. (The inscription above the vessel, which the author reads πλοῦ[τος] ἡρήνης [i.e., εἰρήνης] "wealth of peace," is surely πλοῦ[s] ἡρήνης "voyage of peace").

[10] The fundamental article is Wallinga, "Nautika." Unless otherwise qualified, the tonnages I give are of a ship's burden, the weight it could carry, reckoned in tons of 1,000 kg. (= 2,200 lb.).

[11] One of Wallinga's key contributions was to correct the long-standing misconception that the talent was the standard unit for measuring a ship's capacity.

[12] Cf. Wallinga, "Nautika" 2-3, 24. See Lane, "Tonnages" for an excellent survey of the various tonnages in vogue from the Middle Ages on. In the 13th century, for example, the Genoese used a measure of volume (*baril* = 1.69 cu. ft. [ca. .05 m³]) for wine cargoes and a measure of weight (*cantar* = 104¾ lb. = 47.6 kg.) for cargoes of alum, while the Venetians conformably used the *botta* (15.9 cu. ft. [ca. .45 m³]) for wine and the *milliarium* (= 1049 2/5 lb. = 477 kg.) for salt (Lane 219, 221).

[13] Cf., e.g., the examples in EIGHT, notes 36, 40, 45.

[14] Wallinga, "Nautika" 7-8, 25.

[15] *Ibid.* 6.

[16] *Ibid.* 4-7.

170

container, the content of a parallelepiped whose height and width equaled the height and maximum width of the jars involved.[17] Since the size of shipping jars varied widely,[18] and cargoes were more often than not composed of jars of different sizes,[19] this mode of stating a vessel's capacity was admittedly imprecise.[20] Occasionally the talent was used, in the case of cargoes whose weight rather than size counted (e.g., lead or copper) or whose makeup was so miscellaneous that some common denominator, as it were, was called for.[21]

The capacity of seagoing freighters has been consistently and seriously underrated.[22] The smallest craft the ancients reckoned suitable for overseas shipping was 70 to 80 tons burden.[23] From the fifth

[17] *Ibid.* 17-20. The same was true of later ages; cf. Lane, "Tonnages" 219-20, 225, 227-28. Thus, the wine cask which the English called a tun and which had a net volume of 28.6-33.8 cu. ft. was considered to "occupy" 57-67 cu. ft. (Lane 218-19). Colbert established for France a *tonneau de mer* (a ton of gross content) of 42 *pieds cubes*, whose net volume was under 28 *pieds cubes* (Lane 220, 225).

[18] Cf., e.g., V. Grace, "Standard Pottery Containers of the Ancient Greek World," *Hesperia*, Suppl. VIII (Athens 1949) 175-89, esp. 180.

[19] Cf. EIGHT, note 51 (cargo of full-sized and half-sized jars). The wreck found off the Grand Congloué carried jars varying from 75 to 115 liters, gross volume (Wallinga, "Nautika" 19). Even in the more or less homogeneous cargo of the Albenga wreck there was significant variation, from 100 to 128 liters, gross volume (*ibid.* 19).

[20] Cf. Wallinga, "Nautika" 23.

[21] Wallinga offers as examples a contract dealing with cargoes of ruddle, an extremely heavy substance ("Nautika" 8-9), and port regulations, which had to apply to all ships regardless of their cargo (*ibid.* 11-12; cf. note 23 below).

[22] Lefebvre des Noëttes, *De la marine antique à la marine moderne: la révolution du gouvernail* (Paris 1935) 49 and 69-70, puts the maximum size at the absurdly low figure of 60 tons. Torr (25-26) puts the largest merchantmen used commercially at 250 tons. Most recently, Rougé, calculating (79) a cargo of 3,000 jars at 30-odd tons when it was at the very least 100 and quite likely much more (see Demosthenes 35.10 discussed in App., pt. 1), reached the conclusion (492; cf. 226, 330) that, down through the 1st century B.C., only "navires de tonnage relativement faible" were in use.

[23] A fragment of the port regulations of Thasos (*IG* XII Suppl. p. 151, no. 348 as emended in *SEG* XVII 417), dating from the second half of the 3rd B.C., divides the harbor into two sectors and states that "no ship under 3,000 talents burden [= 80 tons] may haul out within the boundaries of the first, and no ship under 5,000 talents [= 130 tons] within those of the second" ($\pi\lambda o\hat{\iota}o\nu$ $\mu\dot{\eta}$ [$\dot{\alpha}$]$\nu\acute{\epsilon}\lambda\kappa\epsilon\iota\nu$ $\dot{\epsilon}\nu\tau\grave{o}s$ $\tau\hat{\omega}\nu$ [$\ddot{o}$]$\rho\omega\nu$ $\tauο\hat{\upsilon}$ $\mu\grave{\epsilon}\nu$ $\pi\rho\acute{\omega}\tau ου$ $\dot{\epsilon}\lambda\acute{\alpha}$[$\sigma\sigma\omega$ $\phi\acute{o}\rho\tau ο\nu$ $\ddot{\alpha}\gamma ο\nu$ $\tau\rho$]$\iota\sigma\chi\iota$[λ]$\acute{\iota}\omega\nu$ $\tau\alpha\lambda\acute{\alpha}\nu\tau\omega\nu$, $\tauο\hat{\upsilon}$ $\delta\epsilon\upsilon\tau\acute{\epsilon}\rho ο$[$\upsilon$] $\dot{\epsilon}\lambda\acute{\alpha}\sigma\sigma\omega$ $\ddot{\alpha}\gamma ο$[ν] $\pi\epsilon\nu\tau\alpha$[κ]$\iota\sigma\chi$[$\iota\lambda\acute{\iota}\omega\nu$] $\tau\alpha$[$\lambda\acute{\alpha}\nu\tau\omega$]$\nu$). The clear implication is that the 80-tonner was a small ship, the smallest the harbor authorities cared to allow to use the facilities. The Roman jurist Scaevola, writing toward the end of the 2nd A.D., states (*Dig.* 50.5.3) that those shipowners are exempt from compulsory public services who "have had built and furnish to the transport service of the

171

century B.C. on, carriers of from 100 to 150 tons burden were in common use[24] while those of from 350 to 500 tons, though obviously considered big, were by no means rare.[25] When it came to passengers, vessels could take as many as 600 on long voyages.[26]

In the Hellenistic Age, the building of superwarships (103-15 above) apparently affected merchant navies: not long after the middle of the third century B.C., the biggest merchant ship known from the ancient world, a brobdingnagian three-masted, three-decked grain carrier of some 1,700 to 1,900 tons burden, came off the ways under the eye of Archimedes himself. In subsequent centuries giant

government of Rome a seagoing vessel no smaller than 50,000 *modii* [= 340 tons] or a number of vessels each no smaller than 10,000 *modii* [= 68 tons]" (*naves marinas fabricaverint et ad annonam populi Romani praefuerint non minores quinquaginta milium modiorum aut plures singulas non minores decem milium modiorum*). Conformably, the emperor Claudius granted (Gaius, *Inst.* 1.32c) certain privileges to whoever "built a seagoing vessel that would hold no less than 10,000 *modii* of grain" (*navem marinam aedificaverint quae non minus quam decem milia modiorum frumenti capiat*). The clear implication of both is that a 70-tonner was the smallest-sized carrier the government considered useful, and this squares very well with the Thasian port regulation.

[24] See App., pt. 1.

[25] The size the Roman Imperial government preferred for its fleet of grain carriers was 340 tons (see note 23). From the 5th B.C. to the 2nd or 3rd A.D., the expressions μυριοφόρος, μυριαγωγός, *aut sim.* "10,000-er" were used to identify, in a general way, a big freighter: cf. Thucydides 7.25.6; Ctesias, *Indica* 6; Strabo 3.151, 17.805; Philo, *de opificio mundi* 38 (113), *de plantatione Noe* 6 (24), *de aeternitate mundi* 26 (138); Automedon in *Anth. Pal.* 10.23.5; Heliodorus, *Aeth.* 4.16.6. Wallinga has demonstrated ("Nautika" 20-23) that the figure refers not to talents, which would make the vessels only 260 tons burden, but to *amphorae* or *medimni*, which would make them at least 400 (a *medimnus* of grain weighs ca. 40 kg.). A wine-carrier discovered off Albenga on the Ligurian coast seems actually to have been a "10,000-er"; see Lamboglia 214, 219. The cargo covered an area 25 m. (82 feet) long and 8-10 (27-33) wide, and consisted of five or more superimposed rows of amphorae which, Lamboglia estimates, would total at least 10,000. Since each jar held 26 liters and weighed empty 17-18 kg. (cf. Benoit 163; Lane, "Tonnages" 218), the cargo would come to 430-40 tons. Conformably, a passage in the Digest (19.2.61.1) dealing with a cargo of at least 250 tons (it included 3,000 *metretae* of wine = at least 200 tons [cf. *IG* II² 903, discussed in App., pt. 1] and 8,000 *modii* of grain = ca. 55 tons) gives no indication whatsoever that this was in any way exceptional in size.

[26] Josephus, *Vita* 15; the vessel, en route from Palestine to Rome, came to grief near the mouth of the Adriatic. There were 276 aboard the grain freighter that St. Paul boarded to go to Rome (Acts 27.37), but it was an off-season sailing and the ship very likely could have accommodated many more; cf. TWELVE App. note 6.

grain freighters of perhaps 1,300 tons burden plied between Alexandria and Rome. The dimensions of one of these has been preserved: 180' x 45' x 43½' (55 m. x 13.72 x 13.25).[27]

Dimensions and tonnage did not have any fixed ratio; they necessarily varied with the cargo for which a vessel was designed. A 250-ton *lithegos* "stone-carrier" would have been shorter and sturdier than, e.g., a *sitegos* "grain-carrier" of the same capacity; a vessel built with no particular cargo in mind could hold more grain, which went either loose or in sacks, than wine, which went in ponderous clay amphorae of shapes not too well adapted for stowing. As it happens, our evidence for dimensions is more or less limited to ships that carried heavy loads such as wine or building stone. These commonly ran 60-100' long and 25-33' broad (19-33 m. x 7-10) while bigger types reached 130' x 33' (40 m. x 10).[28]

II HULL

A FEW representations are available to show what merchantmen looked like in the sixth century B.C. (Figs. 81-82, 97). But then follows a lacuna lasting almost 500 years and covering the very period when the sailing ship must have seen its fullest development, when it grew from the respectable size it had already achieved by the fifth century (68 above) to the three-masted three-decker, the largest type of sailing vessel the ancient world was to know. Yet, all this, our pictorial record omits;[29] it does not resume until well into the first century A.D. We must simply go on the assumption that, seamen being by nature conservative and Romans being by nature little interested in the sea, the craft portrayed on reliefs and wall-

---

[27] See App., pt. 2. [28] See App., pt. 3.

[29] A. W. Persson, "Die hellenistische Schiffbaukunst und die Nemischiffe," *Skrifter utgivna av Svenska Institutet i Rom* IV.1, *Opuscula Archaeologica* I (1935) 129-63, argues that a series of gems (e.g., Furtwängler, *Ant. Gem.* XLVI 48, 50, 51; Moll E III 3401-3404, 7095) pictures Hiero's famous grain carrier. It is an ingenious suggestion, since all clearly portray the same ship, a big sailing vessel whose most prominent feature is six massive towers, and eight massive towers were a prominent feature on Hiero's ship (Athenaeus 208b, cited in App., pt. 4). Only one mast appears, so, to make the identification, we must accept Persson's excuse (148-49) that one mast and only six towers were all the artist could fit into the space at his disposal.

paintings and mosaics of the Roman Imperial Age were not too different from those that plied the Mediterranean in the previous half-millennium.[30]

The evidence is limited not merely chronologically but in scope as well. The design of sailing ships must have displayed the same variety in ancient days as in later. Yet the pictures at our disposal reveal only obvious distinctions.

As it happens, the most obvious is one that goes back to the Bronze Age (35 above) and is found in merchant galleys and small craft to boot (158 above, and FOURTEEN, note 13): certain sailing ships were given a more or less rounded bow of the traditional type; others were given a concave prow that ended in a jutting, ramlike cutwater. So far as we can tell, no difference in function was involved; indeed, the ancient artists seem to go out of their way to show the two types together performing the same work (Figs. 145, 147).[31]

Those with ramlike prow are distinguished by curved figureheads, a scroll (Figs. 145, 174) or the like (e.g., goose-heads in Figs. 147 [ship to the right], 191), which are not often found in their sisters with the traditional bow. The latter can be divided into at least three types. One includes hulls so rounded at stem and stern that they are nearly crescent-shaped. Though rather plain forward, with prow ending in a simple upright stempost, they are adorned aft with a goose-headed sternpost and most often an overhanging stern gallery (Figs. 144-146, 149, 150, 154).[32] The second is a simpler version of the first: hull with the same crescentlike lines, but plain at both ends,

[30] A straw in the wind is provided by a fragment from a Hellenistic relief which shows the stern of a merchantman with the inclined stern gallery attested in later times and, socketed into the sternpost, the very same adornment that appears, e.g., on a Roman merchantman of the 3rd A.D. (Fig. 147, center ship); see *Prak. Ath.* 1956 (Athens 1961) 222 and pl. 107b (J. Kontes, who published the piece, has erroneously interpreted it as part of a warship in a sea battle).

[31] E.g., it was a favored motif in the Foro delle Corporazioni at Ostia; see Becatti, pls. 175, 176, 178-80.

[32] Other examples are Espérandieu 686, 687 (= Moll B IV a 51); the latter is more elaborate than most, with a goose-headed stempost and with a gallery forward as well as aft.

with an unadorned stem- and sternpost (Figs. 142, 143, 148).[33] The third shows distinctly heavier and less rounded lines than the other two, and the stempost is capped with a massive block-shaped adornment (Figs. 151, 156).[34] The first and third include seagoing ships of large size; the second seems limited to somewhat smaller craft.

Ancient ship-pictures, almost always in profile and generally simplified, give the impression that sailing vessels were all beamy and big-bellied. Many in fact were, as we can tell from a number of wrecks whose keel and bottom planks show flat floors with little or no dead rise.[35] But not all were so made, for other wrecks reveal unmistakable signs of the sharp-built hull,[36] a feature that could never have been discerned in the preserved representations.

III HOLD, DECKS, ACCOMMODATIONS

In the hold[37] of a merchantman, the floor was given over to ballast and bilge. Floor timbers had limber holes[38] to allow free pas-

---

[33] For other examples see N. Avigad, "The Beth-She'arim Necropolis" *Antiquity and Survival* 2 (1957) 244-61, fig. 4 = *MM* 48 (1962) 69, fig. 1 (graffito of 2nd-4th A.D.); Chéhab, *loc. cit.* (note 9 above); Alfieri, *op. cit.* (ONE, note 20), fig. 6.

[34] The graffito of the big merchantman *Europa* found at Pompeii seems to belong to this type; see A. Maiuri, "Navalia pompeiana," *Rendiconti della Accademia di Archeologia . . . di Napoli* 33 (1958) 7-34, esp. 18-22 (reproduced in Benoit 124, fig. 73). Also a graffito (2nd-4th A.D.) found at Beth-She'arim; see *Sefunim (Bulletin), The Maritime Museum of Haifa* 1 (1966) pl. 5.1.

A wall-painting found at Dura and dating ca. 230-250 A.D. shows an extreme variety of this angular hull-form; see C. Kraeling, *The Christian Building, Excavations at Dura-Europos, Final Report* viii, pt. ii (New Haven 1967) pls. 18, 36, 37. The shape is remarkably similar to that of a ship figured in an Arabic ms. of the 13th century (*IH* ill. 185). Perhaps both sailed the same waters.

[35] Marseilles (Benoit 145), Titan (Taylor, fig. 35 and pl. 9), Dramont A (Benoit, pl. 28), Kyrenia (Katzev), Yassi Ada Roman (van Doorninck, by letter), Yassi Ada Byzantine (van Doorninck, figs. 10-19), Fiumicino barges (Testaguzza 135-43).

[36] Mahdia (G. Picard and M. de Frondeville in *Bulletin archéologique du Comité des travaux historiques et scientifiques* (1955-56) 129-41, esp. 134; Grand Congloué (Benoit 132); Chrétienne A (Dumas 130, 133-47, 164; cf. 157).

[37] The hold was called in Greek κοίλη ναῦς "the hollow [part of the] ship"; cf., e.g., Polyaenus 4.7.4.

[38] Grand Congloué (Benoit 130, fig. 75), Dramont A (Benoit, pl. 28), Chrétienne A (Dumas 157-58).

sage of bilge water,[39] which was such a problem that often a hand
—generally the least able in the crew—was detailed to stand by and
keep an eye on it.[40] On larger craft, bailing was done by sophisticated
devices such as an Archimedean screw operated by a treadmill;[41]
on smaller, with buckets.[42] Ballast was usually sand or stone.[43] Car-

[39] The same word was used of both the bilge and the water in it, ἄντλος or ἀντλία
in Greek, *sentina* in Latin. Cf., e.g., Athenaeus 5.208f (cited in App., pt. 4) with
Aeschylus, *Septem* 795-96 (καὶ κλυδωνίου / πολλαῖσι πληγαῖς ἄντλον οὐκ ἐδέξατο
"And despite the many blows of the battering sea, it took in no bilge water"). For
the Latin, cf. Cicero, *ad fam.* 9.15.3 (*vix est in sentina locus* "There is scarcely room
in the bilge") with *de sen.* 17 (*alii sentinam exhauriant* "Others drain off the bilge
water").

[40] Paulinus of Nola, *Epist.* 49.1 (the duty was given to an old man, a poor wretch
who was the worst sailor aboard and whom the rest of the crew scorned); Lucian,
Jup. Trag. 48 (a bad captain will put a cowardly, lazy, good-for-nothing [who
should be in the bilge] in a post of command and put a good man in the bilge).
Cf. Artemidorus 1.48 (cited in the next note) and Suetonius, *Tiberius* 51 for mention of duty in the bilge as a form of punishment. On Hiero's superfreighter, bilge
duty required a number of hands (Athenaeus 5.207f, cited in App., pt. 4).

[41] For the use of the Archimedean screw see Athenaeus 5.208f (cited in App.,
pt. 4); for its operation by a treadmill see Vitruvius 10.6.3 (*et ita cochleae homini-
bus calcantibus faciunt versationes* "And so the water-screws make their revolutions
by means of men treading on them"); and for its common use aboard ships, cf.
Artemidorus 1.48: οἶδα δέ τινα, ὃς ἔδοξε τοῦ παντὸς σώματος ἀτρέμα μένοντος τοὺς πόδας
αὐτοῦ μόνους βαδίζειν καὶ προβαίνειν μὲν μηδὲ βραχύ, ὅμως δὲ κινεῖσθαι· συνέβη αὐτῷ
εἰς ἀντλίαν καταδικασθῆναι. καὶ γὰρ τοῖς ἀντλοῦσι συμβέβηκε διαβαίνειν μὲν ὡς βαδί-
ζουσιν, ἀεὶ δὲ μένειν ἐν τῷ αὐτῷ τόπῳ "I know of a case where a man dreamed that
his whole body seemed to remain absolutely still except for his feet, which seemed
to be walking, and, though they kept moving, he did not seem to advance the
slightest. He, it turned out, later was sentenced to the bilge, because it so happens
that the men who bail the bilge move as if walking but always remain in the same
spot."

[42] Called ἀντλητήρια in Greek (cf. Dio Cassius 50.34.3-4: Antony's men at the
Battle of Actium used the *antleteria* to draw up seawater for putting out fires) and
sentinacula in Latin (cf. Paulinus of Nola, *Epist.* 49.3: *sentinaculi haustum umore
destricta siccataque navi* "with a drawing of the bilge bucket, the water was drawn
off and the vessel left dry.") Things did not change for a long while—there is a
graphic description of continuous bailing with buckets during a stormy voyage in
a letter written by an 11th century traveler (Goitein 321).

[43] The Greek word for ballast is *herma*, the Latin *saburra*, which means literally
"sand." For the use of sand as ballast in galleys, see FIVE, note 67. There was a
corpus saburrariorum "guild of sand-heavers" at the port of Rome, and its members
presumably handled ballast; see SIXTEEN, note 46. For the use of rock, cf. Lycophron
618 (τὸν ἑρματίτην νηὸς ἐκβαλὼν πέτρον "heaving overboard the rock ballast"). Ballast stones have been found in the Gelidonya wreck of the 13th b.c. (see Bass, *Gelidonya* 39, 48) and Chrétienne A of the 1st b.c. (Dumas 159). The Mahdia wreck,
dated 1st b.c. (Taylor 41), was carrying some gravestones and inscribed slabs from
Peiraeus of the 4th and 3rd b.c., which must have been part of the ballast; see R.

goes of wine jars, building stone, tiles, ingots, or the like, were loaded right on the lining (Fig. 157), but more perishable wares, such as sacks of grain, were no doubt stored well above the bilges.[44] The commonest form of dunnage was twigs and branches (Fig. 155),[45] which provided a useful elasticity, and stowage methods were employed that skillfully utilized every inch of space (Fig. 157).[46] Once the hatches had been battened down,[47] additional cargo could be loaded on deck (Figs. 148, 155).[48] Deep in the hold was the water supply, at times carried in a single large receptacle rather than tuns or the like.[49] There were some accommodations down in the bowels, the ancient equivalent of steerage.[50] The ship's galley re-

Dain, "Inscriptions attiques trouvées dans les fouilles sous-marines de Mahdia," *REG* 44 (1931) 290-303, 51 (1938) 423-24.

[44] See App., pt. 5.

[45] Branches and twigs, almost certainly serving as dunnage, were found on the planking of the Gelidonya wreck of the 13th B.C. (Bass, *Gelidonya* 49). Very fine twigs had been used in the Chrétienne A wreck (Dumas 159). For Odysseus' use of brush when he built the boat that took him from Calypso's island, see TEN, App. 2.

[46] On methods of stowing amphorae, see Wallinga, "Nautika" 28-36. There were at least five superimposed rows of amphorae in the Albenga wreck (Lamboglia 219).

[47] Hatches were perhaps called in Latin *transtra*, the pieces that "ran across" the openings to the hold. Statius includes among the steps in getting a seagoing ship under way the instruction (*Silvae* 3.2.28) *pars transtra reponat* "Let some replace the *transtra*." Since it has always been assumed that he was talking of an oared dispatch boat, the word here has been translated "rowers' benches," a meaning it can have in certain contexts, and the explanation offered that these had been removed during the unloading of cargo (references in TWELVE, App., note 9). But fast dispatch boats do not carry cargo, and rowing thwarts are normally not removable. As pointed out below (TWELVE, App., note 9), the ship was a big grain carrier, and what needs replacement on any freighter after unloading are the hatches.

[48] For other examples of deck loads of amphorae, see Becatti, pls. 182, 183 (Ostia).

[49] On the *Syracusia* (Athenaeus 5.208a, cited in App., pt. 4) the tank was forward and consisted of a completely enclosed compartment made of wood and lined with waterproof fabric; the capacity was 2,000 *metretae* = ca. 20,000 gallons or 2,750 cu. ft. (78 m.³). On the ship that St. Gregory Nazianzen took from Alexandria to Greece the water supply—apparently located on deck—was also one large receptacle; cf. *Poemata de seipso* 11.145-47 (Migne, *PG* 37, 1039): στροβουμένης / νεὼς, ῥαγεῖσα σπείρετ' ἐν βυθῷ σκάφη, / ἢ τὸν γλυκὺν θησαυρὸν εἶχεν ὕδατος "While the ship was pitching and rocking, the receptacle which held our treasured supply of sweet water was smashed and scattered its contents into the depths." On the Yassi Ada Byzantine ship there was at least one water jar (maximum diameter 23″ = 58.5 cm.) in the after part of the galley (van Doorninck 190; photograph in *AA* (1962) 546, fig. 5.

[50] Lucian, *Jup. Trag.* 48 "not even able to stretch their legs on the bare boards alongside the bilge water" (οὐδὲ ἀποτείναι τοὺς πόδας δυνάμενοι ἐπὶ γυμνῶν τῶν σανίδων παρὰ τὸν ἄντλον); Athenaeus 5.207f (cited in App., pt. 4).

ceived much attention, to judge from the Yassi Ada Byzantine wreck (Figs. 152-153).[51] In this modest-sized coaster, some 70 feet long, just forward of the sternpost a segment running the width of the ship—about 9 feet at this point—and measuring about 5 feet fore and aft was entirely given over to the galley; to provide light and ventilation, its roof rose well above the deck, probably as high as the level of the gunwale, leaving over 7½ feet of headroom. As protection against fire, the roof was completely covered with standard rooftiles, one of which was pierced by a large smokehole. The port half of the segment held the hearth. This consisted of a grid of iron bars fixed in clay, which supported a floor of flat tiles,[52] the whole raised about 2 feet above the galley floor. The starboard half was the work space, and the bulkheads here were fitted with lockers for the cooking utensils and dining service; behind was a large water jar. Cramped as it was, such a galley was almost luxurious compared with the fireboxes in use as late as Columbus' day.[53]

The main deck on freighters of any size ran from stem to stern.[54]

[51] van Doorninck 107-203. For an imaginative reconstruction of the galley, done to scale and based on all available evidence, see *National Geographic*, September 1968, 418-19.

[52] A good many wrecks have yielded tiles which were obviously not part of the cargo: Mahdia (A. Merlin in *CRAI*, 1909, p. 655), Grand Congloué (Benoit 156), Titan (Taylor 83), Albenga (Taylor 54 and *RSL* 18, 1952, pp. 209-11), Anticythera (J. Svoronos, *Das Athener Nationalmuseum* 1, Athens 1908, pp. 78-79), Chrétienne A (Dumas 167), San Pietro wreck (*Archaeology* 18, 1965, p. 209), Punta Scaletta (Lamboglia, "Punta Scaletta" 252), Torre Sgarrata (Throckmorton, "Torre Sgarrata"). These were generally explained as being from the deckhouse roof, but reasons were never offered for the use of such heavy and fragile material. Van Doorninck's careful analysis (107-202) of the tiles from the Yassi Ada Byzantine wreck clears the matter up: they came from the hearth and galley roof. For a photograph of the tiles found at Yassi Ada, see *AA* (1962) 546, fig. 4. Flat tiles, probably from the hearth, were also found in the Torre Sgarrata wreck (Throckmorton, *ibid.*).

[53] S. Morison, *Admiral of the Ocean Sea* (Boston 1942) 178. Throckmorton reports finding teeth and bone fragments of pig, sheep, and cattle in the Torre Sgarrata wreck, which would indicate that at least certain of those aboard ate well.

[54] This was true even of small freighters on the Nile. Cf. *P. Lond.* v 1714.33-34 (A.D. 570): σεσανιδωμένον ἀπὸ πρύμνης εἰς πρῷραν "decked from stern to stem" and *P. Mon.* 4.11-12 (A.D. 581), which uses practically the same phrase; the two vessels involved were 300 artabs (= 7½ tons) and 500 artabs (= 12½ tons) burden respectively, and both were sailing craft since the detailed listing of their gear does not mention oars. In *P. Lond.* III 1164h (p. 164; A.D. 212 = *Select Papyri* 38) a vessel is described as ἐστρωμένον καὶ σεσανιδωμένον, which can only mean "floored

Bigger ships had a lower deck,[55] and the biggest two.[56] Whether the lower decks were complete or partial is not known.[57]

There was never a foredeck but always a poop deck—or rather an ample platform created by the roof of the superstructure at the stern.[58] This sometimes took the form of a shelter open on all sides (Fig. 154) but more often of a deckhouse (Figs. 143, 144, 146-147, 156) with doors to port and starboard.[59] Thus, this raised afterdeck

---

and decked"; cf. Casson, "Note on a Nile Boat," *AJPh* 63 (1942) 333-34.

This main deck on both oared and sailing ships was called in Latin *forus* or *fori*; see Aulus Gellius 16.19.14 (*stansque in summae puppis foro* "standing on the deck at the very stern") and Cicero, *de sen.* 17 (*alii per foros cursent* "others run about the deck"), and cf. Vergil, *Aen.* 6.412, Lucan 3.630, Ammianus 27.5.2. Petronius uses the term *constratum* (100.3: *vox supra constratum puppis* "a voice coming from the deck aft" and 100.6: *super constratum navis occuparemus secretissimum locum* "choose the most secluded spot on the ship's deck [for sleeping]"), which is the exact equivalent of the Greek *katastroma*, first attested as the term for the fighting deck of a man-of-war (87 above) but later used also of the main deck of a merchantman; e.g., Lucian (*Navig.* 5) gives the depth of the *Isis* (see App., pt. 2) "from the main deck [*katastroma*] to the floor," while Synesius remarks (*Epist.* 4.163a) that, during a storm, the passengers "by being on the deck [*katastroma*] give up the ghost to the air [instead of drowning]" (ἐπὶ τοῦ καταστρώματος ὄντας ἔτι πρὸς τὸν ἀέρα τὴν ψυχὴν ἐρυγεῖν). Another Greek term for deck was *sanidoma*, literally "planking"; e.g., Septuagint, *Macc.* 3.4.10: τοῖς ζυγοῖς τῶν πλοίων προσηλω-μένοι . . . , ἔτι καὶ τῷ καθύπερθε πυκνῷ σανιδώματι διακειμένῳ, ὅπως πάντοθεν ἐσκοτισ-μένοι τοὺς ὀφθαλμούς "[prisoners] tied to the crossbeams of the ships . . . , and, what is more, with their eyes in complete shadow because of the thick deck (*sanidoma*) just above." Compare the use of the related term σεσανιδωμένον, literally "planked," to mean "decked" in the passages cited at the beginning of this note. Cf. also SEVEN, note 44.

[55] In Fig. 151 the crew stands on the main deck, and the line of prominent oblong projections marks the ends of the beams of the lower deck.

[56] E.g., the *Syracusia* (Athenaeus 5.207c, cited in App., pt. 4).

[57] Moschion's use of the word *triparodos* (see the passage from Athenaeus cited in the previous note), which means "with triple gangways," i.e., access passageways to the three deck levels, perhaps indicates that the lower decks at least were partial.

[58] Homer calls the decking at stem and stern of his galleys *ikria* (FOUR, note 4), and the word in nonnaval contexts means "raised platform"; applied to merchant ships it can only mean the platformlike poop deck. In the vessel described by Synesius, though the passengers gather on the *katastroma* (see note 54 above), the captain "took his stand on the poop deck" (*ikria*: ἐπὶ τῶν ἰκρίων ἑστώς, *Epist.* 4.161a). Cf. Heliodorus, *Aeth.* 5.24.2: τῶν μὲν εἰς τὰ κοῖλα τῆς νεὼς καταδυομένων, τῶν δὲ προ-μαχεῖν ἐπὶ τῶν ἰκρίων ἀλλήλοις παρακελευομένων "Some hid in the bowels of the ship, others raised the cry to do battle on the poop deck" [*ikria*; as the highest level, a natural point for rallying the defense].

[59] For other examples, see Becatti, pl. 182, *AM* pl. 14b = *IH* ill. 64.

179

did not as a rule cover the full width of a ship; along each gunwale was a corridor on which the deckhouse doors opened and which continued aft to give access to an inclined overhanging gallery that girdled the stern.[60] On larger craft the overhang was great enough to allow for a small shelter suspended over open water behind the sternpost (Figs. 144, 146, 151, 154), perhaps a latrine to judge from its position.[61] On the poop deck in front of the sternpost was the helmsman's station; from here he operated the long tiller bars of the steering oars. On some craft the stern gallery was matched by another at the bow (Fig. 151).[62]

The choicest accommodations were in the deckhouse. Here were the quarters for the owner or his agent, for the captain,[63] and for a few select passengers traveling, as it were, first class.[64] On very special

[60] In the Yassi Ada Byzantine wreck, the aftermost part of the deckhouse, roofed with standard roof-tiles, formed the roof of the galley. It very likely extended at this point from gunwale to gunwale. The forward part of the deckhouse, where the helmsman had his station, probably included the captain's quarters (van Doorninck 195, and cf. note 63 below). This part could very well have been separated by narrow passageways from the gunwales.

[61] Venetian pilgrim galleys of the 15th century had a latrine in precisely this spot (Lane 19, note 32). The elegant galleries that girdled the sterns of late 16th century warships served as officers' latrines; cf. IH 98.

[62] Espérandieu 687 (cf. note 32 above).

[63] In the Yassi Ada Byzantine wreck, all the coins found came from a locker just forward of the galley on the starboard side (van Doorninck 115-17); since a captain was responsible for the cash aboard, this should mark the site of the captain's cabin. In the big grain carrier Isis (App., pt. 2), the cabins were also at the stern (αἱ κατὰ τὴν πρύμναν οἰκήσεις "the accommodations toward the stern," Lucian, Navig. 5). The exact location of the handsome owner's cabin on the Syracusia is not given but, since the cookhouse was aft (Athenaeus 5.207c, cited in App., pt. 4), the likelihood is that it was nearby. The captain's cabin (diaeta magistri) of the ship in Petronius' Satyricon (115.1) was no doubt aft. The last two passages show that the term for cabin in Greek was diaita and that this was borrowed by the Romans in the Latinized form diaeta; cf. diaetarius discussed in THIRTEEN, note 83.

[64] E.g., Achilles Tatius assigned quarters here to the protagonists of his novel. His hero reports (5.15.3) that "we had a shelter all our own on the ship, completely closed in" (ἰδία . . . καλύβη τις ἦν ἐπὶ τοῦ σκάφους περιπεφραγμένη) and (5.16.4) that "my present circumstances seem to be so many symbols of marriage. Our yoke [zygos, whose second meaning is "crossbeam"; cf. TEN, App. 3, s.v. beams] is this which is suspended over my head. Our bonds are those which stretch to the yard. . . . Beneath the yoke is our bridal chamber [thalamos, whose second meaning is "cubicle aboard ship"] and our ties [literally, "ropes made fast"]. And the rudder is near our chamber" (ἐμοὶ μὲν γὰρ δοκεῖ τὰ παρόντα γάμων εἶναι σύμβολα. ζυγὸς μὲν οὗτος ὑπὲρ κεφαλῆς κρεμάμενος, δεσμοὶ δὲ περὶ τὴν κεραίαν τεταμένοι . . . ὑπὸ ζυγὸν

ships, particularly royal yachts, these accommodations could offer genuine luxury: lavishly fitted cabins, promenades, exercise areas, baths, lounges, chapels.[65] Most passengers, as in mediaeval times, camped on deck, living either in the open or under tiny temporary shelters.[66] There were also steerage accommodations, as mentioned above, in the bowels of the ship.[67]

As we might expect, religion was also accommodated aft. The monster freighter built by King Hiero boasted a special, magnificently decorated chapel dedicated to Aphrodite, presumably the

---

ὁ θάλαμος, καὶ κάλῳ δεδεμένοι. ἀλλὰ καὶ πηδάλιον τοῦ θαλάμου πλησίον). The mention of the steering oar shows that the spot was aft. It was not below decks because some of the running rigging—the braces to the yardarms are specifically mentioned—was made fast nearby. It follows that the crossbeams overhead could only be those supporting the poop deck.

[65] E.g., on Hiero's *Syracusia*; see Athenaeus 5.207c-f (cited in App., pt. 4). Suetonius (*Caligula* 37) describes the craft Caligula had built for his pleasure cruises; they were equipped with baths, shaded promenades, dining salons, even vine and fruit arbors. Plutarch's statement (*Lucullus* 7.5) that Mithridates built ships "without gilded cabins, baths for his concubines, sumptuous women's quarters," implies that such were generally found on royal yachts. The Nemi barges, built by Caligula probably to serve as a floating shrine (Ucelli 285-92), amply bear out all the above. Floors were paved with figured mosaics, walls were revetted with colored marbles, bronze figures of superb workmanship were added for decorative effect all over the ships (Ucelli 199-231). G. Kapitän has reported (by letter) a wreck of Roman date off Ognina in Sicily that has a cabin floor paved with mosaic. Handsome bronze lamps, lampholders, dishes, and pitchers that have been found in a number of wrecks very likely were fittings from well-furnished cabins; cf. G. Kapitän and A. Fallico, "Bronzi tardoantichi dal *Plemmyrion* presso Siracusa," *Bollettino d'Arte* (1967) 90-97.

[66] Cf. Achilles Tatius 2.33.1 (ἔτυχε δέ τις ἡμῖν νεανίσκος παρασκηνῶν "Some young man happened to set up his shelter alongside us") and Petronius 100.6 (cited in note 54 above). Lucian (*Toxaris* 20) tells how a man who fell overboard was saved because his best friend who "was naked [and hence stripped for swimming] in his bed" (γυμνὸν ἐν τῇ εὐνῇ ὄντα) heard the cries; the bed must have been on deck near the rail. As late as the 11th century, Mediterranean seagoing vessels still offered no passenger staterooms (Goitein 315), though those operating in the Indian Ocean did (Goitein 315; cf. *IH* ill. 185. This may have been due to Chinese influence; seagoing junks had passenger cabins centuries before they were known in the West).

[67] A regulation of St. Pachomius governing the behavior of his monks when aboard ship just about sums up all the possibilities: "No brother is to sleep in the area of the bilge or the inside of the ship if the others are sleeping on the thwarts or deck planks" (*in loco sentinae et interiore parte navis, caeteris fratribus super transtra et tabulata quiescentibus, nullus dormiat, Regula Pachomii* cxix [Migne, *PL* 23, p. 80]).

181

guardian deity of the ship.[68] The average shipowner contented himself with an altar on the poop (Fig. 146) or the afterdeck (Fig. 151); here thanks were formally offered when a vessel arrived safely home.[69] On some craft the arrangements were more elaborate, with the altar set in a niche.[70]

[68] Athenaeus 5.207e, cited in App., pt. 4.

[69] Fig. 146 shows the thanksgiving ceremony being carried out on a ship entering its home port. The altar has a fire ablaze on it. Three people stand about it: a woman holding the *acerra*, or incense box; on her right a man who is sprinkling grains of incense on the fire; on her left another who holds a *patera* and bowl for a libation. One of the men must be the *naukleros*, the owner or charterer (see 315 below), and the woman his wife; the other may be the captain. On this scene, see D. Wachsmuth, ΠΟΜΠΙΜΟΣ Ο ΔΑΙΜΩΝ: *Untersuchung zu den antiken Sakral-handlungen bei Seereisen* (Berlin dissertation, 1967) 145-47. Wachsmuth calls the altar portable; perhaps, but see the following note.

[70] A stone altar 60 cm. (23 5/8″) high was found in the Spargi wreck along with a colonnette and other bits of worked stone that seem to have formed elements of a niche or shrine; see *CIAS* II 156 and *RSL* 30 (1964) 261-62. A graffito of a big sailing ship (Maiuri, *loc. cit.* [note 34 above]) shows something on the after deck which may be a niche with an altar in it.

TONNAGE AND DIMENSIONS OF ANCIENT FREIGHTERS

Part i: Tonnage of Freighters of Average Size

THE PORT regulations from Thasos (note 23 above), which testify that at least by the third century B.C. vessels under 3,000 talents (= 80 tons burden) were of negligible size, at the same time indicate that those of 5,000 talents (= 130 tons burden) were at least of average size. The same conclusion can be deduced from certain roughly contemporary inscriptions which honor merchants, or on occasion rulers, either for having made outright gifts of grain to a given city or for having sold it at less than the current price. In a number of these the exact amount of grain involved is stated:

	Amount			
Reference and Date	*(Medimni)*	*(Tons)*	*Given To*	*Given By*
IG xi.4.627 (1st half of 3rd b.c.)	500	20	Delos	Dealer of Byzantium
*Syll.*³ 354 (ca. 300 B.C.)	2,333	95	Ephesus	Rhodian dealer
Insc. de Délos 442A 100-105 (179)	2,800	115	Delos	Massinissa
SEG i 361 (end of 4th)	3,000	120	Samos	Dealer of Torone
IG ii² 360 (ca. 325/4)	3,000	120	Athens	Cypriote dealer
IG ii² 363 (324/3)	3,000	120	Athens	Dionysius, ruler of Heracleia
IG ii² 398 (320/19)	3,000	120	Athens	Dealer from the Hellespont[1]
IG ii 408 (ca. 330)	4,000 (also certain amount of barley)	165	Athens	Two dealers of Heracleia

[1] The text (stoichedon) now reads (lines 12-14): [ἀπέστει]λεν πυρῶν μ[εδίμνους . . . 'Aθήναζε] κτλ. In view of the other inscriptions, the restoration μ[εδίμνους XXX seems almost certain. Cf. the similar restoration in *SEG* xxi 298.14 (323/2 B.C.).

IG II² 400 (320/19)	4,000	165	Athens	lost
IG II² 845 (208/7;				
cf. *SEG* xvi 71)	8,000	330	Athens	an Aetolian; cf. *REG* 51 (1938) 428

The figures present a consistent picture: each of the persons honored had donated, as seems *a priori* likely, a shipload of grain, while a few, doubly generous, had donated two. And the commonest size was 3,000 *medimni* = 120 tons.

The evidence concerning cargoes of liquids, though less precise, points in the same direction. The inscription *IG* II² 903 (176/5) honors a man who loaded 1,500 *metretae* of olive oil intending to sell it abroad and bring back a cargo of grain to the Peiraeus, but returned with his original load when he learned of a scarcity of oil at Athens. The weight of the oil is about 52.6 tons (the *metretes* held 39 liters, and a liter of oil weighs .9 kg.), to which we must add the weight of the jars. At this period these commonly varied between 20 and 40 liters.[2] Assuming that all were of 20 liters, 3,000 would be needed , and since a 20-liter jar weighs some 17-18 kg.,[3] this would add another 51 to 54 tons, bringing the total weight of the cargo to a little over 100 tons. If, as is most likely, there was a miscellany of various sizes, this would not change the figure significantly. Again, one of Demosthenes' speeches on a bottomry case mentions (35.10) a ship that was carrying a load of 3,000 jars from Mende to Scione. The capacity is not stated, but since in Demosthenes' time also the smallest shipping jars in common use held 20 liters, the cargo must have been at least as large as the one just described. Lastly, the wreck of the Grand Congloué, dating about the middle of the second century B.C., had aboard an estimated 3,000 jars for a total weight of a little over 100 tons.[4]

PART 2: OVERSIZE FREIGHTERS

As POINTED out above (172), freighters of 350-500 tons, though considered large, were not out of the ordinary. There were some, however, much larger than this that were very much out of the ordinary, the *Queen Marys* of their day, as it were. We know of at least three such. All operated out of Alexandria which, requiring a sail of weeks or months over open water to reach western Mediterranean ports (TWELVE, notes 82, 86), had a special use for oversize freighters, even as the north Atlantic has had for oversize liners in our own century.

---

[2] Grace, *op. cit.* (NINE, note 18) 180.
[3] Benoit 163; Lane, "Tonnages" 218.   [4] Benoit 163-64.

One, the *Syracusia* (or *Alexandris*, as she was later called), was the largest merchantman built in antiquity, a distinction that induced a certain Moschion to draw up a detailed description of her, which Athenaeus centuries later included in an account of sundry remarkable ships.[5] The vessel was ordered by Hiero II of Syracuse, probably about 240 B.C.,[6] to use for transporting grain, Sicily's chief export at the time. He employed Archimedes himself as supervising architect.

Moschion, unfortunately, gives no dimensions. He does, however, itemize the cargo the vessel carried on her maiden voyage: 60,000 measures of grain, 10,000 jars of pickled fish, 20,000 talents of wool, 20,000 talents of miscellaneous items. Estimates of the weight of all this have ranged from 3,650 tons to as high as 4,200.[7] Yet even the lower figure is inconceivable: in the subsequent history of shipping, vessels capable of carrying this amount were not designed until the end of the nineteenth century, and then only after the use of iron elements or steel hulls had been introduced.[8] What brings the total so high is the grain. We are told that 60,000 measures were aboard, but what unit of measurement is intended, as so often happens, is not stated.[9] Since the ship was loaded at Syracuse, it has always been taken for granted that Greek *medimni* were meant, and since a *medimnus* of grain weighs about 40 kg., this part of the cargo was computed at 2,400 tons.[10] However, if we assume that during the interval of more than four centuries between Hiero's launching of the ship and Athenaeus' transcription of Moschion's account, the original figure in *medimni* was somehow converted to *modii*, the measure used

[5] Athenaeus 5.206d-209; see pt. 4 of this Appendix. Older writers such as Torr (28-29) were inclined to dismiss much of the account as fanciful, but, save perhaps for some occasional exaggeration (e.g., the numbers of the crew), it seems perfectly sound. Indeed, recent evidence has confirmed many of Moschion's details: we know now, for example, that the use of lead sheathing over pitch-impregnated fabric was a well-established practice (210 below). Even the 10-lb. and 15-lb. spikes are not apocryphal; a wreck of quite average size yielded one that weighed 4.05 kg. or 9 lb. (*Klio* 39, 1961, p. 306).

[6] Cf. note to *OGIS* 56.17; the Ptolemy involved is more likely Euergetes (246-221 B.C.) than Philadelphus (283-246).

[7] Torr 27, 3,650 tons; *ESAR* I 105, 4,000 tons; B. Graser, *De veterum re navali* (Berlin 1864) 48-49, 4,200 tons (followed by Beloch, *Griechische Geschichte*[2] IV.1.299 and W. Tarn, *Hellenistic Civilization*[3], London 1952, p. 250).

[8] E.g., Donald McKay's monster *Great Republic*, rated 4,556 tons, was the largest wooden ship ever built (334' 6" long, 53' 6" broad, 38' deep); the size was made possible by the use of iron braces.

[9] Cf., e.g., Polybius 5.89.7-9; Livy 43.6.3, 11.

[10] E.g., Torr 27. Köster (163-64) takes the 60,000 units to represent jars, which is impossible since grain was not shipped that way (see App., pt. 5). Consequently the figure he arrives at for the total cargo (3,310 tons) is worthless (yet repeated uncritically by Miltner 922).

by the Roman administration after it completed its reorganization of the island in 210 B.C., we arrive at a total that makes far better sense:

	(tons)
60,000 *modii* of grain	400
10,000 jars of pickled fish at,	
e.g., 50 kg. per jar	500
20,000 talents of wool	520
20,000 of misc. cargo	520
	1,940

This still makes the ship—as well it should be—extraordinarily large for that age, but not impossibly large. Moreover, the weight assigned above to the fish is a pure guess, deliberately set on the heavy side. It could easily be half that, which would bring the weight aboard to under 1,700 tons.[11]

The second superfreighter known is the *Isis*, one of the large grain ships used on the Alexandria-Rome run in the second century A.D. Blown off her course during one voyage, she put in at Athens, where Lucian, seizing the chance to get a look at such an unusual sight, paid the vessel a visit and reported some details about it. He unfortunately omits the capacity but does give the dimensions: "She was 120 cubits [180' or 55 m.] in length, the ship's carpenter said, beam more than a quarter of that [45' plus or 13.72 m. plus], and from the deck to the bottom, to the deepest point in the bilge, 29 cubits [43½' or 13.25 m.]."[12] Both classical scholars and nautical experts have tried to work out the vessel's capacity from these figures; the variation in their conclusions is so great it is hard to believe they are all dealing with the same ship. Here are, in rough chronological order, the estimates that have been put forth so far: Jal[13] ca. 1,500

---

[11] Some comparative figures may help: the *Constitution*, 175' long on the gun deck x 43' 6" wide x 14' 3" deep in the hold, was rated at 1,576 tons burden; Nelson's *Victory*, 186' long on the gun deck x 51' 6" wide x 21' 6" deep in the hold, was rated at 2,162 tons burden.

[12] εἴκοσι καὶ ἑκατὸν πήχεων ἔλεγε τὸ μῆκος ὁ ναυπηγός, εὖρος δὲ ὑπὲρ τὸ τέταρτον μάλιστα τούτου, καὶ ἀπὸ τοῦ καταστρώματος ἐς τὸν πυθμένα, ᾗ βαθύτατον κατὰ τὸν ἄντλον, ἐννέα πρὸς τοῖς εἴκοσι *Navig*. 5. The episode may be fanciful (cf. J. Bompaire, *Lucien écrivain* [*Bibl. des écoles françaises d'Athènes et de Rome* 190, Paris 1958] 534-36), but this does not affect the verisimilitude of the details provided.

[13] A. Jal, *Arch. nav.* II 154 (cf. 150-51). Jal based his figures on the rather close resemblance of the *Isis'* dimensions to those of a French third-rate (for plans of a standard French third-rate, see F. Paris, *Souvenirs de Marine*, pt. 5, Paris 1892, no. 249).

tons, Graser[14] 1,575 (repeated in some of the handbooks),[15] Smith[16] 1,100-1,200, Breusing[17] 2,672, Assmann[18] 2,000, Köster (165) 3,000-3,500, Miltner (922) 3,250.

Clearly Lucian's three dimensions are in themselves not enough for determining the *Isis'* capacity. An additional figure is necessary, namely the length of her keel. If she were a stubby bluff-bowed craft like a British East Indiaman, her keel would be very long and she would have roomy holds fore and aft as well as amidships. But if her keel were much shorter than her overall length, she would have overhangs fore and aft, like the ships of the sixteenth century, and the sole area for accommodating sizable amounts of cargo would be amidships. The only way to determine the length of her keel is to find a comparable craft from a later age whose complete dimensions are known. It need not be a ship with exactly the same dimensions; one in which the relation between the dimensions is the same will do. We can assume that the *Isis* looked in general like any big Roman merchantman (Figs. 144, 149, 156). An eminently comparable ship is a Venetian man-of-war of the sixteenth century.[19] Its shape is very like that of Roman merchantmen and, though a much smaller ship than the *Isis*, it has proportions astonishingly similar:

	Isis (feet)		*Venetian Man-of-War* (feet)	
Overall Length	180		ca. 119	
Beam	45	= 25 per cent of length	29½	= 25 per cent of length
Bottom of hold to (presumably) highest deck	43½	= slightly less than beam	28½	= slightly less than beam

[14] *Op. cit.* (note 7 above) 47.

[15] E.g., J. Marquardt, *Das Privatleben der Römer*[2] 406 and note 9; L. Friedländer, *Darstellungen aus der Sittengeschichte Roms*[10] I, 425 and note 10.

[16] 147-50. Smith arrived at his conclusion by comparing the *Isis* with two British ships of known dimensions and tonnage. However, he chose his ships rather arbitrarily: he took one of one shape and a second of another, compared the *Isis* with both, derived two possible figures, and chose the mean between the two.

[17] Breusing 157. A. C. Johnson in *ESAR* II, 402, note 42, gives 2,700, a figure arrived at by a colleague in the engineering department to whom he had submitted the problem and who apparently used the same approach as Breusing.

[18] Assmann 1622.

[19] Casson, "Isis," 54, fig. 3, after Paris (*op. cit.*, note 13 above) pt. 3, no. 172.

The keel of the Venetian ship is 75½' long or ca. 63½ per cent of the overall length. The length of the *Isis'* keel works out conformably to ca. 114 feet.

The next step is to employ a formula involving length of keel which has been traditionally used for computing the burden of sailing ships. A standard one is the following:[20]

$$\frac{\text{length of keel x beam x one-half of beam}}{94}$$

Applying this to the *Isis* (114' x 45' x 22½ / 94), her tonnage works out to 1,228, a figure far more in line with the technology of the age than the astronomical estimates mentioned above. It makes her, properly, smaller than the *Syracusia*, but still an impressively large ship. After the fall of Rome, merchantmen of this size did not appear in any considerable numbers until the end of the eighteenth and beginning of the nineteenth centuries, when the British East Indiamen grew from 800 to 1,200 tons. Rome imported 150,000 tons of grain yearly from Egypt; had the fleet of carriers all been as large as the *Isis*, there need only have been 80 units to do the job.[21]

The third superfreighter was the vessel in which Caligula had the obelisk now standing in front of St. Peter's transported from Alexandria to Rome.[22] The obelisk weighs 322 tons, and it was brought over along with the four pieces forming its pedestal, which account for another 174

[20] See David Steel, *The Elements and Practice of Naval Architecture* (London 1822[3]) 210, and cf. 212-13; cf. Smith 148. From 1677 on, the British navy estimated the burden of the vessels it hired by the following formula: length on deck—3/5 of beam x beam x ½ beam / 94 (Lane, "Tonnages" 228-29); this is the formula Graser used [*loc. cit.* note 7 above]. This, while useful for the boxlike craft of the 17th and 18th centuries, would exaggerate the capacity of the crescent-shaped Roman merchantman.

[21] Twenty million *modii* yearly; see Aurelius Victor, *Caes.* 1.6: *ex Aegypto urbi annua ducenties centena milia frumenti inferebantur* "Each year 20,000,000 measures of grain were brought into the city from Egypt." Each vessel could make only one and a half round trips during the sailing season; cf. TWELVE, App.

[22] Pliny, *NH* 16.201: *nave, quae ex Aegypto Gai principis iussu obeliscum in Vaticano Circo statutum quattuorque truncos lapidis eiusdem ad sustinendum eum adduxit, qua nave nihil admirabilius visum in mari certum est. CXXX modium lentis pro saburra ei fuere* ". . . the ship, which, at the orders of the Emperor Gaius, brought over from Egypt the obelisk that stands in the circus on the Vatican Hill along with the four blocks of the same stone that serve as its base. Certainly nothing more wonderful has ever been seen on the water than this ship. There were 130,000 *modii* of lentils aboard as ballast."

tons.[23] These probably were able to be fitted into the hold, but the obelisk necessarily was carried on deck; the ship, therefore, needed considerable ballasting, and this was provided for by a load of about 800 to 900 tons of lentils.[24] Thus the total weight aboard was 1,300 or so tons—a figure that agrees nicely with that arrived at for the *Isis*. Indeed, the natural assumption is that the vessel was modeled on the great Alexandria-Rome grain carriers. It must have been bigger than the *Isis* because it apparently turned out to be commercially impractical: after discharging its unique load, it was kept on display until Claudius had it filled with concrete and sunk to form part of a mole of the new harbor he was building for Rome.[25]

PART 3: DIMENSIONS OF FREIGHTERS OF AVERAGE SIZE

THE dimensions of the *Isis* were, until very recently, the only known of any freighter. Thanks to underwater archaeology, which has accounted for the discovery and examination of scores of wrecks, a good many dimensions are now known, at least approximately. They are all of vessels that were hauling building stone or shipping jars, for only such cargoes have been able to survive and mark the spot of an ancient wreck.

The wrecks and their dimensions are as follows (since wooden freighters in all ages have had a general length to beam ratio of 3 or 4:1, approximations of missing dimensions are given in brackets):

Wreck and Reference	Date	Length (meters)	Beam (meters)	Remarks
Kyrenia Katzev	End of 4th B.C.	15	[4-5]	

[23] D. Fontana, *Del modo tenuto nel trasportare l'obelisco Vaticano* (Rome 1590) 9, 23: the obelisk weighs 963,537 lb. and the four pieces 165,464, 67,510, 179,826, 110,778 respectively; a ton = ca. 2,996 of the pounds Fontana reckons in.

[24] Torr (26) reckons that Egyptian lentils, packed in a hold, would average ca. 15 lb. per *modius*.

[25] See the passages from Pliny cited in SIXTEEN, note 23. Similarly, the vessel in which Augustus had the so-called Flaminian obelisk (now in Piazza del Popolo) brought over was not used commercially but was put on display at Puteoli (Pliny, *NH* 36.70). The three pieces of this obelisk, according to Fontana (70), weigh 1,322,938 lb. = ca. 440 tons, or about 50 less than the Vatican obelisk with its pedestal.

Testaguzza has identified a part of the mole that almost certainly reveals traces of the forward part of this vessel (107, 112-13). However, his identification of the stern and starboard side, which leads him to assign a size of 104 m. x 20.30 = 343' x 66½' (109, 116-19) is by no means as sure.

189

Grand Congloué Benoit 164	Mid-2nd B.C.	23	6.80	cargo of ca. 3,000 jars = 108 tons
Mahdia Taylor 48, 51	1st B.C.	30-[40]	10	cargo of at least 230 tons keel 26 m. long
Albenga Lamboglia 219	1st half of 1st B.C.	40	8-10	cargo of ca. 10,000 jars = 450 tons
Spargi CIAS ii 156, 161	End of 2nd B.C.	25-30	8-10	
Chrétienne A Dumas 163-64	1st half of 1st B.C.	[24-32]	8	mast-step to deck 3.56 m.
Titan Taylor 88	Mid-1st B.C.	25	[6-8]	keel 17 m. long
Dramont A Taylor 97	End of 1st B.C.	20	7-8	
Marzamemi I Klio (1961) 292, 298	3rd A.D.	[21-32]	7-8	cargo of at least 200 tons
Marseilles Benoit 145-46	3rd-5th A.D.	17	[4.25-5.5]	
Yassi Ada Byzantine AA (1962) 552, van Doorninck 84	7th A.D.	19-21	5.2	1,000 jars
Isola delle Correnti Klio (1961) 284, 286	undatable	[30-40]	10	cargo of at least 350 tons

The last three were carrying building stone, the others chiefly amphorae.

PART 4: HIERO'S SUPERFREIGHTER (ATHENAEUS 5.206d-209b)

Since many of the technical points in this account have become clear only in the light of recent evidence, the older translations are unsatisfactory. Even the latest, Gulick's in the Loeb Classical Library (1928), has a number of serious mistakes and, in general, is not precise enough for the student of nautical antiquities. I have used Kaibel's text (Teubner 1887). The Hiero involved is Hiero II, 306-215 B.C.

περὶ δὲ τῆς ὑπὸ Ἱέρωνος τοῦ Συρακοσίου κατασκευασθείσης νεώς, ἧς καὶ Ἀρχιμήδης ἦν ὁ γεωμέτρης ἐπόπτης, οὐκ ἄξιον εἶναι κρίνω σιωπῆσαι, σύγγραμμα ἐκδόντος Μοσχίωνός τινος, ᾧ οὐ παρέργως ἐνέτυχον ὑπογυίως. γράφει οὖν ὁ Μοσχίων οὕτως· Ἱέρων ... ἦν ... περὶ ναυπηγίας φιλότιμος, πλοῖα σιτηγὰ κατα-
206f σκευαζόμενος, ὧν ἑνὸς τῆς κατασκευῆς μνησθήσομαι. εἰς ὕλην μὲν ξύλωσιν ἐκ τῆς Αἴτνης παρεσκεύαστο ἑξήκοντα τετρηρικῶν σκαφῶν [τὸ] πλῆθος ἐξεργάσασθαι δυναμένην. ὡς δὲ ταῦτα ἡτοιμάσατο γόμφους τε καὶ ἐγκοίλια καὶ σταμῖνας καὶ τὴν εἰς τὴν ἄλλην χρείαν ὕλην τὴν μὲν ἐξ Ἰταλίας, τὴν δ' ἐκ Σικελίας, εἰς δὲ σχοινία λευκέαν μὲν ἐξ Ἰβηρίας, κάνναβιν δὲ καὶ πίτταν ἐκ τοῦ Ῥοδανοῦ ποταμοῦ καὶ τἆλλα πάντα τὰ χρειώδη πολλαχόθεν. συνήγαγε δὲ καὶ ναυπηγοὺς καὶ τοὺς ἄλλους τεχνίτας καὶ καταστήσας ἐπὶ πάντων Ἀρχίαν τὸν Κορίνθιον ἀρχιτέκτονα
207 παρεκάλεσε προθύμως ἐπιλαβέσθαι τῆς κατασκευῆς, προσκαρτερῶν καὶ αὐτὸς τὰς ἡμέρας. τὸ μὲν οὖν ἥμισυ τοῦ παντὸς τῆς νεὼς ἐν μησὶν ἓξ ἐξειργάσατο ... καὶ ταῖς ἐκ μολίβου ποιηθείσαις κεραμίσιν ἀεὶ καθ' ὃ ναυπηγηθείη μέρος περιελαμβάνετο, ὡς ἂν τριακοσίων ὄντων τῶν τὴν ὕλην ἐργαζομένων τεχνιτῶν χωρὶς τῶν ὑπηρετούντων. τοῦτο μὲν οὖν τὸ μέρος εἰς τὴν θάλασσαν καθέλκειν προσετέτακτο, τὴν λοιπὴν κατασκευὴν ἵν' ἐκεῖ λαμβάνῃ. ὡς δὲ περὶ τὸν καθελκυσμὸν αὐτοῦ τὸν εἰς τὴν θάλασσαν πολλὴ ζήτησις ἦν, Ἀρχιμήδης ὁ μηχανικὸς
207b μόνος αὐτὸ κατήγαγε δι' ὀλίγων σωμάτων. κατασκευάσας γὰρ ἕλικα τὸ τηλικοῦτον σκάφος εἰς τὴν θάλασσαν κατήγαγε. πρῶτος δ' Ἀρχιμήδης εὗρε τὴν τῆς ἕλικος κατασκευήν. ὡς δὲ καὶ τὰ λοιπὰ μέρη τῆς νεὼς ἐν ἄλλοις ἓξ μησὶ κατεσκευάσθη καὶ τοῖς χαλκοῖς ἥλοις πᾶσα περιελήφθη, ὧν οἱ πολλοὶ δεκάμνοοι ἦσαν, οἱ δ' ἄλλοι τούτων ἡμιόλιοι — διὰ τρυπάνων δ' ἦσαν οὗτοι ἡρμοσμένοι τοὺς σταμῖνας συνέχοντες· μολυβδίναις δὲ κεραμίσιν ἐπεστεγνοῦντο πρὸς τὸ ξύλον, ὑποτιθεμένων

ὀθονίων μετὰ πίττης — ὡς οὖν τὴν ἐκτὸς ἐπιφάνειαν ἐξειργά-
207c σατο, τὴν ἐντὸς διασκευὴν ἐξεπονεῖτο. 41. ἦν δὲ ἡ ναῦς τῇ μὲν
κατασκευῇ εἰκόσορος, τριπάροδος δέ· τὴν μὲν κατωτάτω ἔχων
ἐπὶ τὸν γόμον, ἐφ' ἣν διὰ κλιμάκων πυκνῶν ἡ κατάβασις ἐγίνετο·
ἡ δ' ἑτέρα τοῖς εἰς τὰς διαίτας βουλομένοις εἰσιέναι ἐμεμη-
χάνητο· μεθ' ἣν ἡ τελευταία τοῖς ἐπὶ τοῖς ὅπλοις τεταγμένοις.
ἦσαν δὲ τῆς μέσης παρόδου παρ' ἑκάτερον τῶν τοίχων δίαιται
τετράκλινοι τοῖς ἀνδράσι, τριάκοντα τὸ πλῆθος. ἡ δὲ ναυκλη-
ρικὴ δίαιτα κλινῶν μὲν ἦν πεντεκαίδεκα, θαλάμους δὲ τρεῖς
εἶχε τρικλίνους, ὧν ἦν τὸ κατὰ τὴν πρύμναν ὀπτανεῖον. ταῦτα
δὲ πάντα δάπεδον εἶχεν ἐν ἀβακίσκοις συγκείμενον ἐκ παν-
207d τοίων λίθων, ἐν οἷς ἦν κατεσκευασμένος πᾶς ὁ περὶ τὴν Ἰλιάδα
μῦθος θαυμασίως· ταῖς τε κατασκευαῖς καὶ ταῖς ὀροφαῖς, καὶ
θυρώμασι δὲ πάντα ἦν ταῦτα πεπονημένα. κατὰ δὲ τὴν ἀνωτάτω
πάροδον γυμνάσιον ἦν καὶ περίπατοι σύμμετρον ἔχοντες τὴν
κατασκευὴν τῷ τοῦ πλοίου μεγέθει, ἐν οἷς κῆποι παντοῖοι θαυ-
μασίως ἦσαν ὑπερβάλλοντες ταῖς φυτείαις, διὰ κεραμίδων
μολυβδινῶν κατεστεγνωμένων ⟨ἀρδευόμενοι⟩, ἔτι δὲ σκηναὶ
κιττοῦ λευκοῦ καὶ ἀμπέλων, ὧν αἱ ῥίζαι τὴν τροφὴν ἐν πίθοις
εἶχον γῆς πεπληρωμένοις, τὴν αὐτὴν ἄρδευσιν λαμβάνουσαι
207e καθάπερ καὶ οἱ κῆποι. αὗται δὲ αἱ σκηναὶ συνεσκίαζον τοὺς
περιπάτους. ἑξῆς δὲ τούτων Ἀφροδίσιον κατεσκεύαστο τρίκλινον, δάπεδον ἔχον ἐκ λίθων ἀχατῶν τε καὶ ἄλλων χαριεστάτων
ὅσοι κατὰ τὴν νῆσον ἦσαν· τοὺς τοίχους δ' εἶχε καὶ τὴν ὀροφὴν
κυπαρίττου, τὰς δὲ θύρας ἐλέφαντος καὶ θύου· γραφαῖς ⟨δὲ⟩
καὶ ἀγάλμασιν, ἔτι δὲ ποτηρίων κατασκευαῖς ὑπερβαλλόντως
κατεσκεύαστο. 42. τούτου δ' ἐφεξῆς σχολαστήριον ὑπῆρχε πεντάκλινον, ἐκ πύξου τοὺς τοίχους καὶ τὰ θυρώματα κατεσκευασ-
μένον, βιβλιοθήκην ἔχον ἐν αὐτῷ, κατὰ δὲ τὴν ὀροφὴν πόλον
207f ἐκ τοῦ κατὰ τὴν Ἀχραδίνην ἀπομεμιμημένον ἡλιοτροπίου. ἦν
δὲ καὶ βαλανεῖον τρίκλινον πυρίας χαλκᾶς ἔχον τρεῖς καὶ
λουτῆρα πέντε μετρητὰς δεχόμενον ποικίλον τοῦ Ταυρομενίτου
λίθου. κατεσκεύαστο δὲ καὶ οἰκήματα πλείω τοῖς ἐπιβάταις καὶ
τοῖς τὰς ἀντλίας φυλάττουσι. χωρὶς δὲ τούτων ἱππῶνες ἦσαν
ἑκατέρου τῶν τοίχων δέκα· κατὰ δὲ τούτους ἡ τροφὴ τοῖς ἵπποις
208 ἔκειτο καὶ τῶν ἀναβατῶν καὶ τῶν παίδων τὰ σκεύη. ἦν δὲ καὶ
ὑδροθήκη κατὰ τὴν πρῷραν κλειστή, δισχιλίους μετρητὰς δεχο-
μένη, ἐκ σανίδων καὶ πίττης καὶ ὀθονίων κατεσκευασμένη.
παρὰ δὲ ταύτην κατεσκεύαστο διὰ μολιβδώματος καὶ σανίδων

κλειστὸν ἰχθυοτροφεῖον· τοῦτο δ᾽ ἦν πλῆρες θαλάττης, ἐν ᾧ πολλοὶ ἰχθύες [εὖ] ἐτρέφοντο. ὑπῆρχον δὲ καὶ τῶν τοίχων ἑκατέρωθεν τρόποι προεωσμένοι, διάστημα σύμμετρον ἔχοντες· ἐφ᾽ ὧν κατεσκευασμέναι ἦσαν ξυλοθῆκαι καὶ κρίβανοι καὶ
208b ὀπτανεῖα καὶ μύλοι καὶ πλείους ἕτεραι διακονίαι. ἄτλαντές τε περιέτρεχον τὴν ναῦν ἐκτὸς ἐξαπήχεις, οἳ τοὺς ὄγκους ὑπειλήφεσαν τοὺς ἀνωτάτω καὶ τὸ τρίγλυφον, πάντες ἐν διαστήματι ⟨συμμέτρῳ⟩ βεβῶτες. ἡ δὲ ναῦς πᾶσα οἰκείαις γραφαῖς ἐπεπόνητο. 43. πύργοι τε ἦσαν ἐν αὐτῇ ὀκτὼ σύμμετροι τὸ μέγεθος τοῖς τῆς νεὼς ὄγκοις· δύο μὲν κατὰ πρύμναν, οἱ δ᾽ ἴσοι κατὰ πρῶραν, οἱ λοιποὶ δὲ κατὰ μέσην ναῦν. τούτων δὲ ἑκάστῳ παρεδέδεντο κεραῖαι β΄, ἐφ᾽ ὧν κατεσκεύαστο φατνώματα, δι᾽ ὧν ἠφίεντο λίθοι πρὸς τοὺς ὑποπλέοντας τῶν πολε-
208c μίων. ἐπὶ δὲ τῶν πύργων ἕκαστον ἀνέβαινον τέτταρες μὲν καθωπλισμένοι νεανίσκοι, δύο δὲ τοξόται. πᾶν δὲ τὸ ἐντὸς τῶν πύργων λίθων καὶ βελῶν πλῆρες ἦν. τεῖχος δὲ ἐπάλξεις ἔχον καὶ καταστρώματα διὰ νεὼς ἐπὶ κιλλιβάντων κατεσκεύαστο· ἐφ᾽ οὗ λιθοβόλος ἐφειστήκει, τριτάλαντον λίθον ἀφ᾽ αὑτοῦ ἀφιεὶς καὶ δωδεκάπηχυ βέλος. τοῦτο δὲ τὸ μηχάνημα κατεσκεύασεν Ἀρχιμήδης. ἑκάτερον δὲ τῶν βελῶν ἔβαλλεν ἐπὶ στάδιον. μετὰ δὲ ταῦτα παραρτήματα ἐκ τροπῶν παχέων συγκείμενα
208d διὰ ἀλύσεων χαλκῶν κρεμάμενα. τριῶν δὲ ἱστῶν ὑπαρχόντων ἐξ ἑκάστου κεραῖαι λιθοφόροι ἐξήρτηντο β΄, ἐξ ὧν ἅρπαγές τε καὶ πλίνθοι μολίβου πρὸς τοὺς ἐπιτιθεμένους ἠφίεντο. ἦν δὲ καὶ χάραξ κύκλῳ τῆς νεὼς σιδηροῦς πρὸς τοὺς ἐπιχειροῦντας ἀναβαίνειν κόρακές τε σιδηροῖ [κύκλῳ τῆς νεὼς], οἳ δι᾽ ὀργάνων ἀφιέμενοι τὰ τῶν ἐναντίων ἐκράτουν σκάφη καὶ παρέβαλλον εἰς πληγήν. ἑκατέρῳ δὲ τῶν τοίχων ἑξήκοντα νεανίσκοι πανοπλίας ἔχοντες ἐφειστήκεσαν καὶ τούτοις ἴσοι περί τε τοὺς
208e ἱστοὺς καὶ τὰς λιθοφόρους κεραίας. ἦσαν δὲ καὶ κατὰ τοὺς ἱστοὺς ἐν τοῖς καρχησίοις οὖσι χαλκοῖς ἐπὶ μὲν τοῦ πρώτου τρεῖς ἄνδρες, εἶθ᾽ ἑξῆς καθ᾽ ἕνα λειπόμενοι· τούτοις δ᾽ ἐν πλεκτοῖς γυργάθοις διὰ τροχιλίων εἰς τὰ θωράκια λίθοι παρεβάλλοντο καὶ βέλη διὰ τῶν παίδων. ἄγκυραι δὲ ἦσαν ξύλιναι μὲν τέτταρες, σιδηραῖ δ᾽ ὀκτώ. τῶν δὲ ἱστῶν ὁ μὲν δεύτερος καὶ τρίτος εὑρέθησαν, δυσχερῶς δὲ ὁ πρῶτος εὑρέθη ἐν τοῖς ὄρεσι τῆς
208f Βρεττίας ὑπὸ συβώτου ἀνδρός· κατήγαγε δ᾽ αὐτὸν ἐπὶ θάλατταν Φιλέας ὁ Ταυρομενίτης μηχανικός. ἡ δὲ ἀντλία καίπερ βάθος ὑπερβάλλον ἔχουσα δι᾽ ἑνὸς ἀνδρὸς ἐξηντλεῖτο διὰ κοχλίου,

Ἀρχιμήδους ἐξευρόντος. ὄνομα δ᾽ ἦν τῇ νηὶ Συρακοσία· ὅτε δ᾽ αὐτὴν ἐξέπεμπεν Ἱέρων, Ἀλεξανδρίδα αὐτὴν μετωνόμασεν. ἐφόλκια δ᾽ ἦσαν αὐτῇ τὸ μὲν πρῶτον κέρκουρος τρισχίλια τάλαντα δέχεσθαι δυνάμενος· πᾶς δ᾽ ἦν οὗτος ἐπίκωπος. μεθ᾽ ὃν χίλια πεντακόσια βαστάζουσαι ἁλιάδες τε καὶ σκάφαι πλείους. ὄχλος δ᾽ ἦν οὐκ ἐλάττων . . μετὰ τοὺς προειρημένους ἄλλοι τε ἑξακόσιοι παρὰ τὴν πρῷραν ἐπιτηροῦντες τὰ παραγγελλόμενα.

209 τῶν δὲ κατὰ ναῦν ἀδικημάτων δικαστήριον καθειστήκει ναύκληρος, κυβερνήτης καὶ πρῳρεύς, οἵπερ ἐδίκαζον κατὰ τοὺς Συρακοσίων νόμους. 44. σίτου δὲ ἐνεβάλλοντο εἰς τὴν ναῦν μυριάδας ἕξ, ταρίχων δὲ Σικελικῶν κεράμια μύρια, ἐρίων τάλαντα δισμύρια, καὶ ἕτερα δὲ φορτία δισμύρια. χωρὶς δὲ τούτων

209b ὁ ἐπισιτισμὸς ἦν τῶν ἐμπλεόντων. ὁ δ᾽ Ἱέρων ἐπεὶ πάντας τοὺς λιμένας ἤκουεν τοὺς μὲν ὡς οὐ δύνατοί εἰσι τὴν ναῦν δέχεσθαι, τοὺς δὲ καὶ ἐπικινδύνους ὑπάρχειν, διέγνω δῶρον αὐτὴν ἀποστεῖλαι Πτολεμαίῳ τῷ βασιλεῖ εἰς Ἀλεξάνδρειαν· καὶ γὰρ ἦν σπάνις σίτου κατὰ τὴν Αἴγυπτον. καὶ οὕτως ἐποίησε, καὶ ἡ ναῦς κατήχθη εἰς τὴν Ἀλεξάνδρειαν, ἔνθα καὶ ἐνεωλκήθη.

206d I cannot refrain from mentioning the ship built by Hiero of Syracuse, the one supervised by Archimedes the mathematician, since there is an account of it published by a certain Moschion which I recently read over carefully. Here is what Moschion writes:

206e "Hiero, . . . eager to gain a reputation in the field of shipbuilding, had a number of grain-carriers built, the construction of one of

206f which I shall describe. For the materials, he collected timber from Mt. Etna, enough to build 60 quadriremes; then, partly from Italy and partly from Sicily, the wood for treenails, the upper and lower parts of the frames, and other elements;[26] for cordage, esparto from Spain and hemp and pitch from the Rhone Valley; and the rest of his needs from a variety of places. He recruited carpenters and other craftsmen, chose one of them,[27] Archias of Corinth, to be

207 foreman, pressed him to set right to work, and gave the project his personal attention daily. Since there were three hundred craftsmen, exclusive of assistants, working on the materials, half of the ship

[26] Etna, with its forests of fir and pine (cf. Diodorus 14.42.4), provided the long straight lengths needed for planks and keel. The *gomphoi* would require a hardwood such as oak (TEN, note 54), while the *stamines* and *enkoilia*, the upper and lower parts of the frames (TEN, App. 3, s.v. frames), would require bentwood.

[27] Reading, with Gulick, the ms. ἐκ instead of the restoration ἐπί.

was finished in six months, down to the sheathing of each area, as it was completed, with lead sheets.[28] He gave orders to launch this portion so that the rest of the work could be carried out afloat. After much discussion of how the launching should be done, Archimedes, the engineer, carried it out by himself along with a handful 207b of assistants. He constructed a screw-windlass which drew that huge craft down to the sea.[29] (Archimedes was the first to discover and construct the screw-windlass.) The rest of the ship took another six months; the whole hull was pinned together with copper spikes, most of which weighed ten pounds and some fifteen[30] (these were made fast to the ribs, being fitted into holes bored through,[31] and then were covered over by the underlayer of tarred fabric and overlayer of lead sheeting that protected the planking).[32]

"Once the exterior was finished, work began on the interior. The

[28] See note 32 below.

[29] The hull had probably been planked up to a point just above the waterline. Launching at this early stage helped hasten the day when the vessel could be put to use, since it was customary to let new construction stand a while until the joints and seams closed up as a result of the swelling of the wood in water (cf. TEN, note 21); this could now take place while work went ahead on the superstructure. Another reason may have been the fear that, if complete, the weight might be too great even for the genius of an Archimedes to launch.

Another launching by the celebrated scientist, perhaps in origin a variant on that of the *Syracusia*, is described by Plutarch (*Marcellus* 14.8): "[Archimedes] took one of the royal three-masted merchantmen which much muscle and hard work had succeeded in hauling ashore, loaded aboard it a good number of passengers and the usual amount of cargo, and, seated off to the side, with no great effort—simply setting in motion with one hand a multiple block and tackle—drew it toward him smoothly and without any jerking, just as if it were gliding through the water" (ὁλκάδα τριάρμενον τῶν βασιλικῶν πόνῳ μεγάλῳ καὶ χειρὶ πολλῇ νεωλκηθεῖσαν, ['Αρχιμήδης] ἐμβαλὼν ἀνθρώπους τε πολλοὺς καὶ τὸν συνήθη φόρτον, αὐτὸς ἄπωθεν καθήμενος, οὐ μετὰ σπουδῆς ἀλλὰ ἠρέμα τῇ χειρὶ σείων ἀρχήν τινα πολυσπάστου, προσηγάγετο λείως καὶ ἀπταίστως καὶ ὥσπερ διὰ θαλάττης ἐπιθέουσαν).

[30] On these figures, see App., pt. 2, note 5.

[31] The procedure conforms to what we know of the construction of merchant ships (203-207 below): first, completion of a shell of planks (here, for special reasons, it was done in two stages; see note 29); then insertion of frames, which were secured to the planking by copper spikes driven in from the outside. A literal rendering of the Greek would be "(the spikes) through the use of augers were fitted holding together the ribs," which may be an elliptical way of saying that holes were bored through planking and ribs, the two were joined by a dowel being passed through the hole, and then the spike was driven through the dowel—which was the ancients' usual way of pinning ribs to planking (Fig. 161).

[32] A number of wrecks have yielded evidence of lead sheathing set over a layer of pitch-impregnated fabric; see TEN, App. 1.

207c   vessel, though built after the model of a 'twenty-er,'[33] had three
levels of gangways. The lowest, reached by numerous companion-
ways, was for working cargo. The next was designed to give access
to the cabins. The next—the highest—was for the men-at-arms
aboard. At the middle level, along both sides of the ship, were the
cabins for the men, thirty in all and each of 4-couch size.[34] The
owner's cabin[35] was of 15-couch size with three cubicles of 3-couch
size; the kitchen aft was for these. All had floors done in multi-
colored mosaic; in these was worked, in amazing fashion, the whole

207d   story of the *Iliad*. Trim, overheads, and doors were all carefully
worked. Along the uppermost gangway was a gymnasium, as well
as promenades built to suit a vessel of this size. These had mar-
velously flourishing plant beds of all kinds,[36] which were watered
through covered lead tiles; there were also arbors of white ivy and
grapes whose roots, planted in big jars filled with soil, were watered

207e   in the same way as the flower beds. The arbors provided shade
along the promenades. Alongside these was a chapel to Aphrodite,[37]
three-couch size, whose floor was paved with agate and other of the
most decorative stones found on the island of Sicily. The bulkheads
and overhead were of cypress, and the doors of ivory and aromatic
cedar. It was beautifully fitted out with paintings, statues, and
utensils for libations. Next to it was a reading room, five-couch
size, with bulkheads and doors of boxwood, a library, and, in the
overhead, a circular concavity made to look like the sundial at

207f   Achradina.[38] There was also a bath, three-couch size, with three
copper tubs and a 50-gallon basin made of the colored stone of
Tauromenium. There were, in addition, accommodations for pas-

---

[33] On the word *eikosoros* "twenty-er," see NINE, note 5.

[34] A standard way of expressing size of rooms at this time whether in houses (e.g.,
P. Cairo Zen. 59445 [= *Select Papyri* 171]. 8, 13; 3rd B.C.) or aboard ships (e.g.,
P. Cairo Zen. 59054.34-35, 257 B.C., cited in EIGHT, note 41). Cf. the Japanese method
of reckoning room size by the number of mats needed to cover the floor.

[35] ναυκληρική "of the *naukleros*," the owner or charterer of a freighter. Since the
vessel in this case was state-owned, the *naukleros* would have been Hiero's agent if
the ship was being operated in the king's name, and a charterer if it was not; cf.
THIRTEEN, note 69.

[36] Cf. the practice on French East Indiamen of carrying flats in which greens and
fresh vegetables were raised.

[37] Presumably the guardian deity of the vessel; cf. FIFTEEN, note 14.

[38] Achradina is a part of the city of Syracuse. There the famed tyrant Dionysius I
(405-367 B.C.) had set up a tall and conspicuous sundial (Plutarch, *Dio* 29.2).

sengers and the bilge-watchers.[39] Furthermore, there were ten
stables along each side, which also held the fodder for the horses
208   and the gear of the riders and grooms. There was a sealed water-
tank in the bows with a capacity of 20,000 gallons, made of planks
and waterproofed fabric. Alongside it was a sealed fishtank made
of planks and sheets of lead; it was filled with seawater, and a good
supply of fish was kept alive in it. On each side of the vessel, beams
jutted out, spaced the same distance apart;[40] these supported wood-
208b   bins, ovens, stoves, millstones, and other services. All around the
ship ran a series of exterior *atlantes*, 9 feet high and spaced equally
apart, which supported the uppermost parts of the deck structures
and the triglyph;[41] moreover, the whole ship was decorated with
suitable paintings.

"There were eight towers, of the same height as the deck-struc-
tures of the ship:[42] two aft, the same number forward, and the rest
amidships. Each was fitted with two booms ending in open platforms through which stones could be hurled down on any enemy
208c   that sailed underneath. On each tower were stationed four heavily
armed young marines[43] and two archers. The insides of the towers
were crammed with stones and missiles. Across the ship ran a bat-
tlemented parapet surrounding a raised fighting deck that rested
on pillars; on this was set a catapult capable of throwing a 180-lb.
stone or an 18-foot dart; the instrument had been designed by
Archimedes. Its range with either missile was 200 yards. Nearby
were protective screens joined by thick leather straps and hung
208d   from bronze chains. Each of the three masts was fitted with two
booms for dropping missiles; from these, grappling hooks or chunks

[39] These were a form of steerage; see NINE, note 50.

[40] Most likely the deck beams which, on ancient merchantmen, were often brought
through the side; cf. 210 below. Apparently on this ship they were extended further
than usual to support a series of outhouses.

[41] These 9-foot *atlantes* (male caryatids) supported the "*onkoi* and the *triglyph*."
Elsewhere the author uses *onkos* of the whole mass of a vessel above the waterline
(5.204e); here, in the plural, it must refer to "masses" above the deck, the superstructure or the deck houses (cf. the following note). These, it seems, were in two
levels: the first, 9 feet high, was adorned by a series of *atlantes* supporting an architrave which, as in a Greek temple, was decorated with triglyphs; the architrave in
turn supported the uppermost portions of the deck structures.

[42] The word translated "deck structures" is *onkoi*; see the previous note. The towers, in other words, were the same height as the superstructure.

[43] The word translated "young marines" here and just below (208d) is *neaniskoi*;
on this term, see THIRTEEN, note 33.

of lead could be hurled down on an enemy. All around the ship ran an iron palisade as protection against boarding attempts; there were also grappling irons fired by catapult, which could seize an enemy ship and pull it alongside for a mortal blow. Sixty fully armed young marines were stationed along each side of the vessel and an
208e   equal number around each mast with its booms for dropping missiles. There were also men in the mast-tops, which were made of bronze: three were assigned to the main [literally "first mast"], and two and one to the other two respectively. Slaves hauled up stones and darts to the mast-top parapets by means of baskets on lines running over blocks.

"There were four anchors of wood and eight of iron.[44] When it came to the timber for the masts, the fore and mizzen were found easily enough, but the main was located only with great difficulty
208f   in the mountains of Bruttium by a man from Sybotis; Phileas of Tauromenium, the engineer, had it hauled down to the shore. Although the bilge was extraordinarily deep, it was bailed by only one man using a screw-pump, one of Archimedes' inventions.[45]

"The name of the ship was the *Syracusia* ["The Syracuse"] but, when Hiero sent it off, he changed it to the *Alexandris* ["The Alexandria"]. As ship's boats it had, first a *kerkouros* of 3,000 talents burden [= 78 tons][46] with a full complement of oars, then some *haliades* of half that burden, and, in addition, a good number of small boats. The total complement was no less than [*number lost*]; over and above all mentioned so far, there were 600 stationed for-
209   ward to carry out any orders. Crimes committed aboard ship were brought before a court consisting of the owner, captain, and first mate,[47] who judged in accordance with the laws of Syracuse.

---

[44] The Nemi ships also carried both types of anchor; see ELEVEN, note 115.

[45] The so-called Archimedean screw, still in use in Egypt; see L. Sprague de Camp, *The Ancient Engineers* (New York 1963) 152.

[46] Arab and Jewish shippers of the 11th A.D. saw to it that on their Mediterranean routes "normally, a larger ship was accompanied by a smaller vessel, belonging to the same proprietor or to one of his relatives or friends," a "maidservant boat" as it was called; see S. Goitein, *Studies in Islamic History and Institutions* (Leiden 1966) 304-305 and cf. 348 for the use of the same procedure on voyages in the Indian Ocean. Could not this 78-ton *kerkouros*, manifestly too large to be carried aboard or in tow, have been a "maidservant boat"? Sailing craft of the 11th A.D. were not too different from those of ancient times, and we would expect to find similar practices in both periods.

[47] *naukleros, kybernetes, proreus*. The *naukleros* in this case would be Hiero's agent or a charterer; cf. note 35 above. On these terms see 315 below.

"The vessel was loaded with 60,000 measures of grain, 10,000 jars of pickled Sicilian fish, 20,000 talents of wool, and 20,000 talents of miscellaneous cargo;[48] in addition, there were the provisions for the crew. When Hiero heard that, of all the harbors it was to call at, some would not accommodate the ship at all and others were risky, he decided to send it as a gift to King Ptolemy in Alexandria, since there was a shortage of grain in Egypt at the time. He did so, the ship arrived at Alexandria, and docked there."

209b

PART 5: THE PACKING AND STOWING OF CARGO

IN THE ancient Mediterranean world, where wood was relatively scarce and expensive, the clay jar and not the barrel was the shipping container par excellence (cf. SIXTEEN, note 43). In addition to wine and oil, the two chief commodities, amphorae were used for shipping the widely used fish sauce *garum* (*Gallia* 16, 1958, pp. 34-37; 18, 1960, p. 45; 20, 1962, p. 159, 164), preserved fish (*Gallia* 16, 1958, p. 6; 20, 1962, p. 148; *CIAS* 11.216; Taylor 81), olives (*Gallia* 18, 1960, p. 45; Taylor 95), nuts (*Gallia* 20, 1962, p. 164; *National Geographic* June, 1970, 848), pitch (*Gallia* 18, 1960, p. 43; 20, 1962, pp. 157-59). Nuts were also transported in sacks or the like (*National Geographic* June, 1970, 848).

Tiles were shipped stacked on edge in layers, and stacked so carefully that apparently a minimum of dunnage was needed (Frost 217-19 and pl. 28; cf. *Gallia* 20, 1962, p. 162, fig. 33). Architectural elements such as columns, bases, etc., and pieces of building stone, some weighing as much as 40 tons, frequently went by boat (*Klio* 39, 1961, pp. 276-318, esp. 284, 290); these must have been stowed with equal care, each piece lashed solidly into place and well protected with dunnage (cf. Taylor 51). Metals were shipped in ingots (cf. Fig. 191). Divers have recovered lead ingots of 30-34 kg. off Marseilles (*Gallia* 18, 1960, pp. 55-56); 66 kg. off Fos (*Gallia* 16, 1958, pp. 36-37); 31-35 kg., 17 kg., and small lozenges of but 1.3 kg., in the Mahdia wreck (A. Merlin, "Lingots et ancres trouvés en mer près de Mahdia," *Mélanges Cagnat*, Paris 1912, pp. 383-97, esp. 383-85). Lozenges of lead have also turned up off Agde, as well as a monster block weighing 103 kg. (*RSL* 30, 1964, p. 271). Copper oxhide ingots averaging 20 kg. were found in a wreck of the thirteenth century B.C (Bass, *Gelidonya* 52-57), and disk-shaped ingots of a wide range of sizes in wrecks of later periods (10.5 kg. to 95.5 kg. at Marseilles and $22\frac{1}{2}$-24 kg. in the

[48] Cf. App., pt. 2.

Straits of Bonifacio, *Gallia* 20, 1962, pp. 154-56, 175-76; 44-62 kg. off Agde, *RSL* 30, 1964, p. 274); rectangular bars have also been found (off Agde, and weighing ca. 5-7 kg.; *RSL* 30, 1964, p. 274).

Grain was transported in sacks. Papyri from Egypt attest the presence of σιτομετροσακκοφόροι "grain-measurers-and-sack-carriers"; their job was to measure the grain into sacks and transport it first into the silos and then from the silos into the ships that took it down the Nile (*P. Berol. Frisk* 1, col. 22, lines 3-4; A.D. 155). A wall-painting found at Ostia (*AM* 14b = *IH* ill. 64) illustrates the same procedure going on there. Both, to be sure, concern inland waters, but it is reasonable to assume that grain could also cross open water the same way. Grain was also shipped in bulk, poured into the hold; cf. *Dig.* 19.2.31, a passage dealing with loss "when a number of shippers have poured their grain in common into a . . . ship" (*in navem . . . cum complures frumentum confuderant*). Sometimes separate bins were available (*quod si separatim tabulis aut heronibus aut in alia cupa clusum uniuscuiusque triticum fuisset, ita ut internosci posset etc.* "but if each shipper's wheat was set off by planks or partitions, or each was in its own container, so that it could be identified, etc.").

CHAPTER TEN

Shipbuilding

THERE ARE basically two ways to construct a wooden boat. The one that has been most widely used, particularly in the west, begins with the setting up of a sturdy skeleton of keel and frames; then, to this, is fastened a smooth skin of planks. The other, favored by southern Asia and some parts of northern Europe, is just the reverse: first a sturdy shell of planks is erected by pinning each plank to its neighbor, and then a certain amount of framing is inserted to stiffen the shell.[1]

The shipwright who starts with a shell of planks may join these to each other in any of three ways. One, best known to us because it was used in northern Europe and can be traced from pre-Viking days to our own, produced the clinker-built boat: each plank overlaps the one below for a certain distance, and pegs or nails or rivets driven through where the thickness is double hold the two securely together.[2] In the other two methods, the planks are set edge to edge to make a smooth skin; what differs is the means for fastening. In the waters from the east coast of Africa to India and in many other areas as well, the planks are sewn to one another with twine made from coconut husk or split bamboo or whatever fiber is available. This was, for example, the favored practice of Persian and Arab shipbuilders: we hear of boats so made as early as the first century A.D.; Marco Polo reported seeing them; and the technique was widely practiced in the Indian Ocean, Red Sea, Persian Gulf, and elsewhere until the coming of westerners brought in the more flexible keel-and-frames system.[3] In other parts of India, notably eastern Pakistan, in

[1] Cf., e.g., Hornell 188-98.

[2] Hornell 195-206. In certain clinker-built types of eastern Pakistan, the lapstraking is reversed: the upper edge of the lower plank lies outside the lower edge of the upper; see B. Greenhill, "The Boats of East Pakistan," *MM* 43 (1957) 124, 127.

[3] Hornell 192-93, 234-37; Greenhill, *op. cit.* (previous note) 211-12; Hourani 87-98.

parts of Egypt, and in certain other regions, the planks are pegged together or stapled together or nailed together.[4] Whichever of the two systems of fastening is used, the frames, if any, are always inserted later. They are not necessarily of any great strength: in some makes of boat the number depends simply upon how much money the shipwright wants to spend; he often adds frames only after the boat has been in long use and the shell has started to weaken.[5]

As far back as primitive times the Mediterranean knew the sewn boat (9-10 above), and ancient Egypt had learned by the third millennium B.C. to build craft of edge-joined planks (13-15 above). The great forward step, however, was taken by Greece and Rome. The Greco-Roman shipwright,[6] as we are now aware through the discovery of numerous ancient wrecks, created his own form of shipbuilding, one so refined that it more resembles cabinet work than carpentry. In a sense he combined the two basic ways of putting together a planked boat. Like the Egyptian (from whom he may have gotten the idea), he started with a shell of planks. But instead of using casual joinery he locked these together with mortises set as close as the joints in a piece of furniture;[7] and into this tightly knit structure

[4] Greenhill, "The Karachi Fishing Boats," *MM* 42 (1956) 54-66; Greenhill, *op. cit.* (note 2 above) 108-109, 113-16; Hornell 215-25 (Egypt), 208-209 (Polynesia).

[5] Cf. Greenhill, *op. cit.* (previous note) 59 and *op. cit.* (note 2 above) 113, 121, 122; Hornell 221-23.

[6] The term for shipwright was *naupegos* in Greek, *faber navalis* in Latin. Shipyard was *naupegion* in Greek (ναυπηγεῖον in Diodorus 19.58.4), *textrinum* in Latin (cf. Servius in *Aen.* 11.326: *loca in quibus naves fiunt, graece* ναυπήγια, *latine textrina dici* "places where ships are built are called *naupegia* in Greek, *textrina* in Latin." He cites Ennius [*ROL* 1, p. 52]: *campus habet textrinum navibus longis* "The plain has a shipyard for war galleys").

[7] The knowledge that boats were edge-joined by mortises and tenons has cleared up a number of hitherto puzzling passages in literature. On the building of Odysseus' so-called raft (*Od.* 5.244-57), see App. 2. Hiero's great freighter, it is now clear, was built in this fashion (cf. Athenaeus 5.207 a-b, NINE, App., pt. 4, and note 31). And Ovid's hitherto enigmatic *cunei* "wedges" in *Met.* 11.514-15 (*iamque labant cunei, spoliataque tegmine cerae / rima patet, praebetque viam letalibus undis* "And now the wedges work loose, and the seam, stripped of its covering wax, lies open and furnishes a path for the death-dealing waters") is now explainable: the *cunei* are tenons; the storm was so violent that the mortise-and-tenon joints holding plank to plank came loose, the seams opened up and broke the coat of wax over the hull (see 211 below), and the water came in. (This interpretation was first offered by J. Vars in *CRAI*, 1896, pp. 386-87.) In the Yassi Ada Byzantine wreck, the tenons were in-

he inserted framing as complete as that in hulls made with a pre-erected skeleton. So far as we can tell, the building of ships in this impressive fashion was peculiar to the Greco-Roman world, and the sole method it used for every form of vessel from seagoing freighters to lake skiffs.[8] Very likely it goes back at least to Homer's day,[9] and it was still employed well into the Byzantine period.[10] The technique of pinning a skin to a pre-erected skeleton, which was to produce the European craft of the great days of sail, came into widespread use no earlier than the Middle Ages.[11]

The Greek or Roman shipwright's first step was to set up the stocks[12] and lay the keel. He then added stempost and sternpost, probably scarfing them to the keel (Fig. 158), and, very likely, set up a number of temporary mold frames to guide him in giving the shell the proper lines.[13] His very next step was to begin on the planking, fixing the garboard strakes into a rabbet in the keel and edge-joining the succeeding planks with an appropriate number of mor-

---

deed wedge-shaped (van Doorninck 90), and some on the Grand Congloué were slightly wedge-shaped toward the tip (Frost 244, fig. 48.3).

[8] Every ancient Mediterranean wreck examined so far has turned out to be made with edge-joined planks (App. 1; see also the wreck off Fos described in *Gallia* 16, 1958, p. 37). Harbor boats (the Fiumicino barges) and a riverboat found in the Thames (the County Hall boat) were so made. Even three small boats found with the Nemi ships were so built (Ucelli 233-34), as well as a fishing skiff, 5.25 m. long, unearthed on the site of Rome's harbor (Testaguzza 132). The sole exceptions are two Thames riverboats, whose construction must reflect local practice (338-39 below).

[9] See App. 2.

[10] See note 37 below.

[11] Its earliest appearance is in certain local craft, dating to the 2nd and 3rd centuries A.D., found in the Thames; see 338-39 below. The earliest skeleton-first construction so far discovered in the Mediterranean area dates from the 11th century A.D.; see Alfieri, *op. cit.* (ONE, note 20) 206 and fig. 7 (riverboat found in the Po Valley, 10.50 m. long and 2.50 wide). O. Hasslöf, "Wrecks, Archives and Living Tradition," *MM* 49 (1963) 162-77, esp. 170-72, citing instances among northern European hulls of the past few centuries in which edge-joined shell construction and skeleton-first construction were combined, shows that the two are not mutually exclusive and suggests that the latter may have developed out of the former.

[12] Cf. App. 3, s.v. *stocks*, and see Fig. 163.

[13] The Persian Gulf Arabs today lay the keel, add stempost and sternpost, then the garboards and other bottom strakes, and then set up mold frames on the *outside* of the hull; see T. Johnstone and J. Muir, "Portuguese Influences on Shipbuilding in the Persian Gulf," *MM* 48 (1962) 58-63.

tises and tenons. Sometimes he planked up to the gunwales before turning to insert the frames;[14] sometimes he planked only a certain distance up the sides and then inserted frames, finishing off the planking later.[15] The strakes ran in thickness from 3.5 to 10 cm. (1 3/8″ to 4″) as a rule. In some cases they were astonishingly thin, as little as 2 cm., so that the mortises take up half the available thickness at least; on occasion certain strakes along the turn of the bilge were made double the thickness of their neighbors in order to project beyond the surface of the hull and serve as bilge keels.[16] A wide variety of widths appear in the same vessel, reflecting no doubt whatever could most economically be sawn from the available logs.[17] The garboard strakes were mortised to the keel rabbet (Figs. 161, 167), and often, to make up for the thinness of the wood, they and other bottom planks were doubled, the outer layer being spiked to the inner (Fig. 161).[18] For edge-joining strakes, tenons ca. 5 cm. (2″) broad were most common, but they could run as wide

[14] See Figs. 162-163; both shipwrights planked up to the gunwales before inserting frames.

[15] This is how the Yassi Ada Byzantine wreck was built: 16 strakes, up to the waterline just below the first wale, were laid (Fig. 167); then frames were inserted; then wales were bolted to the frames; then planking between the wales was nailed to the frames up to the gunwales, without edge-joints (van Doorninck 92-98). The Pantano Longarini wreck, roughly contemporary (A.D. 500-700), was constructed in the same fashion (Throckmorton-Kapitän 187).

[16] See App. 1. Ancient writers liked to talk of the scant thickness of wood that separated a sailor from death, only three fingers wide (Dio Chrysostom, Orat. 64.10) or four (Diogenes Laertius 1.103; Juvenal 12.58-59), at the most seven (Juvenal 12.59), i.e., between 2½″ and 5¼″. For examples of particularly thin strakes, see the wrecks of the Isle de Porquerelles and Monaco in App. 1.

On the Kyrenia wreck, eight strakes from the keel to the turn of the bilge are 7-8 cm. thick. The ninth is 14-16 cm. thick, just double the thickness of the others. The tenth is regular thickness, the eleventh again double thickness (Katzev).

[17] E.g., in Dramont A, the garboard strakes are 22 cm. wide while the other strakes run as narrow as 7 (Benoit 143).

[18] Other examples of the mortising of the garboards to the keel or keel rabbet are the Grand Congloué (Benoit pl. XXII) and Mahdia (Benoit 142, fig. 77 and 77bis = Taylor 50, fig. 10). The Mahdia wreck had double planking with bands of impregnated cloth between the layers (Taylor 48). Other examples of double planking are: Grand Congloué (Benoit pls. 21-23), Dramont A (Benoit pl. 28). The Punta Scaletta wreck was apparently entirely double planked (Lamboglia, "Punta Scaletta" pl. 3). The bow area of the Kyrenia wreck was double-planked with lead sheathing between the layers (Katzev).

as 10 cm. (4″); in depth they could penetrate halfway into the plank (Figs. 159-161). On seagoing ships the mortises seem never to have been more than ca. 25 cm. (10″) apart, most often they were but 5 or 10, and sometimes by being staggered so that one was nearer the outboard surface and the next nearer the inboard, there was hardly any space at all between them (Figs. 165-166).[19] The carpenter probably cut mortises into the upper edge of a plank already in place, and then clamped the plank to come next temporarily in place and marked where the corresponding half of each joint was to go.[20] The wood for the strakes, because of the many joints each had to take, was selected with great care, not too green and not too dry;[21] too green, on drying out, would leave the joints with too much play; too dry would involve the risk of splitting when the plank was knocked into place and the joints driven home. As a matter of fact, there is some evidence that the tenons were greased to make them

[19] For statistics on the tenons and their spacing, see App. 1.

[20] Cf. App. 2, note 4.

[21] Cf. Theophrastus, *Hist. Plant.* 5.7.4: ναυπηγικῇ δὲ διὰ τὴν κάμψιν ἐνικμοτέρᾳ ἀναγκαῖον · ἐπεὶ πρός γε τὴν κόλλησιν ἡ ξηροτέρα συμφέρει · ἵσταται γὰρ καινὰ τὰ ναυπηγούμενα καὶ ὅταν συμπαγῇ καθελκυσθέντα συμμύει καὶ στέγει, πλὴν ἐὰν μὴ παντάπασιν ἐξικμασθῇ · τότε δὲ οὐ δέχεται κόλλησιν ἢ οὐχ ὁμοίως "In shipbuilding, because bending is necessary, wood which is rather green must be used. Of course, where joining [*kollesis*] is involved, somewhat drier wood is in order. For new naval construction is let to stand; when it has set, it is launched and then it [sc. the wood in it] closes up and becomes watertight. If, however, it was bone dry [sc. to start with], in that case it will not take the joining, or will not take it so well." The term *kollesis*, translated "joining" above, is usually rendered here with its basic meaning of "gluing" (e.g., by A. Hort in the Loeb translation, 1916), but that hardly suits the context, for (1) glued seams are never used in ordinary shipbuilding, and (2) the use of glue runs directly counter to Theophrastus' statement that the planking, once in water, συμμύει "closes" (that is, by swelling), for a glued joint is, in the nature of the case, a fully closed joint from the outset. A good parallel is offered by Procopius, *Bell. Goth.* 4.22.10, where the closely related word *kollema* is used to mean "piecing together" with a scarf or lap joint. Reporting on a boat of hoary antiquity that was kept as a monument in Rome, he expresses astonishment that the planks were all of one piece (μονοειδῆ ξύμπαντα), that there was not one single instance of "*kollema* of planks," that is, piecing together (ξύλων δὲ κόλλημα οὐδὲ ἓν τὸ παράπαν ἐνταῦθά ἐστιν), nor were there any of broken lengths held together by some kind of iron fastening (οὐδὲ σιδήρων ἄλλη τινὶ μηχανῇ . . . ἐρήρεισται); since each strake was one unbroken length, the only iron fastenings in the vessel were those pinning ribs to planking (4.22.14: σανίς τε . . . ἑκάστη . . . κέντρα σιδηρᾶ τούτου ἕνεκα προσλαβοῦσα μόνον, ὅπως δὴ ταῖς δοκοῖς ἐναρμοσθεῖσα τὸν τοῖχον ποιῇ "Each . . . strake has . . . iron fastenings solely for this one purpose: that, pinned to the ribs, it form the wall [of the hull]").

slip easily into their slots.²² Once they were driven home, each was transfixed by a wooden peg to make sure it would never separate (Figs. 159-160, 165-166).²³

When the shell had been completed or nearly so, the shipwright turned to the insertion of frames (Figs. 162-163),²⁴ scoring lines on the inner surface of the shell to show where each was to go.²⁵ On seagoing craft he usually placed them not over 25 cm. (10″) apart, and most were often much closer (Figs. 157, 161, 164, 168, 190).²⁶

²² Cf. Plutarch, *Mor.* 321d: [a newly made ship] "must stand and set an appropriate length of time until the fastenings [*desmoi*] take hold and the treenails [*gomphoi*] knit. If it is launched while the joints [*harmoi*] are still damp and slippery, they will all loosen up" (στῆναι δεῖ καὶ παγῆναι σύμμετρον χρόνον ἕως οἵ τε δεσμοὶ κάτοχοι γένωνται καὶ συνήθειαν οἱ γόμφοι λάβωσιν· ἐὰν δὲ ὑγροῖς ἔτι καὶ περιολισθάνουσι τοῖς ἁρμοῖς κατασπασθῇ, πάντα χαλάσει).

²³ Cf. App. 1.

²⁴ The relief pictured in Fig. 163 decorated the tombstone of P. Longidienus, a *faber navalis* "shipwright," who died in Ravenna sometime in the late 2nd or early 3rd A.D. We see Longidienus using an adze to give a curve to a massive timber for a vessel whose hull is complete. Those who have dealt with this relief hitherto, unaware of the Greco-Roman way of building a boat, have been hard put to explain what is going on. H. Blümner, *Technologie und Terminologie der Gewerbe und Künste bei Griechen und Römern* (Leipzig 1875-86) II 341, misled by a faulty line-drawing of the scene, thought that Longidienus was making a ladder to serve as gangplank; E. Kornemann, *RE* s.v. *fabri* 1896 (1909), says that he is working on a plank—which makes little sense since all the planks are obviously in place; H. Gummerus, "Darstellungen aus dem Handwerk auf römischen Grab- und Votivsteinen," *JDI* 28 (1913) 63-126, esp. 92, offers the desperate solution that the piece is for a second ship that Longidienus is just beginning to construct. Longidienus is simply shaping a frame to insert in his completed shell of planks. As a matter of fact, the tips of a number which he has already put into place are clearly visible. Fig. 162, a photograph taken in 1929 in Sweden, where the shell-of-planks technique is still practiced (for clinker-built boats), shows a perfect parallel, a Swedish boatwright at precisely the same stage in construction as the Roman.

²⁵ Such lines have been found on the Chrétienne A wreck (Dumas 159) and the Yassi Ada Byzantine wreck (van Doorninck 92-93).

²⁶ On the terminology for frames, see App. 3; for the size and spacing, see App. 1.

In skeleton-first ships, e.g., those of the 17th century, the thickness of the frames and the "rooms"—the spaces between them—were equal; the framing was thus consistently stronger than that used in these edge-joined ancient hulls. On the other hand, the *Grace Dieu* of 1417, a huge clinker-built vessel ca. 140 feet long and therefore comparable to Roman ships in being a large craft built on the shell-of-planks technique, had framing much heavier even than the skeleton-first ships: her frames averaged 11″ in width and were set only 5″ apart; some were as thick as 19″ and sometimes there was no space at all between frames. See R. C. Anderson in *MM* 20 (1934) 158-70. Similarly, in the Woolwich ship of just a century later, originally clinker-built and hence with the framing that goes with this kind of con-

The spacing was very commonly far from regular (Fig. 164),[27] for reasons we can only guess at.[28] Some vessels had frames that were all one unit, a floor timber spanning the bottom of the vessel scarfed to futtocks (Figs. 161, 167);[29] other vessels mingled this type with frames that had no floors, descending on either side to rest by the keelson (Fig. 190).[30] The shipwright secured each frame to the planking with treenails, and then transfixed the treenails with bronze spikes (Fig. 161); on occasion spikes longer than the thickness to be spanned were used so that the part protruding through the inner face of the frame could be clenched.[31] The frames which included floors were sometimes pinned to the keel and keelson.[32] Over the frames, made fast to them by nails, was laid a lining (Figs. 157, 168).[33] A keelson rested on the keel, appropriately notched to pass

---

struction, the frames were 14″ in width and spaced 5″ apart; see W. Salisbury, "The Woolwich Ship," MM 47 (1961) 81-90, esp. 86.

[27] Particularly noticeable in the Chrétienne A wreck (see App. I).

[28] E.g., in the Yassi Ada Byzantine wreck, there are certain wide gaps to allow for passage of through beams, and, where particular strength was desirable, the frames were set closer than usual (van Doorninck 68, 80-81, 93-94).

[29] The Titan wreck (Taylor 88), the Nemi barges (Ucelli, Tables II, VI), the Yassi Ada Byzantine wreck (van Doorninck 94).

[30] Dramont A (Benoit 143; his pl. XXVIII.3 shows the arrangement clearly), Chrétienne A (Dumas 157), Port-Vendres (Chevalier 264; see plan there), Marseilles (personal examination). Probably this was the system used on the Grand Congloué as well; Benoit's reconstruction (fig. 75), as Dumas (157) points out, is unconvincing. The same mingling of frames with floors and frames without floors is found in the Arab craft of the Persian Gulf; see Johnstone-Muir, op. cit. (note 13 above) 60.

[31] Cf. also Throckmorton figs. 10 and 15 (Anticythera), Benoit pl. XXVIII.I (Dramont A). The spikes on the Grand Congloué wreck (Benoit 151) and the Nemi barges were clenched (Ucelli 152 and figs. 153, 154). The description of the building of Hiero's superfreighter may contain a reference to this treatment of the spikes; see NINE, App., note 31.

[32] In the Titan wreck, a thick treenail ran through the keelson and floor timber deep into the keel; see Taylor 88 and Fig. 161. In the Monaco wreck, a bronze bolt ran through keel, floor, and keelson, secured by riveting on top of the keelson; see Benoit fig. 79. In the Yassi Ada Byzantine wreck, iron bolts ending in a key that passed transversely through a slot in the tip held frames where particular strength was required (e.g., around the main hatch), whereas the others were pinned with iron nails (van Doorninck 94-95).

[33] A papyrus (see NINE, note 54) describes a Nile boat as "floored and decked," i.e., it was substantial enough to have at least the floor timbers covered by a lining.

over the floor timbers (Fig. 161).[34] On some ships, at the point where the mainmast was stepped, the keel had laid over it a bulky piece of timber with a socket hollowed out in it to receive the heel of the mast.[35] Below the keel was a false keel,[36] of resistant beech if it had to withstand the wear and tear of hauling on to shore or across shipways.

The details given above are based on remains dating from the fourth century B.C. to the fourth A.D. Wrecks of later date, the sixth and seventh A.D., reveal that by this time the painstaking workmanship attested heretofore had given way to methods more simple and materials more massive.[37] Tenons, instead of standing cheek by jowl,

---

[34] Other examples: Benoit pl. xxviii (Dramont A) and fig. 78 (Marseilles). See also the photograph of the Titan wreck published with his article "Une nouvelle étape de l'archéologie: Les recherches sous-marines," *Atti del Settimo Congresso Internazionale di Archeologia Classica* i (Rome 1961) 59-69, pl. 1.2. Cf. the following note.

[35] Dumas, photograph 7 and fig. 14 on p. 124. The massive timber of the Dramont A wreck, identified as a keelson by Benoit (143), is probably part of the mast-step complex; cf. Dumas 154-56. In the Chrétienne A wreck, the mast-step timber, averaging a breadth of 22-24 cm. and a height of 27 cm. (8 5/8- 9½″ x 10 5/8″) swells out to 48 cm. (19″) and stays this wide for 2.31 m. (ca. 7½′). In other words, under the mast the width was doubled.

An astonishingly similar method of stepping the mast is attested by the remains of an English ship of the early 16th century; see Salisbury, *loc. cit.*, note 26 above. The keelson, averaging in section 9″ x 13″, swells out under the mast to 12″ x 26″—a doubling of the width (Salisbury 85-86). As Salisbury points out (86), from the next century on, the keelson was a much heavier timber, more or less equal in section to the keel.

The interior of the mast-step of the Chrétienne A wreck (best seen in Dumas, photograph no. 11) is carved in an intricate fashion with one of the fore-and-aft surfaces carefully curved. Exactly the same arrangement is found in the mast-step of the Kyrenia wreck (Katzev). Dumas offers the cogent suggestion (156) that this type of step was for a retractable mast. In the Kyrenia wreck, the curved surface faces aft. Dumas (166-67) had found tiles and pottery at the north end of the Chrétienne A wreck and, since these most probably came from the ship's galley, had suggested that this end was the stern; since the curved surface of the mast-step faces this way, his suggestion is confirmed.

The slots that appear on either side of the mast-step of the Kyrenia wreck might very well be for the sides of a tabernacle.

The mast-step is preserved in the Port-Vendres wreck (Chevalier 265). It has an inclined plane, rather than a curved surface, on one of the fore-and-aft faces.

[36] Cf. App. 3, s.v. The false keel has been preserved on the Kyrenia wreck. It is on the *outside* of the lead sheathing: the latter was tacked in place over hull and keel, and then the false keel was added (Katzev).

[37] The Yassi Ada Byzantine and the Pantano Longarini wrecks; see App. 1 and

were here placed 30 to 45 cm. (11¾-17¾″) apart, while some were as much as 90 cm. (35½-39½″) apart. They were wedge-shaped and so carelessly fitted that a vast amount of play was left in each joint and pinning them with pegs was out of the question. Frames were heavier than they had previously been on boats a good deal larger, were separated by wider distances than had been usual earlier, and were fastened to the planking by means of iron nails alone.

With planks edge-joined by close-set mortises, caulking was hardly needed, and, on the wrecks examined so far, evidence of it has yet to be found.[38] When the joints were relatively widely spaced, as in the late wrecks just described, caulking was essential, and it was done with whatever was locally available.[39] Most hulls—but not all

cf. van Doorninck 90-92 and Throckmorton-Kapitän 186-87. The date of the first is fixed to the reign of Heraclius (A.D. 610-641) by 16 gold and over 50 copper coins found in the hull (van Doorninck 2). The second dates between A.D. 500 and 700 (cf. Throckmorton-Kapitän 185).

A statement by Procopius (*Bell. Pers.* 1.19.23) implies that by this time iron fastenings had replaced the mortise and tenon as the common way to edge-join planks. Remarking on the sewn boats characteristic of the Arabian Gulf, he says, "They are not made in the same manner as other ships. For they are not smeared with pitch or with any other substance whatsoever, and, furthermore, the planks are not pinned to each other with iron going right through them but are tied together with some sort of loops" (οὐ τρόπῳ τῷ αὐτῷ ᾧπερ αἱ ἄλλαι νῆες πεποίηνται. οὐδὲ γὰρ πίσσῃ οὐδὲ ἄλλῳ ὁτῳοῦν χρίονται, οὐ μὴν οὐδὲ σιδήρῳ διαμπερὲς ἰόντι ἐς ἀλλήλας αἱ σανίδες ξυμπεπήγασιν, ἀλλὰ βρόχοις τισὶ ξυνδέδενται). For another reference in Procopius to the use of iron fastenings in planks, see note 21 above. Those described in this passage may be some form of staple, such as is used today in East Pakistan for edge-joining planks (cf. Greenhill, *op. cit.* [note 2 above] 108-109, 115-16). The absence of pitch on Arab craft was noted also by Marco Polo (1 xix, trans. H. Yule): "The ships are not pitched, but are rubbed with fish-oil." They are still so treated; cf. R. Bowen, "Arab Dhows of Eastern Arabia," *The American Neptune* 9 (1949) 111.

[38] The absence of caulking has been noted in the Titan wreck (Taylor 88), the Nemi barges (Ucelli 152-53), the County Hall boat (Marsden, "County Hall" 111). It is implicit in Strabo's remark (4.195) about the shipbuilding practice of the Veneti on the Bay of Biscay, who "do not bring together the joints of the planks but leave gaps. These they caulk with seaweed" (οὐ συνάγουσι τὰς ἁρμονίας τῶν σανίδων, ἀλλ' ἀραιώματα καταλείπουσι. ταῦτα δὲ βρύοις διανάττουσι). On the other hand, the two local craft found in the Thames that were built by nailing planks to frames were both caulked (see FOURTEEN, note 56).

[39] Vergil mentions *stuppa* "tow" on Aeneas' ships (*Aen.* 5.681-82), and there was a guild of *stuppatores* "tow-men," i.e. caulkers, at the port of Rome (*CIL* XIV 44). Papyrus was used in Egypt (Herodotus 2.96, cited in Two, note 15), seaweed on

—were protected against marine borers by a sheath of lead sheets set over a layer of tarred fabric and held in place by multitudinous large-headed copper tacks (Fig. 184).[40]

Since the sea bottom has for the most part yielded only the floors of ancient wrecks, for the construction of the upper part of the hull we must turn to the far less conclusive evidence of ancient literature and representations. Sometimes all the deck beams[41] were brought through the sides (Fig. 151), a practice that is attested in Egypt from the third millennium B.C. on (Figs. 14, 18) and remained charac-teristic of Mediterranean shipbuilding up to at least Renaissance times;[42] sometimes only massive through-beams, inserted aft and amidships and forward to add to the lateral stiffening,[43] came through the sides (Figs. 144, 156, 163). Powerful waling pieces[44] girdled the

---

the Bay of Biscay (Strabo, cited in the previous note), and Pliny mentions the use of certain reeds "where they grow hard and retain a measure of viscosity" (*NH* 16.158, reading, with J. André, editor of the Budé edition, 1962, the ms. *ubi limo-siore induruit callo*; Pliny explains that these reeds, "pounded and inserted in the seams of ships, solidify the structure, being more tenacious than glue and, for filling cracks, more reliable than pitch" [*contusa et interiecta navium commissuris fer-ruminat textus, glutino tenacior, rimisque explendis fidelior pice*]). In comparing them to glue he must mean that they have a stickiness which keeps them well in place, not that they served to "glue" the seams together; glued seams have no need of caulking with pounded reeds.

[40] See App. 1. The Nemi barges were sheathed right up to the gunwales (Fig. 184), but ordinary merchant ships probably had only the underwater surface pro-tected, like the coppering on sailing ships of the 18th century and later. The fabric interlining on the Nemi ships was of wool (Ucelli 153), on the Grand Congloué wreck of linen (Benoit 152; cf. Frost 248). The description of Hiero's superfreighter mentions both a lead sheathing and fabric underlay (Athenaeus 5.207b; see NINE, App., pt. 4).

[41] Cf. App. 3, s.v. *beams.*

[42] For another clear instance, see *DS* s.v. *horeia*, fig. 3882, taken from a mosaic of the 5th A.D. found in Tunis. For the continuous history of the practice, see Throckmorton-Kapitän 187 (6th/7th A.D.), *IH* ill. 98 (14th century).

[43] In the Yassi Ada Byzantine wreck, wider spacing of frames at certain points (Fig. 164) reveals at least three such through-beams: two aft and one amidships. Probably there was another forward, but the bow area is missing. Cf. van Door-ninck 70-73. The ship in Alfieri (*op. cit.*, ONE, note 20) fig. 6 has two, one fore and one aft.

[44] Cf. App. 3. On the Yassi Ada Byzantine wreck, the wales were half-logs at least 10 cm. (4") thick, and were matched by a stringer of roughly the same size on the inside (van Doorninck 96-97). A bolt, secured by a washer and key passing

hull horizontally (Figs. 144, 146, 151, 156, 167), held in place by spikes or even long bolts that passed through the frames and the lining; matching these on the inside were heavy stringers. Bulwarks ran along the amidships section outside the gunwale, ending in the winglike projections (226-27 below) for housing the steering oars (Figs. 142, 144-146, 149; in Fig. 151 the projection runs over the butt ends of the lower deck beams). Massive cables were kept aboard for undergirding the ship during emergencies.[45]

It was usual to smear the seams or even the whole hull with pitch or with pitch and wax.[46] Ship's paint was encaustic, i.e., wax melted to a consistency that could be applied with a brush and to which color had been added.[47] The colors available, mostly mineral deriva-

---

transversely through a slot in the tip, ran through wale, frame, and stringer; see Fig. 167. For another example of waling pieces, see Throckmorton-Kapitän 187.

[45] The *hypozomata*. They were a highly important part of the gear on warships, and their precise function has given rise to much controversy; see 91-92 above. There seems to be little question that on merchant ships they were cables to pass under the hull. We know from St. Paul's narrative (Acts 27.17, cited in FIVE, note 73) that a number were carried aboard for use in emergencies to secure planking that had loosened and was letting in water. Smith (65-67) offers numerous illuminating modern parallels.

[46] *Anth. Pal.* 11.248: μιν ἅπασαν ἐπὶ ζυγὰ γομφωθεῖσαν / ἤλειφον πεύκης τῇ λιπαρῇ νοτίδι "And when it [the ship] had been pinned together up to the thwarts, they smeared it with the glistening sap of the pine"; Procopius 1.19.23, cited in note 37 above. Vergil's phrase *uncta carina* (*Aen.* 4.398) must refer to hulls so anointed. The Nemi barges seem to have had a coat of pitch, with perhaps some slight admixture of bitumen, plus some substance containing iron, possibly minium, as coloring matter (Ucelli 179-80). For pitch and wax, cf. Pliny, *NH* 16.56: *zopissam vocari derasam navibus maritimis picem cum cera* "The pitch with wax scraped off seagoing ships is called 'live pitch'"; Vegetius 4.44: *unctasque cera et pice et resina tabulas* "planks smeared with wax and pitch and resin." The hull of the Yassi Ada Byzantine wreck had been smeared with pitch, inside and out, up to the waterline (van Doorninck 100).

[47] Mention of covering a ship with wax may refer either to encaustic paint or to a protective coating, since examples of Egyptian material show that wax was used in both ways (Lucas, *op. cit.* Two, note 1, p. 352). Cf. Arrian, *Periplus* 5 (in the Black Sea, where naval supplies of all kinds were short, "even the wax was scraped off" [καὶ ὁ κηρὸς ἀπεξύσθη]); Valerius Flaccus 1.478-80 (the Argo had to venture where it would be impossible "to repair any wounds with pitch or soft wax" [*vel pice vel molli conducere vulnera cera*]); Lucian, *Dial. Mort.* 4 (κηρόν, ὡς ἐπιπλάσαι τοῦ σκαφιδίου τὰ ἀνεῳγότα "wax to pay any open seams on the skiff"); Vegetius 4.37 (*cera etiam, qua ungere solent naves, inficitur* "And the wax with which hulls are smeared, is also colored" [i.e., with a tint of the same shade as the sea to serve

tives, were purple, white, blue, yellow, brown, green, red;[48] the
brighter shades, such as red or blue or purple, were used for bow-
patches and decorative effects.[49] The hull could either be painted or
left the black color it took on from its coat of tar.[50]

The woods preferred for planking, frames, and keel were fir, cedar,
and pine,[51] but cypress, elm, alder saw service as well,[52] the choice

---

as camouflage; cf. ELEVEN, note 47]); Ovid, *Met.* 11.514-15, cited in note 7 above.
The classic description of painting with encaustic is given by Pliny, who refers
specifically to its use on ships (*NH* 35.149: *quae pictura navibus nec sole nec sale
ventisque corrumpitur* "This kind of painting on ships is not damaged by either
sun, wind, or salt water"). See Lucas 352-53 and the bibliography he cites in 352,
notes 13 and 14.

[48] These are the colors Pliny mentions (35.49), save brown, for which see the
following note. Craft used for scouting were painted to resemble the sea (cf. ELEVEN,
note 47), and so were pirate craft (Philostratus, *Imagines* 1.19.3: γλαυκοῖς μὲν
γέγραπται χρώμασι "[The pirate craft] is painted a blue-gray color").

[49] The original of Fig. 97 has a black hull, and the prominent lateral stripe is
dark red. The original of Fig. 154 has what appears to be a gilded stern gallery,
a goose-necked sternpost figure in alternate stripes of brown and white, a sea monster painted on the stern in light blue against a dark blue background, underneath
which is a panel in dark red with a bright red oculus within. The famous Etruscan
tomb-painting of a fishing scene shows a skiff with a dark blue bow-patch, and the
mural of a Tiber rivercraft has a dark brown hull with a dark blue oculus on the
sternpost. See *IH* ills. 62-64. Homer mentions red, purple, and blue bow-patches
(FOUR, note 18).

Wall-paintings in a building of the early 2nd A.D. near Rome's docks on the Tiber
depict a series of local small boats whose hulls are all gaily painted and some almost
completely covered with pictures and patterned devices done in a multitude of
colors; see G. Jacopi, "Scavi in prossimità del porto fluviale di S. Paolo," *Monumenti
Antichi* 39 (1943) 45-96 and pls. 3-12 (including three in color). They could well
be ceremonial craft in festive dress, though we cannot rule out the possibility that
the wall-painter's whim was responsible for the gorgeous decoration.

[50] Thus Homer calls his galleys black (FOUR, note 15). The hulls Procopius has
in mind in the passage cited in note 37 above were obviously the color of the pitch
with which they had been coated.

Pliny (*NH* 35.49) makes the statement that encaustic colors "are now [being
used] even for merchant ships" (*iam vero et onerariis navibus*); presumably he
means that, in the flourishing times he lived in, many more merchant shipowners
than ever before, not content with merely a pitched hull, went in for decorative
paint jobs.

[51] Theophrastus, *Hist. Plant.* 5.7.1: ἐλάτη μὲν οὖν καὶ πεύκη καὶ κέδρος, ὡς ἁπλῶς
εἰπεῖν, ναυπηγήσιμα. τὰς μὲν γὰρ τριήρεις καὶ τὰ μακρὰ πλοῖα ἐλάτινα ποιοῦσι διὰ κουφό-
τητα, τὰ δὲ στρογγύλα πεύκινα διὰ τὸ ἀσαπές· ἔνιοι δὲ καὶ τὰς τριήρεις διὰ τὸ μὴ εὐπορεῖν
ἐλάτης. οἱ δὲ κατὰ Συρίαν καὶ Φοινίκην ἐκ κέδρου· σπανίζουσι γὰρ καὶ πεύκης. οἱ δ᾽ ἐν
Κύπρῳ πίτυος· ταύτην γὰρ ἡ νῆσος ἔχει, καὶ δοκεῖ κρείττων εἶναι τῆς πεύκης "The shipbuilding timbers, in a word, are silver fir, fir, and cedar. For triremes and other

most likely being determined by what was available locally.[53] Oak
was freely used for frames and false keel, and particularly for tenons
and treenails, but apparently not too often for strakes.[54]

---

warcraft, silver fir is used because of its lightness; for merchant ships, fir because
of its resistance to rot. Some also use it for triremes because silver fir is not available. In Syria and Phoenicia, cedar is used because of the lack of fir. In Cyprus,
pine is used; this is found on the island, and it seems to be better than the local
fir."

For the keel, cf. Theophrastus 5.7.2: τὴν δὲ τρόπιν τριήρει μὲν δρυΐνην ἵνα ἀντέχῃ
πρὸς τὰς νεωλκίας, ταῖς δὲ ὁλκάσι πευκίνην—ὑποτιθέασι δ' ἔτι καὶ δρυΐνην ἐπὰν νεωλκῶσι
—ταῖς δ' ἐλάττοσιν ὀξυΐνην· καὶ ὅλως ἐκ τούτου τὸ χέλυσμα "[They make] the keels
of triremes out of oak to withstand hauling out, of merchantmen out of fir (but they
put on an underlayer of oak if these haul out), and of small craft out of beech; false
keels are totally of beech."

The remains of wrecks show a wide use of pine for strakes: Grand Congloué
(Benoit 149), Titan (Taylor 88), Marseilles (Gallia 16, 1958, p. 11), Yassi Ada
Byzantine (van Doorninck 100), Nemi (Ucelli 152), Kyrenia (Katzev). However,
for keel and frames, a variety of woods was used: Grand Congloué had keel and ribs
of pine (Benoit 150) but floors of oak (Benoit 149); Yassi Ada Byzantine had a keel
of cypress and frames of elm (van Doorninck 100); Albenga had frames of oak
(Taylor 54); Kyrenia had mostly frames of pine with a few selected frames forward
of beech (Katzev).

[52] Vegetius (4.34) mentions the use of cypress, Plato (Laws p. 705c) implies it,
and the Yassi Ada Byzantine wreck shows a keel, sternpost, through-beams, wales,
and other heavy timbers of cypress (van Doorninck 100) and the Pantano Longarini wreck wales, stringers and planking of cypress (Throckmorton-Kapitän
187). Elm was commonly used: the Mahdia wreck was made of elm (Taylor 51),
and so were the frames of the Yassi Ada Byzantine wreck (van Doorninck 100),
and the planking of the Anticythera wreck (Throckmorton 41). The use of alder
is attested by the way Roman poets use alnus, like abies "fir" and pinus "pine,"
by metonymy for a ship (e.g., Vergil, Geor. 2.451: torrentem undam levis innatat
alnus "the buoyant alder swims over the rushing waves"; Lucan 3.520: emeritas
repetunt navalibus alnos "They search the dockyards for ships [literally "alders"]
past service"). One of the Fiumicino barges had planking of larch (Testaguzza 132).

[53] E.g., all the woods used in the Nemi barges—pine, fir, oak—were certainly
local; cf. Ucelli 143. For woods used in local craft, see 339-41 below.

[54] Theophrastus points out that it cannot take salt water (5.4.3; ἐν δὲ τῇ θαλάττῃ
σήπεται "It rots in salt water"). It was used for floors in the Grand Congloué (Benoit
149), for frames in the Albenga (Taylor 54) and Pantano Longarini wrecks
(Throckmorton-Kapitän 187) and in three of the Fiumicino barges (Testaguzza
130-32), for the deck in the Nemi barges (Ucelli 159). Oak treenails have been
found in the Grand Congloué (Benoit 152), Monaco (Benoit 146), Anticythera
(Throckmorton 41), Nemi (Ucelli 152), and oak tenons in Yassi Ada Byzantine
(van Doorninck 100), Kyrenia (Katzev), and Anticythera (Throckmorton 41). The
Punta Scaletta wreck had oak planking (Lamboglia, "Punta Scaletta" 240), and so
did two of the Fiumicino barges (Testaguzza 130-31).

HULL CHARACTERISTICS OF ANCIENT WRECKS

Name of wreck, date, and reference	length (in meters)	mortises: breadth x depth (in cm.)	mortises: distance apart (in cm.)	keel: height x width (in cm.)	frames: general height x width (in cm.)	frames: distance apart (in cm.)	frames: arrangement	strakes: thickness (in cm.)	lead sheathing
Kyrenia 4th B.C. M. Katzev, by letter; cf. *Nat'l Geog.* 1970, 841-57	ca. 15	3.5 - 4.3 x 6.1	ca. 6 - 10	20 x 13	ca. 8 x 10	ca. 15	alternation of frames with floors and frames without	7 - 8	yes
Punta Scaletta 2nd B.C. Lamboglia, "Punta Scaletta," pls. 2, 3	35				? x 7.5 - 9	most 9 -15 some 21 - 24		6	
Grand Congloué 2nd B.C. Benoit 149-52; Frost 244	23	5 -7 x 6 - 8 pinned	8 - 10	17 x 12	9 - 10 x 7 - 9	10 - 14		3 - 6	yes
Spargi 2nd B.C. *CIAS* II 155-56	25 - 30	4.5 x 3 pinned	7		12 x 12	22		3.6	yes
Anticythera 1st B.C. Throckmorton 41, 43		8 -10 x 7 pinned	.5 - 2			25 - 30 on centers		7 - 9	yes
Mahdia 1st B.C. Benoit 142; Taylor 48; author's data	30	10 x ? pinned	16	29 x 23		19 - 21 on centers			yes
Albenga 1st B.C. *RSL* 1952, 202-209; Taylor 54	40	5 x 3	4 - 5		10 x 10 -12			4	yes

Titan 1st B.C. Taylor 87-89; Benoit 144, 150	25	8 × 10 pinned	2-10	28 × 26	10 × 9-11	15	floors joined to futtocks		yes
Chrétienne A 1st B.C. Dumas 111-12, 124, 135-37, 144, 153, 157; Frost 266	24 - 32	5.5 × 9 pinned	3.5	? × 9.5	10 × 8-10	in pairs: two 6.75 apart, then two 17.5 apart	mingling of frames with floors and without floors	4.5 - 6	no
Dramont A 1st B.C. Benoit 143-44; Taylor 98-99	20	5 × 7-8 pinned	8-10	22 × 20	7 × 12-16	15	alternation of frames with floors and frames without	7 (double-planked)	no
Nemi 1st A.D. Ucelli 153-55, 373-74	73 (beam 24) 71.3 (beam 20)	10 × 10 pinned	10	40 × 30	30 × 20	50	floors scarfed to futtocks	10	yes
Torre Sgarrata 2nd-3rd A.D. Throckmorton, "Torre Sgarrata"	30	12-14 × 8-10 pinned	3 - 7.5		8.5 × 15	8 - 18		6.5 - 7	probably no
Elba 3rd A.D. Archaeology 21 (1968) 219	18								
County Hall 3rd A.D. Marsden 111	18 - 21	12.5 × 6.25 pinned	15	16.5 × 21.5	16.5 × 11.5	14 - 42	floors joined to futtocks	5 - 7.5	no
Yassi Ada Roman 3rd-5th A.D. van Doorninck, by letter		7-9 × 5-5.5	ca. 25	22 × 12.2	12 × 12	12	alternation of frames with floors and frames without	garboards 5.3, others 4.2	
Fiumicino barge 3rd-5th A.D. Testaguzza 130, 136, and author's data	18.7 (beam 6.6)	10 × 6 pinned	53	16 × 16	15-17 × 7-8	30	mingling of floors with frames and without	4 - 7.5	no

(continued on next page)

Name of wreck, date, and reference	length (in meters)	mortises: breadth x depth (in cm.)	mortises: distance apart (in cm.)	keel: height x width (in cm.)	frames: general height x width (in cm.)	frames: distance apart (in cm.)	frames: arrangement	strakes: thickness (in cm.)	lead sheathing
Fiumicino barge 3rd-5th A.D. Testaguzza 131-32, 138, and author's data	16.6 (beam 6.4)	6 x 3.25 pinned	57		10-12 x 8-10	25	mingling of floors with frames and without	4	no
Marseilles 3rd-5th A.D. Benoit 145-46; author's data	17	7.5-9 x 6.5	6-8	25 x 35	floors 24 x 20, futtocks 24 x 13	15-16	alternation of frames with floors and frames without	garboards 7, others 5	
Monaco 3rd-4th A.D. Benoit 145-46, 150, 151		5 x ?		22 x 11	15 x 7			2	
Port-Vendres 4th A.D. Chevalier 264, and by letter		pinned		32 x 29	20-30 x 10-15	6-19; most are 13-15 apart			no
Pantano-Longarini 6-7th A.D. Throckmorton-Kapitän 186-87	30-40	not pinned							
Yassi Ada Byzantine 7th A.D. van Doorninck 84, 87, 90, 93, 94	19-22 (beam 5.2)	5 x 3.5 not pinned	30-90 amidships 30-45 sternwards	35.5 x 22	14 x 14	21	floors joined to futtocks	3.5	no
Ile de Porquerelles undated Dumas, by letter			4.5					2	

APPENDIX 2

ODYSSEUS' BOAT (*Od.* 5.244-57)

WHEN Odysseus at long last received celestial permission to leave Ogygia, Calypso helped him get started on building some sort of craft (σχεδία sc. ναῦς literally "improvised boat") by taking him to a stand of selected trees and giving him tools: ax, adze, and drill. He felled twenty trees (244); adzed them into planks (245); "bored them all and fitted them to each other" (247); "hammered it [sc. the craft] with pegs and joints" (248), making his craft as wide "as a good shipwright will lay out the lines for the bottom of a beamy merchantman" (249-51); then "worked away setting up decks by fastening them to close-set frames" (252-53); and "finished up with the long pieces" (253). This completed the hull; he then turned to the fittings: mast and yard, rudder, latticed bulwarks, "plenty of brush" (254-57).

Though *schedia*, like its Latin synonym *ratis*, is frequently used in poetry in the sense of "boat," many translators, particularly the earlier ones, have taken it here to mean a raft.[1] This damns their rendering from the start, since rafts are made of logs, whereas Odysseus took the trouble to adze his logs into planks, and, more important, rafts, being simple platforms, do not have frames (ribs) or decks, whereas Odysseus provided his craft with both.

Of those who realized that Odysseus was making a boat and not a raft, some made no real effort to understand what was going on,[2] while those who did were thwarted by lines 248 and 252: what could Homer mean by "hammer the craft with pegs and joints"? And why bring in the frames after the craft has been hammered together?[3] These have been

[1] E.g., Pope, Butcher and Lang, Palmer, Bryant, Mackail; among more recent translators, A. T. Murray in the Loeb Classical Library version (1919), T. E. Shaw (1932), S. O. Andrew (1948), F. L. Lucas (1948), R. Lattimore (1967), A. Cook (1967). Although we have no ancient translation to aid us, we have the next best thing, an imitation of the passage by Nonnus in his *Dionysiaca* (40.446-62). He clearly takes the whole as the description of the building of a boat, although he obviously was ignorant of the meaning of a number of Homer's terms (e.g., in lines 459-62 he transforms 5.256-57 into a procedure for caulking the hull).

[2] W. Rouse (1937, reprinted as a paperback by Mentor Books) tosses about such terms as "spars," "a bluff ketch," and "copings," which simply make no sense. Berard in the Budé edition (1924) solved things for himself by condemning the last half of line 247 and the first half of 248.

[3] E. V. Rieu (Penguin Books, 1946) assumes that Odysseus crisscrossed the planks to make a floor of double thickness, cutting in lapped joints where two planks

puzzles only because we think in terms of traditional Western methods of boatbuilding. If we assume that Odysseus put together his craft by starting with a shell of planks, each made fast by pinned mortises and tenons to its neighbors, and then inserted framing, everything falls into place. After adzing his logs into planks, he "bored them all" (247: τέτρηνεν δ᾽ ἄρα πάντα), i.e., drilled first into the two edges to make the slots for the mortises and tenons (carpenters traditionally rough out such slots with the drill and finish off with the chisel) and then across the slots for the dowels that would transfix each joint; then he "fitted them to each other" (247: καὶ ἥμοσεν ἀλλήλοισιν), i.e., matched the planks to make sure every mortise coincided with its tenon; then he "hammered the craft with pegs and joints" (248: γόμφοισιν δ᾽ ἄρα τήν γε καὶ ἁρμονί-ῃσιν ἄρασσεν), i.e., knocked the planks together, thereby driving the tenons into the mortises, and hammered the transfixing dowels into their holes;[4] he kept building the shell up, plank upon plank, until it had the ample lines a good shipwright will lay out for "the bottom of a beamy merchantman"; then he "worked away setting up decks by fastening them to close-set frames" (252-53: ἴκρια δὲ στήσας, ἀραρὼν θαμέσι σταμίνεσσιν, / ποίει), i.e., inserted a strong set of frames into his shell and fastened the decking (no doubt just at prow and stern) to them; then he "finished up with the long pieces" (253: ἀτὰρ μακρῇσιν ἐπηγκενί-δεσσι, τελεύτα), i.e., put on the gunwales,[5] and the hull was complete.[6]

---

crossed and transfixing each crossing with a dowel—which is not only an unusual way to build a boat but a far cry from what Homer says. I mistranslated the passage myself (*AM* 39) by, among other things, taking the "joints" to refer to battens.

[4] This is precisely what Homer's great editor took the lines to mean: ὁ δὲ Ἀρίσταρ-χός φησι διὰ τοῦ πρώτου τὸ μὲν τέλειον τῆς ἁρμογῆς μὴ εἶναι, ἀλλ᾽, ὡς ἄν τις εἴποι, ἁρμόζοντα κατεσκεύασε, καὶ πρὸς ἄλληλα συγκαταγαγὼν ἐσκέψατο εἰ ἁρμόζει ἀλλήλοις. τῷ δὲ ἐξῆς συνέκλεισε καὶ κατεγόμφωσε. διὰ γὰρ τοῦ ἄρασσε τὸ τέλος τῆς ἁρμογῆς παρέστησε "Aristarchus says that the words of the first line [247] indicate that there was no completion of the joinery but that he, so to speak, prepared the pieces that were to join and, lining them up together, gauged whether they would fit into each other. In the next line, he locked them together and doweled them. The words 'knocked them home' indicate completion of the joinery" (schol. to lines 247, 248).

[5] That the μακραὶ ἐπηγκενίδες are gunwales is only a conjecture, but it is hard to see what else they can be. Eustathius made the same conjecture (p. 1533, 41: ἔστι δὲ ἐπηγκενὶς . . . καθ᾽ ἣν οἱ σκαλμοὶ πήγνυνται "the epenkenis . . . is [the plank] in which the tholepins are fixed"). The term certainly meant "gunwale" in Byzantine times; see SEVEN, note 49. Morrison (*GOS* 48) takes the *stamines* and *epen-kenides* of lines 252-53 to refer to the uprights and railings about the stern pictured on Geometric vases.

[6] Some of the commentators were on the right track. As long ago as 1876, Merry and Riddell (*Homer's Odyssey* I, pp. 226, 538) suggested that Odysseus had fastened

One final minor crux. As the last step, before turning to sails and rigging, Odysseus "piled on plenty of brush" (257: πολλὴν δ' ἐπεχεύατο ὕλην). The scholiast suggests the brush was ballast, but ballast surely would be sand or stone. Odysseus must have strewn a layer of twigs or branches over the bottom, the Homeric Age's equivalent of a duckboard, to keep his feet out of the water that inevitably collects in the bilge of a wooden boat. Just such brush has been found under the cargo of a wreck dating from the thirteenth century B.C.[7]

---

plank to plank with mortises and tenons, and they offered a translation which is substantially correct (p. 538). In 1926, F. Brewster ("The Raft of Odysseus," *HSCP* 37, 49-55) drew a parallel between Odysseus' technique and that used today in certain parts of Egypt. Though incorrect in a few details, basically Brewster was right: both the ancient Egyptians and ancient Greeks started with a shell of planks. W. B. Stanford, author of a recent commentary (1961), is unaware of the contradiction involved in extending approval to Brewster and, at the same time, quoting in full Rouse's translation (cf. note 2 above).

[7] See Bass, *Gelidonya* 49.

APPENDIX 3

THE TERMINOLOGY FOR STRUCTURAL MEMBERS OF THE HULL

Beams. In the open galleys of early times, the thwarts on which the rowers sat served as the beams. In Greek these were called *zyga*: see *Od.* 9.99, 13.21, and cf. Four, note 19; Sophocles, *Ajax* 249 (θοὸν εἰρεσίας ζυγὸν ἑζόμενον [as oarsman,] "seated on the swift bench of rowing"); Theognis, cited under *frames* below. In Latin they were called *transtra* (e.g., Lucan 10.494-95 [cited in Eleven, App. 1, note 13]; Vergil, *Aen.* 4.573). The same terms were applied to the beams of sailing ships; see *Septuagint, Macc.* 3.4.10, cited in Nine, note 54; Caesar, *Bell. Gall.* 3.13, cited in Fourteen, note 58. Moschion uses the general term *tropoi* (cf. *LSJ*, s.v., vii) for the beams on Hiero's superfreighter (Athenaeus 5.208a; cf. Nine, App. pt. 4), and Procopius (cited in note 21 above) the general term *dokoi*.

Writers of the fifth century and earlier use the term *selmata* of the thwarts. The clearest instance is Aeschylus, *Pers.* 358-59, where a messenger refers to the Greeks "rushing to the *selmata* of their ships" (σέλμασιν ναῶν ἐπανθορόντες) to flee to safety (Morrison, *GOS* 151, forces the sense in trying to make it mean poop deck here). And Euripides uses *selma*, like *zygon*, of the beam of a sailing ship; see *Cyclops* 505-506: ὁλκὰς ὥς, γεμισθεὶς / ποτὶ σέλμα γαστρὸς ἄκρας "like a merchantman, loaded to the beam at the top of its belly" (not "wale," as Torr [41] suggests; Euripides is picturing the ship from the inside, its capacious hold filled to the crossbeams that ran from gunwale to gunwale). The word is also used of the planks of the decking at stem and stern; the clearest instance is Euripides, *Helen* 1563-66 which tells of the loading of a bull onto the *selmata* at the bow (1566: ταῦρον φέροντές τ᾽ εἰσέθεντο σέλματα [sc. of the prow; cf. line 1563]), which can only be the deck planks there. Aeschylus, *Ag.* 1442-43 may possibly refer to the planks of the poop deck, but the text is too unsure for any hard and fast conclusions; see the extensive note *ad loc.* in E. Fraenkel's commentary (Oxford 1950). However, in Philostratus, *Vita Ap* 3.35 the *selmata* clearly do mean the deck planks aft (the Egyptians, creating a special type of merchantman for the India trade, added "numerous cabins of the kind that set upon the stern decking" [πλείους οἰκίας, οἵας ἐπὶ τῶν σελμάτων;

220

on such cabins, see 179-80 above]). For a similar use of *zygon* to mean "poop," see *GOS* 197.

Cutwater. *steira* in Homeric Greek (*Il.* 1.482 = *Od.* 2.428), *stereoma* in later Greek (Theophrastus cited in Five, note 45).

False Keel. Called in Greek *chelysma*, literally "turtle shell," since it acts as a sort of carapace for the keel. The definition is given by Pollux (1.86): τὸ δ' ὑπὸ τὴν τρόπιν τελευταῖον προσηλούμενον, τοῦ μὴ τρίβεσθαι τὴν τρόπιν, χέλυσμα καλεῖται "That which is nailed to the underside of the keel as a finishing piece so that the keel will not get rubbed is called the false keel [*chelysma*]." This is supported by Theophrastus' statement, in one place that the *chelysma* was all of beech to withstand hauling out (5.7.2, cited in Ten, note 51), and, in another, that beech was placed "under the keel" (5.8.3: "[Beech grows so tall in Italy that] it runs in unbroken lengths under the keel on Tyrrhenian ships" [ὥστ' εἶναι διανεκῶς τῶν Τυῤῥηνίδων ὑπὸ τὴν τρόπιν]).

Frames. Both Greek and Latin, using the same figure of speech as English, could refer to frames as "ribs"; see, for Greek, Theognis 513 (where he talks of stowing things νηός τοι πλευρῇσιν ὑπὸ ζυγά "by the vessel's ribs [*pleurai*], under the thwarts"), and, for Latin, Pliny, *NH* 13.63, cited below. More technical terms were *stamines* in Greek (e.g., *Od.* 5.252, cited in App. 2 above; Athenaeus 5.207b, cited in Nine, App., pt. 4) and *statumina* in Latin (Caesar, *Bell. Civ.* 1.54, cited in One, note 12, referring to the light ribs of British coracles), both of which contain the root *sta-* "stand." Theophrastus speaks at one point of the *enkoilia*, literally "pieces that go into the hollow" (*Hist. Plant.* 4.2.8: ἡ δὲ μέλαινα [ἄκανθα] ἰσχυροτέρα τε καὶ ἄσηπτος, δι' ὃ καὶ ἐν ταῖς ναυπηγίαις χρῶνται πρὸς τὰ ἐγκοίλια αὐτῇ "but black [acacia] is stronger [*sc.* than the white] and resistant to rot, so they use it in shipbuilding for the *enkoilia*"), which Pliny translates "ribs" (*NH* 13.63: *dumtaxat nigra, quoniam incorrupta etiam in aquis durat, ob id utilissima navium costis* "But black acacia, since it lasts without rotting even in water, is for this reason the most useful for the ribs [*costae*] of ships"), and a passage in Athenaeus (5.206f; cf. Nine, App., pt. 4 and note 26) makes it clear that a big ship had both *stamines* and *enkoilia*. So, the first, the "standing ribs," must be the part of the frame that stood along the vessel's side, the top and middle futtocks in our terminology, while the second, the "ribs in the hollow" would be the part that fitted into the curve of the bilge, the ground futtock.

221

The floor timbers, the part of the frame that spanned the floor of the ship, were possibly called *metrai*. This word, meaning basically "womb," had numerous derived senses, one of which was "bolt of a lock"; see *Corp. Gloss. Lat.* II 173.45: *repagulum* . . . μήτρα θύρας; *BGU* 1028.19-20 (2nd A.D.): ἤλων εἰς . . . θυρώματ(α) καὶ μήτ(ρας) χελωνίων τῶν πλαγίω(ν) "nails for . . . the doorways and bolts of the locks set sideways" [i.e., vertically along the door to fix into a hole in the threshold, like a modern door-check?]). In other words, the term was applied to the long narrow strip that ran from door to jamb or from door to adjoining door. A compound of amphi- and *metra, amphimetrion*, means "ship's floor" (Pollux 1.87; Dain 5.2.8; Hesychius, s.v.), from which we may infer that *metrai* came to mean the "long narrow pieces spanning a ship's bottom" or floor timbers. This sense suits very well a passage in Strabo (15.691) describing the boats of India and Ceylon. He says that they are not only poorly rigged but "made, on both sides, without bilge spanning pieces [*metrai*]" (κατεσκευασμένας δὲ ἀμφοτέρωθεν ἐγκοιλίων μητρῶν χωρίς), i.e., without floors or futtocks—precisely the way the traditional craft of the area are still made, for they are of sewn planks with so little framing that they can be taken apart for resewing yearly (Hornell 236). It would not be surprising for Strabo to single out for mention this feature so strikingly at variance with the multiple-framed Mediterranean craft familiar to him.

The terminology used in the Mediterranean in later ages bears out my interpretation of *stamines* and *metra*. Italian shipwrights of the sixteenth century called the futtocks *stamenali* and the floor timber *matera* (Jal, *Gloss. naut.* s. vv.); the first is beyond any doubt a derivative of *stamines*, and the second could well come from *metra*.

Herodotus uses *nomeis* literally "regulators," of the withes that formed the ribs of skin-covered coracles (1.194.2, cited in ONE, note 10) and in referring to the absence of ribs in Egyptian craft (2.96, cited in TWO, note 15), but, save for a deliberate reference to Herodotus' usage in Procopius (*Bell. Goth.* 4.22.12), the term appears nowhere else.

KEEL. *tropis* in Greek (e.g., see Pollux cited under *false keel* above), *carina* in Latin (e.g., see Caesar cited in ONE, note 12).

MORTISE and TENONS. *subscus* in Latin (see Pacuvius cited in ONE, note 26); cf. Ovid's use of *cuneus* "wedge" in the passage cited in TEN, note 7. Greek apparently was content with nontechnical terms such as *harmoi, harmoniai, desmoi* (cf. the passages cited in TEN, notes 21, 22, and *Od.* 5.248 cited in App. 2).

Ribs. See *frames*.

Stocks. In Greek *dryochoi* "oak-holders," since galley keels were traditionally of oak (Ten, note 51). Homer describes the row of axes through which Odysseus made his celebrated shot as being set up "like stocks" (*Od.* 19.574); for a convincing reconstruction, see A. Wace and F. Stubbings, *A Companion to Homer* (London 1962) 534-35. The expression ἐκ δρυόχων *aut sim.* "from the stocks," used literally in, e.g., Polybius 1.38.5 (αὖθις ἔγνωσαν ἐκ δρυόχων εἴκοσι καὶ διακόσια ναυπηγεῖσθαι σκάφη "They decided to build afresh 220 vessels from the stocks up"), could also mean figuratively "from scratch" (e.g., Plato, *Timaeus* 81b). Procopius, *Bell. Goth.* 4.22.12, mistakenly equates *dryochoi* with *nomeis*, which are ribs (see under *frames* above), and Torr (39-40) repeats the error.

Strakes. *sanides* in Greek (see Procopius cited in Ten, note 21), *tabulae* in Latin (see Vegetius cited in Ten, note 46). Both mean basically "planks," "boards."

Thwarts. See *beams*.

Treenails. *gomphoi* in Greek (see the passages cited in Ten, note 22 and *Od.* 5.248 cited in App. 2), possibly *pali* in Latin (see Plautus, *Men.* 403-404 and cf. the remarks of W. Sedgewick, "De re navali quaestiunculae duae," *Mnemosyne* 4, 1951, pp. 160-62).

Wales. *zosteres* "girdles" in Greek. Cf. Heliodorus, *Aeth.* 1.1.2: τὸ γὰρ ἄχθος ἄχρι καὶ ἐπὶ τρίτου ζωστῆρος τῆς νεὼς τὸ ὕδωρ ἀνέθλιβεν "Its load pressed the ship down in the water right up to its third wale." Torr mistakenly suggests *selma* as another term; see under *beams* above.

Rudder, Rigging, Miscellaneous Equipment

I RUDDER

LIKE THE VIKINGS, the ancient Mediterranean seaman knew only the side rudder; unlike them, he insisted on using one to port as well as starboard. The side rudder has often, through ignorance, been condemned as inefficient. Quite the contrary: it is not a whit inferior in performance to the stern rudder (which replaced it only by offering advantages of another kind).[1]

The ancient version consisted of an oversize oar hung in slanting fashion on each quarter in such a manner that it was able to pivot (Figs. 128, 170). It was operated by a tiller bar socketed into the upper part of the loom (Figs. 146, 179).[2] Pushing or pulling the tiller

---

[1] See O. Crumlin-Pedersen, "Two Danish Side Rudders," *MM* 52 (1966) 251-61, esp. 256-58, where the author describes the remarkable performance of a modern replica of a Viking ship (the Ladby ship, 20.6 m. long) steered by a nearly exact replica of a Viking side rudder (the Vorsa rudder): "It was possible to hold the helm in position with a single finger in spite of energetic efforts on the part of the rowers to turn the ship the other way" (257). Thus Lucian hardly exaggerates when he says of the *Isis* (*Navig.* 6; cf. NINE, App., pt. 2) "And all that [the enormous ship and its cargo] one tiny little old man just now brought through safely, pivoting those huge steering oars with a fragile steering bar" (κἀκεῖνα πάντα μικρός τις ἀνθρωπίσκος γέρων ἤδη ἔσῳζεν ὑπὸ λεπτῇ κάμακι τὰ τηλικαῦτα πηδάλια περιστρέφων.) Probably the most glaring example of ignorance of the true capacities of the side rudder is the book Lefebvre des Noëttes devoted to it (see NINE, note 22).

[2] The steering oar was called πηδάλιον in Greek and *gubernaculum* in Latin (although both terms could be used in the more generalized sense of "helm"). The blade was πτέρυξ literally "wing" (Apollonius Rhodius 4.929: a goddess rose from the deep and ὄπιθε πτέρυγος θίγε πηδαλίοιο "seized the blade [*pteryx*] of the steering oar from behind"; *IG* ii² 1607.74-75: [πηδάλια δύο, τοῦ] ἑτέρο(υ) ἡ πτέρυξ ἀδόκι[μος] "two steering oars, one with blade not fit for use"). The loom was αὐχήν literally "neck" (see Polyaenus cited in note 7 below). The tiller bar was in Greek κάμαξ (see Lucian, cited in note 1) or, more frequently, οἴαξ (Aristotle, *Mech.* 850 b: διὰ τί τὸ πηδάλιον, μικρὸν ὂν καὶ ἐπ᾽ ἐσχάτῳ τῷ πλοίῳ, τοσαύτην δύναμιν ἔχει ὥστε ὑπὸ μικροῦ οἴακος καὶ ἑνὸς ἀνθρώπου δυνάμεως . . . μεγάλα κινεῖσθαι μεγέθη πλοίων; "How is it that the steering oar, so small and set at the rearmost part of a vessel, yet has so much power that a ship's vast bulk can be moved by the power of one man

bar made the loom pivot within its fastenings, and this put the blade at an angle to the hull and thereby directed the ship.[3] The loom formed the turning axis, and the proper distribution of the amount of blade forward of and abaft this turning axis was of crucial importance.[4] In some representations (Figs. 105, 108) an unbalanced division is clearly apparent; in most, the turning axis seems to bisect the blade, but this may be because artists were unaware of such niceties or disregarded them for aesthetic effect. There were, in addition, other asymmetries that were important for effective performance.[5]

using a little tiller bar [*oiax*]?"; Vitruvius 10.3.5: *navis onerariae maximae gubernator, ansam gubernaculi tenens qui* οἴαξ *a Graecis appellatur* "the helmsman of a huge merchant ship, holding the handle of the steering oar, called the *oiax* by the Greeks"). In Latin it was *clavus* (see Ammianus cited in note 10, and Jerome cited in note 19 below). All three of these terms for tiller were also used in the more generalized sense of "helm." On the possibility of a single tiller yoking the two oars, see note 14 below.

[3] Thus Seneca (*Epist.* 90.24) compares the action of a steering oar to the tail of a fish.

[4] Cf. Crumlin-Pedersen 255-56: in the Vorsa rudder, the larger area is forward of the turning axis; in more modern forms of rudder, abaft it. Possibly this is why ancient helmsmen personally supervised the making of their steering oars (Plato, *Crat.* 390d: τέκτονος μὲν ἄρα ἔργον ἐστὶν ποιῆσαι πηδάλιον ἐπιστατοῦντος κυβερνήτου, εἰ μέλλει καλὸν εἶναι τὸ πηδάλιον "So, is it not the work of a carpenter to make a steering oar under the supervision of a helmsman, if the steering oar is to be a good one?"). The asymmetry of ancient steering oars was observed long ago (cf. e.g., Admiral Serre, *Les marines de guerre* II, Paris 1891, p. 356) but never correctly explained.

[5] Numerous representations of rudders show a heel-like extension at the bottom of the after edge (Figs. 108-110, 114), and the Vorsa rudder has the identical feature (Crumlin-Pedersen fig. 2). Other asymmetries (cf. Crumlin-Pedersen 253-55) may have existed but could hardly be reproduced in the small-scale reliefs or two-dimensional paintings and mosaics that constitute the bulk of our pictorial evidence.

Morrison argues (*GOS* 291-92) that the steering oars moved laterally as well as pivoting about their axis. The demonstrably asymmetrical construction of ancient rudders makes sense only if they pivoted; moreover, a steering oar fixed in brackets as in Fig. 170 could only have pivoted. He interprets Plato, *Alc.* 117c (τὸν οἴακα εἴσω ἄγειν ἢ ἔξω "to move the tiller in or out") as meaning that the oar was thereby pushed and pulled laterally, but Plato's words could just as well mean that the tiller bar was pushed outward in an arc toward the gunwale and pulled inward in a reverse arc toward the helmsman. He takes Polyaenus 5.43 (Calliades, when overtaken by a swifter enemy galley, "again and again eased [ἔσχαζε, literally "let go"] the helm on whatever side the enemy sought to attack so that the pursuer, hitting the cheek of his outrigger against the steering oar, could not strike, since his ram was against the sternmost thranite rowers" [τὸ πηδάλιον ἔσχαζε συχνῶς, καθ' ὁπότερον ἂν ἐμβάλλειν μέλλοι, ἵνα ὁ διώκων προσκρούων ταῖς ἐπωτίσι πρὸς τὸ πηδάλιον ἐμ-

On warships, when there was no outrigger or oar-box, the steering oars could be made fast by means of lines or thongs[6] directly to the hull; when there was an outrigger or oar-box (Figs. 108, 114, 129, 170) or a similar longitudinal projection (Figs. 122-123, 131, 133), these were extended to form a housing for the steering oars, which emerged from the after face.[7] Since a war galley when not in use had to be drawn up on a beach or into a slip, the blades were fitted with pennants by which they could be hoisted safely above the line of the keel (Fig. 108).[8]

The steering oars of sailing ships, far heavier and called upon to withstand stronger pressures, were much more solidly fixed. On all seagoing merchantmen, the side planking was extended into power-

βαλεῖν μὴ δύνηται τῷ τὴν ἐμβολὴν εἶναι κατὰ τὰς πρώτας θρανίτιδας]) to mean that Calliades let the steering oar on the threatened side swing up to its horizontal position of rest (as in Fig. 108) and thereby fended the attacker off. But the single blade of an oar could hardly unaided hold off the massive bulk of an *epotis*. Tarn ("Warship" 140) is surely right in suggesting it was some trick of steering. Calliades was perhaps constantly working the tiller to steer an evasion pattern in which he went off at an angle to his pursuer, deliberately offering his quarters as a tempting target. Then, at just the right moment, he must have eased the helm to put his ship on an exactly parallel course so that the menacing ram was no longer pointing toward his hull but toward the blade of the stroke oar. The total effect would be to get the enemy to go in for frequent spurting (cf. TWELVE, note 37) and thereby wear him out.

[6] Orpheus, *Argonautica* 276-77: ἐπὶ δ᾽ αὖτ᾽ οἴηκας ἔδησαν / πρυμνόθεν ἀρτήσαντες ἐπεσφίγξαντο δ᾽ ἱμᾶσιν "And then they attached the rudders, hanging them from the stern, and bound them tightly with thongs." Vegetius tells (*Re mil.* 4.46) how, in the height of an engagement, well-trained hands can sneak up in small boats to an enemy's stern and "without being observed cut the lines with which the enemy's steering oars are made fast" (*secreto incidunt funes quibus adversariorum ligata sunt gubernacula*).

[7] Polyaenus (3.11.14) reports that Chabrias (an Athenian commander of the 4th B.C.) introduced the use of supplementary steering oars forward of the regular set for stormy weather, viz., a second pair which "he passed through the outrigger on each side, alongside the thranite oars, with their looms and tiller bars reaching above the deck; as a result, even when the stern rose clear of the water, by means of these the ship could still be steered" (θάτερα διὰ τῆς παρεξειρεσίας κατὰ τὰς θρανίτιδας κώπας παρετίθει, τοὺς αὐχένας ἔχοντα καὶ τοὺς οἴακας ὑπὲρ τοῦ καταστρώματος ὥστε ἐξαιρομένης τῆς πρύμνης τούτοις τὴν ναῦν κατευθύνεσθαι).

[8] Euripides, describing how a penteconter was made ready for sea, after mention of the setting of sail and the running out of the oars, adds (*Helen* 1536) πηδάλιά τε ζεύγλαισι παρακαθίετο "And the steering oars were let down alongside [the quarters] by the pennants." For other instances of the "letting down" of the helm on galleys, see Germanicus, *Aratea* 355; Avienus, *Aratea* 767.

ful winglike projections on the quarters. These formed the housing for the steering oars (Figs. 142, 144, 146-147, 149, 151, 154, 156); well protected within were the fittings inside which the oars pivoted.[9] On large craft, whose steersmen were stationed high on the poop deck, the tiller bar had to run upward and inward (clearest in Figs. 144, 146),[10] being socketed at an appropriate angle through the head of the loom.[11] Both rudders were almost always operated by one man.[12] On big ships the distance from his station over the centerline to the gunwale was too great to span with a single pole (e.g., on the *Isis* it would have meant tiller bars 20 feet long; cf. NINE, App., pt. 2). The problem was solved by using a tiller bar that ran athwartship for a certain length and ended in a sort of universal joint. This connected with a lighter extension that continued, more or less in a fore-and-aft direction, to the helmsman, who gripped both bars as one does the bannisters of a narrow stairway (clearest in Fig. 154; also visible in Fig. 179, where the change

[9] In the representations, the housing screens what lies behind, so we can only guess at the precise way the oars were fastened. A rather conventionalized decorative stone model shows the oars pivoting in a slanted sleeve in the housing; see J. Le Gall, "Un 'modèle réduit' de navire marchand romain," *Mélanges Charles Picard* (Paris 1949) 607-17, esp. 613-16. In Fig. 170—perhaps only an artist's conception of what a Homeric ship looked like—they are fixed firmly in a pair of brackets. A finely done relief of a vessel of the 14th A.D. (*IH* ill. 98), though of a different age, belongs to the same tradition and is most instructive. Here the stern gallery embraces the upper part of the loom at gunwale level and a wooden collar projecting from the hull embraces it just above the blade. Thus, as in Fig. 170, the only motion possible for the oar is pivoting within these two collars (or sliding up and down within them; cf. note 17 below). A papyrus document (see note 135 below) refers to the "brackets" (ὄγκοι) of the steering oars; they may have been collars such as these.

Seagoing ships in some places (e.g., the Celebes Islands) still use side rudders and, interestingly, on these a winglike housing still appears; see *IH* ill. 196 and Hornell, pl. 34a.

[10] Since the bar slanted down markedly from helmsman to gunwale, the ancients could speak not only of "turning" the tiller, but of "raising" and "lowering" it (cf. Ammianus 21.13.10: *cautus navigandi magister clavos pro fluctuum motibus erigens vel inclinans* "a careful captain at the helm, raising or lowering the tiller bars as required by the motion of the waves"); e.g., "raising" the starboard tiller would head the vessel to starboard.

[11] Similarly, the Vorsa rudder had a socket so cut as to take an oblique tiller; see Crumlin-Pedersen 254.

[12] See the passage from Lucian cited in note 1, from Ammianus in note 10 above, and from Lucretius in note 15 below.

of angle shows that the shaft running from the gunwale is the tiller bar and not the loom of the oar; see also Fig. 177).[13] Depending on circumstances, a helmsman could use both steering oars at once (Figs. 131, 179)[14] or just one (Figs. 128, 151).[15] As on warships, the steering oars could be raised and lowered, but their weight required heavy pennants probably fitted with tackles.[16] These served to lift the oars when at anchor and so keep them from banging about,[17] and perhaps, when under way, to adjust their immersion in the water and equalize it by, e.g., the lowering of the windward oar and the raising of the leeward.[18] In heavy weather the tiller bars were fitted with relieving tackles to take the strain off the helmsman's arms.[19]

[13] A similar arrangement appears on a heavy galley, which, despite its ram, may be a merchant ship, figured in a mosaic of the 3rd A.D. found in Tunis (Foucher, figs. 9, 10).

[14] Cf. Ammianus 21.13.10 (cited in note 10). It is often suggested that there was some gear yoking the two rudders (GOS 291, Torr 77, de Saint-Denis 51), but it never appears in the preserved representations, and the written evidence offered is either misinterpreted (de Saint-Denis suggests the "ropes" mentioned in Jerome, Epist. 100.14, on which see note 19 below) or inconclusive (the passages cited by Torr and Morrison with "tiller" in the singular and "steering oars" in the plural need only mean that, in general, one oar was in operation at a time; as the evidence from the Viking Age amply demonstrates, this was all that was needed).

[15] Lucretius 4.903-04: et manus una regit quantovis impete euntem / atque gubernaclum contorquet quolibet unum "and one hand controls [a huge ship], no matter how great its speed, and a single steering oar turns it in any direction."

[16] One form this rig could take is to be seen in the 14th century vessel mentioned in note 9: the pennants consist of heavy tackles attached to either side of the blade and running, one to the gunwale and the other to the stern gallery; taking up on one of these would cause the loom to slide upward in the collars. Another form can be made out in Fig. 144: here the pennants run from the gunwale down along the hull and pass through two holes, set alongside each other, in the blade.

[17] Thus Statius, in celebrating a friend's departure (Silvae 3.2) from Puteoli on an Alexandrian grain ship (lines 21-24), invokes sea goddesses to take care of the necessary steps for getting under way and among these includes the injunction (line 29) to "let down into the water the rudder that guides the curved vessel" (demittat aquis curvae moderamina puppis). And on St. Paul's ship, after the uneasy night moored off the coast of Malta, the crew cut away the anchors and (Acts 27.40) "loosening the pennants of the steering oars" (ἀνέντες τὰς ζευκτηρίας τῶν πηδαλίων), got under way; these must have been brought up tight when the vessel anchored. In Fig. 146 the sailor in the ship's boat seems to be hauling on the pennants preparatory to docking.

[18] I owe this suggestion to Mr. O. Crumlin-Pedersen.

[19] Cf. Jerome, Epist. 100.14, a translation from the Greek of one of Bishop Theo-

II RIGGING

Ropes, Spars, Sails

A MAST and sail, whether ancient or modern, is fitted with half a dozen or so different ropes, each of which has a specific function.[20] An ancient square-rigger (for other rigs, of far less importance, see 243-45 below) carried the following:[21]

---

philus' paschal letters; the passage in question is a simile to underline the point that, in the event of some great strain, one can weather it by dividing it: *sicut enim gubernatores magnarum navium, cum viderint inmensum ex alto venire gurgitem . . . spumantes fluctus suscipiunt, eosque prorae obiectione sustentant, flectentes in diversum gubernacula, et prout ventorum flatus et necessitas imperarit, stringentes funiculos vel laxantes; cumque unda subsederit, ex utroque navis latere laborantia clavorum vincla dimittunt, ut parumper quiescentia venturo gurgiti praeparentur; qui cum rursus advenerit, stringunt clavorum capita et palmulas dilatant, ut huc atque illuc scissis flatibus, aequalis sit utriusque lateris labor, et quod simul non poterat sustineri, divisum tolerabilius fiat* "Similarly, the helmsmen on large ships, when they spy a tremendous wave coming out of the deep at them . . . meet the foaming water and support its rush by thrusting the prow at it, turning the rudders in opposed directions [i.e., first in one direction, then in the opposite] and, as the gusts of wind and necessity dictate, tightening or slackening the lines. When the wave has subsided, they release the straining bonds of the rudders on each side of the vessel in order, during the moment's quiet, to be ready for the next wave. When it comes, they tighten their grip on the heads of the tiller bars and turn the blades broadside to the water so that, by splitting the gusts, some to one side and some to the other, the strain is equal on both sides, and what could not be supported all at once becomes endurable when divided." In other words, helmsmen take big seas not directly head on but first on one bow and then the other; to do this they must first steer in one direction and then another (*flectentes in diversum gubernacula*), which automatically brings the wind from one side to the other (*huc atque illuc scissis flatibus*); at the same time, depending upon the direction of the wind (*prout ventorum flatus . . . imperarit*), they tighten and slack off on certain lines (*stringentes funiculos vel laxantes*)—and these lines can only be relieving tackles run from each gunwale (*ex utroque navis latere*) to each tiller bar, which were loosened or tightened according as the helmsman pushed the bar away from himself or brought it toward himself. Once the sea has passed, they right the helm, slacking off both tackles (*laborantia clavorum vincla dimittunt*), and wait to see which side to bring the wind on for meeting the next wave. The movements described are no doubt what Theophilus witnessed as a passenger, whatever we make of his interpretation of them.

[20] In the Athenian naval records the technical term for these ropes is τοπεῖα "cordage" (e.g. *IG* ii² 1627.148-53). They were part of the σκεύη κρεμαστά "hanging gear"; for the other equipment included under this term, see the passage cited in App. 2, note 3.

[21] For the problems involved in the identification of these terms and for citation of sources, see App. 1.

Running Rigging (Fig. 171 and cf. Figs. 144, 146), the lines for maneuvering the sail

GREEK	LATIN	
himantes; kerouchoi or *keroiakes*	*ceruchi*	lifts (basically a pair of lines running from masthead to yardarms to keep the yard level or cocked as need be)
ankoina	*anquina*	halyard (line that raises yard and sail into place on mast)
chalinos		? parral (collar of twisted cords holding yard against mast)
hyperai		braces (pair of lines from yardarms to deck permitting lateral adjustment of yard)[22]
podes	*pedes*	sheets (pair of lines controlling lower corners of the sail)
kaloi, kaloes	*rudentes*	brails (lines for shortening sail; see 70 above and 259 below)

Standing Rigging (Fig. 172 and cf. Figs. 144, 146), the fixed lines supporting the mast

protonos		forestay, running from masthead to a point well forward
epitonos		backstay, running from masthead to a point well aft

One important term, that for shrouds (stays bracing the mast laterally; see Fig. 172), is known neither in Greek nor Latin.[23] This and other gaps[24] are not unnatural, since our knowledge comes solely

---

[22] Some ships were also rigged with "back-braces," i.e., preventer braces; see App. I.

[23] Morrison (*GOS* 300-301) suggests that *ankoina* = "shroud," but see App. I on this term.

[24] Foucher (14-15) and Rougé (54) have identified certain lines in a few ship-pictures as bowlines (the lines run, when sailing on the wind, from the leading edge to a point forward to keep the leech flat; cf. *IH* ill. 82). They look, if anything,

from haphazard mention in literature and documents (see App. 1).
The terminology, it is clear, is basically Greek, the Latin being often
but borrowed versions.

Cordage was made from papyrus,[25] hemp,[26] flax,[27] and esparto
grass, particularly the variety grown in Spain.[28]

Blocks (pulleys) were in shape and construction very much like
those we use today (Figs. 144, 146). Some had metal sheaves and
axles like modern blocks, but others were made entirely of wood,
sheave and axle as well as the shell.[29]

Belaying pins were used for making lines fast.[30]

THE MAST (*histos* in Greek, *malus* or *arbor* in Latin) was generally composite, girdled with wooldings at fixed intervals,[31] though

---

like topmast stays. In "Un voilier antique," *Antiquités africaines* 1 (1967) 83-98,
Foucher identifies as a possible bowline a line that, he says (94-95), is shown attached to the tip of the yard of the artemon; as the photographs he publishes show
(p. 91, nos. 17, 18), the line runs *behind* the *artemon* sail, presumably to the mast.

[25] *Od.* 21.390-91; Herodotus 7.25.1, 34.1, 36.3 (Egyptians furnish ropes of papyrus,
and the Phoenicians ropes of white flax, for Xerxes' bridges); Theophrastus, *Hist.
Plant.* 4.8.4 (ἐκ τῆς βίβλου ἱστία τε πλέκουσι . . . καὶ σχοινία . . . γίνεται δὲ καὶ ἐν
Συρίᾳ . . . ὅθεν καὶ Ἀντίγονος εἰς τὰς ναῦς ἐποιεῖτο τὰ σχοινία "[The Egyptians] weave
sails and rope from papyrus. . . . It grows in Syria also, and Antigonus used it for
the cordage on his ships").

[26] Athenaeus 5.206f (cited in NINE, App., pt. 4); Varro, *re rust.* 1.23.6 = Pliny,
NH 19.29.

[27] See the citation from Herodotus in note 25; Euripides, *IT* 1043 (a ship is moored
with χαλινοῖς λινοδέτοις "flaxen bridles"); Ovid, *Fasti* 3.587 (sail is lowered by
[ropes of] *torto lino* "twisted flax"). The ropes of tow (σχοινία στύππινα) in *P. Cairo
Zen.* 59755 (3rd B.C.) were probably of coarse flax, since the plant was widely grown
in ancient Egypt.

[28] Pliny, *NH* 19.30: *verumtamen conplectatur animo qui volet miraculum aesti-
mare quanto sit in usu omnibus terris navium armamentis etc.* "But, to appreciate
what a miracle [the Spanish esparto plant is], one must realize how great its use
is all over the world for ship's cordage, etc." Cf. Athenaeus 5.206f (cited in NINE,
App., pt. 4); Livy 22.20.6 (in 217 B.C. Hasdrubal had collected a massive amount
of esparto grass to supply his navy).

[29] On ancient blocks, see J. Shaw, "A Double-Sheaved Pulley Block from Ken-
chreai," *Hesperia* 36 (1967) 389-401 and *National Geographic*, June 1970, 854.
For the ancient terminology, see note 136 below.

[30] Apollonius Rhodius 1.566-67: ἐπ' ἰκριόφιν δὲ κάλωας / ξεστῇσιν περόνῃσι
διακριδὸν ἀμφιβαλόντες "On deck they wrapped each line about its smooth pin";
Nonnos, *Dionys.* 4.229 (cited in App. 1, note 12).

[31] Alternate horizontal bands of two different colors consistently appear in mosaics

certain special masts were solid.[32] It was socketed into the keel (208 above) and, following a custom that has lasted till today, a coin was often put in the mast-step for luck.[33] On seagoing vessels the yard, which might at times be as long as the mast was high, was most often of two pieces fished together (Fig. 90); this is why it is referred to in the plural as well as the singular (*keraia* or *keraiai* in Greek, *antemna* or *antemnae* in Latin).[34] The yardarm was called *keras* in Greek and *cornu* in Latin, both words meaning literally "a horn."[35] Along the uppermost part of the mast were fittings to receive the lifts (Fig. 144), while somewhat lower was the fitting for the halyard, a heavy collar with eyes affixed to port and starboard to carry the double lines of the halyard; this fitting seems to have been

---

(Figs. 140, 191; also Foucher figs. 2, 9, 12, all 3rd A.D.), while reliefs (Figs. 147, 151) show horizontal lines at fixed intervals. All seem intended to indicate wooldings holding together a composite mast. (For composite masts on earlier ships, see 69 above). They are a common feature of later Mediterranean craft: see *IH* ill. 94 (15th century square-rigger), ill. 95 (15th century galley); Felix Fabri's description of the ship he sailed on in 1483 (quoted in Lane 19); Jal, *Gloss. naut.* s.v. *rouster* (with illustration).

[32] Cf. Athenaeus 5.208e, cited in NINE, App., pt. 4 (difficulty of finding trees for the masts of Hiero's superfreighter).

[33] Three examples have been uncovered so far: the Chrétienne A wreck (Dumas 121 and pl. 52), Blackfriars wreck (Marsden, *Blackfriars* 36-37), Port-Vendres wreck (Chevalier 267-69). For the practice in later ages, see *MM* 51 (1965) 33-34, 205-10.

[34] The Athenian navy records always use the plural (e.g., *IG* ii² 1604.17, 18; 1606.1; 1608.25). Moreover, they consistently treat the yards as composed of two separate pieces: e.g., in 1612.50-57, 462 main yards are reckoned as enough for 231 ships, and 167 *akateion*-yards as enough for 83 ships, with 1 yard left over. It sounds as if, for the sake of convenience, yards were stored with their two pieces apart and were assembled only just before use. For the singular, cf. Lucian, *Navig.* 5: ἡλίκος μὲν ὁ ἱστός, ὅσην δὲ ἀνέχει τὴν κεραίαν, οἵῳ δὲ προτόνῳ συνέχεται "What a mast [the *Isis*; cf. NINE, App., pt. 2] had! what a yard it carried! what a forestay held it up!"

Similarly, Latin used both the plural (e.g. Ovid, *Met.* 11.483, cited in TWELVE, note 27) and the singular (cf. Tertullian cited in the following note).

Artemidorus (1.35) uses the word ἱστοκεραία, perhaps the technical term for the main yard as against the fore or mizzen yards.

[35] Tertullian, *Adv. Marc.* 3.18: *in antemna . . . extremitates cornua vocantur* "On a yard . . . the tips are called 'horns.'" For the Greek *keras*, see Achilles Tatius 3.1.1-2 (cited in TWELVE, note 24); here and elsewhere it is used to mean "yard" (cf. Lucian, *Amores* 6; *Anth. Pal.* 5.204; *OGIS* 674.30 [1st A.D.]; *P. Oxy.* 2136.6 [3rd A.D.]), but the occurrences are all late and the sense no doubt is a transfer from an original meaning of "yardarm." The same transfer is found in Latin; cf. Ovid, *Met.* 11.482: *demittite cornua* "Lower the yard."

the same on galleys and sailing ships during all of Greco-Roman antiquity (Figs. 91, 98, 144, 149). It most likely was made of bronze. On larger craft of both types, the mainmast was stout enough to support a main-top (*karchesion* in Greek, Latinized to *carchesium*),[36] girdled by a protective railing (*thorakion*).[37] Pine and fir were the woods preferred for masts and yards.[38]

The sail was *histion* (or the plural *histia*) in Classical Greek, *armenon* (*armena*) in post-Classical Greek, *velum* in Latin.[39] Sails today, in order to set properly and have the requisite strength, are made of long strips of cloth sewn together, have the edges surrounded by a bolt-rope, and are often reinforced at the corners or where wear is

[36] Pindar, *Nemea* 5.51: ἀνὰ δ' ἱστία τεῖνον πρὸς ζυγὸν καρχασίου "and stretch the sail up to the crosspiece of the mast-top"; Plutarch, *Them.* 12.1: γλαῦκα δ' ὀφθῆναι ... τοῖς καρχησίοις ἐπικαθίζουσαν "An owl was seen ... to alight on the mast-top"; Cinna, fr. 4 (Baehrens): *lucida cum fulgent summi carchesia mali* "when the shining top at the summit of the mast gleams." The gleam referred to in the last passage probably came from the bronze fitting for the halyard, although other parts of the top may have been of bronze as well (cf. Athenaeus 5.208e [cited in NINE, App., pt. 4]).

[37] Athenaeus 11.474f-75a, quoting Asclepiades of Myrlea (2nd-1st B.C.), provides a convenient summary of the terminology for the parts of the mast: "The lowest part of the mast, that which fits into the socket, is called the 'heel' (*pterna*); the central part, the 'neck' (*trachelos*); the part toward the end, the 'top' (*karchesion*). ... Upon this is set the so-called breastwork (*thorakion*). ... Above the 'breastwork,' reaching up high and ending in a point, is the so-called spindle (*elakate*)" [the last would be that part of the mast from the mast-top to the tip] (τοῦ γὰρ ἱστοῦ τὸ μὲν κατωτάτω πτέρνα καλεῖται, ἣ ἐμπίπτει εἰς τὸν ληνόν· τὸ δ' οἷον εἰς μέσον, τράχηλος· τὸ δὲ πρὸς τῷ τέλει καρχήσιον ... καὶ ἐπίκειται τὸ λεγόμενον αὐτῷ θωράκιον ... ἐπὶ δὲ τοῦ θωρακίου εἰς ὕψος ἀνήκουσα καὶ ὀξεῖα γιγνομένη ἐστὶν ἡ λεγομένη ἡλακάτη).

[38] Theophrastus, *Hist. Plant.* 5.1.7: ἔστι δὲ καὶ μακρότατον ἡ ἐλάτη καὶ ὀρθοφυέστατον. δι' ὃ καὶ τὰς κεφαλὰς καὶ τοὺς ἱστοὺς ἐκ ταύτης ποιοῦσιν "And silver fir grows the tallest and straightest as well, and this is why it is used for yards and masts." Homer also mentions the use of silver fir (FOUR, note 21). The mast of the ship that brought over the Vatican obelisk (NINE, App., Pt. 2) was of fir (Pliny, *NH* 16.201). For masts of pine, cf. Apuleius, *Met.* 11.16, Lucan 2.695-96.

[39] For ἄρμενον "sail," see *P. Lond.* 1164 h 7 (A.D. 212; cited in note 135 below); *P. Oxy.* 2136.6 (A.D. 291); *P. Lond.* 1714.31 (A.D. 570); *P. Mon.* 4.12 (A.D. 581); Porphyrogenitus, *De caerimoniis* 2.45, p. 389 (ἀρμένων ἐνέα ... τῶν θ' καραβίων "nine sails for the 9 *karaboi*"). Cf. Polybius 1.44.3 (λαβὼν δ' οὔριον καὶ λαμπρὸν ἄνεμον, ἐκπετάσας πᾶσι τοῖς ἀρμένοις "catching a favorable wind and spreading all sail") and 21.43.13 (the treaty of 188 B.C., in which Antiochus III surrendered his navy to Rome, stipulated: ἀποδότω δὲ καὶ τὰς ναῦς τὰς μακρὰς καὶ τὰ ἐκ τούτων ἄρμενα καὶ τὰ σκεύη which can only mean "He must hand over the warships and their sails [or possibly 'rigging'] and gear").

likely by patches of leather. They were treated in much the same manner in antiquity: they were made up of oblong blocks of cloth sewn together (clearest in Fig. 142), had the edges secured by a bolt-rope (clearest in Fig. 144), and had their corners reinforced with leather patches.[40] Furthermore, there continued the practice, in vogue since the sixth century b.c. at least (68-69 above), of running light lines evenly spaced horizontally across the front surface; these, intersecting at right angles with the brails, made a distinct checker-board pattern (Figs. 145, 147, 154). Sometimes, instead of lines, strips, perhaps of leather,[41] were sewn across the sail (clearest in Fig. 144). The brails were made fast to the foot, traveled up the forward surface through fairleads[42] sewn in vertical rows (Fig. 144), passed over the yard (Fig. 150), and came down to the deck aft.

Sailcloth was generally made of linen.[43] Usually it was left white,[44] though it could be dyed various colors for various reasons: purple

[40] Plutarch, *Mor.* 664c: τὸ δέρμα τῆς φώκης . . . καὶ τὸ τῆς ὑαίνης οἷς τὰ ἄκρα τῶν ἱστίων οἱ ναύκληροι καταδιφθεροῦσι "seal-hide and hyaena-hide, which ship-owners use to leather over the corners of their sails." These particular hides were chosen, Plutarch explains, because of the seaman's superstition that they kept away lightning; there can be little doubt that others were used as well.

[41] Lucian (*Navig.* 4) reports that he and his friends gazed up at the vast main-sail of the *Isis* (cf. NINE, App., pt. 2) "counting the rows of hides" (ἀριθμοῦντες τῶν βυρσῶν τὰς ἐπιβολάς); the expression applies perfectly to, e.g., the horizontal strips visible on the mainsail of the big merchantman pictured in Fig. 144.

[42] The fairleads were called κρίκοι; cf. Herodotus 2.36 (among the numerous things Egyptians do in just the opposite way from the rest of the world is to fasten τῶν ἱστίων τοὺς κρίκους καὶ τοὺς κάλους . . . ἔσωθεν "the fairleads and brails of the sail on the inside" [i.e., on the after, instead of the fore, side]).

[43] A number of terms occur in Greek: λίνον (Aeschylus, *Prom.* 468: λινόπτερ' ὀχήματα "linen-winged wheels," i.e., sails; Apollonius Rhodius 1.565: κὰδ δ' αὐτοῦ λίνα χεῦαν "From it [the mast] they spread the linen [sails]") or ὀθόνη (Lucian, *Amores* 6: ἀθρόας κατὰ τῶν κάλων τὰς ὀθόνας ἐκχέαντες, ἠρέμα πιμπλαμένου τοῦ λίνου "spreading all the canvas from the brails, with the linen gradually filling"; *Anth. Pal.* 12.53.8: οὔριος ὑμετέρας πνεύσεται εἰς ὀθόνας "A favorable wind will blow on our canvas") or σινδών (*Anth. Pal.* 11.404.4: διαπλεῖ σινδόν' ἐπαράμενος "Raising canvas, he will sail through") or βύσσος (Athen. 5.206c: Ptolemy IV's great pleasure barge had a βύσσινον ἱστίον "sail of linen"). Cf. the use in poetry of λαῖφος "cloth" for "sail" (e.g., Euripides, *Med.* 524: ἄκροισι λαίφους κρασπέδοις "with the very tips of the sail"); the cloth could be only linen. In Latin, linen sails were *lintea* (Vergil, *Aen.* 3.686) or *carbasa* (*Aen.* 4.417).

[44] E.g., *Od.* 2.426 ἱστία λευκά, Catullus 64.235 *candida . . . vela* "white sails." In colored representations the sails are most often white (e.g., the originals of Figs. 81, 141, 154; cf. *IH* ills. 50, 62).

234

to distinguish a royal vessel or a flagship,[45] black for mourning,[46] sea green or blue for camouflage,[47] other colors for decorative effect.[48] The front face of a sail could bear devices or inscriptions or both (Fig. 144).[49]

III THE WARSHIP RIG

WHEN the Mediterranean galley first came into being, its rig consisted of a retractable mast stepped amidships on which hung a single broad square sail. We see it in numerous representations,[50] and it is described in Homer's verses.[51] It apparently filled its role satisfactorily, for it lasted without substantial change until the end of the ancient world.[52]

During a fight warships worked under oars alone; with the rapid maneuvering required, sails were a hindrance rather than a help. Consequently, the practice arose of stowing the sailing gear away

[45] Athenaeus 12.535d and Plutarch, *Alc.* 32.2 (Alcibiades' flagship); Procopius, *Bell. Vand.* 1.13.3, cited in note 82 below (Belisarius' flagship); Plutarch, *Ant.* 26.1 (Cleopatra's barge); Pliny, *NH* 19.22 (Cleopatra's ship at Actium. Pliny notes that "this [a purple sail] was the mark of the emperor's ship" [*hoc fuit imperatoriae navis insigne*]).

[46] Plutarch, *Theseus* 17.4 (the ship carrying the Minotaur's intended victims had black sails).

[47] Vegetius 4.37: *ne tamen exploratoriae naves candore prodantur, colore veneto, qui marinis est fluctibus similis, vela tinguntur et funes* "And, so that no white should betray the presence of ships on scouting duty, their sails and lines are dyed Venetian blue, which is the same color as seawater."

[48] Topsails could be red: see Seneca, *Medea* 328 (cited in note 71 below). Caligula enhanced the look of his yachts with particolored sails (*versicoloribus velis* Suetonius, *Cal.* 37).

[49] Trajan's flagship on the Tigris bore "at the top of the sail the imperial name and other royal titles, lettered in gold" (ἐπ' ἄκρῳ τῷ ἱστίῳ τὸ βασιλικὸν ὄνομα, καὶ ὅσοις ἄλλοις βασιλεὺς γεραίρεται, χρυσῷ ἐγκεχαραγμένα) Arrian, *Parth.* fr. 67 = Jacoby, *FGH* no. 156, fr. 154, vol. II B, p. 876.

[50] Figs. 18, 26-30, 37-41, 48, 50, 52-53, 57-59, 61, 66, 78, 81-82, 90-91, 93. Figs. 91 and 98 and most clearly *GOS* pl. 21e show details of the mast tip. A heavy collar with hoops to port and starboard sat about the mast just below the tip; through the hoops—perhaps they were fitted with sheaves—passed the lines of the double halyard. The whole fitting was most likely of bronze; cf. note 36 above.

[51] *Il.* 1.432-34, *Od.* 2.424-26 = 15.289-91.

[52] Cf., e.g., the pictures on coins of Hadrian's yacht proceeding under sail (*BM Empire* III.1393).

before going into action,[53] or even, if convenient, leaving it ashore.[54] Yet there were moments during battle when a galley could well use the wind's aid—to escape from overwhelming odds or to pull out of danger if crippled.[55] The solution was a second rig for emergencies, one that could be kept on board at all times and be put into operation swiftly and easily.

Very likely this important addition was introduced during the fifth century B.C., when the warship saw so many improvements.[56] There is no question that, by the early fourth, it was standard equipment on Athenian triremes, where it went under the name of *akateion*,[57] the "boat" rig (cf. 159 above) as against the regular gal-

[53] Cf. Livy 36.44.2-3 (191 B.C.): *quod ubi vidit Romanus vela contrahit malosque inclinat, et, simul armamenta componens, opperitur insequentes naves* "When the Romans caught sight [of Polyxenidas' fleet], they furled sail, lowered the masts, and, while still stowing away the gear, awaited the ships coming against them." Since the Roman contingent had been constantly in motion—it had started from Delos, stopped at Chios and Phocaea, and was headed for Corycus (Livy 36.43.11-13)—there had been no chance to leave the sailing gear ashore. When the Carthaginian fleet ran into the Roman off the Aegates Islands in 241 B.C., the latter had the advantage of being stripped for action; the Carthaginians had to stow away sail before going into action (καθελόμενοι τοὺς ἰστοὺς "taking down the masts" Polybius 1.61.1).

[54] Thus Conon, escaping from Aegospotami in 405 B.C., in order to avoid pursuit stopped for a moment at Lampsacus and ἔλαβεν αὐτόθεν τὰ μεγάλα τῶν Λυσάνδρου νεῶν ἰστία "seized there the big sails of Lysander's ships" (Xenophon, *Hell.* 2.1.29); obviously Lysander had left these there when he took his fleet into action. For other references to leaving the working sails ashore, see Thucydides 7.24.2 (the Syracusans, with the capture in 413 B.C. of an Athenian fort used as storehouse, fell heir to the sails and other gear of 40 triremes); Xenophon, *Hell.* 6.2.27, quoted in note 57, below; Plutarch, *Ant.* 64.2 (at Actium, Antony's captains wanted to leave the sailing gear ashore but he forced them to take it aboard); Dio Cassius 50.33.5 (Augustus' ships, stripped for action and hence without sails, were unable to pursue Antony when he fled).

[55] Cf. the passages cited in note 63 below.

[56] When Xenophon reports (*Hell.* 1.1.13 [410 B.C.]) that Alcibiades, readying for action, ordered the ships "to strip off the big sails and follow him" (διώκειν αὐτὸν ἐξελομένοις τὰ μεγάλα ἰστία), the clear implication is that there were also smaller sails aboard; cf. the passage cited in the next note.

[57] Cf. Xenophon, *Hell.* 6.2.27: Iphicrates, setting off in 373 B.C., πάντα ὅσα εἰς ναυμαχίαν παρεσκευάζετο · εὐθὺς μὲν γὰρ τὰ μεγάλα ἰστία αὐτοῦ κατέλιπεν, ὡς ἐπὶ ναυμαχίαν πλέων · καὶ τοῖς ἀκατείοις δέ, καὶ εἰ φορὸν πνεῦμα εἴη, ὀλίγα ἐχρῆτο "made all possible preparations for a naval battle. He started by leaving the big sails behind, as if expecting to go into action. On top of that, he used the boat sails [*akateia*] sparingly, even when there was a favorable wind." For the *akateion* sail on merchantmen, see 241 below.

ley rig. It consisted of a "boat" yard and a "boat" mast[58] fitted with special partners.[59] That mast was stepped in place of the regular mast,[60] and to it was bent some of the extra canvas aboard.[61] By the next century or so, the emergency rig had changed its name from *akateion* to *dolon*—perhaps "sneaker," that which enabled a craft to sneak out of harm's way[62]—and, under this name, can be traced

---

[58] This can be deduced from certain entries in the Athenian navy yard records for the years from 377 to 341 B.C. (*IG* ii² 1604-22). The earliest fragment (1604) mentions "boat yards" (κεραῖαι ἀκάτειοι, lines 17, 18, etc.), "boat masts" (ἱστὸς ἀκάτειος, lines 64, 85, 92), "big masts" (ἱστὸς μέγας aut sim., lines 23, 48, 50), "big yards" (κεραῖαι μεγάλαι, line 23). A fully equipped trireme was issued both sets, as is clear from the many entries that expressly assign all four items to given ships: e.g., 1607.105-12 (373/2); 1609.49-53 (370/69); 1616.46-49 (358/7); 1622.342-50 (342/1). Other entries, which do not specifically mention all, yet imply their presence: e.g., the trireme *Megiste* in 358/7 is listed as having on board a big mast, big yard, and boat mast, all three of which were unserviceable (1616.50-67); obviously the boat yard had been lost altogether.

Navies, whether ancient or modern, rarely succeed in keeping all units up to par. The Athenians were no exception: in 356/5, for example, one warehouse had on hand "big yards" for 231 ships but "boat yards" for only 83 (1612.49-56).

[59] The *parastatai*, literally "standbys" (cf., e.g., 1604.52, 69, 80, etc.), a pair of objects which form part of the wooden gear of triremes. They appear on the lists along with the boat yard and mast, and disappear from them precisely when the latter do (see App. 2). They are usually identified (e.g., Torr 83) as supports for the regular mast, but, to judge from their correlation with the emergency gear, they were, on triremes at least, partners for the boat mast, propping it so securely that it could stand with a minimum of standing rigging. The only ships that continued to be issued *parastatai* were triaconters (1632.6-11 [323/2]), whose regular mast was no doubt relatively short. Morrison (*GOS* 293) suggests that they were stakes for propping the galleys when pulled up on shore; if this were so, their disappearance as a regular piece of equipment becomes inexplicable. Moreover, *parastatai* are mentioned by Cato the Elder (*inc. lib.*, fr. 18) in a context that connects them with the mast: *malum deligatum, parastatae vinctae* "The mast was made fast, the *parastatae* lashed." Isidore, who cites the line (*Orig.* 19.2.11) comments: *parastatae stipites sunt pares stantes quibus arbor sustinetur* "The *parastatae* are a pair of standing posts that support the mast."

[60] See App. 2.

[61] There is no mention ever of a "boat sail" or of "boat ropes" in the naval records, a point that has proved puzzling (cf. Torr 84) but is not inexplicable. Every trireme had, as part of its regular issue, a number of canvas awnings (*katablema, hypoblema, pararrhymata*; cf., e.g., 1611.299-301 [357/6]); any one of these could have been cut and fitted to double as a small sail, using the lines from the regular gear. Venice's galleys, for example, used to press the deck awning into service as a sail (Lane 22-23). In the storehouses, the canvas *pararrhymata* were actually kept in the same boxes as the sails; cf. *Syl.*³ 969.86 [347/6].

[62] In the sense "dagger," which the word also has, it is clearly derived from δόλος "trick"; cf. H. Frisk, *Griechisches etymologisches Wörterbuch* i (Heidelberg 1960)

237

right through to the Byzantine period. The *dolon*, too, consisted of a small-scale mast, yard, and sail.[63]

In the Roman period some galleys appear with a sail taken over from the merchant rig of the age, the *artemon*, the ancient version of the bowspritsail (240 below). It seems to have been used as an aid to steering when cruising under oars.[64]

---

408. Why not the nautical sense as well? (Conformably, its meaning in Artemidorus 2.14, where it is some form of fishing gear, would be "lure" rather than "fishing-rod"; the latter, given in *LSJ*, probably stems from the common identification of the *dolon* mast as one that slants over the bows [see App. 2]). Frisk rightly points out that there is little to be said for a suggested derivation from δέλτος.

[63] In 307 B.C., a Carthaginian admiral, anticipating capture, committed suicide—senselessly, since "his ship, catching a favorable wind, raised the *dolon* and escaped from the battle" (ἡ γὰρ ναῦς φοροῦ πνεύματος ἐπιλαβομένη τοῦ δόλωνος ἀρθέντος, ἐξέφυγε τὸν κίνδυνον Diodorus 20.61.8; the word *dolon* here may reflect the usage of Diodorus' own day rather than of the time he is describing). At the Battle of Lade in 201, "One [Rhodian] ship, because it had been rammed and was going down, raised the *dolon* . . . and escaped" (μιᾶς νηὸς ἐπαραμένης τὸν δόλωνα διὰ τὸ τετρωμένην αὐτὴν θαλλατοῦσθαι . . . ἀποχωρεῖν Polybius 16.15.2). In the encounter with Polyxenidas' fleet in 191, the Romans, after striking their regular masts (see note 53 above), in order to straighten their line of battle, had the left wing "step the *dolons* and head for the open sea" (*dolonibus erectis altum petere intendit* Livy 36.44.3; *erigere* is used of masts, not sails, so clearly a complete second rig was carried). Polyxenidas' ships were similarly equipped since, when things turned against him, he "raised the *dolons* and headed into hurried flight" (*sublatis dolonibus effuse fugere intendit* 36.45.1). The following year his ships again sought safety in their *dolons* (37.30.7). In 36 B.C. some Roman galleys escaped from a defeat by "raising their short sails" (ἀράμεναι τὰ βραχέα τῶν ἱστίων Appian, *Bell. Civ.* 5.111); the ships, then, had two sets of sails, the "short" and the regular. The fleet Belisarius led against Africa in A.D. 533 carried both large and small sails, and Procopius pointedly calls the latter *dolons*: χαλάσαντας τὰ μεγάλα ἱστία τοῖς μικροῖς, ἃ δὴ δόλωνας καλοῦσιν, ἕπεσθαι "lowering the big sails, to follow using the small, which they call *dolons*" (*Bell. Vand.* 1.17.5). P. Lond. Inv. 2305 (3rd B.C.) includes in a list of items (pitch, paint, deck planks) for finishing up the hull of a *kybaia* (see 166 above) a τράπεζα δολωνική (cf. *LSJ* s.v.). This could be a plank with a hole in it that fitted over the mast-hole in the deck to adapt it to the reduced circumference of the *dolon*.

Torr (87) makes much of the change in name, taking the *akateion* to be distinct from the later *dolon*. Their function clearly was identical, and the names were created by seamen, not lexicographers; the one need be no different from the other than, say, a *spanker* is from a *driver*.

[64] In Fig. 127 a number of galleys are shown approaching a dock and the largest, a trireme, carries an *artemon*. Coins issued by Hadrian depict on the reverse the galley that served as his yacht during his numerous voyages; we see it cruising along under oars and the *artemon* (Fig. 122 and cf. *BM Empire*, Hadrian nos. 243-47 = pl. 51.9, 10; 1394-1414 = pl. 85.1-7; all A.D. 119-38).

IV THE MERCHANTSHIP RIG

Throughout ancient times, merchantmen of all sizes used one chief sail, a broad square sail set amidships. A fore and then a mizzen came to be added, but both always remained secondary; the mainsail was *the* sail,[65] and in most representations of ancient freighters (Figs. 144, 146, 149, 151, 154, 156) it dwarfs the rest of the canvas.[66] The main yard was an enormous spar, nearly as long as the ship in some cases, that needed multiple lifts to support it; when squared, its arms reached so far beyond the hull that they could be fitted for dropping stones or other heavy weights on enemy galleys that tried to come alongside.[67] A massive forestay ran from mast-top to the base of the bowspritlike foremast, where it was made fast. The shrouds were set

[65] Thus, ancient writers invariably refer to the mast of a ship rather than its masts. E.g., Cicero observes (*de orat.* 3.180) that the key elements of a ship are its "prow, poop, yard, sails, and mast"; Pliny observes (*Epist.* 9.26.4) that the true test of a helmsman comes when "the rigging sings in the wind and the mast bends"; Apuleius observes (*Florida* 23) that a bad helmsman can destroy a beautiful ship despite its "well-hung tiller, stout ropes, lofty mast." Even more significant, Roman legal writings mention only one mast: *Dig.* 21.2.44: "all that is connected with a ship, such as helm, mast, yard, sail" (*omnia autem quae coniuncta navi essent, veluti gubernacula, malus, antemnae, velum*); cf. *Dig.* 14.2.6, which cites the case of a ship "with its gear, mast, and yard burned by a stroke of lightning" (*ictu fulminis deustis armamentis et arbore et antemna*). The vessel that figures in Petronius' story had but one mast (cf. 109.7: *per antemnam pelagiae consederant volucres* "gulls perched on the yard"), and it was a big ship (101.9). Even the huge *Isis* (cf. Nine, App., pt. 2) had but one mast (Lucian, cited in note 34 above). Correspondingly, Philostratus (*Vita Ap.* 3.35) describes the oversize craft specially designed for the Egypt-India trade as "made very high with respect to sides and mast" (τοίχοις τε ὑπεράραντες καὶ ἱστῷ).

[66] This remained the case right up to the 15th century; see *IH* 59-84.

[67] Yards so fitted were called "stone-bearers" (λιθοφόροι) or "dolphin-bearers" (δελφινοφόροι; "dolphins" must have been sailor's jargon for the special lead weights dropped from the yardarms). Cf. Thucydides 7.38.2-3, 41.2 (in 413 B.C. the Athenians closed the mouth of the bay of Syracuse by anchoring a line of merchantmen with κεραῖαι . . . δελφινοφόροι "yards fitted as dolphin-bearers"); Diodorus 13.78.7, 79.3: οἱ δ' ἐπὶ τῶν μεγάλων πλοίων ἐφεστῶτες ἐπέρριψαν ταῖς τῶν πολεμίων ναυσὶ τοὺς ἀπὸ τῶν κεραιῶν λίθους . . . πλεῖστοι δ' ὑπὸ τῶν λιθοφόρων κεραιῶν ἔπιπτον "Those stationed on the large ships hurled rocks on the enemy craft from the yards. . . . A large number of the enemy personnel fell as victims to the rock-bearing yards" (this was in 407 B.C.). Cf. Athenaeus 5.208d (see Nine, App., pt. 4), where the use of grapnels as well as lead weights is mentioned; the wording may imply the use of special booms fitted as extensions to the yardarms).

up with tackles so they could be adjusted; since ratlines cannot be used when shrouds are so set up, there was a rope ladder abaft the mast for getting aloft (Fig. 151), a feature that marked Mediterranean sailing vessels for the next fifteen hundred years.[68]

As early as the sixth century B.C., a foresail makes an appearance (70 above). In its first form it was a rather large sail, perhaps one half the size of the main, set on a foremast with a distinctly forward rake (Fig. 97). Lack of pictures for the next half millennium (cf. 173 above) prevents us from following its progress, but the description of Hiero's three-masted superfreighter (NINE, App., pt. 4) is proof that the foresail not only lived on, but was joined by a mizzen. When, in the first century A.D., pictorial evidence is once again available, it reveals that the foresail underwent two lines of development. On most ships it became a headsail pure and simple, one very much like the bowspritsail of latter ages, a small square of canvas on a mast slanting low over the bows (Figs. 144, 145, ship to right, 147, 149, 151, 156). On others, however (Figs. 142, 145, ship to left), it remained as it had been five centuries earlier, a sail of fair size hung on a mast often nearly as high as the mainmast. The foremast never lost its strong forward rake[69] until close to the beginning of the fourth century A.D., when some ships finally stepped it almost upright (Fig. 169). The name for the foresail was *artemon* in Greek, which was Latinized to *artemo*.[70]

---

[68] Cf. *IH* ill. 98.

[69] Other examples: Rougé, pls. III b, IV a (3rd A.D.); Rostovtzeff *SEHRE*², pl. 62.3 (3rd A.D.); *DS* s.v. *horeia*, fig. 3882 (5th A.D.).

[70] The word occurs once in Greek (Acts 27.40: the crew of St. Paul's ship, having abandoned the mainsail during the gale, tried to nose into Malta by "raising the *artemon*" [ἐπάραντες τὸν ἀρτέμωνα]) and several times in Latin (e.g., Paulinus of Nola, *Epist.* 49.2; Seneca, *Controversiae* 7.1.2); in all these there is no way of telling whether a bowspritsail or just a jury-sail is meant. However, a passage from Augustine shows the close connection with the steering of a ship that only a bowspritsail could have. Suppose, Augustine argues (*Enarratio in Psalmum* XXXI.4 [Migne 36, p. 259]), a man is an accomplished helmsman but yet has lost the course: "What good is his superb skill at managing the *artemo*, manipulating it, keeping the bow to the waves and the ship from broaching to" (*quid valet quia artemonem optime tenet, optime movet, dat proram fluctibus, cavet ne latera infligantur*). Possibly Augustine has here confused the *artemon* with the tiller; if so, it can only be because

When sailing over open water, vessels raised a topsail (Greek *sipharos* = Latin *siparum*;[71] perhaps also *akateion*[72] in Greek) over

---

both were involved in steering the ship. Smith (153-63) supports the identification by tracing the later history of the word. He demonstrates that the *artemo* of Mediaeval Latin and the *artimone* of early Italian all unquestionably referred to a foresail, at that time a sail of triangular shape. When, later, a sail of such shape was used on square-rigged ships as a mizzen, the traditional name traveled aft with the sail —hence the modern French *artimon* "mizzen."

A passage from the Digest (50.16.242) has been offered as proof (e.g., by Rougé 58-59) that the *artemon* must have been a jury-rig: *malum navis esse partem, artemonem autem non esse, . . . quia pleraeque naves sine malo inutiles essent, ideoque pars navis habetur; artemo autem magis adiectamento quam pars navis est* "The mast is part of the ship; the *artemo* however is not . . . ; this is because many ships are useless without a mast and therefore it is considered a part of the vessel, whereas the *artemo* is rather an addition than a part." The Digest would scarcely bother to make mention of a jury-rig, whereas it would perforce of the foresail, which Roman ship-pictures reveal was a standard element. The passage means that, on a sailing ship, operation without a mast is inconceivable, so a mast is an essential part; operation without a foremast is perfectly possible (cf. Figs. 143-144, 151), so the latter is but an adjunct.

[71] Seneca, *Epist.* 77.1-2: *subito nobis hodie Alexandrinae naves apparuerunt, quae praemitti solent et nuntiare secuturae classis adventum; tabellarias vocant. gratus illarum Campaniae aspectus est; omnis in pilis Puteolorum turba consistit et ex ipso genere velorum Alexandrinas quamvis in magna turba navium intellegit. solis enim licet siparum intendere, quod in alto omnes habent naves. . . . siparum Alexandrinarum insigne est* "Today the Alexandrian ships suddenly made their appearance, the ones that are usually sent ahead to announce that the fleet [i.e., of big grain ships from Egypt] is behind and will be arriving; they are called 'dispatch boats' [*tabellariae*]. To the Campanians they are a welcome sight. The whole mob from Puteoli stands on the dock, and, even in a big crowd of ships, they can pick out those from Alexandria by their sails, since they are the only vessels allowed to keep the topsails raised, although all use them on the open sea . . . The topsail stands out on the ships from Alexandria." Cf. Seneca, *Medea* 327-28: *alto rubicunda tremunt/sipara velo* "The red topsails flutter on top of the [main]sail"; Lucan 5.428-29: *summaque pandens/sipara velorum perituras colligit auras* "Spreading the topsails, highest of the sails, he picks up breezes that would otherwise be lost"; similar expression in Statius, *Silvae* 3.2.27. The Greek occurs only in Epictetus (*Diss.* 3.2.18): ἐπαίρεις τοὺς σιφάρους "You raise your topsails," though the word occurs elsewhere in the sense of "awning" (see *LSJ*, s.v. σείφαρος).

[72] Lucian, *Quomodo historia conscribenda sit* 45: ἀνέμου ἐπουριάσοντος τὰ ἀκάτεια καὶ συνδιολίσοντος ὑψηλὴν καὶ ἐπ' ἄκρων τῶν κυμάτων τὴν ναῦν "a wind blowing fair on the *akateia* and carrying along the vessel high over the tips of the waves"; *Jup. Trag.* 46: ἔφερε μὲν ὑμᾶς τότε . . . ἄνεμος ἐμπίπτων τῇ ὀθόνῃ καὶ ἐμπιπλὰς τὰ ἀκάτεια "And then the wind bore you along . . . falling on the canvas and filling the *akateia*"; *Lexiphanes* 15: ὁλκάδα τριάρμενον ἐν οὐρίῳ πλέουσαν, ἐμπεπνευματωμένου τοῦ ἀκατείου, εὐφοροῦσάν τε καὶ ἀκροκυματοῦσαν "a three-masted merchant ship bowling along and skimming over the waves before a favorable wind, the breeze filling its *akateion*." Since *akateion* is etymologically a smallish "boat" sail, and on warships the

the main. In form it was like a flattened isosceles triangle; the base
was bent to the main yard and the apex ran to the masthead (Figs.
143, 149). Sometimes, in order to clear the forestay, the topsail was
cut vertically into two flattened right triangles (Fig. 144); the vertical sides slid on short staffs running from main yard to masthead
on either side of the mast.[73]

Artemon, mainsail, and main topsail was the standard seagoing
rig.[74] The largest freighters afloat, however, were three-masters, carrying a mizzen (*epidromos*)[75] hoisted on a short mast set midway

---

small sail of the jury-rig (236 above), the meaning that best suits the contexts here
is "topsail," the lofty little sail that gave a vessel extra drive. Lucian mentions it
in the same way as we, say, will talk of a square-rigger rushing along under royals
or t'gallants.

The phrase "Hoist the topsail and flee" was used by Epicurus (φεῦγε τἀκάτιον
ἀράμενος; cf. H. Usener, *Epicurea* [Leipzig 1887] fr. 163) and is repeated by Plutarch
(*Mor.* 15d, 1094d). According to Torr (86) this is a reference to the jury-sail of a
warship. But Plutarch at another point (662c) puts it this way: τὰ ἱστία ἐκάτερ'
ἐπαράμενοι . . . φεύγοιμεν "Flee, raising both sails," which is surely to be taken as
a reference to a merchantman's escaping by clapping on all sail, main and topsail.
Aristophanes, *Lys.* 64, which Torr offers as the earliest example of the expression,
may have nothing to do with sails; see B. B. Rogers' note *ad loc.* On παράσειρον,
which Torr offers as the Greek for topsail, see note 86 below.

[73] This must be the meaning of *contos*, literally "poles," in Paulinus of Nola,
Epist. 49.3: *exsinuatum in contos suos siparum stabat* "The topsail, unfurled, stood
upright on its poles." The cutting of the topsail into two may explain why the
words for it are sometimes in the singular form and sometimes in the plural; cf.
the two previous notes.

[74] Cf. Synesius, *Epist.* 4.160d-161a ταχὺ μὲν τὴν γῆν ἀπεκρύπτομεν, ταχὺ δὲ μετὰ
τῶν ὁλκάδων ἦμεν τῶν διαρμένων (διαρμενίων codd.) "Soon we are out of sight of land,
soon we are among the seagoing merchantmen, the two-masters." Conformably,
pictures of the harbor of Portus (cf. SIXTEEN, note 22) consistently show it filled
with sailing vessels of this type; see Meiggs, pl. 18.

[75] Cf. Pliny, *NH* 19.5: *iam vero nec vela satis esse maiora navigiis, sed quamvis
amplitudini antemnarum singulae arbores sufficiant, super eas tamen addi velorum
alia vela, praeterque in proris et alia in puppibis pandi* "By this time sails larger
than the ships are not enough for us. Even though single trees are not big enough
for the yards, yet other sails are added above the yards, and still others are spread
on prow and poop." The earliest three-masters we know of are Hiero's superfreighter, built about 240 B.C. (NINE, App., pt. 4), and the others in his merchant
marine (see Plutarch quoted in NINE, App., note 29). The grain carriers that plied
between Alexandria and Rome included particularly large units, some of which
were three-masters: cf. Lucian, *Navig.* 14 (a reference to five such ships, τριάρμενα
πάντα "all three-masters"); *Pseudol.* 27 (a quack was beaten up in Alexandria by
ναύτῃ τινὶ τῶν τριαρμένων "some sailor from the three-masters," in other words,
from a crack crew and hence a well-muscled specimen); *Lexiphanes* 15 (cited in

between the mainmast and the sternpost (Fig. 145, ship to left). Both *artemon*-foresail and mizzen were large enough to need full running rigging: lifts, halyards, brails, braces, sheets (Figs. 145, 147, 174). At the same time, they were distinctly subordinate to the main; this was the sail par excellence.[76]

V FORE-AND-AFT RIGS

Fore-and-aft rigs are those in which the sail, instead of lying athwartship as a square sail does, is set parallel with the line of the keel. Less efficient than the square-rig before the wind, they are considerably more efficient than the other when going into the wind. Fig. 173 illustrates several types that have been at home in Mediterranean and Near Eastern waters to the present day: the triangular lateen; the quadrilateral ("Arab") lateen; the spritsail; the gaff-headed sail; the lugsail.

Until a few decades ago there was near unanimity of opinion: the ancients, it was claimed, used no sail other than the square sail.[77]

---

note 72 above). Philostratus mentions (*Vita Ap.* 4.9) a three-master in the harbor at Smyrna, which could well be a member of the Alexandrian fleet.

The word *triarmenos*, literally "trebly equipped," strictly speaking means "with three sails" (cf. note 39 above); since ancient sailing ships consistently carried their working sails one on each mast, the distinction between a "three-sailer" and "three-master" is academic.

For *epidromos* "mizzen," see Pollux 1.91.

The mast-step found in the Chrétienne A wreck was perhaps for a mizzenmast. It is definitely not amidships but well toward one end of the vessel (Dumas 166-67), most likely the after end (cf. Ten, note 35).

[76] E.g., on Hiero's great three-master, a tree for the mainmast was found only after long and hard search, whereas the other two offered no difficulty (Athenaeus 5.208e, cited in Nine, App., pt. 4). The mainmast on this vessel was called the "first" mast, and the others the "second" and "third." This cannot mean that the foresail was the largest, for that would contradict every ship-picture that has come down to us. "First" here must mean the no. 1 mast, the mainmast.

[77] E.g., Böckh 141-43, 153; Torr 78-91; Assmann, *Segel* 1052; Köster 167-76; Moll 21, 26 (where he goes to elaborate lengths to disprove the existence of other kinds of sail). For a full-scale discussion of the problem, see Casson, "Studies" 43-49.

Since the lateen has been a favorite among Arab seamen, there has been a tendency to involve them with its spread, if not its invention, and to place its introduction into the Mediterranean sometime after the Arab Conquest. H. H. Brindley, for example, in "Early Pictures of Lateen Sails," *MM* 12 (1926) 14, holds that the lateen was probably invented by Arabs in the east and that the Mediterranean very likely

As we now can tell from representations that have come to light in the last half-century, they knew at least three types of fore-and-aft rig. The best attested is the sprit, which was in use from, at the latest, the second century B.C. on, chiefly in the north Aegean (where it has remained a favorite until today)[78] but also as far westward as Rome (Figs. 175-179). The most detailed picture (Fig. 179) shows a stout mast stepped far in the bows, with the sail made fast to it very loosely, as is generally the case with this rig. The sprit, a long spar running diagonally across the windward side of the sail, supports the peak, and a double-ended vang made fast to its upper end permits trimming of the peak; no vertical brails are visible since they have no place in such a rig.

Both types of lateen were known. A gravestone of the second century A.D. (Fig. 181)[79] shows an example of the quadrilateral type, the so-called Arab lateen, while a mosaic of the fourth A.D. (Fig. 182)[80] and a graffito (Fig. 180), which is indeterminate in date—it may be as early as the Hellenistic Age or as late as the Byzantine—show the triangular type that was to become standard in the Mediterranean from the Middle Ages on. One of Synesius' letters, dated

---

got it from there. G. S. Laird Clowes, *Sailing Ships: Their History and Development* I (London 1932) 4, holds that it arose in the Mediterranean about the 7th A.D. and was carried westward by the Arabs. R. H. Dolley, "The Rig of Early Medieval Warships," *MM* 35 (1949) 54-55, holds that it was introduced after the time of Justinian but probably was not an Arab discovery. J. Poujade, *La route des Indes et ses navires* (Paris 1946) 158-59, holds that it was invented in India and from there made its way, on the one hand, to the Pacific and, on the other, to the west, being brought there by the Arabs (but see Brindley, *op. cit.* 22, who points out rightly that the development of the lateen of the Far East was totally independent). P. Paris, "Voile latine? Voile arabe? Voile mystérieuse?" *Hesperis: Archives berbères et bulletin de l'institut des hautes études marocaines* 36 (1949) 96, offers no hard and fast conclusions but clearly favors a late date. The fullest case for Arab discovery and dispersion of the lateen is given by G. Hourani, *Arab Seafaring in the Indian Ocean in Ancient and Early Medieval Times* (Princeton 1951) 100-105.

[78] *IH* 198.

[79] For a discussion of earlier scholarly opinion about this relief, see Casson, "Studies" 47-49.

[80] On the date of this mosaic, see Rostovtzeff in *Röm. Mitt.* 26 (1911) 152. He compares it to the pictures in Vat. Vergil Codex 3225, which is generally assigned to the 4th A.D.; see *Codices e Vaticanis Selecti* I, *Fragmenta et Picturae Vergiliana Codicis Vaticani Latini 3225* (Rome 1945³) 12.

A.D. 404, describes a voyage that probably took place on a lateener,[81] and a passage from Procopius, referring to events in A.D. 533, almost certainly contains an allusion to lateen sails.[82]

For a way in which the lateen may possibly have come into being, see 277 below.

VI MISCELLANEOUS EQUIPMENT

NAVIGATIONAL aids were rudimentary. Good handbooks were available, which supplied brief notes on distances, landmarks, harbors, anchorages, and so on,[83] but there is no evidence for the use

[81] See App. 3.

[82] *Bell. Vand.* 1.13.3: to distinguish the three vessels that led the fleet, Belisarius "colored with red paint the sails from the upper angle to [a point] about one-third [down]" (τὰ ἱστία ἐκ γωνίας τῆς ἄνω καὶ ἐς τριτημόριον μάλιστα ἔχρισε μίλτῳ). A glance at Fig. 173 will show that these words most naturally fit a lateen. The passage was noticed over half a century ago by Torr, who cited it (99, note 214) as evidence for the use of colored sails. Torr, to be sure, believed the ancients knew only the square sail, but the passage raised no difficulties for him since he misunderstood the meaning of γωνίας. (Citing Herodotus 8.122.2 as a parallel, he stated that the word here must mean "masthead." This is a mistake: both in Herodotus and here, the word has its usual meaning "angle.") J. Sottas, "An Early Lateen Sail in the Mediterranean," *MM* 25 (1939) 229-30, first offered the passage as evidence for the lateen. Paris (*op. cit.* [note 77 above] 79) attempted to explain the passage away by suggesting that Procopius was referring to a spritsail, which is possible but unlikely, or to a lugsail—which is a wild shot in the dark inasmuch as we as yet have no evidence for the lugsail in the ancient world. Hourani (104, n. 104) suggested that Procopius was referring to topsails, the Roman version of which were triangular. Against this suggestion, however, are the words τὰ ἱστία which, when used without qualification, as here, can only refer to the mainsails. Besides, Belisarius and his staff almost certainly traveled on the galleys in the fleet, not the transports, and galleys carried no topsails.

In the Kyrenia wreck, the mast-step is located in the forward part of the vessel, about one-third of the way abaft the prow (Katzev). Although single-masted square-rigged vessels may have had the mast stepped somewhat forward of amidships (e.g., the Humber keel; see H. Warington Smyth, *Mast and Sail in Europe and Asia*, London, 2nd ed. 1929, 143), it is never stepped that far forward, for this drastically reduces the ship's efficiency on any but downwind courses (cf. Bowen in *MM* 45, 1959, pp. 332-37 and 46, 1960, pp. 303-306). It is entirely possible that the Kyrenia ship carried a lateen or a spritsail.

[83] *Periploi*, to give them their Greek name. See F. Gisinger in *RE* s.v. *Periplus* (1937). A few examples of those drawn up for use by merchant skippers are extant. The *Periplus of the Mediterranean and Black Seas*, completed ca. 225 B.C. and attributed to a certain Scylax, and the anonymous *Stadiasmus Maris Magni* of the 4th A.D. are true coast pilots. The *Periplus of the Erythraean Sea* of the first A.D. is more a trader's handbook. The texts of the first two are in Müller, *GGM*; the

of charts.[84] Every ship carried a leadline to sound depths, and the lead had a hollow on the underside which, filled with tallow or grease, brought up samples of the bottom.[85]

Signaling and identification equipment consisted of flags and lights. In addition to the pennants at the stern (348 below), merchantmen often flew one at the masthead (Figs. 145, 147, ship to the right) and occasionally at each yardarm (Fig. 149);[86] these probably served for both identification and signaling. Warships flew some standard (*semeion*) to indicate the country, or at least the fleet, they belonged to,[87] flagships flew the admiral's pennant,[88] and there were flags of

---

best text of the third is H. Frisk's *Le Périple de la Mer Érythrée* (Göteborgs Högskolas Årsskrift 33, Göteborg 1927).

[84] Cf. Gisinger, *op. cit.* (previous note) 842.

[85] The *katapeirates* (Latinized to *catapirates*) or *katapeirateria*. E.g., Herodotus 2.5.2: κατεὶς καταπειρητηρίην πηλόν τε ἀνοίσεις καὶ ἐν ἕνδεκα ὀγυιῇσι ἔσεαι "And heaving the sounding lead, you will bring up mud and find yourself in eleven fathoms"; Lucilius in Isidore, *Orig.* 19.4.10 (*ROL* III, p. 378): *hunc catapiratem . . . plumbi pauxillum rodus linique metaxam* "this sounding lead . . . a little lump of lead and a line of spun flax." "To heave the lead" in Greek was βολίζω; e.g., Acts 27.28: καὶ βολίσαντες εὗρον ὀργυιὰς εἴκοσι, βραχὺ δὲ διαστήσαντες καὶ πάλιν βολίσαντες εὗρον ὀργυιὰς δεκαπέντε "And they sounded and found it twenty fathoms, and when they had gone a little farther they sounded again and found it fifteen fathoms." A number of ancient sounding leads have been recovered; see, e.g., Benoit 179 and pl. 31.16 (cone-shaped, 25.5 cm. high, 3.5 cm. across at top, 6 cm. across at bottom, weight 5 kg. [9 7/8, 1 3/8, 2 3/8"; 11 lb.]), 31.17 (pyramidal, 21 cm. high, 8 cm. at the base, weight 4.3 kg. [8 1/4, 3 1/8"; 9 1/2 lb.]), fig. 95 (square shaft flaring into wide cone, 20 cm. high, 9 cm. across top, 16 cm. diameter of bottom, weight 13 kg. [7 7/8, 3 9/16, 6 5/16"; 28 lb. 9½ oz.]); the last has four cavities on the underside, apparently for tallow or the like. Gargallo (35 and fig. 13) found a very similar one, but much smaller (12 cm. high, 9 cm. across the bottom, weight 3 kg. [4¾, 3½"; 6½ lb.]) and with five cavities on the underside. A bell-shaped variety was found by G. Kapitän (*Klio* 39, 1961, pp. 308-11; 10 cm. high, diameter 7.5 cm. at top and 20 cm. at bottom, weight 13.4 kg. [3 15/16, 2 15/16, 7 7/8"; 29½ lb.]); there is a markedly deep hollow on the underside.

[86] Pollux (1.91) calls the pennant at the mast-head ὁ ἐπισείων, while Lucian lists τοῦ ἱστίου τὸ παράσειον πυραυγές (*Navig.* 5) as part of the striking decoration of the *Isis* (cf. NINE, App., pt. 2). The word *paraseion* is generally translated "topsail" (see *LSJ*, s.v.), but it is hardly natural for Lucian to say "the red topsail of the mainsail." If the *episeion* is the mast-head pennant, why cannot a *paraseion* be a yardarm pennant? Athenaeus, 5.206c, as emended mentions that on Ptolemy Philopator's sumptuous barge the sail was adorned "with a red *paraseion*" (ἀλουργεῖ παρασείῳ). It may be, however, that the ms. reading παρασείρῳ is correct, and the sail was adorned with a "red edge."

[87] Artemisia, the wily Queen of Caria, "had not only a Persian flag but a Greek one as well. If she herself gave chase to a Greek ship, she flew the Persian flag, but

various kinds for signaling[89] as well as burnished shields for helio-graphs.[90] When the mast was lowered, signal flags could be flown from the *stylis* (346 below). Every ship carried lights at night to

---

if she was herself chased by a Greek ship, she flew the Greek flag" (οὐ μόνον τὸ τῶν βαρβάρων, ἀλλὰ καὶ τὸ τῶν Ἑλλήνων σημεῖον εἶχεν. εἰ μὲν ἐδίωκεν αὐτὴ ναῦν Ἑλληνίδα, τὸ βαρβαρικὸν ἀνέτεινε σημεῖον, εἰ δὲ ὑπὸ Ἑλληνίδος νεὼς ἐδιώκετο, ἀνέτεινε τὸ Ἑλληνικόν Polyaenus 8.53.3). Chabrias once had the flags on his ships lowered, leaving the enemy in doubt as to their nationality "since they were not carrying the Athenian flag" (διὰ τὸ μὴ ἔχειν αὐτὰς τὸ Ἀττικὸν σημεῖον Polyaenus 3.11.11).

[88] Also called *semeion* in Greek; *vexillum* in Latin. Cf. Herodotus 8.92.2: ὡς δὲ ἐσεῖδε τὴν νέα τὴν Ἀττικὴν ὁ Πολύκριτος, ἔγνω τὸ σημήιον ἰδὼν τῆς στρατηγίδος "When Polycritus spied the Athenian ship, he caught sight of the pennant (*semeion*) and recognized the flagship"; Appian, *Bell. Civ.* 5.55: πλησίον τε ἦσαν ἀλλήλων ἤδη, καὶ αἱ ναυαρχίδες ἐκ τῶν σημείων ἐφαίνοντο "They [Antony and Ahenobarbus, in 40 B.C.] were by now near each other, and the flagships were revealed by their pennants"; Tacitus, *Hist.* 5.22: *praetoriam navem, vexillo insignem* "the flagship, marked by its pennant." The general Latin term for "device," *insigne*, could be used of the ad-miral's pennant; cf. Caesar, *Bell. Civ.* 2.6: "[In 49 B.C. at Marseilles, Brutus' ship] could easily be recognized by its pennant" (*ex insigni facile agnosci poterat*). The pennant might be flown from the mast when cruising but, since the sails were lowered in action, it was then necessarily carried elsewhere. Pictures of galleys on Roman coins show a pennant on the *artemon* mast (Fig. 121) or a square *vexillum* aft (Fig. 128; cf. *BM Empire* III, Hadrian no. 1391, pl. 84.13).

[89] A purple flag, the *phoinikis*, was a standard signal for going into action. Cf. Diodorus 13.46.3: καὶ τοῖς μὲν Λακεδαιμονίοις οὐδὲν ἐφαίνετο σύσσημον, τοῖς δ' Ἀθηναίοις Ἀλκιβιάδης μετέωρον ἐποίησεν ἐπίσημον φοινικοῦν ἀπὸ τῆς ἰδίας νεώς, ὅπερ ἦν σύσσημον αὐτοῖς διατεταγμένον "Among the Spartans there was no sign of a signal, but, for the Athenians, Alcibiades hoisted on his own ship a purple device that had been agreed on as the signal" (this was in 410 B.C.); Diodorus 13.77.4: ὁ Κόνων . . . ἦρεν ἀπὸ τῆς ἰδίας νεὼς φοινικίδα · τοῦτο γὰρ σύσσημον ἦν τοῖς τριηράρχοις "Conon . . . hoisted the purple flag (*phoinikis*) on his own ship; this was the signal for the commanding officers of the triremes" (cf. Polyaenus 1.48.2). For general signaling, ships carried flags of different shapes and colors for hoist-ing as well as semaphore flags; cf. Leo *peri thal.* 46 (Dain, p. 30): "Signal with the flag by holding it upright, dipping it to right or left and then shifting it back again to right or left, waving it, raising it, lowering it, removing it from sight, changing it, switching it around by orienting the head now in one direction and now another, or using flags of different shapes and colors as was the practice among the ancients" (τὸ δὴ σημεῖον ὑποσημαινέτω, ἢ ὀρθὸν ἱστάμενον, ἢ ἐπὶ δεξιὰ ἢ ἐπὶ ἀριστερὰ κλινόμενον καὶ ἐπὶ δεξιὰ πάλιν ἢ ἐπὶ ἀριστερὰ μεταφερόμενον, ἢ τινασσόμενον, ἢ ὑψούμενον, ἢ ταπεινούμενον, ἢ ὅλως ἀφαιρούμενον, ἢ μετατιθέμενον, ἢ διὰ τῆς ἐν αὐτῷ κεφαλῆς ἄλλοτε ἄλλως φαινομένης ἀλλασσόμενον, ἢ διὰ σχημάτων, ἢ διὰ χρωμάτων, οἷόν ποτε τοῖς παλαιοῖς ἐπράττετο).

[90] The Alcmeonidae were suspected of having "held up a shield for the Persians, by now back in their boats, to see" (τοῖσι Πέρσῃσι ἀναδέξαι ἀσπίδα ἐοῦσι ἤδη ἐν τῆσι νηυσί Herodotus 6.115). In 405 B.C., Lysander arranged for his scouts to report by "raising a shield when they were half way back" (ἆραι ἀσπίδα κατὰ μέσον τὸν πλοῦν Xenophon, *Hell.* 2.1.27). Plutarch adds the detail that they were to "raise a bronze

identify itself when traveling in company[91] and flagships carried special lights on the stern (Figs. 114, 128) to show the way during night movements.[92] Whether there were also conventionalized running lights is not known.

Sailing ships and warships had at least one ship's boat,[93] towed astern; merchantmen kept a hand stationed in it (Fig. 144).[94] Large

---

shield at the prow" (ἀσπίδα χαλκῆν ἐπάρασθαι πρῴραθεν Lysander 11.1). In 307 B.C., "Demetrius, when he was about one-third of a mile from the enemy, raised the signal agreed upon for going into action, a gilded shield, and this was flashed by each ship to the next" (Δημήτριος μὲν οὖν, τῶν ἐναντίων ἀποσχὼν ὡς ἂν τρεῖς σταδίους, ἦρε τὸ συγκείμενον πρὸς μάχην σύσσημον ἀσπίδα κεχρυσωμένην, φανερὰν πᾶσιν ἐκ διαδοχῆς Diodorus 20.51.1).

[91] Scipio's operating procedure for the overseas expedition against Carthage in 204 B.C. included the instruction that "warships were to carry one light each, transports two, and the flagship would be marked at night by carrying three" (lumina in navibus singula rostratae, bina onerariae haberent: in praetoria nave insigne nocturnum trium luminum fore Livy 29.25.11). Cf. Polyaenus cited in note 133 below.

[92] Xenophon, Hell. 5.1.8: νυκτὸς δ' ἐπιγενομένης φῶς ἔχων, ὥσπερ νομίζεται, ἀφηγεῖτο, ὅπως μὴ πλανῶνται αἱ ἑπόμεναι "When night came on, he [an Athenian admiral in 388 B.C.] led the way showing a light, as is customary, so the ships behind would not lose the course"; Appian, Bell. Civ. 2.89: περὶ ἑσπέραν ἀνήγετο ἐπαγγείλας τοῖς λοιποῖς κυβερνήταις πρὸς τὸν λαμπτῆρα τῆς ἑαυτοῦ νεὼς . . . εὐθύνειν "Toward evening, he [Caesar in 48 B.C.] set sail, instructing the rest of the captains to steer . . . by the lantern of his ship"; cf. Dio Cassius 49.17.2, Florus 4.8.9. For the placing at the stern, cf. Procopius, Bell. Vand. 1.13.3: κοντοὺς τε ὀρθοὺς ἀναστήσας ἐν πρύμνῃ ἑκάστῃ ἀπεκρέμασεν ἀπ' αὐτῶν λύχνα, ὅπως ἔν τε ἡμέρᾳ καὶ νυκτὶ . . . ἔκδηλοι εἶεν "[Belisarius, sailing to Africa in A.D. 533, on the three ships that carried himself and his staff] set poles upright on the poop of each and hung lanterns from them so that [the flagships] . . . would stand out both day and night." R. Lantier (Monuments Piot 46, 1952, pp. 75-76) describes an elaborate bronze ship's lantern found off the coast of Monaco in a wreck of the 1st A.D., and Gargallo a modest clay one found off Syracuse (Archaeology 15, 1962, p. 194 and fig. 4).

[93] The technical term in Greek was epholkion or epholkis, literally "for towing" (e.g., Plutarch, Dem. 17.3; Pomp. 40.5; Strabo 2.99, cited in note 95 below; Athenaeus 5.208f [cited in NINE, App., pt. 4]; Heliodorus, Aeth. 5.24.5; Philostratus, Vita Ap. 4.9 [cited in THIRTEEN, note 88] and 4.32; Achilles Tatius 3.4.1), though it was often loosely called a lembos (e.g., Demosthenes 32.6, 34.10), skaphos (Heliodorus, Aeth. 5.24.2), akatos (Agathias cited in note 96 below), etc. In Latin the technical term was scapha (cf. Dig. 33.7.29: scapha navis non est instrumentum navis "The ship's boat is not part of the ship's equipment") and it usually is called this (Plautus, Rudens 75; Caesar, Bell. Civ. 3.24, 3.101 and Bell. Alex. 46; Petronius 102.1; Paulinus of Nola, cited in note 96 below; cf. Acts, cited in the same note, where the Greek form is used), though other, looser, terms are found (e.g., Statius, Silvae 3.2.31: secuturam religent post terga phaselon "Make fast the skiff [phaselos; cf. 167-68 above] that will follow behind").

[94] In Petronius' Satyricon (102.5) one character reminds another that they cannot escape in the ship's boat because "there is a sailor lying on duty in the ship's boat

sailing ships could have more than one,[95] and could, by rigging lines from the masthead, haul them aboard.[96] Ship's boats could step their own mast with square sail (Fig. 143).[97]

The extra canvas and leather gear carried by Athenian warships is recorded in the naval records.[98] Each trireme had two coverings of some heavy material, most probably hide, which were hung, one over each side, as protection in action against missiles or grapnels; two lighter coverings of canvas for protection against spray; and others, almost certainly of canvas, that served as awnings.[99] Mer-

---

at all times, night and day" (*unum nautam stationis perpetuae interdiu noctuque iacere in scapha*).

[95] Eudoxus, in preparing for the circumnavigation of Africa, had "two ship's boats (*epholkia*) resembling pirate *lemboi*" ἐφόλκια δύο λέμβοις λῃστρικοῖς ὅμοια Strabo 2.99; on *lemboi* see 162 above. Hiero's superfreighter had a 78-ton *kerkouros*, some *haliades* half that size, as well as smaller boats (Athenaeus 5.208f [cited in NINE, App., pt. 4; on the *kerkouros*, see 163-66 above, on the *haliades* FOURTEEN, notes 35 and 52]).

[96] On St. Paul's ship, since the ship's boat was giving trouble under tow during the storm, it was hauled aboard (Acts 27.16). Off Malta, the crew tried to lower it in order to escape in it (27.30: χαλασάντων τὴν σκάφην εἰς τὴν θάλασσαν προφάσει ὡς ἐκ πρῴρας μελλόντων ἀγκύρας ἐκτείνειν "lowering the ship's boat [*skaphe*] into the sea on the pretext that they were going to stretch anchors from the prow"), but the soldiers aboard cut the lines and let it drop into the water (27.32). Paulinus of Nola (*Epist.* 49.1) tells how, when the anchor cables snapped, "The terrified sailors launched the ship's boat" (*nautae exterriti scapham dimiserunt*). Cf. Agathias 3.21 (Dindorf, *Historici Graeci Minores* II, p. 274): νῆες δὲ φορτίδες μεγάλαι . . . μετεώρους εἶχον τὰς ἀκάτους, καὶ ἀμφ' αὐτὰ δήπου τὰ καρχήσια τῶν ἱστῶν ἀνιμηθείσας καὶ βεβαιότατα αἰωρουμένας "Big merchantmen carried their boats (*akatoi*) suspended, hoisted right up to the mast-tops and hanging there in perfect safety" (this was a device to mount an attack on a high-walled city from the sea).

[97] See Nicander cited in TWELVE, note 19. The clearest pictorial example is the graffito of a big merchantman found in Pompeii (see NINE, note 34). Despite the absence of a towline, the small boat in Fig. 143 is probably intended as the ship's boat and not another freighter of a smaller class as has been suggested by E. Pfuhl in *Eph. Arch.*, 1937, pp. 92, 94.

[98] Included under the rubric "hanging gear"; see the passage cited in App. 2, note 3.

[99] The résumé of gear for triremes and quadriremes (cf. App. 2, note 3) includes *hypoblema, katablema, pararrhymata leuka* ("white *pararrhymata*"), *pararrhymata trichina* ("hairy *pararrhymata*"); the last two are always in the plural. All had been standard issue since ca. 370 B.C. (cf. *IG* II² 1609.40-50). That a trireme had two of each kind of *pararrhyma* is clear from 1611.244-53.

The name suggests that the *pararrhymata* were hung over the side, and this is confirmed by Xenophon, *Hell.* 1.6.19 (in 406 B.C., Conon moved his marines from deck to hold and hung the *pararrhymata* to conceal them; the only place they could chantmen unquestionably had spray-screens and extra canvas for awnings and replacement sails, but few details are available.[100] On the rope gear called *hypozomata* that was standard on both galleys and sailing ships, see 91-92 above.

Warships needed lighter mooring lines and ground tackle than merchantmen, since they were much less heavy in construction and most often were hauled up on the beach or put into slips instead of lying at anchor. In the Athenian navy, each trireme and quadrireme had two sets of lines, the one consisting of four lines of ca. 4.75 cm. (slightly under 2″) in diameter, the other of four lines of ca. 3.65 cm. (slightly under 1½″) in diameter; one set, probably the heavier, were the mooring lines, the other the anchor cables.[101] Each galley

---

possibly have been visible from were the side apertures in the outrigger [cf. Figs. 99, 101-102]. See also 2.1.22, where the sidescreens—here called *parablemata*, which must be a synonym—are again used for concealment). The "white *pararrhymata*" were of sailcloth, since they were stored in the same boxes as the sails (see note 61 above). The "hairy *pararrhymata*" must be of hide; cf. Polyaenus cited in FIVE, note 58. Similarly, Roman warships of the 1st B.C. had hide shields for protective use (which could be rigged to catch dew when water ran short; see Caesar, *Bell. Civ.* 3.15: *difficilioribus usi tempestatibus ex pellibus quibus erant tectae naves nocturnum excipere rorem cogerentur* "[Ships on blockade] experiencing rather bad weather, were forced to catch the night moisture with the skins with which they had been covered"). The triangular patch on the bow palisade of the galley in Fig. 81, brown in color in the original, may be a hide *pararrhyma*. The *pararryseis* that Danaüs in Aeschylus' *Suppliants* (715; cited in FOUR, note 49) sees on Aegyptus' ship as it approaches must be spray shields.

The *hypoblemata* and *katablemata* should be awnings. Down to 325 B.C. triremes had been equipped with one of each (1628.576-91, 326/5; 1627.440-53, 330/29; 1624.110-24, between 336/5 and 331/0; and compare 1611.238-43 with 244-53, 357/6); in 330 and 323 they were issued only a *katablema* (1627.454-72, 1631.257-68). Quadriremes, introduced into the navy just before 330 B.C., carried only *katablemata* from the very beginning (1627.459-72, 330/29; 1628.592-608 [cited in App. 2, note 3], 326/5; 1629.1068-85, 325/4; 1631.268-78, 323/2). The simplifying of this canvas gear took place at the same time as the attempt to simplify the rig; see App. 2.

[100] Cicero, *ad Att.* 4.19.1, refers to a crossing Atticus made in weather so rough they had to use δέρρεις "skins"; presumably these were the equivalent of a warship's "hairy *pararrhymata*." For the awnings on a Nile boat, see EIGHT, notes 41 and 49.

[101] Cf. the passages from the naval records cited in App. 2, note 3. The mooring lines and anchor cables are called *schoinia* "ropes," i.e., heavy lines as against the *topeia* or lighter cordage of the running and standing rigging (see note 20 above). The ropes described as "eight-finger" and "six-finger" must be 6″ and 4½″ (15 and 11.5 cm.) respectively in circumference rather than diameter, since lines of

carried two anchors with arms and shank of iron that weighed less than 50 lb. (22.7 kg.) but with a stock of lead that supplied added weight.[102] The anchors were carried hanging from catheads on either bow.[103] Each also had two landing ladders (cf. Figs. 90, 106, 117),[104] and three poles—one longer and two shorter—for fending off or for sounding in the very shallow waters that galleys alone could risk.[105]

Sailing ships had to have considerably heavier gear,[106] and large

---

such a diameter would be impossibly bulky for the purpose. One set was called ἐπίγυα "on-land [lines]," i.e., mooring lines; the other, ἀγκύρεια "anchor [lines]," i.e., anchor cables (e.g., 1611.254-57: σχοινία· . . . Ναυκράτιδι ἐπίγυα ΙΙΙΙ "lines: . . . the *Naukratis*, four mooring lines"; 1611.392-93: τὴν Ἡδίστην σχοινία ἀγκύρεια ΙΙΙΙ "the *Hediste*, four anchor cables"). Other terms for mooring lines are "off-shore [lines]" (ἀπόγαια Polybius 33.9.6; ἀπόγεια Lucian, *VH* 1.42 and *Dial. Mort.* 10.10, Polyaenus 6.8) or "stern [lines]" (πρυμνήσια *Od.* 15.498; Athenaeus 15.672c; Polyaenus 4.6.8; Achilles Tatius 2.31.6; Xenophon, *Eph.* 1.10.8). In Latin, they are called *orae*, literally "shores" (Livy 22.19.10, 28.36.11; Quintilian 4.2.41), and the anchor cables are *ancoralia* (Livy 22.19.10, 37.30.10; Pliny, *NH* 16.34, cited in note 133 below).

[102] For the number issued to each ship, see the passages from the naval records cited in App. 2, note 3. On the size and makeup, see note 131 below.

[103] Cf. Appian, *Syr.* 27: "When a Rhodian galley rammed one from Sidon, there was a jarring blow, and the Sidonian ship's anchor fell on the Rhodian craft and stuck fast" (Ῥοδίας νεὼς ἐς Σιδονίαν ἐμβαλούσης, καὶ τῆς πληγῆς εὐτόνου γενομένης, ἄγκυρα ἐκπίπτουσα τῆς Σιδονίας ἐς τὴν Ῥοδίαν ἐπάγη). Galleys could also be anchored by the stern; see Six, note 99.

[104] Called *klimakis*; cf. 1611.28-32: "total number of ladders 465, sufficient for 232 ships, with one ladder left over" (κλιμακίδων ἀριθμὸς ΗΗΗΗ Γ ΔΓ· αὗται γίγνονται ἐπὶ ναῦς ΗΗΔΔΙΙ καὶ μία κλιμακίς). Elsewhere the simple term *klimax* was used: cf. Euripides, *IT* 1351; Arrian, *Anab.* 1.19.5. Another common term was *apobathra* "gangway": cf. Herodotus 9.98.2; Thucydides 4.12.1; Lucian, *Dial. Mort.* 10.10 and *Toxaris* 20 (cited in note 133 below); Polyaenus 4.6.8. The Latin equivalents were *scala* (Livy 28.36.11; Vergil, *Aen.* 10.653-54) and *pons* (Statius, *Silvae* 3.2.54-55). See also DS s.v. *scala*, fig. 6148 = MM 27 (1941) 40 and pl. 3.

[105] Called a *kontos*. Cf. 1611.33-37: "total number of poles 677, sufficient for 225 ships with 2 poles left over" (κοντῶν ἀριθμὸς ΓΗ Γ ΔΔ ΓΙΙ· οὗτοι γίγνονται ἐπὶ ναῦς ΗΗΔΔ Γ καὶ κοντοὶ δύο). Several entries refer to a κοντὸς μέγας "big pole" (e.g., 1606.51, 1607.21), so the other two presumably were smaller (cf. the restoration κοντοὺ[ς μικροὺς ΙΙ] in 1606.93-94). In the battle of Patras in 429 B.C., when Phormio forced the enemy ships into an ever narrowing circle, "Ship fell against ship and were shoved apart with the boat poles" (ναῦς τε νηὶ προσέπιπτε καὶ τοῖς κοντοῖς διεωθοῦντο Thucydides 2.84.3). Latin took over the Greek term: cf. Vergil, *Aen.* 5.208-209; Tacitus, *Ann.* 14.5; Suetonius, *Caligula* 32.1.

[106] Cf. Aristophanes, *Peace* 36-37: σχοινία / τὰ παχέα συμβάλλοντες εἰς τὰς ὁλκάδας "braiding ropes, the heavy ones for merchant ships."

251

vessels were equipped with winches and capstans forward[107] to handle it; the machines presumably performed other tasks (e.g., hoisting the main yard) which might be hard on human muscle. There were multiple mooring lines and multiple cables (Fig. 150) to serve the rather large number of anchors carried; anchor chain was not unknown but its use was strictly exceptional.[108] For mooring at a quay, merchantmen carried landing ladders and gangways (Fig. 174).[109] For mooring off shore and for emergencies they carried a number of anchors, kept at the ready both fore and aft (Fig. 150),[110] including the "sacred" anchor[111] or what we call the sheet anchor.

Ancient anchors, once almost a complete mystery, are now probably the best-known piece of marine equipment, thanks to the recovery by divers and underwater archaeologists of hundreds of specimens of at least the parts made of lead or stone. Some seagoing ships still were equipped, as in Homer's day, with anchors all of stone, now in the form of a trapezoidal slab with a hole in the narrower upper part to take the cable (Fig. 187).[112] At the other end

[107] Cf. Lucian, *Navig.* 5 (describing the *Isis* [see NINE, App., pt. 2]): πρὸ τούτων αἱ ἄγκυραι καὶ στροφεῖα καὶ περιαγωγεῖς "in front of these [that is, the mast and sail, therefore on the foredeck], the anchors and winches and capstans." For pictures of marine capstans see Casson, "Harbor Boats," pls. 2.3 (= *IH* ill. 69), 5.1 and, for a sample of the kind used on land, Fig. 139.

[108] At the siege of Tyre in 332 B.C., after defending divers had cut the anchor cables time and again, "The Macedonians lowered the anchors with chains instead of ropes" (οἱ δὲ ἀλύσεσιν εἰς τὰς ἀγκύρας ἀντὶ σχοίνων χρώμενοι, οἱ Μακεδόνες, καθίεσαν Arrian, *Anab.* 2.21.6). Caesar, in describing the ships of the Veneti, points out as one of their peculiarities that their "anchors are held by iron chain instead of rope" (*Bell. Gall.* 3.13, cited in FOURTEEN, note 58).

[109] See also Fig. 144; Casson, "Harbor Boats" pls. 2.3, 5.1.

[110] Two graffiti in the Terme Stabiane at Pompeii show ships at anchor with a pair of anchors hanging from an end which Maiuri, who published the pictures (*op. cit.* NINE, note 34, pp. 26-28), identifies as the stern. Examination of the original reveals that in each a figure seated near the other end is unmistakably grasping a tiller bar, making that end the stern.

[111] Lucian, *Fugitivi* 13: τὴν ὑστάτην ἄγκυραν, ἣν ἱερὰν οἱ ναυτιλλόμενοί φασιν, καθιέναι "to lower the last anchor—the sacred one, as seafarers call it"; cf. *Jup. Trag.* 51.

[112] On anchors in Homer, see 48 above. Trapezoidal stone anchors were found in a wreck of the 4th B.C. off Taranto; see A. McCann's forthcoming article in *Archaeology* (1972). For other types of stone anchor, some of considerable complexity, see Frost 37-51 and K. Nikolaou and H. Catling, "Composite Anchors in Late

of the scale was the sophisticated anchor totally of iron and with a stock that could be removed, enabling the apparatus to lie comfortably and out of the way on deck—a type that disappeared at the end of the ancient world and was not in widespread use again until the nineteenth century. A particularly well-preserved example reveals that the whole—arms, shank, stock—could be cased in wood (Fig. 183).[113] Another type of iron anchor had a stock of different material, either wood or lead.[114] Most anchors, however, as the majority of finds amply demonstrate, were not of iron at all but two-thirds of wood and one-third of a heavy substance, either lead or stone.[115] The arms were of wood with metal tips—but no flukes[116]—and the

---

Bronze Age Cyprus," *Antiquity* 42 (1968) 225-28. Many are still in use today.

G. Kapitän is completing a definitive study of ancient anchors based chiefly on the hundreds of specimens of stocks and other anchor elements recovered from the sea floor.

[113] The anchor with removable stock, the so-called admiralty anchor, was reinvented by the Dutch in the eighteenth century and adopted by the British navy in the nineteenth. Another excellent specimen of iron anchor with removable stock was found near Pompeii; see Ucelli 239-40. There was no trace of its wood casing, although it may have had one. The iron anchors of the Yassi Ada Byzantine wreck probably had no wood casing (van Doorninck 231).

[114] Ucelli (241-242) mentions iron anchors found off North Africa, Italy, Sardinia, Sicily; for the south coast of France, see Benoit 172; for Sicily, see Gargallo, fig. 11; for Turkey, see Bass, "Yassi Ada" 542-46. None had their stocks preserved, indicating that these were made of perishable material, probably wood. On the Yassi Ada wreck, only two stocks were found for the 11 iron anchors aboard, and van Doorninck (237-41, 262-63) suggests that most of the stocks must have been of wood. For lead stocks on iron anchors, see note 131 below.

The arms of iron anchors were set in different fashions, which changed over the years, as G. Kapitän's studies have revealed (kindly communicated to me by letter). In Roman Republican times, the arms were straight and formed more or less an isosceles triangle with the shank. In Imperial times, they were curved to form a crescent (Fig. 183, Ucelli 239-40). In late Imperial times they became straight again, sticking out at right angles to the shank for most of their length, with but a slight upward curve at the ends (*AJA* 39, 1935, pp. 74-75, fig. 20; *Not. Sc.* 57, 1932, pp. 434-36, figs. 1, 2, anchor from Pisa, no exact date; Gargallo, fig. 11; the Yassi Ada Byzantine wreck [van Doorninck, figs. 77-88]).

[115] The Nemi barge had one of iron and another of wood with lead stock (Ucelli 235-47); Hiero's superfreighter had eight of the first type and four of the second (Athenaeus 5.208e, cited in NINE, App., pt. 4). Some of the anchors dedicated at Delos were of this second type: *Inscr. de Délos* 1417 A col. 1, lines 165-66, 155 B.C.: ἄγκυραν ξυλίνην τοῦ μολύβδου ἀποκεκομμένην "anchor of wood with its lead cut off."

[116] Flukes do not appear until late in the Roman Imperial period and only rarely then. The iron anchor from Pisa mentioned in note 114 above has flukes, and en-

shank was of wood; sometimes the two were held together by
wooden pins (Fig. 184), sometimes by a lead collar (Fig. 185).[117]
The essential weight the ancients put into the stock—the reverse of
subsequent practice, which was to put it principally into the arms
and shank. The cheapest and most primitive form of heavy stock was
of stone; it had a channel cut where the shank lay, and the two were
lashed permanently together.[118] As early as the sixth century B.C.,
stocks were made of lead,[119] and this eventually became standard.
There were several basic types of lead stock, including one that was
removable. Of those permanently fixed to the shank, some consisted
of a wooden casing filled with a core of lead (Fig. 186);[120] some were
just the reverse, a lead casing surrounding a core of wood;[121] still
others were of solid lead. All generally had a boxlike opening to
receive the shank (Fig. 184). In many instances there is a bar span-

---

gravings on sealstones of the 3rd and 4th A.D. seem to show them (Moll E II a, nos.
e 1, 9, 14, 15, 39), but no earlier examples are attested (Moll's line-drawings showing
anchors with flukes that he attributes to coins from Paestum minted between 282-
267 B.C. [E II a, nos. b 10, 11] are inaccurate; cf., e.g., the specimens illustrated in
P. Garucci, *Le monete dell'Italia antica*, Rome 1885, pl. 122.17, 33, 38).

[117] For examples of such collars found off Sicily, see Gargallo 33, figs. 5 and 8.

[118] Examples from Sicily, Gargallo 32, fig. 2; from Marathon Bay, F. Braemer
and J. Marcadé, "Céramique antique et pièces d'ancres trouvées en mer à la pointe
de la Kynosoura (Baie de Marathon)" *BCH* 77 (1953) 139-54, esp. 151 and fig. 12.

[119] The earliest example of a lead stock was recovered from a wreck of the
sixth B.C. found off Antibes (Benoit 170). This squares with the Greek tradition
that the anchor with arms (as against, e.g., a stone slab) was invented by Anacharsis,
whose *floruit* was ca. 600 B.C.; see Strabo 7.303 (εὑρήματά τε αὐτοῦ . . . καὶ τὴν
ἀμφίβολον ἄγκυραν "And his [Anacharsis'] inventions . . . [include] the two-armed
anchor") and cf. Pliny, *NH* 7.209 (addidit . . . bidentem Anacharsis "Anacharsis
. . . added [the invention of] . . . the two-toothed [type of anchor]").

[120] Gargallo 34, fig. 12; G. Kapitän in *Archaeology* 21 (1968) 63. The lead bars,
trapezoidal in section, described by Braemer-Marcadé (150 and fig. 11) could very well
be lead cores from this make of stock. The type fits perfectly the anecdote Diodorus
(5.35.4) tells about the Phoenician merchants who were so greedy that, when their
ships were fully loaded and there was still silver to be put aboard, they "hacked out
the lead in the anchors and used silver instead, to do the work of the lead" (ἐκκόπτειν
τὸν ἐν ταῖς ἀγκύραις μόλιβδον, καὶ ἐκ τοῦ ἀργύρου τὴν ἐκ τοῦ μολίβδου χρείαν ἀλλάτ-
τεσθαι); cf. the following note. General David Lewis has shown me specimens of
lead cores which were deliberately lightened by being poured about bits of light
volcanic stone.

[121] Braemer-Marcadé 146 and fig. 9. The authors, among others, have mistakenly
assumed that this is the type described in Diodorus' anecdote.

ning the box horizontally (Fig. 186). The bar was formed by the pouring of the lead or lead core with the shank in place—the shank had previously been bored through at the point that was to lodge in the box; when the stock was cast, the molten lead ran through the hole and hardened into a bar that wedded stock and shank inseparably.[122] The removable type of lead stock was relatively thin, flat save for a short ridge across the center, which formed a stop, and with a hole at the base of the ridge. A pin running through the shank into the hole held the two pieces together.[123] Stocks varied somewhat in shape. Some were straight, some straight along the lower edge but bowed along the upper, in some there were gentle lateral curves.[124] Stocks were occasionally decorated, the simplest with a series of astragals, the more elaborate with Medusa heads or dolphins, etc.[125] They sometimes bear inscriptions, very often pious.[126]

The anchors were big as well as numerous.[127] The wreck found off Mahdia, a good-sized ship of no less than 230-tons burden, had at least five, the largest of which, presumably the "sacred" anchor, had a stock that was just under 8 feet (2.35 m.) long and weighed over 1,500 lb. (695 kg.); the next had a stock even longer, about 8½ feet (2.46 m.), but weighing less (ca. 1380 lb. = 628 kg.).[128] Monsters

[122] Cf. Benoit 173. For examples of the boxlike aperture and transverse bars formed by the lead passing through the hole in the shank, see Braemer-Marcadé, figs. 9-10; Torr, fig. 45.

[123] Gargallo, fig. 9.

[124] Ucelli, fig. 276; Braemer-Marcadé, fig. 10; Torr, fig. 45; Benoit 171, 173.

[125] Gargallo 32 (astragals, dolphins), Benoit 170 (Medusa), FA 1959.4282 (caduceus). An unusual form of decoration has been found on a stock from a wreck of the 3rd or 2nd B.C., four tiny human faces with half-closed eyes (Gallia 18, 1960, p. 45 and fig. 13); perhaps they were apotropaic devices.

[126] Ucelli, fig. 276 = Torr, figs. 46-47: Ζεὺς Ὕπατος "Zeus the Highest," Ἀφροδίτη Σώζουσα "Aphrodite our Rescuer," Σώτειρα "Saviour"; SEG xi 18: Ἀφροδίτα Ἐπιλιμενία "Aphrodite of the Harbor"; FA 1959.4282: Veneri, "to Venus" on one face and Iovi "to Juppiter" on the other. Some Latin inscriptions seem to refer to the boat's owner: e.g., T. Fulvi Euti [for Eut(ych)i], FA 1959.4282; C. ACASICI [for C. Ac(ilius) Asici(us)?], Gallia 16 (1958) 36; see also F. Moll in AA 44 (1929) 267-70. The iron anchor from Nemi has its weight inscribed (1,275 Roman lbs.); see Ucelli, fig. 271.

[127] Braemer-Marcadé (147) lists some sample sizes and weights of lead stocks recovered from various points all around the Mediterranean. Gargallo notes (35) an iron anchor with shank 5 m. long.

[128] A. Merlin, op. cit. (NINE, App., pt. 5) 392-93.

such as these could have been handled only by a capstan. St. Paul's ship dropped 4 anchors astern to keep from being driven ashore on Malta, and had others ready forward.[129] The Yassi Ada Byzantine wreck, a small ship only 60-70 odd feet (ca. 19-22 m.) long, carried no fewer than 11 anchors.[130]

The warship's anchor was a variation on the merchant ship's. To cut down on bulk, arms and shank were of iron; the total weight of this part of a trireme's anchor could be less than 50 pounds (22.7 kg.). To this a stock of lead was clamped on with iron clamps to supply the needed weight. Examples of stocks weighing ca. 83 lb. (38 kg.) and 240 (109 kg.) are recorded; since the date is the second century B.C., the latter figure could be for a ship a good deal bigger than a trireme.[131]

[129] Acts 27.29 (ἐκ πρύμνης ῥίψαντες ἀγκύρας τέσσαρας "casting four anchors from the stern") and 27.30 (cited in note 96 above). The ship pictured in Fig. 150 carries an anchor on the starboard quarter.

[130] van Doorninck 211-59. They were iron with movable stocks; some of the stocks were iron, while others may have been of wood with a thin iron bar as core (240-41). They were large (length of shank roughly 2-2.5 m., width across arms roughly 1.2-1.5 m.) but light: one group of three, the heaviest, weighed ca. 306 lbs. [139 kg.], a second group of six, the lightest, weighed ca. 175 lbs. [79.5 kg.]. Two were kept at the ready by each gunwale, and the rest carefully stacked on deck over the amidships line forward of the mast, with the heaviest group on the bottom (248-58).

[131] An entry in the Athenian naval records (*IG* ii² 1627.282-86, 330/29 B.C.) reads as follows: ἀγκύρας σιδηρᾶς, σταθμὸν μναῖ ΔΔ[..] δεσμὰ σιδηρᾶ δόκιμα τὰ ἐκ τῶν λίθων ἐγλυθέντα σὺν τῷ μολύβδῳ ἀριθμὸς ΗΗΗΔΔΔΓ "iron anchor, weight [exact figure lost; between 20 and 50 lb.]; total number of iron clamps in usable condition removed from the 'stones' along with the lead, 335." This must be taken together with certain entries in the temple accounts of Delos listing galley anchors that had been dedicated. E.g., *Inscr. de Délos* 443 Bb 92 (178 B.C.): ἄγκυρα σιδηρᾶ καὶ λίθος μολυβδοῦς ὁλκὴ τάλαντα ΙΙΙΙ μναῖ Δ Δ "iron anchor and 'stone' of lead, weight 4 talents, 20 mnas [ca. 250 lb. = 113.5 kg.]," and 96: ἄγκυρα σιδηρᾶ λίθον οὐκ ἔχουσα "iron anchor without a 'stone.'" In other words, the first was complete, iron shank and arms plus a 250 lb. lead stock; the second consisted only of the iron shank and arms. The first, it would seem, came from a big galley, for a stock that weighed only 1 talent, 26 mnas, or about 8.3 lb. (38 kg.) occurs in another entry (421.17, circa 190 B.C.). That these anchors at Delos must be from galleys is rendered almost certain by the nearby entry νεὼς ἔμβολον "ship's ram" (443 Bb 90). Still another entry (1417 A col. i, line 167) mentions an "iron anchor complete" (ἄγκυραν σιδηρᾶν ἐντελῆ).

The word "stone" that appears in some of the above entries must be a technical term for the stock—which, in the early versions of the anchor with arms, was very likely of stone; cf. note 118 above.

The grapnel anchor, though it very likely was used earlier, is not attested until the seventh century A.D. [132]

Pieces of cork were a regular part of a ship's equipment, serving both as marking buoys for anchors or the like and as life preservers.[133] Tools were carried for the ship's carpenter.[134]

This section can fittingly be closed by a relevant passage from a business document drawn up in A.D. 212 that has emerged perfectly preserved from Egypt's perennially dry sand.[135] It is a lease assigning for long-term use a boat and all that went with it. Other similar documents are extant but none gives such welcome detail; and though the boat is a rivercraft of a scant 10 tons burden (400 artabs), it carried the same basic gear as any seagoing type. The vessel was

[132] One was found in the Yassi Ada Byzantine wreck (van Doorninck 231). Its small size—the span of the arms is but 16-16½" (40-42 cm.)—would indicate that it was for the ship's boat.

[133] For marking buoys, see Pausanias 8.12.1 (ἀπ' αὐτοῦ καὶ ἐν θαλάσσῃ ποιοῦνται σημεῖα ἀγκύραις καὶ δικτύοις "From it [the cork oak] they even make things for use on the sea, markers for anchors and nets"); Pliny, NH 16.34 (usus eius ancoralibus maxime navium piscantiumque tragulis "It [cork oak] is used particularly for ships' anchor cables and fishing nets"); Polyaenus 6.11 (a commander, wishing to elude chase at night, tricked the enemy into following a will o' the wisp by dousing his ship's lights and "setting adrift others anchored to big pieces of cork" [ἑτέρους καθῆκαν ἐς τὴν θάλατταν φελλοῖς μεγάλοις ἐφηρμοσμένους]). For life buoys, see Lucian, Toxaris 20 (φελλούς τε γὰρ πολλοὺς ἀφεῖναι αὐτοῖς καὶ τῶν κοντῶν τινας ὡς ἐπὶ τούτων ἀπονήξαιντο, εἴ τινι αὐτῶν περιτύχοιεν, καὶ τέλος καὶ τὴν ἀποβάθραν "They threw a lot of cork floats to them [i.e., two men overboard] and some of the boat poles so that, if either could get hold of one, he could swim off on it; finally they even threw over the gang-plank").

[134] The Yassi Ada Byzantine wreck has yielded a complete set—hammers, adze, scraper, file, chisel, gouge, wood-boring bit, etc.; see M. Katzev and F. van Doorninck, Jr., "Replicas of Iron Tools from a Byzantine Shipwreck," Studies in Conservation 11.3 (August 1966) 133-42. An ax and adze found in a wreck of the 1st B.C. off Anthéor; see F. Benoit, "Jas d'ancre et pièces d'outillage des épaves de Provence," RSL 21 (1955) 117-28, esp. 127-28.

[135] P. Lond. 1164 h (= Select Papyri, no. 38) lines 7-11: σὺν ἱστῷ καὶ κέρατι καὶ λιναρμένῳ καὶ σχοινίοις καὶ κάδοις καὶ κρίκοις καὶ μαγγάνοις καὶ πηδαλίοις δυσὶ σὺν οἴαξι καὶ ὄκνοις (= ὄγκοις) καὶ κώπαις τέσσαρσι καὶ κοντοῖς πέντε σὺν θηλαῖς σιδηραῖς καὶ θυραβάθραις καὶ διαστῆρι κλιμακίῳ καὶ ἐργάτῃ καὶ ἀγκύραις σιδηραῖς δυσὶ σὺν σπάθαις σιδηραῖς καὶ μονοβόλῳ ἑνὶ καὶ σχοινίοις σεβενίνοις καὶ παρόλκῳ καὶ σχοινίοις ἀπογλοις καὶ ἐμβόλια τρία καὶ μέτρῳ ἑνὶ καὶ ζυγῷ καὶ Κιλικίῳ καὶ κατώτιον κατὰ γευστρίδα σὺν κώπαις δυσὶ ἐξηρτισμένον πᾶσι τοῖς ἀνήκουσι καὶ ὀβελίσκῳ σιδηρῷ.

leased along with "mast, yard, linen sail, ropes, jars, rings, blocks,[136] two steering oars with tiller bars and brackets, four oars, five boat poles tipped with iron, companionway ladder, landing plank, winch,[137] two iron anchors with iron stocks, one one-armed anchor,[138] ropes of palm fiber, tow rope, mooring lines, three grain chutes, one measure, one balance yard, Cilician cloth,[139] cup-shaped two-oared skiff fitted with all appropriate gear and an iron spike."[140]

[136] Some special kind of jar must be meant, perhaps a type used for bailing. The rings (κρίκοι) would be fairleads of various sizes, etc.; such rings, generally made of lead, are commonly found in ancient wrecks (Benoit 176; Gargallo in *Archaeology* 15, 1962, p. 195, fig. 5; *National Geographic*, June 1970, 853-54). The word for block here and in other similar documents is μάγγανον (cf. *P. Mon.* 4.14 [A.D. 581]). More literary terms are τροχός "sheave" (Synesius, *Epist.* 4.163c, cited in App. 3, note 2), τροχιλεῖον "block" (Athenaeus 5.208e, cited in NINE, App., pt. 4), and τροχιλεία (*trochlea* in Latin) "block and tackle" (see *LSJ* s.v.).

[137] "Companion-way ladder" is *thyrabathra*, literally "opening-ladder," probably a ladder to get from the deck-hatch to the hold; "landing plank" is *diaster klimakios*, literally "ladder-spanner"; "winch" is *ergates*, which occurs in literary passages as well (see *LSJ* s.v.; for other terms, see note 107 above).

[138] The two iron anchors have iron σπάθαι, literally "swords"; these must be the stocks, whose material is specified since on certain iron anchors they could be of wood or lead (see note 114 above). For one-armed anchors, see Breusing 108 and Ucelli, fig. 274; such anchors must be set out by hand, and hence are useful only on rivercraft or the like. They apparently were common on the Nile, since other documents mention them (*Sammelb.* 9683, 4th A.D.; *P. Lond.* 1714.32, A.D. 570).

[139] A coarse cloth, perhaps here used for tarpaulins.

[140] The iron spike perhaps was for driving into the bank to serve as a stake for tethering the skiff. Such stakes, called *tonsillae*, are mentioned by Latin writers (Festus 538.28: *palum dolatum in acumen et cuspide praeferratum . . . quem configi in litore navis religandae causa* "a stake, hewn to a point and tipped with iron . . . which is stuck into the shore for tying a boat to." He cites examples of use from Pacuvius and Accius [*ROL* II, pp. 250, 522]).

APPENDIX 1

THE TERMINOLOGY FOR RIGGING

A NUMBER OF terms have been identified with a fair degree of certainty. Of the running rigging, there is little doubt that *podes* "feet" was the Greek and *pedes* the Latin for sheets,[1] *hyperai* "uppers" the Greek for braces,[2] *kaloi* or *kaloes* the Greek and *rudentes* the Latin for brails.[3]

[1] Odysseus, after finishing the hull of the boat that was to take him from Calypso's island, "fitted it with braces, brails, and sheets" (ὑπέρας τε κάλους τε πόδας τ᾽ ἐνέδησεν ἐν αὐτῇ *Od.* 5.260). Cf. Euripides, *Or.* 706-707: καὶ ναῦς γὰρ ἐνταθεῖσα πρὸς βίαν ποδὶ / ἔβαψεν, ἔστη δ᾽ αὖθις ἢν χαλᾷ πόδα "for a ship, with the sheet close-hauled and straining, heels in the water, but rights itself when the sheet is slacked off"; Lucian, *Charon* 3: ἐνδοῦναι ὀλίγον τοῦ ποδός "ease the sheet a bit [i.e., when caught by a sudden squall]"; Cicero, *ad Att.* 16.6.1 *pedibus aequis* (cf. TWELVE, note 43) and Ovid, *Fasti* 3.565 *pede aequo* "with equal sheet(s)," i.e., with the wind dead astern so that the yard was at right angles to the keel, and the port and starboard sheets, therefore, slacked off the same amount. See also the passages cited in TWELVE, note 13.

The term *propes*, literally "forward sheet," cited from Turpilius (2nd B.C.) by Isidore of Seville (*Orig.* 19.4.3) may possibly mean "tack-rope."

[2] The braces, attached to the yardarms would naturally be the "upper" ropes. The conclusive evidence is the proverb quoted by Hyperides (Frag. 181): ἀφεὶς τὴν ὑπέραν τὸν πόδα διώκει "Drop the *hypera* to go after the *pous*." In adjusting sail, the only ropes that are always and of necessity handled—and therefore the only ones to which this proverb could apply—are the brace and the sheet. Since we know that the *pous* is the sheet, the *hypera* must be the brace. Harpocration, who cites the saying (s.v. ἀφείς), explains it as referring to "those who disregard what is the more important and waste time on trifles." In adjusting a square sail, the first rope to be set was the brace, which swiveled yard and sail horizontally into position to catch the wind; the trimming of the sheet came second—and the trimming did precious little good if the yard and sail had not first been placed at the right angle to the wind by the brace.

The Latin is unknown; whenever braces call for mention Latin authors refer to them by vague terms such as *funis* or *rudens*, both meaning "rope" (cf. Ovid, *Met.* 3.615-16 *conscendere summas / ocior antemnas prensoque rudente relabi* "[none] swifter at climbing to the top of the yard and, seizing a rope [most probably a brace], sliding down"). The word *versoria* has been suggested (de Saint-Denis 32-33, Rougé 53) but with no good reason. It occurs but twice in Latin literature, both times in the same phrase *cape vorsoriam* "Stand by the *versoria*" (Plautus, *Merc.* 875, *Trin.* 1026). The context makes it clear that the command has to do with reversing course—a maneuver that involves helm and sheets as well as braces, and, what is more, *both* braces and not just one.

[3] Plato, *Protagoras* 338a: πάντα κάλων ἐκτείναντα, οὐρίᾳ ἐφέντα, φεύγειν εἰς τὸ πέλαγος τῶν λόγων "slacking off on every brail, driven by a fair wind, to flee into the open waters of words"; similar metaphorical use occurs in Aristophanes, *Knights* 756, Euripides, *Medea* 278, Oppian, *Hal.* 2.223 (slacking off on the brails allowed the sail to belly out; see Fig. 90). For the Latin, cf. Lucan 5.426-27: *totosque ru-*

Of the standing rigging, *protonos* was the Greek for forestay,[4] and *epitonos* for backstay.[5] And we can add, thanks to the papyri, *opisthyperai* "back-braces," i.e., preventer braces.[6]

Many attempts have been made to pin down still others but with no great success. The evidence, for the most part by-the-way remarks in literature or explanatory notes furnished by late (and often lubberly) compilers and commentators, is vague or inconclusive or downright contradictory. The most reliable is what can be culled from the records of the Athenian Navy Yard (see FIVE, note 1). One particularly useful entry, repeated again and again in identical form, is that listing the rigging of a quadrireme.[7] Each ship had

καλῳδίων μηρύματα	ΔΓ III	"18 coils of lines"
ἱμάντες	II	"2 *himantes*"
ἄγκοινα διπλῆ		"double *ankoina*"
πόδες	II	"2 sheets"
ὑπέραι	II	"2 braces"
χαλινός		"*chalinos*"

The quickest—and most graphic—way of demonstrating the total lack of agreement about these terms (save, of course, *podes* and *hyperai*) is by means of a chart showing the translations that have been offered:

---

dentes / *laxavere sinus* "The brails let loose all the folds [of the sail]"; 3.44-45: *legere rudentes* / *et posito remis petierunt litora malo* "They took up on the brails [to furl sail] and, lowering the mast, made for shore under oars"; Pliny, *Epist.* 8.4.5: *immitte rudentes* "Slack off the brails!" (a command in getting under way).

[4] For Homer, see FOUR, notes 31, 32; Lucian, *Navig.* 5 (cited in ELEVEN, note 34); *P. Col. Zen.* 100, *P. Cairo Zen.* 59754 (cited 261 below). *LSJ* erroneously offers "halyard" as a second meaning, misunderstanding its use in the passages cited in TWELVE, note 26.

[5] For a long while only *hapax legomenon* in Homer (*Od.* 12.423), but now attested as well in *P. Col. Zen.* 100 (cited below, 261).

[6] *P. Col. Zen.* 100, *P. Cairo Zen.* 59756, cited below, 261. Several ship-pictures (Fig. 154; *IH* ill. 50) show yardarms fitted with two braces each: the regular brace maintaining tension toward the stern, and another which leads back from the regular brace to maintain tension in the opposite direction; the first is the *hypera* and the other must be the *opisthypera*. In a relief (2nd A.D.) of a seagoing ship found at Palmyra, the starboard preventer brace is prominently indicated; see H. Ingholt, *Gandhāran Art in Pakistan* (New York 1957) pl. VI.2 = Kraeling, *op. cit.* (NINE, note 34) pl. 38.3.

[7] E.g., *IG* II² 1627.148-153. For a detailed discussion of these entries, see Böckh 144-58, Torr 82-85, Assmann, *Segel* 1051 (who is followed by Miltner 944).

	Böckh	*Torr*	*Assmann-Miltner*
καλῴδια	stays, esp. shrouds	brails	stays
ἱμάντες	lifts	halyards	halyards
ἄγκοινα	parral	forestay	parral
χαλινός	halyard	backstay	[none offered]

To the evidence of the inscriptions can now be added that of three inventory lists of ship's rigging found on papyri. The documents in question all date from the third century B.C. Presumably they refer to Nile sailing craft; however, since these carried squaresails, they were fitted with the same ropes as any square-rigger. The entries which concern us particularly are the following:

P. Col. Zen. 100		*P. Cairo Zen.* 59756		*P. Cairo Zen.* 59754
ἄγκοινα	α	ἀγκοίνη π(ηχῶν) κβ	α	
ὑπέραι	β			
ὀπισθοπέραι	β	ὀπισθυπέρα π(ηχῶν) ιβ	α	
ἱμάντες	β	ἱμάντες παλαιοί	β	ἱμάντες
πρότονος	α			πρότονος
ἐπίτονοι				

The most important piece of information to be derived from these documents is a precious detail contained in *P. Cairo Zen.* 59756: the length of the *ankoina* and the *opisthypera*, 22 cubits as against 12. Torr had identified the *ankoina* (= Latin *anquina*) as the forestay—but a forestay, running from the tip of the mast to a point forward, would be only a bit longer than a brace, not almost twice as long. Moreover, the other two papyri reveal that the standard nautical term for forestay was still what it had been in Homer's day—*protonos*. Böckh and Assmann had identified the *ankoina* as the parral[8]—but a parral, a collar of relatively light line which carries no weight (the weight of a yard is borne by the halyard and lifts) is neither "double" nor twice the length of a brace. Morrison (*GOS* 300-301) has suggested "shrouds," but this would mean that the shroud was a single piece of line run from gunwale to masthead to the other gunwale, and shrouds are not usually so rigged.

There is only one rope that completely fits the requirements—and fits

[8] Followed by de Saint-Denis, who, in treating the Latin equivalent *anquina* (7), identifies it as the parral.

261

them perfectly: the halyard. A halyard, which must reach from the deck to a block at the top of the mast and down again, in the nature of the case will be nearly twice as long as a brace, which must reach only from the deck to the yardarm. Moreover, the halyard neatly satisfies the etymology of the word: ἄγκοινα is akin to ἀγκάλη, it is that which is "bent."[9] A halyard is the one rope that is at all times "bent": going up, as it does, to the masthead and then down again, part of it bends, hairpin fashion, over the sheave in the block at the masthead. The ἄγκοινα διπλῆ, the "double halyard" of the naval records, offers no problem. On all sailing ships of any size, modern as well as ancient, raising a heavy sail is a back-breaking job. The standard way of making it easier was by introducing a tackle—an arrangement whereby the halyard was doubled, looping around two blocks at the top of the mast (clearest in Fig. 144).[10] Alcaeus, in describing a storm, mentions that the "ankoinai work loose";[11] anyone who has been on a sailing craft in a blow will remember that one of the lines that generally works loose is the halyard. Alcaeus' use of the plural also points to the halyard, since halyards were double even in his day (cf. Figs. 81, 90).[12]

[9] *LSJ* s.v. defines the word as "halyard." Warmington so takes it (*ROL* III, p. 193) in a preserved verse of Lucilius: *funis enim praecisus cito atque anquina* [ms: *anchora*] *soluta* "the cable was quickly cut and the halyard cast off" (the poet is describing how a skipper saved mast, sail, and all the other tackle when a storm suddenly hit his moored ship; he must have cut the anchor loose as his first step in order to get into open water, and, as his next, lowered the massive, wildly swinging yard to the deck by releasing the halyard). A line preserved from Cinna (Frag. 5, Baehrens) by Isidore (19.4.7) reads *atque anquina regat stabilem fortissima cursum* "and that the mighty halyard maneuver a stable course," a reference to loosening the halyard in order to lower the sail and so make a vessel ride on a more even keel (275 below); the halyard is "mighty" since, having to carry the weight of a vast spar nearly as long as the ship, it was no doubt the stoutest line of the running rigging.

[10] For a photograph of the detail, see Meiggs, pl. 21, and for a line-drawing, Torr, fig. 30.

[11] The manuscript reading is χόλαισι δ' ἄγκυραι. Since "the anchors grew slack" makes no sense, ἄγκονναι has been suggested instead (so Page, *op. cit.* [FOUR, note 33] 148.9; for the identical scribal error, see the passage from Lucilius cited in note 9 above).

The suggestion has been made (Rougé 52) that the *chalatorii funes* in Vegetius, *re mil.* 4.46 are halyards. The term appears nowhere else, the reading is very doubtful (the manuscripts read *collatorio* or *collatorios*), and Vegetius makes it fairly clear that he is talking not of halyards but lifts ([*funes*] *quibus antemna suspenditur* "[ropes] from which the yard hangs").

[12] On the double halyard, cf. Nonnos, *Dionys.* 4.228-29: κόλπωσεν ἀχείμονι λαῖφος ἀήτῃ / διχθαδίους δὲ κάλωας ἐφαψάμενός τινι γόμφῳ "He bellied the sail before the mild breeze and made the double halyard (*kaloes*) fast to a pin."

The *himantes*, inasmuch as they are mentioned in all three papyri as well as in the naval records, must be an essential part of the rigging. A still further clue is the fact that, as the listings show, a vessel carried more than one, generally a pair. Since we know the names for sheets, braces, and the halyard, we can by elimination identify the *himantes* as the last pair of ropes so far unaccounted for, the lifts, the ropes that run from masthead to yardarm. In later documents the lifts are called *keroiakes* or *kerouchoi* (= *ceruchi* in Latin), literally "horn-holders," the word for yardarm meaning literally "horn" (232 above).[13]

Of the six terms listed in the naval records, only the "coils of lines" and the *chalinos*, literally "bit," remain to be identified. For the latter I would suggest the parral. This, an arrangement of twisted cords that held the yard against the mast much as a bit is held against a horse's mouth, perhaps was reminiscent of a bit with the reins attached. Thus the eighteen "coils of lines" are left to apply possibly to the brails, possibly to both brails and standing rigging.

[13] Lucan 8.177: *instabit summis minor Ursa ceruchis* "The Little Bear will stand on the highest point of the lifts"—a natural place, since the lifts were the topmost lines of the rigging, running right to the head of the mast; 10.494-95: [in a fire on board the flames leaped so high that] *tempore eodem / transtraque nautarum, summique arsere ceruchi* "At one and the same time both the rowers' benches and the tops of the lifts were ablaze"; Lucian, *Navig.* 4: ἀνιόντα τὸν ναύτην διὰ τῶν κάλων, εἶτα ἐπὶ τῆς κεραίας ἄνω ἀσφαλῶς διαθέοντα τῶν κεροιάκων ἐπειλημμένον "the sailor swinging himself aloft by the ropes and then running the length of the yard along the top in perfect safety holding on to the lifts" (which offered convenient points of support, running as they did from intervals along the yard to the masthead; see, e.g., the ship in Fig. 144, which has four pairs). Torr (93, note 200) is in error in taking κεροῦχος to be the same as καρχήσιον "mast-top," and *LSJ* in equal error in defining κεροῦχοι as "braces." In all citations given, the meaning "lifts" suits very nicely.

The term κεροῦχος is not attested in Classical Greek, although the Latin loanword *ceruchus* is sure evidence of its existence. *LSJ* cites Pherecrates, *Frag.* 12 (Kock); the word here is not Pherecrates' but a 17th century restoration which makes scant sense (Breusing [67] long ago suggested κεραία, which would be distinctly better).

APPENDIX 2

THE WARSHIP'S EMERGENCY RIG

THE prevailing opinion would have it that the *akation* or *dolon*, the warship's emergency rig, was not a smaller-scale version to replace the working mast and sail, but rather what we see on certain Roman galleys (Figs. 122, 127), a bow-sail like the merchantman's *artemon* (240 above).[1]

If purpose alone is considered, this view has little to recommend it. A bow-sail is chiefly a help to stability or steering: it enables the vessel to sail more efficiently by keeping the bow from digging in, and, when braced about in the appropriate direction, can aid the steering. It also provides drive—but only in proportion to its size, and its size on a Roman galley was strictly limited by its position over the prow, a position it had to have for exercising the functions just described. On the other hand, a sail carried amidships is a driver pure and simple, and, although an emergency rig designed specifically for handling on a crowded warship during action will perforce be smaller than the working rig, it will beyond the shadow of a doubt be considerably larger than any bow-sail; the greater the size the greater the drive, and drive is what a galley in flight needs above all else. Conformably, when ancient writers refer to the emergency sails, it is often with the distinct implication that these are a substitute for the larger working sails; Livy, moreover, reveals that a fleeing warship had to raise and step its *dolon*, which makes it sound far more like an emergency rig than a bow-sail, whose unobtrusive location

[1] Breusing 68-78; Assmann, *Segel* 1050; de Saint-Denis 53, 56, 113; Rougé 59. The two last argue for identification with the bow-sail on the grounds that its mast would be conveniently available since it "sans doute . . . restait debout pendant le combat" (de Saint-Denis 56). No doubt this bow-mast did so remain—but a *dolon*-mast did not, as Livy expressly says (see ELEVEN, note 63); de Saint-Denis failed to see the contradiction, while Rougé saw it but dismissed it. The sole evidence connecting the *dolon* with the prow is the word of the late and bookish lexicographer, Isidore of Seville, in a context that is sadly confused. Isidore reports (*Origines* 19.3.3) that "the *dalum* is the smallest sail and is set at the prow. The *artemo* was devised for directing a ship rather than for speed" (*dalum minimum velum et ad proram defixum. artemo dirigendae potius navis causa conmentatum quam celeritatis*). Isidore has here defined perfectly the function of a small sail set at the bow—but he (or his copyists) so misunderstood matters as to assign the function to a sail of one name and the size and location to a sail of another. And it is only a guess that "*dalum*"—the reading of all the manuscripts—is a mistake for *dolon*.

would presumably allow it to be carried in place at all times.[2] Yet all the above considerations, cogent though they may be, are admittedly inconclusive: if a vessel is reported as having lowered her big sail and raised her smaller, it can still be argued that the latter, as on Roman warships, was set on a bowsprit-like mast forward rather than a jury mast amidships, and that this bowsprit-like mast was unstepped and stored away when a vessel went into action.

There is yet a second problem. For the fourth century b.c. we have at our disposal the invaluable evidence of fragments of the official records of the Athenian navy yard. These reveal that triremes were equipped with "boat mast" and "boat yard" only until shortly after the middle of the century.[3] Was the use of emergency rigs discontinued? If so, why only

[2] Cf. the citations from Xenophon in ELEVEN, notes 56, 57, from Procopius in ELEVEN, note 63, and Epicrates as quoted in Athenaeus 11.782f [Edmonds II, p. 354]: κατάβαλλε τἀκάτεια, καὶ κυλίκια / αἴρου τὰ μείζω "lower the boat-size (cups) and hoist the bigger goblets." For Livy's statement (36.44.3), see ELEVEN, note 63.

[3] Cf. Torr 82-84. The last appearances are IG II² 1622.152, 321-22, 347-48 (342/1); the mention in connection with only 3 triremes out of at least 18 or so seems significant. The records continue from 334 to 322 (1623-32), and it is absolutely certain that, in this later period, boat yards and masts were a thing of the past. Numerous entries spell out precisely what a ship's equipment consisted of, and invariably only a *histos* and *ķeraia,* working mast and yard, are included (1624.105-24; 1627.436-53, 454-72; 1628.576-608; 1629.1050-85; 1631.257-78). A typical entry (1628. 576-608) runs as follows:

"trierarchs booked as having received for a cruise a complete set of [trireme] hanging or wooden gear—
 those [booked as having] hanging gear are in possession of the following: undergirds, sail, running and standing rigging [*topeia*], awnings [a *hypoblema* and a *ķatablema*], white side covers [*pararrhymata leuķa*], side covers of hide [*pararrhymata trichina*], four eight-finger lines, four six-finger, two anchors
 those [booked as having] wooden gear are in possession of the following: set of oars, steering oars, ladders, mast, yard, poles.
"trierarchs booked as having received for a cruise a complete set of quadrireme hanging or wooden gear—
 those [booked as having] wooden gear are in possession of the following: set of oars, steering oars, ladders, mast, yard, poles
 those [booked as having] hanging gear are in possession of the following: undergirds, sail, side coverings of hide, white side coverings, awnings [only *ķatablemata*], running and standing rigging, two anchors, four eight-finger trireme lines, four six-finger."

ὅσοι τῶν τριηράρχων γεγραμμένοι εἰσὶ ἔχοντες εἰς πλοῦν ἐντελῆ σκεύη κρεμαστὰ ἢ ξύλινα,
ὅσοι μὲν κρεμαστά, τάδε ἔχουσιν· ὑποζώματα, ἱστίον, τοπεῖα, ὑπόβλημα, κατάβλημα, παραρύματα λευκά, παραρύματα τρίχι(να), σχοινία ὀκτωδάκτυλα ΙΙΙΙ, ἑγδάκτυλα ΙΙΙΙ, ἀγκύρας δύο
ὅσοι δὲ ξύλινα, ἔχουσιν· ταρρόν, πηδάλια, κλιμακίδας, ἱστόν, κεραίας, κοντούς·

265

at Athens, since they obviously continued elsewhere for many a century? It so happens that there is an answer to these questions, and it happily settles the nature of the rig as well.

Careful study of the entries in the records reveals that the emergency rig did not die out but yielded to a substitute arrangement. Mention of the boat yard and mast ceases precisely when mention of a new type of sail, the "light" sail, begins: boat yards and masts make their last appearance in records drawn up in 342/1; "light" sails make their first in the very next fragment preserved, dated 334/3,[4] and continue until 323/2, the date of the latest fragment we have.[5] Thus, the trireme *Pandora*, whose equipment before 342 specifically included a "big yard" and by implication complete sailing gear of both sizes, in 323 has only a working mast and yard—*but now rigged with a light sail*.[6] The abrupt dropping of the one and introduction of the other can hardly be coincidence. During these very years the navy was simplifying the other canvas gear issued;[7] this looks very much like more of the same, an attempt to do away with the trouble and expense of equipping ships with two sets of gear and to experiment with a single set carrying a sail of finer cloth.[8] The naval au-

---

ὅσοι τῶν τριηράρχων γεγραμμένοι εἰσὶ ἔχοντες εἰς πλοῦν ἐντελῆ σκεύη τετρήρων ξύλινα ἢ κρεμαστά,

ὅσοι μὲν ξύλινα, τάδε ἔχουσιν· ταρρόν, πηδάλια κλιμακίδας, ἱστόν, κεραίας, κοντούς·
ὅσοι δὲ κρεμαστά, τάδε ἔχουσιν· ὑποζώματα, ἱστίον, παραρύματα τρίχινα, παραρύματα λευκά, καταβλήματα, τοπέα, ἀγκύρας δύο, σχοινία τριηριτικὰ ὀκτωδάκτυλα τέτταρα ἐγδάκτυλα IIII.

For details on these items, see 249-51 above. *IG* ii² 1629.1058-59 is exceptional in assigning triremes more than one *hypoblema* and *katablema*; elsewhere both are consistently in the singular.

[4] 1623.44-46, 270-75, 315-20, 330-33. For the wording of the entries, see note 9 below.

[5] 1632.128-30, 143-49, 149-54, etc.

[6] Cf. 1622.231-33 with 1631.479-82. In the first, the *Pandora* involved was the work of the master shipwright Xenocles, and his name has been restored as the master builder of the *Pandora* listed in the second inscription, making the two vessels the same. The editors have maintained this restoration despite the objections of Schmidt (36), who argues that they are different galleys. Even if they are, this does not change matters one whit: a trireme *Pandora* in 342/1 had an *akateion* rig; in 334/3 a trireme with the same name and presumably just about the same in all other respects had no *akateion* rig but did have a "light" sail.

[7] See ELEVEN, note 99.

[8] The traditional sails are described as "thick"; cf. 1631.415-17: κρεμαστὰ τριηριτικά· ἱστία λεπτά II ἀντὶ τούτων παρέδοσαν παχέα δύο "trireme hanging gear: 2 light sails; instead of these, they made restoration of two thick sails."

If the navy intended to fit all triremes with the new sails, it got no further than it had earlier in providing them with a double set of rigs (see ELEVEN, note 58).

thorities may have reckoned that, for a trireme in danger, the loss of time in setting up the rig was less important than the extra drive a sail of normal size would give—particularly a light one, which would be more effective than heavy canvas in the mild weather that alone permitted galleys to come out and fight.[9] Judging by the presence of the *dolon* in the next century, the experiment was ultimately abandoned, and warships returned to the old practice of using two rigs.

What the above makes perfectly clear is that, in the Athenian navy, the emergency rig was, as one would expect, a substitute for the mainsail. There is no reason to think that it was any different elsewhere.

E.g., in 330/29 there were 100 sails in storage on the Acropolis, all the old type, and, in the dockside warehouse, 288, of which 74 were light (1627.58-67); in 326/5 the warehouse had 281, of which 72 were light (1628.242-44); in 325/4 the number of light sails was down to 68 (1629.368-71). No quadriremes or quinqueremes are ever listed as having the new sail, which may simply mean that, having less need of it than their weaker sisters, their priority was low.

[9] It may be that the light sail was issued *in addition* to the regular. The standard entry reads ξύλινα ἐντελῆ, κρεμαστὰ ἐντελῆ, ἱστίον τῶν λεπτῶν (e.g., 1632.147-49, 153-54, 158-59, etc.) which seems to mean "complete set of hanging gear (i.e., lines, awnings, sail), the sail [thereof being one] of the light ones," but might just possibly mean "plus a sail from the light ones."

APPENDIX 3

DID SYNESIUS SAIL ON A LATEENER?

IN 404 A.D., Synesius, later to become well known as Bishop of Ptolemais, traveled by ship from Alexandria along the coast to Cyrene and wrote up his experiences in a lively and amusing letter (*Epist.* 4). Although Synesius was hopelessly ignorant of all matters connected with the sea,[1] certain actions which he describes and could not possibly mistake make it very likely that the vessel he sailed on had a lateen rig.

That it was no galley but a plain sailing ship is clear from the small size of the crew, a mere thirteen in all, including the captain (160a). Moreover, it could not have been of any great size, and it certainly had but a single mast and sail, for at one point Synesius expresses his alarm when he found himself—where he obviously never expected to be—far offshore amid big seagoing "two-masted freighters" (ὁλκάδων διαρμένων 160d-161a; cf. ELEVEN, note 74).

In the middle of the first night out a storm made up. All aboard put their hands to the ropes—but to no avail, since these turned out to be jammed in the blocks.[2] The following day, when the wind slacked off and the sun came out and dried the ropes, a second try was made. At this point Synesius drops a significant remark: ὑπαλλάττειν μὲν οὖν ἱστίον ἕτερον νόθον οὐκ εἴχομεν, ἠνεχυρίαστο γάρ "We weren't able to sub-

---

[1] Cf. Casson, "Bishop Synesius' Voyage to Cyrene," *The American Neptune* 12 (1952) 294-96. Contrary winds forced the captain to tack, i.e., follow a zigzag course, making a "zig" out to sea and doubling back for a "zag" toward shore. This utterly mystified Synesius. He was even more mystified when the captain, to avoid that seaman's *bête noire*, a lee shore, stayed far out at sea for a while. The captain was considerate enough to take time to explain what he was doing, pointing out that this happened "to be the way sailing ships were navigated" (τοιοῦτον τὸ ναυτίλλεσθαι τέχνῃ 161c), but Synesius refused to be convinced, partly through ignorance, partly through deep distrust of the captain because he was a Jew. Synesius everywhere brands him as an incompetent, yet he emerges from even this hostile account as an able and experienced seaman. Modern commentators have been as shortsighted as Synesius. E.g., most recently A.H.M. Jones in his definitive *The Later Roman Empire* (Oxford 1964) devotes nearly half a page (II 842-43) to a résumé of the voyage full of sympathy for Synesius' unfortunate choice of vessel. It is rather the skipper, an able man maligned for properly—and successfully—doing his duty, who deserves the sympathy.

[2] 163c: πᾶσιν ἱστίοις ἡ ναῦς ἐφέρετο, ὑποτεμέσθαι δὲ οὐκ ἦν, ἀλλὰ πολλάκις ἐπιχειρήσαντες τοῖς καλωδίοις ἀπηγορεύκειμεν, τῶν τροχῶν ἐνδακόντων "The ship was swept along under full sail, and it was impossible to stop her; time and again we had attacked the ropes and worn ourselves out, since the blocks bit into them."

268

stitute another, bastard sail since it was in pawn" (163d). In other words, the ship was so rigged that, to shorten sail, one took down the ordinary sail and replaced it with another, presumably smaller, used for extraordinary circumstances such as the present emergency.[3] This is confirmed by his words when the ship ran into another storm a few days later: πάλιν δὲ δυσπειθὲς ἦν τὸ ἱστίον καὶ οὐκ εὔτροχον εἰς καθαίρεσιν "Again the sail was hard to handle, and it couldn't be made to move *for lowering*" (164d).

I have italicized the last two words for they are critical: they describe a procedure precisely the opposite of that used on vessels rigged with squaresails, whether ancient or not. Sailors shorten a squaresail not by lowering it but by *raising* it to the yard; the ancients did this by taking up on the brails (70 above). One of the prime weaknesses of the lateen, on the other hand, is that it allows of no quick and efficient way of shortening sail; it is basically a fair-weather rig. If the wind suddenly makes up, the crew of a lateener has one standard recourse: to lower yard and sail to the deck, and either scud under bare poles, or, stripping off the sail, set another, smaller one—*which is precisely the procedure that Synesius is describing*. The lateen-rigged galleys of Venice's great fleets of the fourteenth and fifteenth centuries carried a whole series of sails of diminishing sizes to be used under varying wind conditions.[4] A tramp such as the craft Synesius was on would have but one for emergencies. For some reason, during his voyage it was not available; we hardly need take seriously his amusing sneer that it was at the pawnbroker's.[5]

[3] *nothos* "bastard" as used here sounds like a sailor's word. The same figure of speech was very common among sailors of later ages to express the meaning "not of normal size." E.g., in the Venetian navy the largest type of galley was called a *bastarda*; the largest sail a galley carried, a *bastardo*; a smaller size of cannon, a *bastardo*; and so on. See Guglielmotti, s.v. *bastardo*. Synesius clearly means by *nothos* a sail of smaller size than the one normally set. Fitzgerald's translation (*The Letters of Synesius of Cyrene*, London 1926, p. 86) "to replace our sail by a new one" is misleading.

[4] Cf. Lane 22-23. For a fascinating description of what happens aboard a lateener when the wind makes up, see Alan Villiers, *Sons of Sinbad* (New York 1940) 272-75.

[5] As some commentators do; see, e.g., Assmann, *Segel* 1052 and Jones, *loc. cit.* (note 1 above).

Seasons and Winds, Sailing, Rowing, Speed

I SEASONS AND WINDS

HESIOD URGED all sailors to stay away from the sea except for the fifty days after the summer solstice, in July and August.[1] Hesiod, to be sure, was a lubberly farmer living in the stony hills of Boeotia, but even Vegetius, who speaks professionally, points out that the sailing season par excellence is from 27 May to 14 September, and that the outside limits are 10 March to 10 November.[2] And, in point of fact, this is the way things were for the whole of the ancient period: during late fall and winter, sailing was reduced to the absolute minimum—the carrying of vital dispatches, the ferrying of urgently needed supplies, seaborne military movement that was impossible to delay.[3] All normal activity was packed into the summer

---

[1] Hesiod, *Works and Days* 663-65: ἤματα πεντήκοντα μετὰ τροπὰς ἠελίοιο . . . ὡραῖος πέλεται θνητοῖς πλόος "The 50 days after the summer solstice . . . is the right time for men to sail the seas."

[2] Vegetius, *re mil.* 4.39: *a die VI. kal. Iunias usque in Arcturi ortum, id est in diem VIII. decimum kal. Octobres, secura navigatio creditur . . . post hoc tempus usque in tertium idus Novembres incerta navigatio est. . . . ex die . . . tertio idus Novembres usque in diem sextum idus Martias maria clauduntur* "From the 6th day before the kalends of June until the rising of Arcturus, that is until the 18th before the kalends of October, is believed to be the safe period for navigation. . . . From then up to the 3rd before the ides of November, navigation is uncertain. . . . From the 3rd before the ides of November to the 6th before the ides of March, the seas are closed." Practically the same sailing season was still being observed in the 11th century, see Goitein 316.

[3] Thucydides mentions several instances of naval movement on the sea during the winter, all the result of exceptional circumstances: in the winter of 430/29 B.C., a squadron of 20 was sent around the Peloponnese to be on station at Naupactus when the sailing season reopened, and six were sent off to collect tribute in Lycia and Caria (2.69); in the winter of 427/6, a naval force of 30 attacked the Lipari Islands, since lack of water prevented a summer expedition (3.88); in the winter of 425/4, a Persian captured with important dispatches was sent back in a trireme from Athens to Ephesus (4.50). E. de Saint-Denis, "Mare clausum," *REL* 25 (1947) 196-214, supplies numerous references from Latin authors to the sailing season and

and a few weeks before and after it; at other times the sea lanes were nearly deserted, and ports went into hibernation to await the coming of spring.[4] It was not merely the severity of winter storms, although these played their part. It was even more a matter of visibility: during the winter a much greater incidence of cloudiness obscures the sun by day and the sky by night, making navigation difficult in an age that did not have the mariner's compass, and more often do scud and mist veil the cliffs, headlands and mountains,

---

a review of out-of-season voyages both military (201-203) and commercial (203-207). J. Rougé, "La navigation hivernale sous l'Empire romain," *REA* 54 (1952) 316-25, adds references from the Church Fathers. *Cod. Theod.* 13.9.3 (A.D. 380), addressed to the shippers of Africa, puts the situation clearly: acceptance and loading of government cargoes shall take place from 1 April to 1 October; transport of such cargoes shall take place from 13 April to 15 October; from November to April, navigation is suspended (*Novembri mense navigatione subtracta, Aprilis, qui aestati est proximus, susceptionibus adplicetur. Cuius susceptionis necessitas ex kal. Aprilib. in diem kal. Octob. mansura servabitur; in diem vero iduuum earundem navigatio porrigetur* "From the month of November, navigation shall be discontinued; the month of April, since it is just before the summer, shall be employed for the acceptance of cargo. The necessity of such acceptance from the kalends of April to the kalends of October shall be preserved permanently; but navigation shall be extended to the day of the ides of the aforesaid months").

The limitation affected every phase of maritime activity: a shipper had to pay increased interest rates on maritime loans if he sailed out of season (Demosthenes 35.10), or had to guarantee to complete contracted voyages within the season (*Dig.* 45.1.122.1); the Emperor Titus would not trust his troops to an overseas crossing in winter (Josephus, *Bell. Jud.* 7.1.3); a prince, presumably able to commandeer the best available transport, refused to go even from Caesarea to Ionia in winter (Josephus, *Ant. Jud.* 16.2.1); imperial enactments made in Europe in the autumn "practically never reached Africa till the following spring or early summer" (Jones *op. cit.* [ELEVEN, App. 3], I 403, cf. III 92). Caligula's threat of death sent to Petronius in Antioch in midwinter took all of three months to get there (Josephus, *Bell. Jud.* 2.10.5); the messengers who followed with word of the assassination [24 January A.D. 41] "had a good voyage" (εὐπλόουν) and so arrived 27 days earlier. The trip in summer could have been made in two weeks (cf. Table 1).

One exception was the run between Rhodes and Alexandria. According to Demosthenes (56.30), "There sailing goes on continuously, so [certain specialists in bottomry loans] were able to put the same money to work two or three times, whereas when residing here [i.e., at Athens] they had to stay put through the whole winter awaiting the suitable season [sc. for sailing]" (ἐκεῖσε μέν γε ἀκέραιος ὁ πλοῦς, καὶ δὶς ἢ τρὶς ὑπῆρχεν αὐτοῖς ἐργάσασθαι τῷ αὐτῷ ἀργυρίῳ. ἐνταῦθα δ' ἐπιδημήσαντας παραχειμάζειν ἔδει καὶ περιμένειν τὴν ὡραίαν).

[4] An important subject that has never been treated is the extent of the economic dislocation that all port towns had to suffer because of the limited sailing season.

which, sighted from far off, gave skippers fair warning to stay clear.[5]

During the heart of this curtailed sailing period, Mediterranean winds are prevailingly northerly. This is particularly true of the eastern basin: "From June to September . . . the windroses between the 30th and 35th parallels, 20th to 35th meridians [i.e., from Egypt to Crete and from Cephallenia to Syria] show almost as steady a northwesterly direction as would be found from the northeast in many parts of the northern trade wind belts of the open ocean."[6] To the north, in the Aegean, the situation is practically the same: in the southeastern portion the northerly winds "reach the exceptional frequency of 80 percent or more in July and August."[7] In the western basin, northerlies still hold sway, the prevailing winds in the Tyrrhenian and Ionian Seas being northwest.[8] The situation changes finally at the Gulf of Lion, where summer winds are often from the southwest (although there is still the hard-blowing northerly Mistral to reckon with).[9] And, between the Balearics and the Strait of Gibraltar, easterlies are most common.[10]

Sailing season and wind direction combined to give a definite pattern to ancient seaborne activity. Ships traveling in most southerly directions—e.g., from Italy or Greece to Africa, Asia Minor, Syria, Egypt—could generally count on a quick and easy downhill voyage. But they paid for this on the return, which had to be made in the teeth of the prevailing wind. And, in both directions, they had to be prepared for fairly stiff breezes. The Mistral in the west has al-

---

[5] Cf. Vegetius' list (4.39) of the dangers of winter sailing: *lux minima noxque prolixa, nubium densitas, aëris obscuritas, ventorum imbri vel nivibus geminata saevitia* "scant daylight, long nights, dense cloud cover, poor visibility, and the violence of the winds doubled by the addition of rain or snow." Storms, it will be noted, are listed last. There is no question, of course, that they were troublesome. Herod's near shipwreck off Pamphylia took place in midwinter (Josephus, *Bell. Jud.* 1.14.2-3); a storm-plagued voyage described in a pair of papyri (*P. Cairo Zen.* 59029 and *P. Mich. Zen.* 10) took place in December or in January; and St. Paul's celebrated disaster happened because his skipper risked sailing from Crete after the season proper had closed (Acts 27.9).

[6] *HO* 154A.32-33.   [7] *HO* 154B.25.   [8] *HO* 152.32-33.
[9] *HO* 152.33, 578.   [10] *HO* 151.41.

ready been mentioned; the Etesians, which ruled the Aegean, are notorious: "In August they attain such violence that sailing vessels for weeks at a time cannot beat against them but have to tie up behind islands."[11] Compared with the winds, other factors were minor: the Mediterranean's currents are in general too feeble, and its tides too faint, to be of significance; the one affected only certain straits (Hellespont, Bosporus) and the other only the ends of deep inlets (the head of the Adriatic, the Syrtes) or certain channels (Euripus, the Strait of Messina).[12]

II SAILING

ANCIENT square-riggers, with their broad mainsail as principal driver, were designed first and foremost for traveling with the wind astern[13] or on the quarters.[14] But, when pressed, the ancient mariner could also sail a close-hauled course with the wind abeam or forward

[11] E. Semple, *The Geography of the Mediterranean Region* (New York 1931) 580.

[12] Semple 582-83.

[13] Nonnos, *Dionys.* 4.231: ἰσάζων ἑκάτερθε νεὼς πόδας "equalizing the sheets on both sides of the vessel"; Cicero, *ad Att.* 16.6.1 and Ovid, *Fasti* 3.565, both cited in ELEVEN, App. 1, note 1; Catullus 4.20-21: *utrumque Iuppiter/ simul secundus incidisset in pedem* "A favorable wind falls on both sheets simultaneously," i.e., they were slacked off the same amount; Lucan 8.193: *cornibus aequis* (cf. the following note) "with equal yards," i.e., with the yard squared, set at right angles to the keel. (In *Aen.* 4.587, however, by *aequatis velis* Vergil means "with sails set the same way," i.e., the ships in the fleet, sailing close-hauled, had all braced the yards about and sheeted the canvas home at exactly the same slant; cf. Mohler 61).

[14] Achilles Tatius 2.32.2: ἡ κεραία περιήγετο "The yard was braced round."

Lucan (8.193-99) gives a nice description of turning from a run to a reach. Pompey, fleeing from his defeat at Pharsalus in August, 48 B.C., left Lesbos and headed southwest for Chios with a following wind (the Etesians are often northeasterly). Off Chios the skipper swung the vessel to the south to negotiate the strait between Chios and the mainland, taking the wind on the port quarter: "The sails, hanging evenly from the squared yards, he twisted round, and headed the ship to port, and, to cut through the waters that Chios and the rocks of Asina make so turbulent, he slacked off the lines toward the prow [i.e., the weather sheets and braces] and took in those toward the stern [i.e., the lee sheets and braces]. The sea sensed the movement and changed its sound as the prow cut a different way through the water and the vessel headed on a new course" (*iusto vela modo pendentia cornibus aequis / torsit et in laevum puppim dedit, utque secaret / quas Asinae cautes et quas Chios asperat undas / hos dedit in proram, tenet hos in puppe rudentes. / aequora senserunt motus aliterque secante / iam pelagus rostro nec idem spectante carina / mutavere sonum*).

of the beam—"tack" as we put it, "make a sheet"[15] as he put it—although, with his relatively inefficient rig he could probably point no closer to the wind than seven points.[16] The yard was braced round till it ran from bow to quarter and slanted toward the wind,[17] thus bringing the windward sheet forward of the mast.[18] When his destination lay well to windward he resorted, as ships willy-nilly did until the age of steam, to tacking, i.e., he pursued a zigzag course with the ship taking the wind first on one bow and then being swung about to take it on the other.[19]

[15] ποδιαῖον ποιεῖσθαι in Greek, facere pedem in Latin; see notes 19 and 24 below.

[16] Cf. Smith 177-78. Square-riggers as late as the 19th century could get no closer than six—i.e., if headed north, they could aim no better than WNW on one leg, ENE on the other.

[17] Lucian, Navig. 9: πρὸς ἀντίους τοὺς ἐτησίας πλαγιάζοντας "They slanted against the foul northerly tradewinds." Vergil, Aen. 5.16: obliquatque sinus in ventum "And [Palinurus] slants the sails toward the wind" (the fleet had left Carthage under the favorable land breeze [Zephyros secundos 4.562] and, heading westward, probably for Corsica, must have gone on a starboard tack when the prevailing northerly set in [Aquilone, 5.2]; then, when the wind suddenly backed to the west [mutati transversa fremunt et vespere ab atro / consurgunt venti "The winds have changed and rage against us, rushing from the black west," 5.19-20], Palinurus had to harden sheets and braces [colligere arma 5.15] and sail as close to the wind as he could to hold his course—without success, as we learn from 5.27). Lucan 5.427-28: flexo navita cornu / obliquat laevo pede carbasa "The crew swivels the yard and slants the canvas on the port tack" (the words translated "port tack" literally mean "left sheet." Lucan is describing Caesar's crossing from Brindisi to Dyrrachium, a course roughly NE. Since the winds were northerly [cf. 5.417], his ships were on a port tack, i.e., with the wind coming over the port bow and the sails on the starboard side of the mast, held slanted toward the wind by the port sheet).

[18] Seneca, Medea 322: prolato pede transversos captare Notos "carry the sheet forward to catch the south winds blowing against [the ship]"; Pliny, NH 2.128: isdem autem ventis in contrarium navigatur prolatis pedibus "By carrying the sheet forward, ships sail in opposite directions on the same wind."

[19] Cf. Vergil, Aen. 5.830-32: una omnes fecere pedem; pariterque sinistros, / nunc dextros, solvere sinus; una ardua torquent / cornua, detorquentque "Together they all tacked; in unison they unfurled the sails, now to port and now to starboard; together they braced the lofty yards about, then braced them round again [on the new course]." Cf. Achilles Tatius 3.1.3-6: κλίνεται δὴ κοῖλον τοιχίσαν τὸ σκάφος καὶ ἐπὶ θάτερα μετεωρίζεται καὶ πάντη πρηνὲς ἦν. . . . μετεσκευαζόμεθα οὖν ἅπαντες εἰς τὰ μετέωρα τῆς νηός. . . . αἰφνίδιον δὲ μεταλλάττεται τὸ πνεῦμα ἐπὶ θάτερα τῆς νηὸς. . . . καὶ τρίτον καὶ τέταρτον καὶ πολλάκις τὸ αὐτὸ πάσχοντες κοινὴν ταύτην εἴχομεν τῷ σκάφει τὴν πλάνην "The ship heels over, laying one side in the water amidships and going high in the air on the other side; it is all aslant. . . . So we all change our position to the high side of the vessel. . . . Suddenly the wind leaps to the other side of the ship. . . . And a third time, a fourth time, many times, we go through the same procedure, keeping up with the gyrations of the ship." It was rather the ship

When the wind was somewhat too strong for normal sailing, the yard was carried lower on the mast to bring down the center of pressure;[20] this maneuver, by keeping the bow from digging in, enabled the vessel to plane better. When that did not suffice, sail was shortened by taking up on the brails.[21] The vertical brailing ropes (70 above) were a unique device for reefing, one that was made possible by the absence on ancient ships of superimposed sails. It had many advantages to offer over the reef points of later ages.[22] For one, it furnished immediate and complete control of the sail from the

---

that was leaping to the other side of the wind as she changed from one tack to another.

Nicander, describing the motion of the horned viper Cerastes, likens it to a "merchantman's ship's boat which, forcing its way to windward in a violent blow, when hit by the blast of the southwesterly dips its whole side in the sea" (τράμπιος ὁλκαίης ἀκάτῳ ἴσος ἤ τε δι᾽ ἅλμης / πλευρὸν ὅλον βάπτουσα κακοσταθέοντος ἀήτεω / εἰς ἄνεμον βεβίηται ἀπόκρουστος λιβὸς οὔρῳ Theriaca 268-70). The Cerastes has a "side-winding movement," it "advances by throwing its body forward in a series of loops, and leaves as its track in dust or sand a series of parallel lines, disconnected and oblique to the direction in which the snake is moving," A. Gow and A. F. Scholfield, Nicander: The Poems and Poetical Fragments (Cambridge 1953) 175. Anyone who has sailed a small boat will recognize the aptness of the simile, for this is precisely the way a small boat advances against a strong wind: as each gust hits it, it heels over sharply; then, as the tiller is eased, it makes a looping turn oblique to its course into the wind and straightens up.

[20] Seneca, Epist. 77.2: quotiens ventus increbruit maiorque est quam expedit, antemna submittitur: minus habet virium flatus ex humili "Whenever the wind makes up and becomes too strong for comfort, the yard is lowered; for the wind exerts less force from low down." The yard could be lowered a third of the way down the mast (Sallust, Hist. 4.3 [Maurenbrecher]: demissis partem quasi tertiam antemnis "lowering the yard about a third") or halfway (Caesar, Bell. Alex. 45.2: antemnis ad medium malum demissis "with yards lowered half-way; Seneca, Medea 323-24: nunc antemnas medio tutas / ponere malo "place the yards safely at midmast"). Cf. Aristotle, Mech. 851a: ὅσῳ ἂν ἡ κεραία ἀνωτέρα ᾖ, θᾶττον πλεῖ τὰ πλοῖα "The higher the yard is carried, the faster ships sail." Plutarch records another reason for carrying sail this way: to escape detection. Lucullus got safely to Rhodes "by sailing with the sails lowered far down during the day and only raising them at night" (μεθ᾽ ἡμέραν μὲν ὑφειμένοις πλέων τοῖς ἱστίοις καὶ ταπεινοῖς, νύκτωρ δ᾽ ἐπαιρομένοις, Luc. 3.3).

[21] ἱστία στέλλειν, aut sim. (e.g., Il. 1.433; Od. 3.10-11, cited in note 28 below; Aristophanes cited in note 23 below; Aristotle cited in note 24 below); contrahere vela (e.g., Cicero, ad Att. 1.16.2; Livy 36.44.2, cited in ELEVEN, note 53; Seneca, Epist. 19.9). The Latin expression means literally "contract the sails."

[22] In its method of operation and efficiency it compares with the highly praised Chinese lugsail; cf. IH 176-77. Both systems, of course, are limited to rigs without superimposed sails.

deck; there was no need to send men aloft. For another, it permitted the shortening of selected areas of the sail. When the wind blew hard over the quarters or astern, the center of the sail was brailed up (Figs. 81, 147).[23] When the wind blew hard while a ship was sailing close-hauled, with its yard braced about to run from bow to quarter, the brails abaft the mast were tightened to reduce the amount of canvas aft; this lessened the tendency of the ship to head into the wind with consequent easing of the pressure on the helm.[24] In this particular maneuver the sail must have been

[23] Cf. Aristophanes, *Frogs* 999-1000: "drawing in sail and using only the tips" (συστείλας ἄκροισι / χρώμενος τοῖς ἱστίοις).

[24] Aristotle, *Mech.* 851b: διὰ τί, ὅταν ἐξ οὐρίας βούλωνται διαδραμεῖν μὴ οὐρίου τοῦ πνεύματος ὄντος, τὸ μὲν πρὸς τὸν κυβερνήτην τοῦ ἱστίου μέρος στέλλονται, τὸ δὲ πρὸς τὴν πρῴραν ποδιαῖον ποιησάμενοι ἐφιᾶσιν; ἢ διότι ἀντισπᾶν τὸ πηδάλιον πολλῷ μὲν ὄντι τῷ πνεύματι οὐ δύναται, ὀλίγῳ δέ, ὃ ὑποστέλλονται. προάγει μὲν οὖν τὸ πνεῦμα, εἰς οὔριον δὲ καθίστησι τὸ πηδάλιον, ἀντισπῶν καὶ μοχλεύον τὴν θάλατταν. ἅμα δὲ καὶ οἱ ναῦται μάχονται τῷ πνεύματι· ἀνακλίνουσι γὰρ ἐπὶ τὸ ἐναντίον ἑαυτούς. "Why is it that sailors, after sailing with a favorable wind, when they wish to continue on their course even though the wind is not favorable, brail up the part of the sail toward the helmsman, yet, as they go close-hauled, leave the part toward the prow unfurled? It is because the rudder cannot produce an effect against the wind when it is strong, but can when it is not, and this is why they shorten [sc., the sail area aft]. The wind moves the ship forward, and the action of the rudder converts it into a favorable breeze, producing an effect against it and using the sea as a fulcrum. At the same time, the sailors join in the fight against the wind, for they lean their bodies in the direction opposite to it." Cf. on this passage, *GOS* 312-13, where it is rightly pointed out that Aristotle's reasoning is perfectly sound. When sailing close-hauled, too much canvas aft will increase the tendency of a vessel to come into the wind, which must be counteracted with the rudder; a stiff enough wind can make this arduous. Reducing the sail area aft corrects the situation.

The same maneuver is described by Achilles Tatius (3.1.1-2): ἐγείρεται δὲ κάτωθεν ἄνεμος ἐκ τῆς θαλάσσης κατὰ πρόσωπον τῆς νηός, καὶ ὁ κυβερνήτης περιάγειν ἐκέλευε τὴν κεραίαν. καὶ σπουδῇ περιῆγον οἱ ναῦται, πῇ μὲν τὴν ὀθόνην ἐπὶ θάτερα συνάγοντες ἄνω τοῦ κέρως βίᾳ (τὸ γὰρ πνεῦμα σφοδρότερον ἐμπεσὸν ἀνθέλκειν οὐκ ἐπέτρεπε), πῇ δὲ πρὸς θάτερον μέρος φυλάττοντες τοῦ πρόσθεν μέτρου καθ' ὃ συνέβαινεν οὔριον εἶναι τῇ περιαγωγῇ τὸ πνεῦμα "A wind arose from low over the water and struck the ship head on. The captain ordered the yard braced around [i.e., in order to go off on a tack]. The sailors quickly braced the yard around, and, on one side they furled the canvas aloft toward one yardarm by main force since the wind had hit too strongly to permit them to haul [sc. on the brails] against it, while, on the other side, toward the other yardarm, they kept just as much of the original spread of sail as would take the wind at the proper angle for bringing the yard around." In this instance, not only was the canvas brailed up to form a triangle, but, be-

276

brailed up into a triangle, with the yard and luff forming the arms and the hypotenuse running from the foredeck up to the after yard-arm (Figs. 188a-b). The weather sheet would be made as tight as possible in order to draw the luff taut, and this would have a tendency to bring the forward yardarm down, to tilt the whole sail (Fig. 188c). A square sail brailed up in this triangular fashion and set aslant is in shape not unlike a lateen, and may possibly have sparked the invention of that all-important sail (244-45 above).

But the brails had uses over and above that of shortening sail. In light breezes they could be so slacked off[25] that the great main-sail would belly out to meet the forestay (Fig. 90).[26] Sail could be completely doused in a twinkling by taking up on all the brails at once.[27] Sail was furled in the same way when a ship snugged down

---

cause of the violence of the wind, sail was at the same time further shortened by making the triangle as small as possible.

[25] κάλως ἐξιέναι, ἐφιέναι aut sim. in Greek (e.g., Plato, *Prot.* 338a [cited in ELEVEN, App. 1, note 3] and *Sisyphus* 389c; Euripides, *Medea* 278 and *Trojan Women* 94; Aristophanes, *Knights* 756), *immittere* or *laxare rudentes* in Latin (e.g., Ovid, *Pont.* 4.9.73; Vergil, *Aen.* 10.229; Pliny, *Epist.* 8.4.5, cited in ELEVEN, App. 1, note 3).

Vergil (*Aen.* 3.267-68; cf. 682-83) gives a full description of the handling of the brails when getting under way: *excussosque iubet laxare rudentis. / tendunt vela Noti* "And he gives orders to shake loose the brails and slack them off. The south wind fills the sails." In other words, the men whip the lines to get them to run through the blocks and fairleads and let the stiff canvas unfurl, and then they let them run out the proper amount. Thus Lucan (2.697-98), in describing a departure that was carried out with maximum secrecy, relates that "the men, hanging [from the yards], lower the furled canvas and do not shake the stout brails so as not to make the wind whistle [sc. by whipping the lines briskly to get them to run]" (*strictaque pendentes deducunt carbasa nautae / nec quatiunt validos, ne sibilet aura, rudentes*).

[26] Euripides, *Hec.* 111-12: σχεδίας / λαίφη προτόνοις ἐπερειδομένας "the ship resting its sail on the forestays"; *Frag.* 773.42: σινδὼν δὲ πρότονον ἐπὶ μέσον πελάσσει "The sail travels to the middle of the forestay"; *Anth. Pal.* 10.2.7: λαίφεα δ' εὐυφέα προτο-νίζετε "Forestay your well-cut sail" (i.e., let it out to the forestay). Synesius, describing one of the storms his ship ran into (ELEVEN, App. 3) says (*Epist.* 4.164c) ᾠόμεθα προτονίζειν τὴν ναῦν "We had in mind to forestay the ship." Torr suggests (94) that he means "tightening the forestay to secure the mast," but Synesius unmistakably says "ship" and not "mast." It is not unlikely that he was trying to use a nautical term he did not understand; cf. ELEVEN, App. 3, note 1.

[27] Ovid, *Met.* 11.483: *antemnis totum subnectite velum* "bind the whole sail to

or docked (Fig. 150);[28] men had to go aloft only to secure the canvas with gaskets (Figs. 144, 151).

III ROWING

As MENTIONED earlier (80, 104 above), one of the trireme's advantages was that it enabled the oarsmen to row from a seated position. In recompense, it required an extraordinarily high degree of skill to ensure that 170 individual rowers would carry out orders with split-second speed and in perfect unison.

A good crew was said to be "beaten together,"[29] i.e., it rowed with a unified beat. Veteran crews had to be kept up to the mark by frequent exercises and racing competitions.[30] New crews had to be broken in by arduous repetitive practice.[31] When galleys were not available for this, the men could be given the elementary lessons seated in temporary platforms erected on land.[32] Here they would

---

the yard"; Lucan 9.328: *antemnae suffixit lintea summae* "had tied the canvas to the yard on high."

[28] Cf. *Od.* 3.10-11: οἱ δ' ἰθὺς κατάγοντο, ἰδ' ἱστία νηὸς ἐίσης / στεῖλαν ἀείραντες, τὴν δ' ὥρμισαν "They straightway made for the shore, quickly raised [i.e., brailed up] and furled the sail of their graceful craft and moored her." Conformably, to unfurl sail is to "lower" it; cf. Achilles Tatius 2.32.2, where he says of a ship getting underway τὸ ἱστίον καθίετο "The sail was lowered."

[29] συγκεκροτημένος; see Polyaenus 3.11.7, cited in note 32 below (same expression in Polybius 1.61.3) and cf. Thucydides 8.95.2: Ἀθηναῖοι . . . ἀξυγκροτήτοις πληρώμασιν ἀναγκασθέντες χρήσασθαι "the Athenians . . . forced to use untrained (*axynkrotetos*) crews."

[30] See *GOS* 308-309. The boat race in Book Five of the Aeneid may very well be based on races held by the Roman navy that Vergil had seen; cf. Mohler 62.

[31] Cf. Herodotus' story (6.12) of the Ionian citizens who volunteered to serve as rowers and, able to take but a week of the rigorous training an experienced naval officer put them through, quit en masse.

[32] Chabrias trained raw Egyptian recruits in the 4th century B.C. this way; see Polyaenus 3.11.7: "He removed the oars from the triremes, laid out long planks along the shore so that the men could sit in single file, issued them the oars, and, putting them in charge of bilingual *keleustai* (302 below), in a few days he taught them to row and filled his ships with trained [*synkekrotemenos* 'beaten together'] oarsmen" (ἐξελὼν τὰς κώπας τῶν τριήρων, ξύλα μακρὰ παραβαλὼν ἐπὶ τὸν αἰγιαλὸν ὥστε ἐφ' ἕνα καθῆσθαι, δοὺς αὐτοῖς τὰς κώπας καὶ κελευστὰς τῶν διγλώσσων ἐπιστήσας ὀλίγαις ἡμέραις ἐλαύνειν αὐτοὺς ἐδίδαξεν [καὶ τὰς ναῦς] ἐρετῶν συγκεκροτημένων ἐπλή-ρωσεν). The Romans trained men for their newly built fleet in 260 B.C. in this way (Polybius 1.21.1-2) and Agrippa for Augustus' newly reconstituted fleet in 37 B.C.

learn all the basics of rowing: to stroke to the "trireme tune" or the beat of a mallet,[33] back water,[34] dip and hold the oars,[35] ship oars,[36] spurt.[37] The next stage in the oarsmen's education was necessarily carried out on the water and in concert with other ships,

---

(Dio Cassius 48.51.5: τοὺς δὲ ἐπ' ἰκρίων ἐρέττειν ἤσκει "the [oarsmen] he taught to row on frames"). Thiel, *History* 172-73, note 345, argues that this was standard practice, which goes a little too far; it was standard emergency practice.

[33] Cf. Athenaeus 12.535d: [Alcibiades, making a grand entrance into Athens' harbor, had a concert flute-player] "pipe the trireme beat" (ηὔλει τὸ τριηρικόν). For time-keeping aboard Roman galleys by pounding with a mallet, see THIRTEEN, note 52.
The general term for what gave the time to the rowers, whether the sound of an instrument or a coxswain's cry, was *keleuma* or *keleusma* in Greek (cf., e.g., Euripides, *IT* 1405: κώπῃ προσαρμόσαντες ἐκ κελεύσματος "fitting [their movements] with the oar to the beat [*keleusma*])," which was borrowed by Latin (cf., e.g., Martial 3.67.4: *lentos tinguitis ad celeuma remos* "You dip the oars lazily to the beat [*celeuma*]" and 4.64.21).

[34] πρύμναν (ἀνα)κρούεσθαι in Greek; see, e.g., Thucydides 1.50.5, 7.40.1; Aristophanes, *Wasps* 399; Diodorus 11.18.6. *inhibere remis puppim (aut sim.)* in Latin; see, e.g., Cicero, *ad Att.* 13.21.3 (*inhibitio . . . remigum motum habet et vehementiorem quidem remigationis navem convertentis ad puppim* "Inhibitio [literally 'holding back'] of the oars involves motion, rather violent motion at that, of the oarage driving the ship toward the stern"); Livy 26.39.12, 30.10.17, 37.30.10.

[35] τὰς κώπας καθεῖναι in Greek; see Thucydides 2.91.4, Dio Cassius 50.31.5. *remos demittere* in Latin; see Livy 36.44.8. The maneuver was used to check way or to steady the ship.

[36] Cf. GOS 310.

[37] Latin authors employ a variety of expressions. Caesar (*Bell. Gall.* 3.14.6, 4.25.1; *Bell. Civ.* 2.6.5) prefers (*navem*) *incitare* "hurry [a ship] along," others *concitare* (Cicero, *de orat.* 1.33.153; Livy 30.25.8; Curtius 4.3.2; Frontinus, *Strat.* 1.5.6). Tacitus (*Ann.* 3.1) uses *alacre remigium* "quick rowing"; cf. *citis remigiis* in Ammianus 24.6.5, *concitatis remigiis* in Seneca, *Dial.* 6.18.7.
In Greek, the technical expression is ῥοθίῳ τῇ εἰρεσίᾳ πλεῖν "to travel with rush-roaring rowing" (Polyaenus 4.7.6). Greek poets of the 5th B.C. use the expression *rhothios* "rush-roaring" of the sound of a galley's oars as they pass through the water (cf. GOS 203, 311) and *pitylos* of the "plash" as the oars are dipped. Both terms seem to have entered the technical vocabulary; for the latter see THIRTEEN, note 53. Aristophanes, *Frag.* 84 (Edmonds I, p. 596), speaks of a "ship, when it makes a rush-roar with its plashes" (ναῦς ὅταν ἐκ πιτύλων ῥοθιάξῃ).
We have no information on how fast an ancient galley could spurt. The galleys of Louis XIV, driven by four or five men to the oar, could do perhaps 6 nautical miles per hour (Masson 206). Conformably, Guglielmotti (s.v. *palata*) reckons that Italian galleys of the 16th century did ca. 5½ to 6½, with a maximum of ca. 8¾ (30 strokes a minute, each producing 9.1 m. of forward movement, and a nautical mile is 1854.96 m.), or even better for very brief dashes. Thus Admiral Rodgers' estimate of better than 7 for ancient galleys may not be far from the mark (W. Rodgers, *Greek and Roman Naval Warfare*, Annapolis 1937, p. 516).

279

namely the fundamental fleet evolutions: to form line ahead,[38] to form line abreast,[39] to go from one to the other.[40] Lastly there would be intensive training in battle maneuvers, notably the *diekplous* "break-through" and the *periplous* "envelopment."[41]

It was only during battle or in emergencies that all oars were manned.[42] Sails were used as much as possible.[43] When rowing was unavoidable, and particularly when it had to be carried on for long periods, the men were divided into squads that rowed in turn, or, if all were kept at the oars, were given regular short spells of rest.[44]

[38] κατὰ μίαν (ναῦν) Thucydides 2.84.1, 2.90.4; ἐπὶ μιᾶς Xenophon, *Hell.* 1.6.29-31. For the Latin terminology, see W. Lacey, "Some Uses of *Primus* in Naval Contexts," *CQ* 51 (1957) 118-22, esp. 118. Livy's word for a file of ships (i.e., in line ahead) is *agmen* (36.43.13, 37.23.8). The head of the file is *prima navis*, the ship in the center is *media navis*, the tail end is *extremum agmen*, and "to bring up the rear" is *agmen cogere* (36.44.4, 37.23.8).

[39] In Greek μετωπηδόν (πλεῖν aut sim.) Herodotus 7.100, Thucydides 2.90.4. Livy's word for a line of ships abreast is *acies* or *frons* (30.10.4, 37.30.6; cf. Lacey, *ibid.*).

[40] Cf. Thucydides 2.90.4: ἐπιστρέψαντες τὰς ναῦς μετωπηδὸν ἔπλεον "Turning the ships [sc. from line ahead], they sailed line abreast"; Livy 37.29.8: *regia classis, binis in ordinem navibus longo agmine veniens . . . aciem . . . explicuit* "The king's fleet, traveling line ahead in a long double column, . . . went into . . . line abreast."

[41] On these terms, see *GOS* 137-39, 314-19.

[42] Cf. *GOS* 309. As Thucydides tersely puts it (7.14.1): "Peak performance of a crew is short" (βραχεῖα ἀκμὴ πληρώματος).

[43] Cf. *GOS* 310-11, Mohler 48-52. Cicero's description (*ad Att.* 16.6.1) of a trip he took in a 20-oared *actuariola* (cf. 16.3.6 and EIGHT, note 11) from Pompeii to Vibo is instructive: *magis commode quam strenue navigavi; remis enim magnam partem, prodromi nulli. illud satis opportune, duo sinus fuerunt quos tramitti oporteret, Paestanus et Vibonensis, utrumque pedibus aequis tramisimus* "I had a comfortable rather than strenuous voyage, for most was under oars and there were no boisterous northerlies (*prodromi*). A bit of luck this, since we had two bays to cross, the Bay of Paestum and the Bay of Vibo, and we crossed both with the yard squared" (*pedibus aequis*; see note 13 above). In other words, in a light galley the rowers went to work not only when there was no wind but on occasion when there was too much—and winds like the *prodromi* "forerunners," i.e., the harbingers of the summer northerlies (Pliny, *NH* 2.123), though fine for heavy sailing craft could often prove too much for a galley with its scant draft. At such times it had to stay out of open water and creep along the coast, perforce under oars. Cicero voyaged comfortably because he was spared both the buffeting a light boat takes in a stiff breeze and the long hours needed to row around the circuit of a bay. Heading straight for his destination, he went along mostly under oars but catching just the right wind for crossing the mouth of the Gulf of Salerno and Gulf of S. Eufemia.

[44] When Athens sent a second trireme rushing to Mytilene in 427 B.C. to countermand the orders for the town's destruction, "the men slept and rowed in turns" (οἱ μὲν ὕπνον ᾑροῦντο κατὰ μέρος, οἱ δ' ἤλαυνον Thucydides 3.49.3). When Conon was caught at Aegospotami in 405 B.C. with the crews wandering ashore, he ordered the ships launched, but "some were double-leveled (*dikrotoi*, i.e., had only two banks

IV SPEED

How fast could ancient ships travel?[45]

The question so put is meaningless: the speed of a sailing ship depends first and foremost on the direction of the wind and varies drastically with it. Against the wind, 100 miles can take as long as 200

---

manned), some single-leveled (*monokrotoi*, i.e., had only one bank manned), and some completely empty" (αἱ μὲν τῶν νεῶν δίκροτοι ἦσαν, αἱ δὲ μονόκροτοι, αἱ δὲ παντελῶς κεναί Xenophon, *Hell.* 2.1.28). Traveling under only one bank of oars was apparently common practice. Polyaenus (5.22.4) reports that Diotimus (4th B.C.), after sending off contingents from his crews to set up an ambush against an enemy city, "held off with his ships [sc., to give the ambush party enough time], ordering the marines on the decks to prepare for battle and the oarsmen to ply the thalamite, the zygite, and the thranite oars in turn" (ταῖς ναυσὶν ἀνεκώχευε παραγγείλας τοῖς μὲν ἐπὶ τῶν καταστρωμάτων διασκευάζεσθαι πρὸς μάχην, τοῖς δὲ ἐρέταις ἀνὰ μέρος ὀτὲ μὲν τὰς θαλαμιάς, ὀτὲ δὲ τὰς ζυγίας, ὀτὲ δὲ τὰς θρανίτιδας κώπας ἀναφέρειν). Vergil (*Aen.* 5.268-81) describes Sergestus' ship as returning ingloriously from the boat race "feeble under one bank" (5.271: *ordine debilis uno*; it had lost a good many oars taking a cliff too closely); the poet only means that the ship's progress was feeble in comparison with the others which had dashed home under full power.

On spells of rest, cf. Statius, *Theb.* 6.799-801: *sic ubi longa vagos lassarunt aequora nautas / et signum de puppe datum, posuere parumper / bracchia: vix requies, iam vox citat altera remos* "Just so, when long wandering over the sea has wearied the rowers and the signal is given from the poop, they rest their arms for a while; a scarce moment of repose, and then a second cry recalls them to the oars"; cf. H. Levy in *Classical Journal* 41 (1945-46) 327.

[45] The topic has been dealt with any number of times. The various treatments, over and above taking no cognizance of wind direction, have been marred by inaccuracy and incompleteness. Here are a few samples. W. Götz, *Die Verkehrswege im Dienste des Welthandels* (Stuttgart 1888) 259 cites *Od.* 15.474ff. as evidence that the trip from the Cyclades around Malea to Ithaca took 7 days. The passage does not say this: 7 days out at sea a killing took place, and some unspecified time later the vessel arrived at Ithaca. Similarly, W. Riepl, *Das Nachrichtenwesen des Altertums* (Leipzig 1913) 164 cites Livy 36.29 to prove that Nicander traveled from Asia Minor to Greece in 12 days. Again, the passage does not say this: Nicander went to King Antiochus, spent an undisclosed amount of time there, then returned to Phalara, a port on the Malian Gulf; the round trip, including whatever time he spent with the king, took 12 days (Livy's words *duodecimo is die quam conscenderat navem* mean "on the 12th day from the day he had boarded ship" at the port from which he left Greece). E. de Saint-Denis, "La vitesse de navires anciens," *Rev. arch.* sér. 6, 18 (1941) 121-38, omits Marcus Diaconus, *Vita Porph.* 6, though citing other passages from the same work. R. van Compernolle, "La vitesse des voiliers grecs à l'époque classique (Vᵉ et IVᵉ siècles)," *Bulletin de l'institut historique belge de Rome* 30 (1957) 5-30, dismisses as exceptional (16) a significant passage from Thucydides (2.97.1; see Table 2), makes (16) an unproven and probably wrong assumption about another (6.1; see note 93 below), and bases his chief argument on a series of totally hypothetical calculations. See also notes 60 and 111 below.

with the wind. Columbus flew to America with the Atlantic trades at his heels; when he met headwinds while working north along the coast of South America he was lucky to log 1 mile forward an hour. A very first step must be to classify voyages according to the winds encountered en route, to sift those done with a wind from some point abaft the beam, enabling the ship to move at its fastest directly toward its destination, from those with a wind from ahead, when a vessel must go through the uncomfortable and wearisome procedure known as tacking.[46] Fortunately, in many cases we are specifically told what the wind conditions were; where we are not, we can often make some sort of guess by using modern hydrographic information.[47] No doubt, in any crossing, the nature of the vessel, whether fast or slow, and the force of the wind had a certain effect. But never as much as wind direction; this sets the basic speed, and other factors cause only variations.

Voyages Made with Favorable Winds

Pliny, in a much-quoted passage,[48] mentions a pair of record voyages and a number of others that he obviously considers examples

[46] This has been overlooked by the many who provide a list of miscellaneous voyages and from it deduce—or let the reader deduce—what the "average speed" of ancient ships was. E.g., Götz (*op. cit.* previous note, 260) concludes that the average speed in Greek times was 4 to 6 knots. Riepl (*op. cit.*, previous note, 160-61) agrees with this but adds (168) an overall average of 5-7½ knots for the whole ancient period. Cedric Yeo (*TAPA* 77, 1946, p. 232) holds for an average speed of about 3 to 4 knots. W. Kroll (*RE*, s.v. *Schiffahrt* 411 [1921]) offers 5.6 knots for the Greek period and up to 7.5 for the Roman.

[47] The same winds prevail today as in the days of the ancients. Cf. e.g., Casson, "Isis" note 4, which points out that Nelson in 1798 met the identical winds encountered in a voyage described by Lucian. Mohler has applied hydrographic information and a knowledge of the effect of the wind on sailing vessels to certain passages in the Aeneid with illuminating results.

[48] *NH* 19.3-4: "[Is there a greater miracle than the flax plant which (sc. made into sails) enabled] Galerius to reach Alexandria on the seventh day from the Strait of Messina, and Balbillus on the sixth . . . and Valerius Marianus . . . from Puteoli on the ninth day with the lightest possible breeze . . . , which puts Gades, near the Pillars of Hercules, within seven days of Ostia, Hither Spain within four, the Province of Gallia Narbonensis within three, Africa within two?" (*ut Galerius a freto Siciliae Alexandriam septimo die pervenerit, Balbillus sexto . . . , Valerius Marianus . . . a Puteolis nono die lenissimo flatu . . . , quae Gades ab Herculis columnis septimo die Ostiam adferat et citeriorem Hispaniam quarto, provinciam Narbonensem tertio, Africam altero*).

of exceptionally fast runs. All, of course, must have been made under favorable wind conditions. The voyages are as follows:[49]

TABLE I

Voyage	Distance (nautical miles)	Length of Voyage (days)	Overall Speed (knots)
Ostia-Africa	270[50]	2	6
Messina-Alexandria	830	6[51]	5.8
Ostia-Gibraltar	935	7	5.6
Ostia-Hispania Citerior	510[52]	4	5.3
Ostia-Provincia Narbonensis	380[53]	3	5.3
Messina-Alexandria	830	7	5
Puteoli-Alexandria	1,000	9	4.6

This list gives a good point of departure. It provides an upper limit: with a wind from the right direction, a speed of between 4½ and 6 knots could be realized. The variation reflects the two factors mentioned above, that some winds are stronger and some ships faster than others, a point that can be best illustrated by the voyage between Ostia and Africa. The record crossing took, as Pliny shows, two days. A more usual but still fast trip took two and a half or three days,[54] while

[49] Distances are based on G. Philip, *Mercantile Marine Atlas*[16] (London 1959) maps 16-18; Reed's *Tables of Distances Between Ports and Places in all Parts of the World*[11] (Sunderland 1947) 62-69; *Table of Distances between Ports*, United States Navy Hydrographic Office, no. 117 (Washington, D.C. 1943) *passim*. In addition, I have used measurements based on the U.S. Navy Hydrographic Office charts of the Mediterranean.

[50] This is the distance between Ostia and a point off Cape Bon. Pliny, who cannot resist exaggeration, adds that this voyage (like the Puteoli-Alexandria run mentioned in note 48) was done "with the lightest possible breeze."

[51] The voyage from Italy to Alexandria was a downhill run with the northwest trades at a vessel's heels all the way. Goitein (326) reports one 11th century run from Palermo to Alexandria that took 13 days and another that took 17, while Ibn Jubayr went from Sardinia to Egypt in 12 (Goitein 325). I estimated that the Marseilles-Alexandria run would take 20-30 days ("Speed" 146); Goitein (325) has subsequently reported an 11th century voyage that took exactly 25.

[52] The distance from Ostia to Tarraco through the Strait of Bonifacio is 510 nautical miles.

[53] To Narbo.

[54] Cato the Elder showed the Senate at Rome a fig that "had been picked at Carthage three days before" (*tertium . . . ante diem . . . decerptam Carthagine*, Pliny, *NH* 15.75; cf. Plutarch, *Cato Maior* 27.1).

Marius, when hastening from the war against Jugurtha to stand for the consulship at Rome needed three and a half or four days despite a favorable wind.[55] Probably he boarded the first available ship for Rome which, as it happened, was slow; possibly the wind, although from the right direction, was not very strong.

The next step is to compare Pliny's record runs with other voyages of which we have knowledge. Here are seven, each one made, we are specifically told, under favorable wind conditions.

TABLE 2

Voyage	Distance (nautical miles)	Length of Voyage (days)	Overall Speed (knots)
Corinth-Puteoli[56]	670	4½	6.2
Abdera-Mouth of Danube[57]	500	4	5.2
Rhegium-Puteoli[58]	175	1½	5
Carthage-Gibraltar[59]	820	7	4.9
Syrtes-Alexandria[60]	700	6½	4.5
Alexandria-Ephesus[61]	475	4½	4.4
Carthage-Syracuse[62]	260	2½	4.3
Phycus-Alexandria[63]	450	4½	4.3
Puteoli-Tauromenium[64]	205	2½	3.4

[55] Plutarch, *Marius* 8.5: "And he crossed the sea [from Utica to Rome] in four days with a favorable wind" (καὶ τὸ πέλαγος τεταρταῖος οὐρίῳ πνεύματι περάσας).

[56] Philostratus, *Vita Ap.* 7.10: "[From Corinth, Apollonius] left . . . toward evening. Happening upon a favorable wind and a current running his way . . . , he arrived at Puteoli on the fifth day" (ἀφῆκεν . . . ἅμα ἑσπέρᾳ. τυχὼν δὲ οὐρίου πνεύματος καὶ τινος εὐπλοίας ὑποδραμούσης . . . ἀφίκετο ἐς Δικαιαρχίαν πεμπταῖος).

[57] Thucydides 2.97.1: "From Abdera to the Pontus Euxinus as far as the Ister— if the wind stands steady at the stern, this is a voyage for a merchantman of four days and as many nights" (ἀπὸ Ἀβδήρων πόλεως ἐς τὸν Εὔξεινον πόντον μέχρι Ἴστρου ποταμοῦ— . . . περίπλους . . . ἦν αἰεὶ κατὰ πρύμναν ἱστῆται τὸ πνεῦμα, νηὶ στρογγύλῃ τεσσάρων ἡμερῶν καὶ ἴσων νυκτῶν). There is some variation in the figures given for the distance covered in this voyage. Whibley, *A Companion to Greek Studies* (Cambridge 1931⁴) 588, gives 460 nautical miles; Assmann (*Segel* 1053) 470; Köster (179) 500; Kroll (*op. cit.*, note 46 above, 411) 520; How and Wells, note to 7.183.3, 600 miles = 520 nautical miles; the same figure in Grundy, "The Rate of Sailing of Warships in the Fifth Century B.C.," *CR* 23 (1909) 107-108. The higher figure presumes that the voyage was made in constant sight of land. For the figure of 500 which I have adopted, see Köster 179, note 1.

[58] Acts 28.13: "After one day [spent at Rhegium] the south wind blew, and we

With the single exception of the last voyage—which will be dealt with in a moment—the figures here agree very well with Pliny's records: with the wind in the right direction, a vessel could log roughly between 4½ and 6 knots. The exception, Apollonius' voyage from Puteoli to Tauromenium, may have been made with a very slow ship or very light breezes or perhaps both. On the second

came the next day to Puteoli" (μετὰ μίαν ἡμέραν ἐπιγενομένου νότου δευτεραῖοι ἤλθομεν εἰς Ποτιόλους).

[59] Scylax, *Per.* 111 (Müller, *GGM* I, p. 90): "From Carthage to the Pillars of Hercules under ideal sailing conditions is a voyage of seven days and seven nights" (ἀπὸ Καρχηδόνος ταύτῃ ἐστὶν ἐφ' Ἡρακλείους στήλας τοῦ καλλίστου πλοῦ παράπλους ἡμερῶν ἑπτὰ καὶ νυκτῶν ἑπτά).

[60] Sulpicius Severus, *Dial.* 1.3.2, 6.1: "Setting forth [from Utica] for Alexandria, with the south wind blowing against us, we were almost carried into the Syrtes. . . . [After some time spent there] the sailors called us back to the sea, we shoved off, and with a favorable voyage arrived at Alexandria on the seventh day" (1.3.2: *Alexandriam petentes, reluctante Austro paene in Syrten inlati sumus. . . . 1.6.1: revocantibus ad mare nautis discessimus, prosperoque cursu septimo die Alexandriam pervenimus*). Since the point on the Greater Syrtes where the landing was made is not precisely indicated, the figure of 700 nautical miles is merely an estimate. H. de Saussure (*Rev. Arch.* Sér. 6, 10, 1937, p. 97) erroneously took the seven days as the length of the voyage from Utica, as did M. Charlesworth, *Trade Routes and Commerce of the Roman Empire*[2] (Cambridge 1926) 247.

[61] Achilles Tatius 5.15.1, 17.1: "And it happened that the wind called us [to leave Alexandria]. . . . After sailing for five days in a row we arrived at Ephesus" (5.15.1: κατὰ τύχην δὲ καὶ τὸ πνεῦμα ἐκάλει ἡμᾶς. . . . 5.17.1: πέντε δὲ τῶν ἑξῆς ἡμερῶν διανύσαντες τὸν πλοῦν ἤκομεν εἰς τὴν Ἔφεσον).

[62] Procopius, *Bell. Vand.* 1.14.8: [In Syracuse, Procopius met a man] "who had, three days before that very day, come from Carthage" (τριταῖόν οἱ ἐκείνῃ τῇ ἡμέρᾳ ἐκ Καρχηδόνος ἥκοντα). Though not stated in so many words, the implication is that this was a fast voyage, hence done with favorable winds.

[63] Synesius, *Epist.* 51 (Migne, *PG* 66, p. 1379): "Leaving Phycus at daybreak, toward evening we made for the bay at Erythra. Spending there as long as it took to quench our thirst and replenish the water supply, . . . we again set forth. Enjoying a rather gentle wind that, however, blew steadily from astern, . . . on the fifth day we saw the torch which they raise from the tower as a signal to new arrivals. Before you could say it, we had disembarked on the island of Pharos" (ἄραντες ἐκ Φυκοῦντος ἀρχομένης ἑῴας δείλης ὀψίας τῷ κατ' Ἐρυθρὰν κόλπῳ προσέσχομεν. ἐνδιατρίψαντες δὲ ὅσον ὕδωρ πιεῖν καὶ ὑδρεύσασθαι . . . αὖθις ἀνήχθημεν· πνεύματι δὲ χρησάμενοι μετρίῳ μὲν, ἀλλ' ἐκ πρύμνης ἀεί, . . . πεμπταῖοι τὸν φρυκτὸν ἰδόντες, ὃν αἴρουσιν ἀπὸ πύργου τοῖς καταγομένοις σύνθημα· θᾶττον ἢ λόγος ἀποβιβασθέντες ἦμεν ἐν τῇ νήσῳ τῇ Φάρῳ). Phycus is the modern Ras-al-Razat in Cyrenaica. A certain amount of time—perhaps as much as half a day—must be allowed for leaving course, beaching, taking on water, and resuming course.

[64] Philostratus, *Vita Ap.* 8.15 [From Puteoli, Apollonius and a friend] "sailed for Sicily with a favorable wind. . . . They arrived at Tauromenium on the third day" (ἔπλευσαν ἐπὶ Σικελίας ἀνέμῳ ἐπιτηδείῳ . . . ἐγένοντο ἐν Ταυρομενίῳ τριταῖοι).

285

long leg of the journey, from Syracuse to the mouth of the Alpheus, the same ship logged but 2.5 knots.[65] We are told nothing about the wind conditions en route but the prevailing winds in this area blow from the NW, which should have been perfect for such a voyage[66]—had he actually encountered them.

In one of the episodes in Xenophon of Ephesus' novel, he has a pirate ship make the 400 miles from Rhodes to Tyre in four days,[67] or an average of 4 knots, somewhat slower than the speeds just given. The wind failed on the second day, and some time was lost in a boarding operation. Presumably the author conceived of the rest of the run as being made under a favorable wind—and the northwesterlies of the area would, in fact, be ideal.[68]

Sulpicius Severus in one of his dialogues mentions that the trip from Narbo to Utica took five days.[69] The distance is about 500 nautical miles, which works out to an average speed of 4.1 knots. For the 70-odd miles through the Gulf of Lion the ship may have had to tack[70] and, if so, made no more than 2 knots.[71] The rest of

---

[65] Ibid.: "Having arrived at Syracuse, they set sail for the Peloponnese . . . [and] arrived on the sixth day at the mouth of the Alpheus" (ἐπὶ Συρακουσῶν κομισθέντες ἀνήγοντο ἐς Πελοπόννησον . . . , ἀφίκοντο δι' ἡμέρας ἕκτης ἐπὶ τὰς τοῦ Ἀλφειοῦ ἐκβολάς). The distance is somewhat over 300 nautical miles.

[66] Cf. Plutarch, Dion 25.1-6 (cited in part in note 98 below) for a graphic picture of the effect of the northwest winds in this area. Dion's fleet was driven before them from Syracuse to the Greater Syrtes. All hands were amazed when a fresh south wind sprang up; they never expected a wind from that direction in those waters.

[67] Xenophon of Ephesus, Eph. 1.12.3, 13.4-5, 14.6: "At first they were carried along by a fair wind . . . and that day and the following night they were borne along. . . . On the second day the wind died down. . . . At first [a galley full of pirates] kept sailing alongside . . . but about midday . . . they leaped on to the ship . . . [and after looting it], completing the rest of the sail in three days, arrived at . . . Tyre" (1.12.3: καὶ τὰ μὲν πρῶτα ἐφέροντο οὐρίῳ πνεύματι . . . κἀκείνην τε τὴν ἡμέραν καὶ τὴν ἐπιοῦσαν νύκτα ἐφέροντο . . . τῇ δὲ δευτέρᾳ ἐπέπαυτο μὲν ὁ ἄνεμος . . . 1.13.4-5: τὰ μὲν πρῶτα παρέπλεον . . . περὶ μέσον ἡμέρας . . . ἀνεπήδησαν ἐπὶ τὴν ναῦν . . . 1.14.6: διανύσαντες ἡμέραις τρισὶ τὸν πλοῦν κατήχθησαν εἰς . . . Τύρον).

[68] 272 above.

[69] Sulpicius Severus, Dial. 1.3.1: "We set sail from Narbo and entered the port of Africa [most likely Utica] on the fifth day; so prosperous a voyage had we made with God's consent" (Narbone navem solvimus, quinto die portum Africae intravimus: adeo prospera Dei nutu navigatio fuit).

[70] The course was southeast and southwest winds are common in this area; see 272 above.

[71] See Table 5 below.

the journey probably saw favorable winds[72] and an average speed of somewhat better than 5 knots.

The following instances are not records of actual voyages but indications of the time it usually took to cross from one point to another. As it happens, all involve sailing with favorable winds.

TABLE 3

Voyage	Distance (nautical miles)	Length of Voyage (days)	Overall Speed (knots)
Ibiza-Gibraltar[73]	400	3	5.5
Epidamnus-Rome[74]	600	4½	5.5
Cape Samonium in Crete-Egypt[75]	310	3 or 4	4.3 or 3.2
Rhodes-Alexandria[76]	325	3½	3.9
Sea of Azov-Rhodes[77]	880	9½	3.9

In the light of this evidence, the lower limit of the rate of speed should be revised from the 4½ knots noted hitherto to 4 knots.

Consider the following voyages:

[72] HO 152.58I and 35, bottom.

[73] Diodorus Siculus 5.16.1: "The Pityuses . . . lie a voyage of three days and three nights distant from the Pillars of Hercules" (Πιτυοῦσσα . . . διέστηκεν ἀπὸ μὲν Ἡρακλέους στηλῶν πλοῦν ἡμερῶν τριῶν καὶ τῶν ἴσων νυκτῶν). The prevailing easterly winds of the area (272) would be favorable.

[74] Procopius, Bell. Goth. 3.18.4: "For those sailing [from Epidamnus] . . . with a favorable wind, it is possible to reach the harbor of Rome in five days" (πλέουσι μὲν ἀνέμου . . . ἐπιφόρου ἐπιπεσόντος πεμπταίοις ἐς τὸν Ῥωμαίων λιμένα καταίρειν δυνατὰ ἔσεσθαι).

[75] Strabo 10.475: "The voyage from Samonium to Egypt takes four days and nights; though some say three" (ἀπὸ δὲ τοῦ Σαμωνίου πρὸς Αἴγυπτον τεττάρων ἡμερῶν καὶ νυκτῶν πλοῦς, οἱ δὲ τριῶν φασί).

[76] Diodorus Siculus 3.34.7: "From the Lake of Maeotis [Sea of Azov], . . . many on merchant ships traveling with a fair wind make Rhodes on the 10th day . . . , and from there make Alexandria on the 4th" (ἀπὸ γὰρ τῆς Μαιώτιδος λίμνης . . . πολλοὶ τῶν πλοϊζομένων οὐριοδρομούσαις ναυσὶ φορτίσιν εἰς μὲν Ῥόδον δεκαταῖοι . . . ἐξ ἧς εἰς Ἀλεξάνδρειαν τεταρταῖοι καταντῶσιν).

[77] See previous note.

TABLE 4

Voyage	Distance (nautical miles)	Length of Voyage (days)	Overall Speed (knots)
Byzantium-Rhodes[78]	445	5	3.7
Byzantium-Gaza[79]	855	10	3.6
Thessalonica-Ascalon[80]	800	12	2.8

The three voyages took place in an area where the prevailing winds would be fair. The slower speed, if they actually did encounter fair winds, may have been because the run involved some island-hopping and coasting, which is more time-consuming than sailing over the open sea. Vessels could not reach their maximum speed until in the waters south of Rhodes.

Combining all the above evidence, it would seem that, under favorable wind conditions, ancient vessels averaged between 4 and 6 knots over open water, and slightly less while working through islands or along coasts.[81]

[78] Marcus Diaconus, *Vita Porph.* 55: "We arrived at Rhodes [from Byzantium] in five days" (ἐφθάσαμεν δὲ τὴν Ῥόδον δι' ἡμερῶν πέντε).

[79] Marcus Diaconus, *Vita Porph.* 27: "I left Byzantium and arrived in 10 days at Gaza" (ἐξεπόρισα ἐκ τοῦ Βυζαντίου καὶ φθάνω δι' ἡμερῶν δέκα τὴν Γαζαίων πόλιν).

[80] Marcus Diaconus, *Vita Porph.* 6: "And going immediately to Ascalon and finding a ship, I set sail and after a good voyage of 13 days, we arrived at Thessalonica. . . . I sailed back, arriving in 12 days at the port of Ascalon" (εὐθέως δὲ κατελθὼν εἰς Ἀσκάλωνα καὶ εὑρὼν πλοῖον ἀνήχθην, καὶ δι' ἡμερῶν δέκα τριῶν εὐπλοήσαντες ἐγενόμεθα εἰς Θεσσαλονίκην . . . ἐπανέπλευσα φθάσας δι' ἡμερῶν δέκα δύο εἰς τὸ ἐμπορεῖον Ἀσκάλωνος). The voyage out was indeed good: though in a direction that would normally encounter foul winds, it took but one day longer than the voyage back.

[81] Cf. Aristides' remark (*Orat.* 36.111 [II, p. 298 Keil]) to the effect that "a vessel, running for a day and a night with a wind from astern—one that whistles, I may add—. . . will make . . . perhaps better than 1,200 stades [120 nautical miles = 5 knots]. I have often made this speed myself during a good crossing" (ναῦς πανημερία θέουσα ὑπ' ἀνέμου κατὰ πρύμναν πνέοντος, προσθήσω δὲ καὶ 'λιγέος', . . . σταδίους ἀνύσει . . . ἴσως μᾶλλον διακοσίους καὶ χιλίους. καὶ ἡμεῖς τοσούτους ἐν εὐπλοίᾳ πολλάκις ἠνύσαμεν). Menippus, who wrote a coast pilot toward the end of the 1st B.C., mentions (Marcianus of Heracleia, *Epitome peripli Menippei* 5 [Müller, *GGM* I, p. 568]) that an ordinary vessel makes 700 stades (70 nautical miles = ca. 3 knots) and a well-made one 900 (90 nautical miles = ca. 4 knots) "in one day" (διὰ μιᾶς ἡμέρας); whether he means 24 hours or just the daylight hours is not certain.

VOYAGES MADE WITH UNFAVORABLE WINDS

The difference in speed when traveling before and against the wind can most graphically be illustrated by several of the voyages of Mark the Deacon. Sailing with a following wind he made from Byzantium to Rhodes in five days and from Byzantium to Gaza in ten (Table 4). The return trip in both cases took just twice as long (Table 5).[82]

Here are the records of voyages made under unfavorable wind conditions:

TABLE 5

Voyage	Distance (nautical miles)	Length of Voyage (days)	Overall Speed (knots)
Cyrene-West Point of Crete[83]	160	2	3.3
Ascalon-Thessalonica[84]	800	13	2.6
Rhodes-Gaza[85]	410	7	2.4
Alexandria-Marseilles[86]	1,500	30	2.1
Puteoli-Ostia[87]	120	2½	2.0
Gaza-Byzantium[88]	855	20	1.8
Rhodes-Byzantium[89]	445	10	1.8
Caesarea-Rhodes[90]	400	10	1.7
Alexandria-Cyprus[91]	250	6½	1.6
Sidon-Chelidonian Isles[92]	350	9½	1.5

[82] Equally graphic are voyages from Rome to the Near East and back. Puteoli to Alexandria could be done in as few as 9 days (Table 1), while the trip back could take over 50 and even 70 (Casson, "Isis" 43-51).

[83] Strabo 10.475: "The voyage from Cyrene to Criumetopon [the west point of Crete] takes two days and nights" (ἔστι δ' ἀπὸ τῆς Κυρηναίας ἐπὶ τὸ Κριοῦ μέτωπον δυεῖν ἡμερῶν καὶ νυκτῶν πλοῦς).

Thucydides mentions that the sail from Carthaginian Neapolis to Sicily, a distance of about 125 nautical miles, took two days and a night (7.50.2: Νέαν πόλιν, Καρχηδονιακὸν ἐμπόριον ὅθενπερ Σικελία ἐλάχιστον δύο ἡμερῶν καὶ νυκτὸς πλοῦν ἀπέχει "Neapolis, a Carthaginian port that lies the closest to Sicily, a voyage of two days and a night"). If he means 36 hours, then the speed works out to ca. 3.5 knots; if

The speed in this list that appears most out of line is the first, but a moment's further study will reveal that it is not really so. From Cyrene to the west point of Crete is NNE. The prevailing wind in these waters is NW. A vessel making the voyage would theoretically travel on a port tack the whole distance. If the wind, however, backed merely a point or so, it would cease being unfavorable. The second,

---

he means 24 hours plus a long day of, say, 15, then it works out to slightly less. The first leg, along the coast of Cape Bon, could take advantage of the favorable land breeze, but the rest would generally be against the prevailing easterlies (HO 151.47). Thus, like the crossing from Cyrene to Criumetopon, it reveals somewhat better time than could be made in trips against the wind all the way.

[84] See note 80 above.

[85] Marcus Diaconus, Vita Porph. 56-57: "Sailing out of Rhodes, the weather being good, we had a good voyage for two days; then a storm suddenly arose.... Toward evening, the wind shifted, and we had good sailing conditions. Putting in four more days on the sea, at dawn on the fifth we sailed up to the beach at Gaza" (πλεύσαντες δὲ ἐκ τῆς Ῥόδου καὶ εὐπλοήσαντες ἐπὶ ἡμέρας δύο εὐδίας οὔσης, ἄφνω κινεῖται χειμών . . . πρὸς ἑσπέραν ἐτράπη ὁ ἄνεμος, καὶ ἐπλέομεν ἐπιτηδείως, καὶ ποιήσαντες ἐν τῷ πελάγει ἄλλας ἡμέρας τέσσαρας, τῇ πέμπτῃ ὄρθρου κατεπλεύσαμεν εἰς τὸ παράλιον μέρος τῆς Γαζαίων).

[86] Sulpicius Severus, Dial. 1.1.3: "There [i.e., at Alexandria] I found a merchant ship that was getting ready to shove off with a cargo for Narbo.... On the 30th day I arrived at Massilia, and from there I came here [to Narbo] on the 10th. So prosperous was the voyage that fell to my pious wish" (navem ibi onerariam inveni, quae cum mercibus Narbonam petens solvere parabat . . . tricensimo die Massiliam adpulsus, inde huc decimo perveni: adeo prospera navigatio piae adfuit voluntati). Indeed it was a prosperous voyage, considering the consistently unfavorable prevailing winds; cf. note 82. Goitein (326) reports an 11th century voyage from al-Kanais, just west of Alexandria, to Palermo that took 25 days, which accords fairly well with the 30 days reported by Sulpicius for the trip from Alexandria to Marseilles. Goitein also reports (325) a voyage from Alexandria to Almeria—the distance sailed would be about the same as from Alexandria to Marseilles—that took 65 days, which accords with Lucian's comment that it could take as much as 70 days to go from Alexandria to Rome (see Casson, "Isis" 43-51).

[87] Philostratus, Vita Ap. 7.16: "Sailing from Puteoli, they arrived at the mouth of the Tiber in three days" (ἀποπλεύσαντες δὲ τῆς Δικαιαρχίας τριταῖοι κατῆραν ἐς τὰς ἐκβολὰς τοῦ Θύμβριδος). Friedländer suggests (Darstellungen aus der Sittengeschichte Roms[10] 1 337-38) that the ship put in for the night at Cajeta and Antium. More likely it was sailing night and day beating into the eye of the northwesterlies.

[88] Marcus Diaconus, Vita Porph. 26: "[At Gaza, Porphyrius] put me aboard, and in 20 days we arrived [at Byzantium]" (ἐπλώισέν με καὶ δι' ἡμερῶν κ ἐφθάσαμεν). Cf. the 11th century voyage reported by Goitein (326) that took 18 days from Alexandria to Constantinople. I had earlier estimated ("Speed" 145) that such a voyage would take 17-20 days. For the route, cf. Josephus, Ant. Jud. 16.17-20 (Herod sailed from Palestine to Byzantium via Rhodes, Cos, Chios and Mitylene. He was held up by contrary winds at Chios).

[89] Marcus Diaconus, Vita Porph. 37: "Leaving [Rhodes] on that very day, we set

the voyage from Ascalon to Thessalonica, is patently exceptionally fast since it took but one day longer than the trip the other way (Table 4), which was in the direction favored by the prevailing winds. The very low average of the last voyage was caused by a heavy storm encountered en route. It would seem therefore that ancient vessels averaged from less than 2 to 2½ knots against the wind.[93]

---

sail and in 10 more days arrived at Byzantium" (ἀναχθέντες ἐκείνῃ τῇ ἡμέρᾳ ἐπλεύσαμεν καὶ δι' ἄλλων ἡμερῶν δέκα ἐφθάσαμεν τὸ Βυζάντιον).

[90] Marcus Diaconus, *Vita Porph.* 34: "Arriving at Caesarea, we set sail . . . and . . . making a good voyage . . . arrived at Rhodes in 10 days" (καταλαβὼν τὴν Καισάρειαν . . . ἐπλεύσαμεν . . . καὶ . . . εὐπλοήσαντες δι' ἡμερῶν δέκα κατήχθημεν εἰς 'Ρόδον).

[91] Lucian, *Navig.* 7: "Setting sail from Pharos with no very strong wind, on the 7th day we sighted Acamas [the western tip of Cyprus]. . . . [On leaving Sidon] we were struck by a great storm and on the 10th day, passing through the Strait of Aulon, we arrived at the Chelidonian Isles" (ἀπὸ τῆς Φάρου ἀπάραντας οὐ πάνυ βιαίῳ πνεύματι ἑβδομαίους ἰδεῖν τὸν 'Ακάμαντα. . . . χειμῶνι μεγάλῳ περιπεσόντας δεκάτῃ ἐπὶ Χελιδονέας διὰ τοῦ Αὐλῶνος ἐλθεῖν). Cf. the 11th century voyage reported by Goitein (321) from Alexandria to Tripoli on the Lebanese coast opposite Cyprus; a constant fight against storm and foul winds, it took 8 days.

[92] See the preceding note.

[93] Wind conditions were often bad enough to force vessels to stop at intermediate ports for days or weeks or even months. It took Cicero three weeks to cross from Patras to Brindisi in 50 B.C. (*ad Fam.* 16.9.1-2), and St. Paul, of course, was forced to spend the winter at Malta (Acts 28.11).

Several voyages have not been included in the tables because it is impossible to determine the winds encountered. There are, e.g., those reported by Herodotus (4.86) that took place in the Black Sea, an area of variable winds. The speed seems to have been 3 to 4 knots, but there is a complication in that the figures Herodotus gives for the distances covered are much greater than they should be. He based them on a ship's average speed, which in this case he overestimated; cf. How and Wells, note to 4.85.2. Strabo (11.498) reports that the voyage from Phasis to Amisus and Sinope took two or three days. Since the distance is 235 nautical miles, this would mean a speed of 3¼ or 5 knots. Then there is the circumnavigation of Sicily to which there are references in Thucydides, Strabo, and Plutarch. On such a course a vessel would theoretically meet both favorable and unfavorable winds. Strabo (6.266) quotes Ephorus as saying that it took five days and nights. Since the distance is roughly 500 nautical miles, the speed works out to 4.2 knots. Plutarch (*Mor.* 603a) says four days, which would mean 5.2 knots. Thucydides (6.1) gives eight days. He means traveling by day only (cf. 2.97.1, cited in note 57 above, where he specifies day and night). If we allow 15 hours a day for sailing, he agrees with Strabo; if less, with Plutarch. In either event, all the figures indicate that, in going around Sicily, vessels met with more fair winds than foul.

Other voyages must be left out of account since there is no certainty that they were nonstop. Vessels traveling along coastal routes frequently put in at a con-

V SPEED OF FLEETS UNDER SAIL

A GOOD deal of attention has been given to how fast ancient warfleets could travel. Aemilius Paulus' record 8-knot dash from Brindisi to Corcyra is quoted again and again.[94] Yet this was surely exceptional speed, unquestionably made with strong favorable winds and very likely helped out by periods of hard rowing.[95] It cannot be used as evidence for a fleet's speed under sail alone.[96] Voyages in which supply ships participated are good evidence, for a fleet's speed is determined by its slowest members, and the swift galleys would have no occasion to put the rowers to work while dawdling alongside

---

venient port for the night, while those threading their way through the Aegean often laid over at whatever island they reached by evening. For example, the 4-day trip from Rome to Stabiae mentioned by Galen (*Methodus Medendi* 5.91-92 [Kuhn, vol. 10, p. 363]) may be a case in point; with any sort of luck with the wind, the voyage would have taken, nonstop, less than two. St. Paul needed 15 days according to the Western text of Acts 27.5 to travel from Sidon to Myra. If the figure can be relied upon, it indicates that his ship probably laid over a good many nights, since the distance is under 400 nautical miles; cf. also Paul's voyages in Acts 20.14-15 and 21.1. A crossing Cicero made from Athens to Ephesus consumed no less than two weeks (*ad Att.* 5.12, 13); the *aphracti* he sailed in must have been very much like the open caiques that still ply between some of the islands today. When Paul went from Philippi to Alexandria Troas (Acts 20.6), he probably laid over at Samothrace (cf. Acts 16.11-12). The 4-day sail from Athens to Rhodes mentioned by Lycurgus (*in Leoc.* 70) probably included stopovers, for the distance is ca. 275 miles and the wind would normally be favorable.

Voyages that took less than a day have also been omitted; they are too short to be of value in computing averages.

[94] Livy 45.41.3: "Departing from Italy, I set sail with the fleet from Brundisium at daybreak; by the 9th hour of the day I reached Corcyra with all my ships" (*profectus ex Italia classem a Brundisio sole orto solvi; nona diei hora cum omnibus meis navibus Corcyram tenui*). Nine Roman hours would be 11½ of our hours, and the distance from Brindisi to the tip of Corcyra is about 90 nautical miles. Cf. Assmann, *Segel* 1054; Köster 180; How and Wells, note to 7.183.3; Mohler 54, note 26.

[95] Cf. Mohler, *ibid.* Grundy's assumption (*loc. cit.* note 57 above) that 8 knots was a reasonable speed was convincingly answered by Tarn ("Fleet-Speeds; A Reply to Dr. Grundy," *CR* 23, 1909, pp. 184-86) who held for 4½ knots or 5 for a fleet in a hurry and 2 when not.

[96] Tarn's estimate of 2 knots (see previous note) includes at least one voyage during which oars very probably were used: Alcibiades took all night up to breakfast to go from Parium to Proconnesus (Xenophon, *Hell.* 1.1.13), about 25 nautical miles—in other words, an average of not over 1½ knots. Very likely this was the speed made while rowing in shifts (cf. 280 above), since in this area the ships were presumably bucking contrary current as well as wind; cf. *The Black Sea Pilot*[9] (Hydrographic Department of the Admiralty, London 1942) 25 and 62.

slow-sailing transports. Almost as good are voyages that lasted several days or more; on such occasions there would be no reason to use the oars, which were intended for battle or emergencies.

TABLE 6

Voyage	Distance (nautical miles)	Length of Voyage (days)	Overall Speed (knots)	Wind
Rhodes-Alexandria[97]	325	3	4.5	Unstated; probably favorable
Greater Syrtes-Heraclea Minoa[98]	475	4½	4.4	Favorable
Sason-Cephallenia[99]	160	1¾	4	Unstated; probably favorable
Troy-Alexandria[100]	550	7	3.3	Favorable
Carales-African Coast[101]	200	2½	3.3	Unstated; probably favorable

[97] Appian, Bell. Civ. 2.89: "[Caesar] left [Rhodes] toward evening . . . and after three days on the sea was off Alexandria" (περὶ ἑσπέραν ἀνήγετο . . . καὶ ὁ μὲν τρισὶν ἡμέραις πελάγιος ἀμφὶ τὴν ᾿Αλεξάνδρειαν ἦν). For the winds in this region, see 272 above.

[98] Plutarch, Dion 25.4-5: "[Off the Greater Syrtes] a land breeze from the south arose; a south wind was hardly what they expected, and they had no faith in the shift. As it little by little freshened and grew strong, they spread all sail . . . and, running swiftly, anchored off Minoa on the fifth day" (αὔραν τινὰ κατέσπειρεν ἡ χώρα νότιον, οὐ πάνυ προσδεχομένοις νότον οὐδὲ πιστεύουσι τῇ μεταβολῇ. κατὰ μικρὸν δὲ ῥωννυμένου τοῦ πνεύματος καὶ μέγεθος λαμβάνοντος ἐκτείναντες ὅσον ἦν ἱστίων . . . θέοντες ἐλαφρῶς πεμπταῖοι κατὰ Μίνῳαν ὡρμίσαντο).

[99] Polybius 5.110.5: "[Philip, from Sason] making a helter-skelter departure and return, arrived at Cephallenia on the second day, having sailed continuously day and night" (οὐδενὶ δὲ κόσμῳ ποιησάμενος τὴν ἀναζυγὴν καὶ τὸν ἀνάπλουν δευτεραῖος ἐς Κεφαλληνίαν κατῆρε συνεχῶς ἡμέραν καὶ νύκτα τὸν πλοῦν ποιούμενος). For the winds, see HO 152.33. Sason is off the mouth of the Aous, not the Achelous as Tarn (loc. cit. note 95 above) has it.

[100] Lucan 9.1004-5: "[After Caesar left Troy] the west wind ever keeping the rigging taut, the seventh night brought in sight the shores of Egypt and the flames of [the lighthouse on] Pharos" (septima nox Zephyro numquam laxante rudentes / ostendit Phariis Aegyptia litora flammis).

[101] Procopius, Bell. Vand. 1.25.21: "Setting sail from there [Carales], with the whole fleet, on the third day they reached the Libyan shore at the point where Nu-

293

Lilybaeum-Cape Bon[102]	65	1	2.7	Generally favorable
Messina-Cephallenia[103]	250	4½	2.3	Unstated; probably favorable
Pisa-Marseilles, via Ligurian Coast[104]	240	4½	2.2	Favorable-unfavorable
Utica-Carales[105]	160	3	2.2	Probably unfavorable
Lilybaeum-Ruspina[106]	140	3½	1.7	Favorable
Lilybaeum (?)-Anquillaria[107]	90	2½	1.5	Unstated
Syracuse-Cape Bon[108]	220	6	1.5	Probably unfavorable
Euripus-Phalerum[109]	96	3	1.3	Variable
Zacynthus-Cape Pachynus[110]	340	12½	1.1	Very light
Lilybaeum-Africa[111]	85	3½	1	Unfavorable
Zacynthus-Mt. Etna[112]	320	15½	.9	Very light

midia borders on Mauretania" (ἄραντες δὲ ἐνθένδε παντὶ τῷ στόλῳ τριταῖοι κατέπλευσαν ἐς τὴν Λιβύης ἀκτὴν ἣ Νουμίδας τε καὶ Μαυριτανοὺς διορίζει). For the winds, see *HO* 151.47.

[102] Livy 29.27.6-8: "They set forth [from Lilybaeum] with a favorable and nicely strong wind. . . . At mid-day fog closed them in. . . . The wind was lighter on the open water. During the night that followed, fog again held them in its grip; at daybreak it dispersed, and the wind gained strength. Soon they sighted land. Not long afterward, the helmsman reported to Scipio that Africa was no more than 5 miles off, that he could see Cape Mercury" (*vento secundo vehementi satis profecti . . . a meridie nebula occepit . . . lenior ventus in alto factus. noctem insequentem eadem caligo obtinuit; sole orto est discussa, et addita vis vento. iam terram cernebant. haud ita multo post gubernator Scipioni ait non plus quinque milia passuum Africam abesse; Mercuri promunturium se cernere*).

[103] Livy 42.48.9: "The praeter Caius Lucretius set out from Naples; after passing through the strait [of Messina], he crossed to Cephallenia in five days" (*C. Lucretius praetor ab Neapoli profectus, superato freto, die quinto in Cephallaniam transmisit*). For the winds, see *HO* 152.33.

[104] Polybius, 3.41.4: "Publius, following the coast of Liguria, arrived off Massalia on the fifth day out of Pisa" (Πόπλιος δὲ κομισθεὶς παρὰ τὴν Λιγυστίνην ἧκε πεμπταῖος ἀπὸ Πισῶν εἰς τοὺς κατὰ Μασσαλίαν τόπους). The northerly and northwesterly winds in this area (cf. *HO* 152.33, 35) would have been unfavorable for the first leg of the voyage as well as part of the last.

[105] Caesar, *Bell. Afr.* 98: "[Caesar] embarked in the fleet at Utica and two days

The first three voyages listed in Table 6 are considerably faster than all the others. This can be easily accounted for. The first involved triremes alone, the third swift *lembi* alone, and the second was made under a wind that was remarkable for its steadiness and

---

later arrived at Carales in Sardinia" (*Uticae classem conscendit et post diem tertium Caralis in Sardiniam pervenit*). For the winds, see *HO* 151.47.

[106] Caesar, *Bell. Afr.* 34: "[The ships, setting out from Lilybaeum], catching a favorable wind, arrived safely . . . on the fourth day at the port of Ruspina (*ventum secundum nactae quarto die in portum ad Ruspinam . . . incolumes pervenerunt*). Ruspina was near the modern Monastir on the east coast of Tunis.

[107] Caesar, *Bell. Civ.* 2.23: "Caius Curio left Sicily for Africa . . . spending two days and three nights sailing, he arrived at the place called Anquillaria" (*C. Curio in Africam profectus ex Sicilia . . . biduoque et noctibus tribus navigatione consumptis appellit ad eum locum qui appellatur Anquillaria*). For the winds, see note 105 above. Anquillaria was somewhere on the tip of Cape Bon.

[108] Diodorus Siculus 20.6.1-2: "[Agathocles and his fleet, leaving Syracuse], sailed for six days and as many nights . . . and, on sighting Libya, a cheer arose among the crews" (ἐξ δ' ἡμέρας καὶ τὰς ἴσας νύκτας αὐτῶν πλευσάντων . . . καθορωμένης δὲ τῆς Λιβύης παρακελευσμὸς ἐγίνετο τοῖς πληρώμασι). The landing was eventually made at the "Quarries" (20.6.3: Ἀγαθοκλῆς ἀποβιβάσας τὴν δύναμιν πρὸς τὰς καλουμένας Λατομίας) which would appear to be those on Cape Bon.

[109] Herodotus 8.66: "They sailed through the Euripus and in another three days came to Phalerum" (ἔπλεον δι' Εὐρίπου καὶ ἐν ἑτέρῃσι τρισὶ ἡμέρῃσι ἐγένοντο ἐν Φαλήρῳ).

[110] Plutarch, *Dion* 25.2: "[From Zacynthus] sailing with a light and gentle breeze for 12 days, on the 13th they were at Pachynus" (ἀραιῷ δὲ καὶ μαλακῷ πνεύματι πλεύσαντες ἡμέρας δώδεκα, τῇ τρισκαιδεκάτῃ κατὰ Πάχυνον ἦσαν).

[111] Caesar, *Bell. Afr.* 2: [About 10 miles out of Lilybaeum Caesar] "aboard a swift ship and carried by a steady wind, three days later arrived with a few warships in sight of Africa; for, except for a very few, all the other merchant ships, scattered by the wind and wandering about, made for shore at different points" (*ita vento certo celerique navigio vectus post diem quartum cum longis paucis navibus in conspectum Africae venit; namque onerariae reliquae praeter paucas vento dispersae atque errabundae diversa loca petierunt*). The passage is not inconsistent, as has been thought (cf., e.g., A. Way's note in the Loeb translation, 1955: "3-4 days' sail in a fast ship seems unduly long for the passage of less than 100 miles. . . . Other apparent inconsistencies are . . . the capriciousness of the wind, which favored the warships but scattered the transports"). When Caesar says "carried by a steady wind" he can only mean that the ships traveled quickly through the water; every schoolboy must have known that Scipio had taken a huge fleet over the same crossing in one quarter the time (see note 102 above). The wind blew steadily but so strongly it drove everyone off course, doing a thorough job of scattering the transports and forcing them ashore at all different points. Assmann (1622) states that this voyage "agrees [with Livy 29.27.6-8] in an average of 2¾ knots." It, of course, does nothing of the sort. The error is repeated by How and Wells, note to 7.183.3.

[112] Procopius, *Bell. Vand.* 1.13.22: "Sailing with a very gentle and languid breeze, on the 16th day they arrived at a deserted spot in Sicily near which Mt. Etna rises" (ἀνέμου δὲ σφίσι μαλακοῦ τε καὶ νωθροῦ κομιδῇ ἐπιπνεύσαντος ἑκκαιδεκαταῖοι κατέπλευσαν τῆς Σικελίας ἐς χῶρον ἔρημον, οὗ τὸ ὄρος ἐγγὺς ἡ Αἴτνη ἀνέχει).

freshness.[113] The rest of the table presents a consistent picture. Before a favorable wind, a fleet could log between 2 and 3 knots.[114] With unfavorable or very light winds, a fleet usually could do no better than 1 to 1½ knots.

[113] With a strong favorable wind Belisarius' fleet made from Malta to Caputvada on the east coast of Tunis, 165 nautical miles, in either a day or a day and a half (Procopius, *Bell. Vand.* 1.14.17: "There [off Malta] a good wind from the southeast sprang up and brought the fleet on the following day to the Libyan shore at the point that in Latin is called Caputvada" [ἔνθα δὴ αὐτοῖς Εὔρου πολύ τι πνεῦμα ἐπιπεσὸν τῇ ὑστεραίᾳ τὰς ναῦς ἐς τὴν Λιβύης ἀκτὴν ἤνεγκεν ἐς χωρίον, ὃ δὴ Κεφαλὴν Βράχους τῇ σφετέρᾳ γλώσσῃ καλοῦσι Ῥωμαῖοι]). His speed must have been anywhere from 4.5 to 7 knots depending upon the exact time of arrival. Similarly, Himilco in 397 B.C. left Carthage "with 100 [of the best] triremes [in the fleet], crossed to Selinus, then doubled the promontory at Lilybaeum, arriving at Motya at dawn of the following day" (ἐκπλεύσας οὖν μετὰ νεῶν ἑκατὸν κατήχθη ἐπὶ τὴν τῶν Σελινουντίων χώραν νυκτός, καὶ περιπλεύσας τὴν περὶ Λιλύβαιον ἄκραν ἅμ' ἡμέρᾳ παρῆν ἐπὶ τὴν Μοτύην Diodorus 14.50.2); since he most likely had left Carthage at dawn, he covered ca. 150 nautical miles in 24 hours for an average speed of slightly over 6 knots.

[114] The expedition sent by Caesar from Lilybaeum to Ruspina, consisting of heavily laden transports, made very slow time.

APPENDIX

THE ALEXANDRIA-ROME
SAILING SCHEDULE

PROBABLY the most ambitious maritime enterprise of the ancient world was the transport of the grain that Rome imported from Egypt, 150,000 tons which traveled annually from Alexandria to Rome during the first three centuries A.D. and required the services of a fleet of the biggest and fastest freighters available.[1] These seem to have fitted their sailings into the curtailed season as follows.

1. Those of the fleet[2] which had wintered at Alexandria loaded up at the beginning of spring with grain that had been stored in the dockside silos, and departed at the earliest possible moment, probably in April. They followed either of two routes: the northerly by way of Cyprus, Myra, Rhodes or Cnidos, south of Crete, Malta, Messina;[3] or the southerly by way of the north African coast to Cyrene.[4] Either, involving beating against the prevailing northwesterlies, could take at least a month,

[1] See NINE, App., pt. 2, and Philo, *In Flaccum* 26 (Caligula advised Prince Agrippa, headed for Palestine, not to make the wearisome journey by short hops from Brindisi to Syria, but to go directly from Puteoli to Alexandria and then backtrack, since that way he could cross on one of the crack Puteoli-Alexandria freighters whose skippers "drive them like racehorses"). Arrian describes the special craft Trajan used on the Euphrates as having "the width and depth of a merchantman big as the biggest Nicomedian or Egyptian" (*Parth.* fr. 67 = Jacoby, *FGH* no. 156, fr. 154, vol. II B, p. 876: εὖρος δὲ καὶ βάθος καθ' ὁλκάδα, ὅσον μεγίστη Νικομηδὶς ἢ Αἰγυπτία). The "Egyptians" he refers to must certainly be the ships of the Alexandria-Rome run.

[2] That they traveled as a fleet is clear—see *P. Oxy.* 1763 (3rd A.D.): οὔπω μέχρι σήμερον τὰ πλοῖα τῆς ἀννώνας ἐξῆλθεν "The ships carrying the grain tribute have not yet sailed"; *W. Chrest.* 445 (cf. note 5 below); and Seneca, *Epist.* 77.1 (cited in ELEVEN, note 71). A. Pelletier, translator of the Éditions du Cerf publication of Philo's *In Flaccum* (1967), overlooks this when he explains (p. 161) that Philo, in reporting Agrippa's passage (cf. previous note), speaks of "captains" because Agrippa and his entourage crossed on numerous small units; the "captains" were the commanders of the several ships in the convoy.

[3] See Casson, "Isis" 43-51.

[4] Cf. *P. Mich.* 490 (2nd A.D.), where a recruit, en route from Alexandria to Rome, dispatches a letter to his family in Egypt through a traveler he met in Cyrene ("Finding someone going your way from Cyrene, I felt I had to let you know I was safe and sound" [ἀπὸ Κυρήνης εὑρὼν τὸν πρός σε ἐρχόμενον ἀνάγκην ἔσχον σοι δηλῶσαι περὶ τῆς σωτηρίας μου]). This southern route was the one used in the 11th century by ships plying between Alexandria and Palermo (Goitein 319-20).

sometimes two, so arrival was at the earliest in May.[5] They dropped their cargoes and immediately set off back to Alexandria, a downhill sail that could be made in two to three weeks or even less (above, Table 1); here they reloaded to squeeze in a second crossing before the season closed.[6] What was all-important was a quick turn-around, but this, unfortunately, could not always be counted on. A papyrus[7] reveals that on one occasion the fleet completed its spring run to Rome on 30 June, unloaded by 12 July—and was still hanging around waiting for clearance on 2 August.

2. Those vessels that had wintered in Rome would leave, probably in ballast,[8] in April,[9] be in Alexandria in May, take on a load of grain, and

[5] For the length of the voyage, see TWELVE, notes 82, 86.

W. Chrest. 445 (= Select Papyri 113, 2nd-3rd A.D.) mentions arrival at Ostia (see Wilcken in Archiv 9, 1930, p. 86) of the grain fleet on 30 June ("I arrived on Epeiph 6, and we unloaded on the 18th of the same month. . . . Day after day we have been waiting for notification of release. Right up to today not one of the grain carriers has been released" [ἐλήλυθα τῇ ς τοῦ 'Επεὶφ μηνὸς καὶ ἐξεκενώσαμεν τῇ ιη τοῦ αὐτοῦ μηνός. . . . καὶ καθ' ἡμέραν προσδεχόμεθα διμισσωρίαν ὥστε ἕως σήμερον μηδέναν ἀπολελύσθαι τῶν μετὰ σίτου]). P. Mich. 490 (see previous note) mentions arrival at Ostia on 20 May ("We arrived at Portus Pachon 25" [ἰς Πόρτον παρεγενάμην Παχὼν κε]; cf. 491.5-6). P. Oxy. 2191 (2nd A.D.) mentions arrival at Puteoli on 29 May ("Having disembarked on Italian soil, I felt I had to write to tell you that I am well . . . we had a slow voyage but not an unpleasant one. . . . Written at Puteoli Pauni 4" [ἐπιβὰς τῆς 'Ιταλικῆς χώρας ἀναγκαῖον ἡγησάμην δηλῶσαι ὑμῖν ὅτι ἔρρωμαι . . . τῷ βραδυπλοίᾳ μὲν χρησάμενος οὐ μὴν δυσπλοίας . . . ἐγράφη ἐν Πυτιόλοις Παῦνι δ]). In none of these is the type of carrier mentioned, but all are evidence for early spring sailings. An inscription from Puteoli (OGIS 594 = Berytus 9, 1948/49, p. 47, A.D. 79) which records a ship that "sailed into Puteoli from Tyre on Artemisios 11 [29 May]" (μηνὸς 'Αρτεμισίου ια κατέπλευσεν ἀπὸ Τύρου εἰς Ποτιόλοις) provides another example of a spring sailing since the journey is about the same length as that from Alexandria and has to be made against the same winds. Cicero, ad Att. 4.10.1, written 22 April, mentions a rumor rife at Puteoli that Ptolemy Auletes had regained his throne; since it most likely came by sea (overland news would have gone via Brindisi and the Appian Way), the ship that carried it clearly had been able to start very early in the year and made excellent time. Titus, for example, when he hastened back from Alexandria to Rome on a merchant ship in the spring of A.D. 71 (Suetonius, Titus 5), did not set sail until after 25 April (cf. P. Oxy. 2725, written 29 April, which mentions his entry into Alexandria on 25 April).

[6] This late summer trip back to Rome is attested by the experience of those ships that, for one reason or another, failed to make it all the way before the season closed. E.g., two notable cases in point are the ship that St. Paul boarded at Myra (Acts 27.6), which only reached Crete by the end of the season (27.9), and the one he completed his voyage on, which had gotten no farther than Malta (28.11).

[7] W. Chrest. 445, cited in note 5 above. The letter was written Mesore 9 = 2 August.

[8] Cf. Strabo 17.793: "The exports of Alexandria are much greater than the imports. Anyone who goes to Alexandria and Puteoli becomes aware of this, watching

be back in Rome by August.[10] This gave them enough time to return to Alexandria and spend the winter there, available for a quick departure the following spring.

---

the ships at arrival and departure and observing how much heavier or lighter they are as they sail in and out" (τὰ ἐκκομιζόμενα ἐξ 'Αλεξανδρείας πλείω τῶν εἰσκομιζομένων ἐστι· γνοίη δ' ἄν τις ἔν τε τῇ 'Αλεξανδρείᾳ καὶ τῇ Δικαιαρχίᾳ γενόμενος, ὁρῶν τὰς ὁλκάδας ἔν τε τῷ κατάπλῳ καὶ ἐν ταῖς ἀναγωγαῖς ὅσον βαρύτεραί τε καὶ κουφότεραι δεῦρο κἀκεῖσε πλέοιεν).

[9] It was one of these sailings that Caligula advised Agrippa to take (cf. note 1). Agrippa was surely in Alexandria by June, since riots that followed in the wake of his appearance took place not long after the death of Drusilla (Philo, In Flaccum 56), who died on 10 June A.D. 38 (Prosopographia Imperii Romani, s.v. "Iulia Drusilla"; the news would have taken some two to three weeks to arrive).

The ship celebrated in Statius, Silvae 3.2, on which Maecius Celer was to sail to Egypt, was not one that had wintered in Rome but one that had arrived in the spring convoy. Statius reports (line 22) that it "heavy laden, was the first to carry the harvest of Pharos [i.e. Egypt] to the shores of Puteoli" (prima Dicarcheis Pharium gravis intulit annum). It is generally assumed, on the basis of a passage from Seneca (Epist. 77.1, cited in ELEVEN, note 71), that what is meant is one of the dispatch boats which heralded the arrival of the convoy (cf. F. Vollmer, P. Papinii Statii Silvarum libri, Leipzig 1898, p. 397; A. Pézard, "Gravis artemo," REL 25, 1947, pp. 215-35, esp. 216; H. Izaac, Stace, Silves, Paris 1944, 1 106, note 1). But Statius' words are far more naturally taken as describing one of the actual carriers; it was prima because it had come in with the first group to arrive that year. For other arguments against the traditional interpretation see NINE, note 47.

[10] E.g., the ship from Alexandria whose passengers according to Suetonius (Aug. 98.2) hailed Augustus off Puteoli shortly before his death (19 August), could have been one that made this sailing.

CHAPTER THIRTEEN

Officers and Men

I NAVY

RANDOM REFERENCES in Homer and remarks dropped by later
poets who purport to describe heroic times reveal in embryo
the officers' ranks that were to become standard in the navies of
Greece and Rome. On the simple vessels of the age, with no room
or need for a complicated organization, there were but three: the
commanding officer (*kybernetes*, literally "steerer"), who manned
the helm;[1] the rowing officer (*keleustes*, literally "orderer"), who
gave the beat to the oarsmen;[2] and the bow officer (*prorates*, literally
"fore-looker"), who stood watch in the bows,[3] an all-important station in an age that knew neither charts nor navigational aids.

---

[1] Cf. Plato, *Republic* 1.341c-d: ὁ ὀρθῶς κυβερνήτης . . . ἐστὶν . . . ναυτῶν ἄρχων . . .
οὐ γὰρ κατὰ τὸ πλεῖν κυβερνήτης καλεῖται, ἀλλὰ κατὰ . . . τὴν τῶν ναυτῶν ἀρχήν ("The
true *kybernetes* is . . . the leader of the crew. . . . He is called *kybernetes* not for
his navigation but for . . . his leadership of the crew"). So Pindar used the phrase
(*Isthm.* 4.71) κυβερνατὴρ οἰακόστροφος "the helm-wielding captain."
   The Homeric heroes, each of whom had brought a fleet to Troy, when on the
water were, so to speak, the admirals of their contingents, and a *kybernetes* exercised
command on each individual unit. Odysseus took the tiller when he had the bag
of winds aboard (*Od.* 10.32-33), but this was exceptional.
   The term *kybernetes* is ambiguous inasmuch as, alongside the specialized sense
of "commanding officer," it retained its original sense of "man at the tiller." Thus
Philostratus, *Vita Ap.* 3.35, describing a special vessel designed for use in the Egypt-
India trade, remarks that "there were many helmsmen (*kybernetai*) aboard this
ship, subordinate to the oldest and the ablest" (πολλοὶ μὲν κυβερνῆται τῆς νεὼς ταύτης
ὑπὸ τῷ πρεσβυτάτῳ τε καὶ σοφωτάτῳ πλέουσι). The latter was the captain, the others
quartermasters, skilled hands to take tricks at the helm.
[2] In Euripides' *Helen*, Menelaus and Helen steal a penteconter to make their escape,
and the rowers start the stroke "when the sounds of the *keleustes*' shout were heard"
(1576: βοῆς κελευστοῦ φθέγμαθ' ὡς ἠκούσαμεν).
[3] Cf. Sophocles, *Assembly of the Greeks* 142, col. II 3 (Pearson I, p. 98), where
Telephus is told to "take a seat at the helm and instruct the officer of the bow
right off to keep an eye on the course taken to Troy by the sons of Atreus"
(πηδαλίῳ παρεδρεύων / φράσεις τῷ κατὰ πρῷραν / εὐθὺς Ἰλίου πόρον / Ἀτρειδᾶν ἰδέσθαι).
Sophocles used the word *prorates* (in a metaphorical sense) in his *Polyxena*, fr.
524 (Pearson II, p. 166). This officer is frequently included in pictures of 6th century galleys (Fig. 90; *GOS*, pls. 11d, 14a, 14b; *Eph. Arch.*, 1912, p. 102, fig. 1 and

When, in the fifth century B.C., the trireme with its highly trained crew and specialized fighting personnel became the capital ship, expansion was inevitable. The three-man staff was increased to five, each of whom wielded authority for the most part, physical execution being carried out by subordinates. The old titles remained in use, but the *keleustes* now left some of the actual time-beating, and the *prorates* the actual lookout, to men under them, while the *kybernetes* took the steering oars only when necessary.[4] And, as their superior, at least formally, there appears a new figure, the *trierarchos* "chief of a trireme." He came into being when Greek states, growing bigger and more sophisticated, were faced with the problem of funding bigger and more sophisticated navies. A favored solution was the appointment of *trierarchoi*, wealthy men who, as a service to their country, had each to undertake the expense of maintaining and running a warship for a year.[5] In the early fifth century, trierarchs as a rule exercised actual command of their vessels, and in certain nations, such as Rhodes, that maintained a proud naval tradition, they continued to do so. In many navies, however, the trierarch, a man chosen for income and not seamanship, became a figurehead who frequently never saw the inside of the ship he paid for and left the command totally in the hands of the *kybernetes*.[6]

---

pl. 6 [= *GOS*, Arch. 95]). Bow officers charged with keeping a sharp lookout (presumably for mudbanks, small boats, driftwood) are attested on Egyptian craft of the Pharaonic period (Boreux 433, note 6).

All three officers—*prorates, keleustes, kybernetes*—are portrayed in a picture of a 6th century B.C. galley (*GOS*, pl. 11d).

[4] Cf. Plutarch, *Mor.* 812c: οἱ κυβερνῆται . . . χρῶνται δὲ καὶ ναύταις καὶ πρῳρεῦσι καὶ κελευσταῖς, καὶ τούτων ἐνίους ἀνακαλούμενοι πολλάκις εἰς πρύμναν ἐγχειρίζουσι τὸ πηδάλιον "Captains . . . have the use of sailors, bow officers, and rowing officers, and frequently they call certain of these to the poop and hand over the steering oar."

[5] The French, for example, in the 15th and 16th centuries, used only chartered galleys belonging to the captains or to entrepreneurs who built up fleets for hire. It took until the 17th century for the Royal Navy to have a fleet that was exclusively its own property (Masson 42-45).

[6] For a good review of the Athenian trierarchy, its history and duties, see *GOS* 260-63. By the end of the 5th B.C., financial stringency forced Athens to appoint two trierarchs per ship. For an example of a trierarch of a ship in the Ptolemaic period using a substitute for the actual command, see *P. Cairo Zen.* 59036 = *Select Papyri* 410 (257 B.C.) and cf. *AM* 148-49.

The earliest information available that is full enough to be useful concerns the Athenian navy of the fifth and fourth centuries B.C. The capital ship was the trireme, and its crew numbered about 200,[7] made up as follows:[8]

OFFICERS

[*trierarchos*]

kybernetes: executive officer when the trierarch had the experience and desire to take command himself, and commanding officer when he did not. Under way, the *kybernetes* took the captain's traditional station on the poop. In emergencies he might handle the tiller himself, but normally he used quartermasters.[9]

keleustes: in charge of the rowing personnel, responsible for their training and morale. His station was near enough to the *kybernetes* to receive orders from him and convert these into shouted commands for the oarsmen.[10]

pentekontarchos: the name means "chief of 50"; most likely it was originally the title of the commanding officer of a penteconter

---

[7] The evidence has been collected by Morrison (*GOS* 254-55). Herodotus "reckons 200 for each ship" in the Persian fleet in 480 B.C. (7.184.1: ὡς ἀνὰ διηκοσίους ἄνδρας λογιζομένοισι ἐν ἑκάστῃ νηί) and Thucydides the same number in 416 (6.8.1: 60 talents as a month's pay for 60 ships or 1 talent [= 6,000 *drachmai*] per month per ship. Since each man was paid a *drachma* a day [6.31.3], there were 200 on each ship [200 men x 1 *drachma* x 30 days = 6,000]).

[8] The best evidence is *IG*² ɪɪ 1951.79-109. These lines list for a trireme: 2 trierarchs, 10 *epibatai*, 1 *kybernetes*, 1 *keleustes*, 1 *pentekontarchos*, 1 *auletes*, 1 *naupegos*, 1 *prorates*, 3 *toxotai*. (For the connection of this inscription with the Battle of Arginusae [406 B.C.], see App., note 4). For other fragments, see *SEG* x 356, and cf. xxɪɪ 53.

[9] On the quartermaster (*pedaliouchos* "steering oar wielder"), see below. Aristophanes, *Knights* 542-44, surveys the high points of a career in the navy as follows: ἐρέτην χρῆναι πρῶτα γενέσθαι πρὶν πηδαλίοις ἐπιχειρεῖν, / κᾆτ' ἐντεῦθεν πρῳρατεῦσαι καὶ τοὺς ἀνέμους διαθρῆσαι, / κᾆτα κυβερνᾶν "You have to be a rower first before putting your hand to the steering oars; then, from there, to bow officer (*prorates*) and keeping an eye on the winds; then to captain (*kybernetes*)." In other words, from rower to quartermaster to bow officer to captain.

[10] For his connection with the oarsmen's morale, see Xenophon, *Oec.* 21.3, with Morrison's translation and comments (*GOS* 267). For his shouted commands, see *GOS* 267, citing Aristophanes, *Frogs* 207-208 (Charon plays *keleustes* as Dionysus sits to the oar) and Xenophon, *Hell.* 5.1.8 (during an operation requiring secrecy, the *keleustes* gave orders by chinking stones instead of shouting). For his station, cf. Silius Italicus 6.360-61: *mediae stat margine puppis, / qui voce alternos nautarum temperet ictus* "At the edge of the poop on the centerline stands the officer to regulate with his voice the successive strokes of the oarsmen."

and then doubled as the name of the subordinate officer on a trireme.[11] Conformably, his station in action may have been by the lower banks, the 54 zygite or thalamite rowers, to pass on the *keleustes'* commands. He had important administrative duties, serving as paymaster, purchasing officer, and recruiting officer.[12]

prorates (or *proretes* or *proreus*): Officer of the bow, stationed on the foredeck[13] and entrusted with keeping a sharp lookout; he had to be particularly sensitive to sudden changes of wind,[14] which could be dangerous for so light and unseaworthy a craft as a war galley. The term *prorates* (or *proretes*) refers only to the naval rank; the alternate form *proreus* can also be used of the first mate of a merchantman (see 318-19 below).[15] The *prorates* ranked well below the other three officers; it was the first commissioned grade a rower, ambitious for promotion, could hope to achieve.[16]

[11] Cf. *GOS* 268. Compare the French use of *capitaine de corvette* for what we call a lieutenant, *capitaine de vaisseau*, etc.

[12] Demosthenes 50.18 refers to a trierarch's dispatching the *pentekontarchos* to certain cities to recruit rowers. In 50.25 there is mention of his services as paymaster and purchasing officer.

[13] See the representations listed in note 3 above.

[14] "Keeping an eye on the winds" is the duty that Aristophanes singles out for mention; cf. note 9 above.

[15] *Prorates* or *proretes*: *IG* ii² 1951.104, 330 (Athens, end of 5th b.c.); *Clara Rhodos* 8 (1936) 228, line 7 (Rhodes, 1st b.c.); *SEG* iv 178 (Rhodes, 2nd b.c.); *ILS* 2827, 2864-66 (Rome, Imperial period). *Proreus*: *Nuova sill.* 5 (Rhodes, 3rd-2nd b.c.); *IG* xii.8.260 (Cos, 1st b.c.). When the word *proreus* is used with no further qualification, we cannot be sure whether the bow officer of a warship or the first mate of a merchantman is intended (e.g., in Aristotle, *Pol.* 1253b 29; Plutarch, *Agis* 1.2).

[16] Cf. the citation from Aristophanes in note 9 above. Conformably, Xenophon, *Ath. Pol.* 1.2, ranks the officers *kybernetes, keleustes, pentekontarchos, prorates*. Plato, *Laws* 4.707a-b, refers to the arts of the captain, *pentekontarchos*, and rowing personnel (which would include the *keleustes*) as the bulwark of a city, omitting any mention of the *prorates*.

These ranks correspond in a general way to those found, e.g., in French or Papal galleys of the 16th-17th centuries (Masson 144, 240, 246-47; Guglielmotti iii 107; Pantera 113-20):

trierarchos (when commanding)	capitaine	capitano
kybernetes	lieutenant + pilote	senior nobile di poppa + piloto
keleustes	comite	comito
pentekontarchos	sous-comite	sottocomito + padrone
prorates	sous-lieutenant	junior nobile di poppa

RATINGS

naupegos: ship's carpenter[17]

auletes or *trieraules*: ship's flutist,[18] who piped the time for the rowers once the *keleustes* had set the stroke.[19]

toicharchoi: literally "side-chiefs," very likely the stroke oars, i.e., the thranite rower nearest the stern on port and starboard respectively.[20]

The above are all that our random sources of information for Athens' navy happen to mention. Undoubtedly there were deckhands, quartermasters, oar-tenders, and others who are attested in the navies of the next few centuries (see 308-309 below).

FIGHTING PERSONNEL

epibatai: marines. Selected from Athens' military class, her highest social body, the marines ranked next to the trierarch (like the *gentilhommes des armes* who manned the French galleys of a later age). The Athenians, favoring the ram as the chief weapon, used a minimal number, generally 10. Other navies that preferred boarding tactics loaded the decks with as many as 40.[21]

---

[17] *IG* ii² 1951.102.

[18] *IG* ii² 1951.100 (*auletes*); Demosthenes 18.129 (*trieraules*, who, as it happens, was in this case a slave).

[19] For the *keleustes* and *auletes* operating as a team, see Polyaenus 5.2.5: Dionysius, to scare a port into submission, sent in a "penteconter . . . loaded with pipers (*auletai*) piping and time-beaters (*keleustai*) time-beating, each marking the time as if for one trireme" (πεντηκόντορος . . . αὐλητὰς αὐλοῦντας ἄγουσα καὶ κελευστὰς κελεύοντας, ὧν (ἂν) ἕκαστος ἀνεκάλει τριήρη μίαν).

[20] Lucian, in a dialogue presumably set in the 4th B.C. (*Dial. Het.* 14.3) has a disappointed lover complain to his mistress that he has given her "whatever a man who's a sailor serving as a mercenary could. And by now I'm chief of the starboard side. . . . If I were a rich man, I wouldn't be rowing for a living (ὅσα ναύτης ἄνθρωπος ἐδυνάμην μισθοῦ ἐπιπλέων. νῦν γὰρ ἤδη τοίχου ἄρχω τοῦ δεξιοῦ . . . οὐ γὰρ ἂν ἤρεττον, εἴ γε πλουτῶν ἐτύγχανον). An oarsman who was "chief of a side" can only be a stroke oar. The position was of undoubted importance as is revealed, e.g., by the favored treatment accorded to the *espaliers*, the stroke oars on French galleys of the 16th and 17th centuries (Masson 60).

[21] *GOS* 263-65. Ten was the standard number during the Peloponnesian War (*GOS* 264), and the Decree of Themistocles gives the same figure for the Battle of Salamis (Jameson 387, lines 24-26: ἐπ[ι]βάτας [δ]έκα [ἐφ' ἑκάστη]ν ναῦν . . . καὶ τοξότας τέτταρας "for each ship 10 marines and 4 archers," a figure more trustworthy than Plutarch's 14 marines, as given in *Them.* 14.1). On the other hand, the Chians

toxotai: archers. These, not from the military class, were included in the ship's company along with the officers and ratings. Their number was generally 4.[22]

ROWERS[23]

thranite oars	62
zygite	54
thalamite	54
	—
	170

A trireme, then, had a nonrowing complement of 30 of whom almost half were the fighting contingent.[24] Since a penteconter apparently used the same number,[25] the conclusion seems to be that

---

in 494 B.C. used 40, while Xerxes swelled the normal allotment of fighting personnel aboard the ships he commandeered with an additional 30 (of whom a few were no doubt archers); see *GOS* 161.

[22] *GOS* 266. There were four per ship at Salamis (Jameson, quoted in the previous note; cf. Plutarch, *Them.* 14.1) and two to three aboard the ships listed in *IG* ii² 1951 (*IG* II² 1951.43-45, two; 106-109, three; 185-87, two; 338-40, two).

[23] For the figures, see 84 above. For *nautai* in the sense of "oarsmen," see *IG* II² 1951.46, 110, 188, 341, and Demosthenes, cited in the following note.

A crew of 150 oarsmen and 30 "puntmen" seems to be indicated for a warship from the Ptolemaic fleet on duty in the Red Sea; see *P. Grenf.* 1 9 (239/8 B.C.), as ingeniously restored by T. Reekmans (*Antidorum W. Peremans sexagenario ab alumnis oblatum*, Studia Hellenistica 16, Louvain 1968, pp. 228-32). The vessel very likely was a trireme. We would expect some word meaning deckhands or sailors instead of puntmen for the 30 nonoarsmen, but Reekmans' restoration of *kontotitai* seems the only one possible.

[24] I have broken down the complement into the familiar categories of officers, ratings, fighting personnel, rowing personnel. The Greek arrangement was somewhat different, distinguishing rowers (*nautai*), marines (*epibatai*), and—at least from the later 5th century B.C. on—*hyperesia* "service group," viz., officers, ratings, and archers (cf. Demosthenes 50.10, 25, 32, 36, where a ship's complement is divided into *hyperesia, epibatai, nautai*; similarly in Athenaeus 5.204b, cited in SIX, note 47). Jameson (389-92) argues that earlier, in Themistocles' decree (*SEG* XVIII 153), *hyperesia* has precisely the opposite meaning, referring to the fighting personnel alone. He does this in order to include the *hyperesiai* within the system of sortition described in the decree for assigning rowers and marines to ships; otherwise they would fall outside it. But the assignment of officers and ratings, trained technical personnel, would necessarily fall outside of any such mechanical scheme of assignment.

[25] Herodotus (7.184.3) reckons 80 men to a penteconter in Xerxes' navy, or 30 above the complement of rowers.

305

navies at this time, for whatever reason, were content to get along with the very minimum in executive and specialized personnel.[26]

IN THE Hellenistic Age, the steady progress in design and armament (97-123 above) was paralleled by a marked development in the man-of-war's complement. Introduction of larger ships and of the catapult as a naval weapon inevitably added a good many new members, but it went beyond that: ships were assigned significantly more specialists than hitherto. Our information for the earlier period was almost exclusively from Athens; now the chief source is Rhodes, which, from the third through the first centuries B.C., maintained a crack navy using the quadrireme as its capital ship.[27] The crew of such a unit included:[28]

---

[26] Far fewer, for example, than on galleys of the 16th and 17th centuries. A French ship of the 17th century, powered by 202 rowers—not too far removed from the 170 of a trireme—had 4 senior officers, 20 junior officers and petty officers and ratings, 29 deckhands, 6 cabin boys, and 8 guards for a total of 67 (there were also 57 oarsmen to man the *canot* and *cayq*), exclusive of fighting personnel, whose number varied widely according to need (Masson 240, 245-52, 313). On a Papal galley of the 16th century, officers, ratings, deckhands, etc. totaled 73, and the fighting personnel 50 (Guglielmotti III 107).

[27] Cf. SIX, note 30.

[28] The chief sources of information are a number of inscriptions set up by Rhodian crews in honor of their officers. The most important are: *Nuova sill.*, no. 5 (3rd-2nd B.C.); M. Segre, "Due nuovi testi storici," *Rivista di filologia* 60 (1932) 446-61, esp. 453-61 (1st B.C.); and "Dedica votiva dell' equipaggio di una nave rodia," *Clara Rhodos* 8 (1936) 225-44 (text on p. 228, 1st B.C.), an article that includes a comprehensive summary of what is known about Rhodian crews.

The fullest inscription is the last. The ranks mentioned are:

line 9	κυβερνάτας		followed by 1 name		
7-8	[the *prorates* was the dedicatee himself]				
11	ναυπαγός	"	"	1	"
13	παδαλιοῦχος	"	"	1	"
15-16	ἐργαζόμενοι ἐν πρῴρᾳ	"	"	5	"
23	ἐν πρύμνᾳ	"	"	5	"
30	καταπελταφέται	"	"	2	"
33	τοξόται	"	"	6	"
	col. II				
9	ἐλαιοχρῆστας	"	"	1	"
11	ἰατρός	"	"	1	"
13	κωποδέτας	"	"	1	"
17	ἐπιβάται	"	"	19	"
				at least	

Nuov. sill. 5 contains part of the listing of two crews. For the first we have, in this

OFFICERS

trierarchos: the trierarch, who in the Rhodian navy generally served as captain. In addition, the trierarch of a smaller unit could command a flotilla of such ships.

epiplous: vice-captain, i.e., the officer assigned as captain when the trierarch chose not to take personal command, or ambitiously assumed the expense of several galleys, or was aboard the flagship of a flotilla of smaller units.[29]

grammateus: literally "clerk," "secretary"; the full form of his title translates "trierarch's secretary and treasurer." An important officer who, by position and duties, had close rapport with the trierarch or *epiplous*.[30]

kybernetes: executive officer and navigating officer, with station on the poop.

prorates: bow officer, who, in the Rhodian navy seems to have ranked just below the *kybernetes*.[31]

---

order: τριήραρχος, γραμματεύς, κυβερνάτας, πρωρεύς, κελευστάς, πεντηκόνταρχος, ναυπαγός, ἐργαζόμενοι (followed by at least three names); for the second: πεντηκόνταρχος, ναυπαγός and probably [ἐργαζόμενοι].

A fragment published in *Clara Rhodos* 2 (1932) 176 (3rd-2nd B.C.), lists, in this order: κυβερνάτας, πρωρεύς, κελευστάς, πεντηκόνταρχος, ἐργαζόμενοι (followed by at least three names). The first four recur, in the same order, in a fragment of an inscription set up by the crew of a Coan quadrireme (*IG* XII.8.260, 82 B.C.), followed by the *iatros* and the marines (at least 20).

[29] Segre, "Dedica" 231-33. C. Blinkenberg, in *Lindos* II: *Inscriptions* (Berlin and Copenhagen 1941) no. 420, note to lines 11-12, argues that the *epiplous* is the captain of the marines, but a papyrus document (see note 6 above) puts Segre's interpretation beyond doubt. This, a letter to a trierarch, refers at one point (line 11) to Ἀντιπάτρῳ τῷ ἐπιπλέοντι ἐπὶ τῆς νεώς "Antipater, who is serving as *epiplous* aboard the vessel" and some lines later (line 21) describes the same man as τῷ παρὰ σοῦ τριηραρχοῦντι τὴν θ' "[Antipater], who is representing you as trierarch of the 'nine.'"

[30] *Nuova sill.* 5 lists, right after the trierarch, γραμματεὺς τριηράρχου καὶ ταμία. A dedication (text in Segre, "Dedica" 240; 1st B.C.) of a Milesian crew lists their *nauarchos* (commodore), *trierarchos, epiplous,* and *grammateus.* A dedication of a Rhodian crew (*IG* XII Suppl. p. 139, no. 317; 2nd-1st B.C.) includes the *grammateus* along with commodores and trierarchs. Similarly, on Papal galleys of the 16th century, the *scrivano* received the same pay as the assistant rowing officer, the *sottocomito* (Guglielmotti III 107), while on French galleys of the early 17th the *écrivain* was paid the same as the chief rowing officer, the *comite* (Masson 144) and by the end of the century his pay was even higher (Masson 245-46).

[31] Cf. the order in the three inscriptions cited in note 28 above.

keleustes: chief rowing officer.

pentekontarchos: probably assistant rowing officer, with no doubt fewer administrative duties than heretofore, since these must have been to a great extent taken over by the *grammateus*.

RATINGS

hegemon ton ergon: literally "chief of the activities," a sort of boatswain heading the *ergazomenoi*, the fore or after deckhands.[32]

naupegos: ship's carpenter.

pedaliouchos: quartermaster (literally "steering oar holder").

elaiochreistes: literally "olive oil anointer"; probably in charge of issuing oil to the crew for rubbing down.[33]

kopodetes: literally "oar binder"; judging from the title, his prime duty was to check the chafing gear and straps of the oars, but he very likely had responsibility for the oars in general.[34]

iatros: ship's doctor; not an officer and generally a foreigner (many came from the island of Cos, the home of Hippocrates).[35]

[32] See note 39 below. Segre, "Dedica" 234, taking the word *ergon* too literally, thinks of this rank as a kind of naval engineer.

[33] *SEG* xv 112 (225 B.C.), a dedication erected by the Athenian citizens aboard an aphract to their trierarch (cf. Six, note 128) mentions (lines 7-9) that he "issued oil to the young men (*neaniskoi*), so that they, by taking proper care of their bodies, could become stronger" (ἔθηκεν δὲ καὶ ἔλαιον τοῖς νεανίσκοις ἵνα ἐπιμελόμενοι τοῦ σώματος δυνατώτεροι γίνωνται). As Jean Pouilloux points out in his discussion of the inscription (*BCH* 80, 1956, pp. 64-66), the *neaniskoi* were youths doing their military service by forming the complement of marines. If the marines needed oil for their bodies, so too did the oarsmen, and the man in charge of distributing it to them would be the *elaiochreistes*. The oarsmen are not mentioned, since this happens to be a dedication set up by, and referring only to, the Athenians aboard, and rowers in this age were generally hired foreigners; cf. App.

[34] Cf. the *remolaio* on Papal galleys, whose duties included cutting new oars, repairing old, adjusting the lead in the handles to achieve the proper balance (Pantera 129). His counterpart on French galleys, the *remolat*, in addition kept careful check on the *galavernes* and *auterelles*, the oak pieces that reinforced respectively the part of the oar where it met the thole and the gunwale about the thole (Masson 60, 247).

[35] Similarly, on French and Papal galleys, the ship's *chirurgien* (*cerusico*) stood low in the hierarchy, receiving the same wage as the *remolat* (*remolaio*); see Masson 144, Guglielmotti III 107.

Seamen

ergazomenoi en prora: literally "those working at the prow," bow deck watch for handling sails and lines; at least 5.

ergazomenoi en prymne: literally "those working at the poop," stern deck watch for handling sails and lines; at least 5.

Fighting Personnel

katapeltaphetai: catapult operators; at least 2.

toxotai: archers; at least 6.

epibatai: marines; at least 19.

Rowers

nautai or

parakathemenoi: literally "the seated ones."[36]

All the above are attested as being part of the crew of a quadrireme. In addition, we may be almost certain that stroke oars (*toicharchoi*), the piper (*auletes, trieraules*), and various grades of guards (*nauphylakes*) were included.[37] For larger or smaller units we have nothing to go on save an indication that, on smaller, the *pentekontarchos* may have been replaced by a *dekatarchos*.[38]

We know a little about promotion in the Rhodian navy. An inscription records a career that was perhaps typical: enlisted man on light units (*triemoliai*) and then on heavier (*cataphracts*); boatswain; bow officer of a *triemolia*; bow officer of a quadrireme.[39]

[36] Cf. Segre, "Due nuovi" 459; A. Maiuri in *Annuario della reale scuola archeologica di Atene* 2 (1916) 136, no. 2b-c, line 10 (Rhodes, 1st B.C.).

[37] At least three types are attested: *archinauphylax* "chief ship's guard" (Maiuri, cited in the previous note, line 5); *nauphylax* or simply *phylax* "ship's guard" (Maiuri, line 7); *skeuophylax* "guard of equipment" (*Sammelb.* 9780, mid-3rd B.C.).

[38] Literally "commander of 10." It occurs once (*SEG* 1 345 = *IG* xii Suppl. p. 108, no. 210, 1st A.D.) as a rank on a Rhodian *triemiolia* (the ship is not specifically identified as such but its name is *Eirena Sebasta*, i.e., the Greek rendition of *Pax Augusta*, and this name—along with *Euandria Sebasta* = *Virtus Augusta*—is attested in a series of inscriptions as a name for Rhodian *triemioliai*; see Six, note 122).

[39] Segre, "Dedica" 228, lines 4-8: στρατευσάμενον ἐν ταῖς τριημιολίαις καὶ ἐν ταῖς καταφράκτοις ναυσί, καὶ ἀγησάμενον τῶν ἔργων, καὶ πρωρατεύσαντα τριημιολιᾶν καὶ τετρήρευς κατὰ πόλεμον. See also *Lindos* ii, no. 707, cited in Six, note 128.

309

THE information next available concerns the Roman Imperial navy of the first three centuries A.D. It reveals that Rome adopted the traditional Greek organization but combined with it some important typically Roman features.[40]

Officers and ratings were taken over with little change—as could be expected in a navy whose ships were largely Greek in design and whose first crews were largely drawn from the Greek-speaking parts of the Mediterranean:[41]

OFFICERS

trierarchus: commanding officer, now permanently assigned to, and always in command of, the ship.[42]

gubernator: the Latin equivalent of the *kybernetes*, executive officer and navigating officer.[43]

proreta: bow officer; probably, as in the Rhodian navy, he ranked just below the *gubernator*.[44]

celeusta: see next entry.

pausarius: *celeusta*, the Latinized form of the Greek *keleustes*, must have been the title of the chief rowing officer; *pausarius* may have been another name for the same rank or the *pausarius* may have been junior to the *celeusta*,[45] being the equivalent of the *pentekontarchos*, a rank not taken over by the Roman navy.[46]

---

[40] The evidence, largely from inscriptions on gravestones, has been collected and analyzed by Starr (43-45, 55-61). L. Wickert, "Die Flotte der römischen Kaiserzeit," *Würzburger Jahrbücher für die Altertumswissenschaft* 4 (1949-50) 100-125, adds very little.

[41] Starr 1-8 (growth of the fleet), 44, 75 (national origin of officers and sailors).

[42] *ILS* 2819-22, 2844, 2846, 2857, 2908, 2910-14.

[43] *ILS* 2828-29, 2853-54.

[44] *ILS* 2827, 2864-66.

[45] *ILS* 2830. Seneca (*Epist.* 56.5) describes the *pausarius* as the one who "gives the stroke to the rowers in a piercing voice" (*voce acerbissima remigibus modos dantem*). The rank is attested in an inscription (*ILS* 2867).

The term *hortator* for the chief time-beater of either a war galley or merchant galley is a literary word found only in poetry (Ennius, *Annales* 480 [*ROL* I, p. 164]; Ovid, *Met.* 3.619; Plautus, *Merc.* 696; cf. Nonius 2.151 s.v. *portisculus*).

[46] The *pentekontarchos* had important administrative duties in the Athenian navy and probably retained some in the Rhodian. Rome, on the other hand, as will appear below, had its own elaborate organization for handling administration.

RATINGS

velarii: "sail-men," i.e., deckhands, the equivalent of the *ergazomenoi* of the Rhodian navy.[47] Some are characterized as receiving double pay; perhaps these were boatswains, the equivalent of the Rhodian *hegemon ton ergon*.[48]

faber: ship's carpenter, the equivalent of the *naupegos*.[49]

subunctor: the equivalent of the *elaiochreistes*.[50]

medicus: ship's doctor, the equivalent of the *iatros*.[51]

pitulus: time-beater? On Roman galleys, in addition to—or at times instead of—the piping of a flute, the measured beating of a mallet, the *portisculus*, gave the time to the rowers.[52] *Pitulus* means literally the "plash" of the oars; it is the Latinized form of the Greek *pitylos*, for which see TWELVE, note 37. I suggest that the rating bearing this name was a time-beater who kept the oars plashing in the proper rhythm.[53]

[47] *Bonner Jahrbücher* 108 (1902) 94; *CIL* XIII 8160.

[48] *velarius duplicarius, ILS* 2878-79.

[49] *ILS* 2831, 2868-70.　　　[50] *ILS* 2877.　　　[51] *ILS* 2898-2900.

[52] The Althiburus mosaic shows an *actuaria* (Fig. 137, no. 13) with three men aboard; one is rowing, another climbing in the rigging, and the third, identified as *"portisculus,"* pounds with a mallet. Fronto, writing to Marcus Aurelius (*De fer als.* 3; Loeb edition, vol. II, p. 4), reminds the emperor that he can "step aboard some vessel to . . . enjoy the sight and sound of time-beaters and rowers" (*aliquam navem conscenderes, ut . . . portisculorum et remigum visu audituque te oblectares*). Originally meaning the instrument, the word apparently came to be applied to its user as well. It is cited by the lexicographers (Festus, s.v. *portisculus*, Nonius 2.151.18) from Ennius (cf. *ROL* I, p. 110) and other writers and may have been a literary rather than technical navy term.

[53] The term is attested twice, *ILS* 2880, the gravestone of the *"pitulus septesemiodialis* of the quadrireme *Dacicus*," and 2881, the gravestone of the *"pitulus* (*septesemiodialis*) of the quadrireme *Vesta*." Note that both these "seven-and-a-half-time" *pituli* are attached to quadriremes. Now, on galleys of the 16th and 17th centuries, which were powered by four or five men to an oar, "the maximum effect of a single stroke with all oars was seven 'bench-distances,' that is, between one stroke and the next, the seventh oar came to rest on the very point in the water where the first oar had struck" (Guglielmotti, s.v. *palata*: il massimo effetto di una sola palata con tutti i remi, era di sette bancate: così che, tra una palata e l'altra, il settimo remo giungeva ad appoggiarsi sull' istesso punto dell' acqua ove aveva battuto il primo"). Could the "seven-and-a-half-time plashers" of these quadriremes have been time-beaters trained to pound out a cadence that would produce seven and a half "bench-distances" per stroke?

symphoniacus: ship's flutist, equivalent of the *auletes* or *trieraules*.[54]

naufylax: ship's guard.[55]

duplicarii: "double pay" men,[56] a term that perhaps covered stroke oars and quartermasters, i.e., rowers and seamen with special skills.

WHEN it came to fighting personnel, Rome abandoned Greek models, for here she had a well-developed tradition of her own to follow—that of the army. The Roman standing navy, founded by Augustus toward the end of the first century B.C., was a late and junior branch of the military establishment. And so it was a natural move to arrange the fighting component on a galley according to a pattern taken from the army. But Rome went even further: she grafted onto each ship a complete army organization.[57] Every crew was treated as a century of the Roman army: these naval centuries, in addition to the strictly naval officers, petty officers, and ratings listed above, had each a centurion (the commanding military officer), *optio* and *suboptio* (his sergeants), *armorum custos* (armorer-sergeant), *bucinator* (bugler), and the rank and file included not only the strictly fighting personnel—marines, archers, catapult operators—but also all the rowers.[58] Since our chief sources of information are gravestones, and since seamen as well as marines recorded on these only their membership as "soldiers" in the century that comprised the ship, we have no way of distinguishing the one from the other,

---

[54] *ILS* 2874. Cf. Cicero's reference (*Div. in Q. Caecil.* 17.55) to an admiral's confiscation of the *symphoniacos servos* "slave *symphoniaci*" of a private person to use in his fleet. As a general term, *symphoniaci* was applied to players of both wind and stringed instruments (see *RE*, s.v.); as a naval term, it must surely mean pipers.

[55] *ILS* 2861.

[56] *CIL* vi 3169, 32771; x 3503-3505, 3507-3508, 3882; xi 343. Cf. Starr 56.

[57] See Starr 57-61 for a full discussion.

[58] E.g., *ILS* 2835 (centurion of the "six" *Ops*), 2855 (*optio* of the trireme *Rhenus*), 2859 (*suboptio* of the quadrireme *Fortuna*), 2886 (*armorum custos* of the liburnian *Virtus*). *CPL* 120 (A.D. 166) is a bill of sale for the purchase of a slave by an *optio* of the trireme *Tigris* from a "soldier" (*miles*) on the same ship. Another "soldier" (*manipularius* in this case) from the trireme *Virtus* supplied an attestation; the witnesses included the bugler (*bucinator*) of the *Virtus*, the *suboptio* of the trireme *Salus*, and the centurion of the trireme *Providentia*; while the *suboptio* of the trireme *Liber Pater* signed for the *manipularius*, who happened to be illiterate.

and consequently no idea of the numbers and nature of the fighting component. A newly discovered tombstone, a handsome and well-carved piece that presumably belongs to someone higher than a common sailor, commemorates a *"dolator* from the liburnian *Satura"*; since the word means "chopper," the *dolatores* aboard a warship may well have been marines assigned to such duties as cutting away enemy grapnels.[59]

The centurion and his subordinates unquestionably had immediate command over, and responsibility for, the fighting component. But their organization seems too elaborate for just that. The naval hierarchy probably was charged with nothing more than the operation and maintenance of the ship, while the military took care of base maintenance and guard, barrack routines, elementary military drill, shore patrol—all matters which, in a peace-time navy such as Rome's, were generally of more concern than naval action.[60]

[59] *dolator de liburna Satura*, inscribed on a tombstone found at Ravenna and dating to the 1st A.D.; see G. Montanari in *Epigraphica* 28 (1966) 155-58 (= *AE* 1967.114). The editor prefers to take *dolator* in the sense of "handler of the *dolon*" (see ELEVEN, App. 2), but this is linguistically impossible; such a rating, if it existed, would have to be called *"dolonarius."* *Dolator* derives from *dolare* "chop," "hew"; in the glossaries—its only attestation so far—it is defined "axe-man" (πελε-κητής). A marine's equipment included, in addition to sword and spears, a grapnel and an axe. Cf. *P. Aberdeen* 70.2-5 (2nd A.D.), a letter in which a marine complains that "I wrote you so many times about the grapnel and spears, and you haven't sent me them" (ἔγραψά σοι ποσάκις περὶ τῆς κόπλας καὶ τῶν λονχαρίων καὶ οὐκ ἔπεμψές μοι αὐτά), and *P. Mich.* VIII 467.19-20 (2nd A.D.), a letter from a marine aboard a liburnian in the Alexandrian squadron who writes to his father asking to be sent "a battle sword . . . and axe and grapnel and two spears of the best quality" (*gladium pugnatorium et . . . dolabram et coplam* [= *copulam*] *et lonchas duas quam optimas*). In a later letter (468.27-29) he asks for another "axe, [since] the one you sent me the sergeant took" (*dalabram* [*sic*] *eam quam mi misisti optionem illan* [*sic*] *mi abstulisse*), so the axe was a significant piece of equipment.

[60] The correspondence of the marine in the Alexandrian squadron cited in the previous note throws some light on naval dress. He requests in one letter "a cloak and belted tunic with trousers" (467.20-21: *byrrum castalinum et tunicam bracilem cum bracis*; the word *castalinum* is obscure), and in another "low-cut leather boots and one pair of felt stockings" (468.24-25: *caligas cori subtalares ed udones, par*). All are items worn by soldiers as well. Cf. M. Bollini, *Antichità classiarie* (Università degli Studi di Bologna, Quaderni di Antichità Ravennati, Cristiane e Bizantine 1 NS, Ravenna 1969) 86-96; using pictures on tombstones as evidence, she concludes (95-96) that, save for certain specialized weapons, navy men were by and large dressed and equipped like army men.

It was the army's practice to issue gear to the troops, charging them for it; they

313

Rome's fleets were far more highly organized than any of the Greek predecessors.[61] This, along with the double organization within each ship, inevitably caused an increase in the administrative personnel. Each vessel had the following:[62]

> *beneficarius*: the trierarch's chief administrative officer.
>
> *secutor*: assistant to the *beneficarius*.[63]
>
> *scriba*: the equivalent of the *grammateus*, the ship's chief secretarial officer. He was assisted by the following yeomen:
>> *adiutor*: chief clerk,
>>
>> *librarius*: keeper of records,
>>
>> *exceptor*: stenographer.[64]

Even the religious side of things felt the Roman penchant for organization. Galleys were assigned a *coronarius* "garlander," a man in charge of placing sacred wreaths on the ship on appropriate occasions, and a *victimarius* "sacrificer."[65] Probably these exercised their specialties only on holidays and had normal seaman's duty at other times.

II MERCHANT MARINE

HEADING the hierarchy[66] on a seagoing merchantman was the

---

could also acquire additional or replacement gear at their own expense (cf. J. Gilliam, "The Deposita of an Auxiliary Soldier," *Bonner Jahrbücher* 167, 1967, pp. 233-43, esp. 237-38). Presumably the navy used the same system. The papyri cited in the previous note, as it happens, involve equipment acquired at the men's own expense.

[61] Cf. Starr, Chapters II, VI.

[62] Cf. Starr 57.

[63] *ILS* 2893 (*beneficarius* of the liburnian *Neptunus*), 2895 (*secutor trierarchi*).

[64] *ILS* 2889 (*scriba* of the liburnian *Varvarina*), 2845 (*adiutor* of the quadrireme *Venus*), 2890 (*librarius*), *CIL* x 3439 (*librarius* of the quadrireme *Dacicus*), *ILS* 2892 (*exceptor* of the quinquereme *Victoria*).

[65] *ILS* 2876 (*coronarius* of the trireme *Danae*), 2875 (*victimarius* of the trireme *Fides*).

[66] Artemidorus 1.35 gives a clear statement of the hierarchy aboard a merchantman: "The *toicharchos* is over a supercargo [*perineos*; see note 83 below], the *proreus* over a *toicharchos*, the captain over a *proreus*, the owner (charterer) over a captain" (ἄρχει δὲ περινέου μὲν ὁ τοίχαρχος, τοιχάρχου δὲ ὁ πρῳρεύς, πρῳρέως δὲ ὁ κυβερνήτης, κυβερνήτου δὲ ὁ ναύκληρος). Artemidorus' hierarchy is in close conformity

owner or charterer (*naukleros* in Greek, *nauclerus* or *navicularius* in Latin),[67] or owners and charterers (*synnaukleroi* in Greek) since there could be more than one.[68] When he carried cargo for his own account, he or his representative was generally aboard.[69] He might

with one that occurs in a simile likening the ranks in the Church to those aboard a ship (*Epistula Clementis* 14.2, in B. Rehm, *Die Pseudoklementinen.* 1, *Homilien* = *Die griechischen christlichen Schriftsteller der ersten Jahrhunderte*, 42, Berlin 1953, p. 16): "Let the owner of this [ship] of ours be God. Let the *kybernetes* be likened to Christ, the *proreus* to a bishop, the sailors to priests, the *toicharchoi* to deacons, the *naustologoi* to catechists, the multitude of our brotherhood to passengers" (ἔστω μὲν οὖν ὑμῖν ὁ ταύτης [νεὼς] δεσπότης θεὸς καὶ παρεικάσθω ὁ μὲν κυβερνήτης Χριστῷ, ὁ πρωρεὺς ἐπισκόπῳ, οἱ ναῦται πρεσβυτέροις, οἱ τοίχαρχοι διακόνοις, οἱ ναυστολόγοι τοῖς κατηχοῦσιν, τοῖς ἐπιβάταις τὸ τῶν ἀδελφῶν πλῆθος). In the Latin version of the passage, the owner is called *dominus*; the captain, *gubernator*; and the *toicharchoi*, *diaconi*. On the *naustologoi*, see note 84 below.
The most complete example of the Latin terminology occurs in a passage of the Digest (39.4.11.2): *magistro vel gubernatore aut proreta nautave* "shipmaster, sailing master, first mate, sailor."
[67] The *naukleros-navicularius* was the man who had the use of a vessel, and thereby the possibility of exploiting it, whether through ownership or charter. If he owned it, he could be called, in Greek *kyrios* or *despotes*, in Latin *dominus*; see the passages in the previous note, and cf. Artemidorus 2.23: "[In dreams] a mast of a ship stands for the owner [*kyrios*], and a figurehead [*antiprosopon*] the *proreus*, while the stern ornament [*cheniskos*] stands for the captain, the rigging the sailors, and the yard the *toicharchos*" (ἱστὸς δὲ τῆς νεὼς τὸν κύριον σημαίνει, ἀντιπρόσωπον δὲ τὸν πρωρέα, καὶ ὁ χηνίσκος τὸν κυβερνήτην, τὰ δὲ ὅπλα τοὺς ναύτας, καὶ τὸ κέρας τὸν τοίχαρχον); the *kyrios* here can be equated with the *naukleros* of Artemidorus 1.35, cited in the previous note. Roman law used the convenient term *exercitor navis* for the exploiter of the ship, whether owner or charterer; see *Dig.* 14.1.1.15: *exercitorem autem eum dicimus, ad quem obventiones et reditus omnes perveniunt, sive is dominus navis sit sive a domino navem per aversionem conduxit vel ad tempus vel in perpetuum* "We define the exercitor as the one to whom all profits and income accrue, whether he be the owner of the vessel or has chartered the vessel from the owner for a lump sum either for a given period or in perpetuity."
[68] For the Greek term, see *Sammelb.* 9571 (cf. E. Seidl's remarks in *Studia et Documenta Historiae et Iuris*, 1958, p. 439); in this document (2nd A.D.), four men are *synnaukleroi* of an *akatos* plying between Ascalon and Alexandria. No parallel Latin term is attested, although joint ownership and chartering certainly existed; cf. *Dig.* 4.9.7.5: *si plures navem exerceant etc.* "when a number of people operate a vessel jointly for profit, etc." (same phraseology in 14.1.1.25, 14.1.4. pr. 1). In the Middle Ages joint ownership was the rule; see W. Ashburner, *The Rhodian Sea-Law* (Oxford 1909) clxiii-xv.
[69] E.g., the *naukleros* was aboard the ship St. Paul sailed on (Acts 27.11), the ship Lucian visited in the Peiraeus (*Navig.* 7, 9), the ship that rescued Pompey after the defeat at Pharsalus (Plutarch, *Pomp.* 73.3, 6), the vessel that took Aristides from Rome to the east (Aristides, *Sermo Sacra* B, p. 305), and was expected to be aboard Hiero's great grain freighter (there was a special cabin for him, and he was

even be his own captain,[70] though most often, and particularly on
large ships, the captain was a hired professional who had full authority over the running of the vessel and full command of its crew.[71]
The Greek term for captain was *kybernetes*.[72] The Latin equivalent,
gubernator, was in popular use given the same sense,[73] but Roman

---

a member of the ship's court along with the captain and first mate; Athenaeus
5.207c, 209a [cited in NINE, App., pt. 4]). In Demosthenes 35, the *naukleros* (35.10-
11, 18) was not aboard, for an affidavit cited as to the cargo taken on was sworn
to only by the captain, supercargo, and miscellaneous witnesses (35.20). Nor was
one aboard the grain ship that Demetrius intercepted when blockading Athens,
for the punishment meted out included only "the execution of the importer and
the captain" (ἐκρέμασε τὸν ἔμπορον καὶ τὸν κυβερνήτην Plutarch, *Dem.* 33.3).

Sometimes a charterer or owner was represented aboard by an agent, whose title
in Greek was *pronaukleros*; cf. *IG* XII.8.585: εὔπλοιά σοι, Ἄρτεμι, ναυκλήρου Εὐτύχου
Μυτιληναίου, προναυκλήρου Τυχικοῦ, κυβερνήτου Ἰουκούνδου "Bon voyage to the ship
Artemis, owner (or charterer) Eutychus of Mytilene, owner's representative Tychi-
kos, captain Jucundus." For another example, see *SEG* XI 1020 (= Sandberg no.
10); also possibly Sandberg no. 9. For a later use of the word, in the Rhodian Sea-
Law of the 7th to 9th century, see W. Ashburner, *op. cit.* [note 68 above] 15 and
cf. 85-86.

Fig. 144 shows a *naukleros* performing a thanksgiving ceremony on the poop of
a homecoming vessel; see NINE, note 69.

[70] In a case of maritime law posed by Cicero (*De inv.* 2.154), he refers to an
owner-captain (*dominus navis, cum idem gubernator esset* "the owner of the vessel,
since he was also the captain"). In Demosthenes 32, a rascally *naukleros*, caught
red-handed attempting barratry, is drowned, and his equally rascally associate then
tries to talk "the *proreus* and sailors (32.7)" into abandoning ship. No captain is
mentioned, which seems to suggest that the drowned man had commanded his own
vessel (cf. Rougé in *Revue de philologie* 39, 1965, p. 93). The term *nauklerokybernetes* "owner-captain" frequently turns up in the business documents from Roman
Egypt (*W. Chrest.* 434; *P. Flor.* 75.8, 29; *P. Lugd. Bat.* 11.1, col. 1 3-5 and col. 11
2-4. All belong to the 4th A.D.). Petronius introduces in his novel a character who
exercised all three functions: he was owner and captain of a vessel which was carrying cargo for his own account (101: *Lichas Tarentinus . . . huius navigii dominus,
quod regit, . . . onus deferendum ad mercatum conducit* "Lichas of Tarentum . . . ,
owner of this ship which he commands, . . . is collecting a cargo to be brought
to market").

[71] Cf. John Chrysostom, *In Epist. II ad Timoth. Cap. 1, Hom.* III 3 (Migne 62,
col. 616): ἐν πλοίοις ὁ κυβερνήτης δικάζει "On board ship, the captain judges cases."
On Hiero's superfreighter cases were decided by a board consisting of the *naukleros*,
kybernetes, and *proreus* (Athenaeus 5.209a, cited in NINE, App., pt. 4).

[72] See the passages cited in notes 66 and 67, and cf. Plato, *Rep.* 1.341c-d (cited in
note 1 above), 342e; 6.488d.

[73] E.g., Plautus, *Rud.* 1014 (*si tu proreta isti navi's, ego gubernator ero* "If you're
going to be first mate of this ship, I'm going to be captain"); Varro, *de ling. lat.*
9.6 (*obtemperare debet . . . gubernatori unusquisque in navi* "Every single person
aboard a ship must obey the captain"); Cicero, *de inv.* 1.58 (the *gubernator* of a

law insisted upon stricter terminology. In legal parlance, the commanding officer of a merchant ship was the *magister navis* "shipmaster" (translated *pistikos* in Greek),[74] who was responsible for fitting out the vessel,[75] including the hiring of a sailing master and presumably the rest of the crew,[76] for maintaining it in good repair,[77] and for all administrative matters, particularly the arrangements concerning cargo and passengers.[78] On certain ships, such as coastal craft and the like, he took care of the operation of the vessel as well.[79] On seagoing craft, however, he usually turned this over

---

ship is compared to the general of an army), 2.154 (cited in note 70 above); Seneca, *Epist.* 85.32 (the *gubernator* undertakes total responsibility for delivery of a vessel to a given port). For a full-scale study of the term with comprehensive examination of the evidence, see C. Moschetti, *Gubernare navem, gubernare rem publicam* (*Quaderni di "Studi Senesi"* 16, Milan 1966) 18-59. Like its Greek equivalent (cf. note 1 above), *gubernator* is somewhat ambiguous, for it may also mean no more than "helmsman" (e.g., Vitruvius, cited in ELEVEN, note 2).

[74] *P. Lond.* 1341.12 and 1433.23, 27, 28, etc. (both 8th A.D.) and cf. Ashburner, *op. cit.* [note 68 above] cxxxi.

[75] Cf. *Dig.* 14.1.1.7: *aut aliquas res emerit utiles naviganti vel si quid reficiendae navis causa contractum vel impensum est* "any purchase [made by the *magister navis*] of use for sailing the vessel or any contracts entered into or expenses incurred for its repair."

[76] Cf. *Dig.* 19.2.13.2: *si magister navis sine gubernatore in flumen navem immiserit et tempestate orta temperare non potuerit et navem perdiderit, vectores habebunt adversus eum . . . actionem* "If a shipmaster [*magister navis*] with no sailing master [*gubernator*] aboard takes a ship into the stream and, a storm springing up, is unable to control the vessel and loses it, the passengers are entitled to bring an action . . . against him." For long this passage has been thought to deal with the hiring of a local pilot to get a ship upstream. As Moschetti's excellent analysis (72-78) reveals, it concerns a vessel that is starting out, that has left the dock for midstream to begin a voyage. A *magister* was responsible for having a sailing master aboard to navigate the ship; if he neglected to do so, even just to get his ship from dock to midstream, he was automatically liable in case of mishap with no consideration of whether an act of god or any of the other usual qualifying circumstances was involved.

[77] *Dig.* 14.1.1.1: *magistrum navis accipere debemus, cui totius navis cura mandata est* "We must define the *magister navis* as the one to whom the care of the entire ship is entrusted." For the interpretation of the passage, see Moschetti 60-72.

[78] *Dig.* 14.1.1.3: *magistri autem imponuntur locandis navibus vel ad merces vel vectoribus conducendis armamentisve emendis* "Moreover, shipmasters [*magistri*] are in charge of chartering vessels, or conveying goods or passengers, or purchasing equipment."

[79] He could, for example, actually handle the helm; see the wall-painting of a harbor craft (*AM* pl. 14b = *IH* ill. 64) where the figure holding the tiller is identified as "Farnaces, the *magister*." In Vergil and other poets, *magister* is actually used

to a *gubernator* "sailing master"; once the ship was under way, the *gubernator* seems to have had fairly wide authority in all matters relating to its handling.[80]

The commander had two principal officers to rely upon, the *pro-*

---

in the sense of "pilot"; see E. de Saint-Denis, "Proreus"—"Proreta," *RPh* 41 (1967) 205-11, esp. 211.

[80] For the presence of a *gubernator* on board and his rank just below the *magister navis*, see the passage from the Digest cited in note 66 above. The *gubernator*'s duties are nowhere spelled out in the corpus of Roman law and can only be guessed at; the guesses run the gamut of opinion, from making him more or less a quarter-master under the *magister* as commanding officer (the majority opinion; cf. Moschetti 15) to making the latter more or less a supercargo under the *gubernator* as commanding officer (first put forth by A. Rocco in 1898 [see Moschetti 17]; its most recent adherent is Rougé [234-38]). Moschetti, reviewing the evidence, argues that the *gubernator* was in command of the total operation of the ship when under way, and assigns to the *magister* largely administrative and commercial duties. Now, Moschetti makes a significant new contribution by demonstrating that the *magister* was the sole person responsible for the maintenance and good condition of the ship (cf. note 77 above). From this the conclusion is unavoidable that he had to exercise considerable command over the crew and to direct much of its activities even when under way. The *gubernator* no doubt shouted the orders for setting sail and course (cf. Seneca, *Epist.* 95.7, where he gives some samples: *move gubernaculum* "Mind the helm!"; *vela summitte* "Take in all sail!" etc.), but it was presumably the *magister* who issued the orders to swab the decks. Moreover, the evidence cited in the previous note proves incontrovertibly that on certain craft he actually handled the tiller.

Dig. 14.1.1.13 states that *si plures sint magistri non divisis officiis, quodcumque cum uno gestum erit, obligabit exercitorem: si divisis, ut alter locando, alter exigendo, pro cuiusque officio obligabitur exercitor* "If there are a number of *magistri* with unspecified duties, whatever commitments are entered into by any one are binding upon the shipowner or charterer; if each has specified duties, e.g., one responsible for chartering, another for collections, only commitments made within the prescribed area are binding upon the shipowner or charterer." This passage is always taken to refer to a single ship and the conclusion drawn that, since there could be more than one *magister* aboard a ship, he could hardly be a commanding officer in any real sense; see Rougé 237, Moschetti 60 and 82. Yet how can there be more than one aboard a ship when the legal definition (cited in note 77 above) of a *magister* makes him the officer in sole charge of the maintenance of the whole vessel? The mistake has lain in taking the "number of *magistri*" mentioned in the passage as being aboard one ship, a wholly gratuitous assumption, since this is nowhere stated or even implied. The *magistri* in question must be those employed on a fleet operated as a unit by one shipowner or charterer; it is only natural that he, for efficiency's sake, would parcel out the duties carried on in port among the various shipmasters available.

The definition of *magister navis* given by D. Maffei (s.v. *armatore, Enciclopedia del diritto* III, Milan 1958, p. 10) is as good as any: "è chi è preposto alla nave, colui che ne cura in particolare l'amministrazione, agendo per l'*exercitor*."

reus for operations and the *toicharchos* for administration;[81] in both instances the duties were different from those performed by the naval officers who bore the same titles (cf. 303-304 above). The *proreus* was, in effect, first mate, the key assistant in the running of the ship and the one who assumed command if anything happened to his superior. Among the *proreus'* major responsibilities was maintenance, to see to it that hull and gear were in perfect order.[82]

The administrative side of things was in the hands of the *toicharchos*, who combined the functions of a purser and supercargo, taking care of both passengers and cargo.[83] Under him was the cargo clerk (*perineos*) and other assistants.[84]

---

[81] The simile in the *Epistula Clementis* (see note 66 above) carefully distinguishes between the personnel connected with the operation of a vessel and those with what it carried. The former (*kybernetes, proreus, nautai*) are equated with the officers, as it were, who guide the ship of the Church (Christ, bishops, priests); the latter (*toicharchoi, naustologoi*) with those (deacons, catechists) in charge of arranging and instructing the "passengers" of the ship, the congregations.

[82] Xenophon, *Oec.* 8.14, in describing a large Phoenician merchantman, mentions "the captain's assistant, who is called first mate [*proreus*] of the ship" (τὸν δὲ τοῦ κυβερνήτου διάκονον, ὃς πρῳρεὺς τῆς νεὼς καλεῖται). Xenophon emphasizes that the *proreus* knew each area of his vessel so well he was aware of every bit of gear there, was sure it was all in place with none missing. For the *proreus'* taking over of command, see Theodoret of Cyrrhus, *Epist.* 78 (Migne 83, col. 1252; *Sources Chrétiennes* 98 [Paris 1964], p. 176): ὅταν ὁ κυβερνήτης τι πάθῃ, ἢ ὁ πρῳρεύς, ἢ τῶν ναυτῶν ὁ πρῶτος τὴν ἐκείνου χρείαν πληροῖ "Whenever anything happens to the captain, the *proreus* or the top man among the sailors fills in for him." For a probable instance, see note 70 above.

[83] The *toicharchos* of a merchantman, unlike his counterpart in the navy, had no connection with the "side" of a vessel, e.g., the command of the port or starboard watch. This is clear from Artemidorus 1.35 and 2.23 (cited in notes 66 and 67 above), which reveal (a) that there was only one aboard a ship, and (b) that his station was not fixed in a given part of the vessel (in dreams he is symbolized not by the prow or poop or sides but by the sailyard). Artemidorus lists him as being "over a *perineos*"—a vital piece of information since it reveals that the *toicharchos'* duties had nothing to do with the sailors, for a *perineos* is just the opposite of these. The word means basically "one who is on a ship but not of the ship's company." Thucydides, for example, uses it (1.10.4, cited in FOUR, note 103) of the chieftains, as against the rowers, on the galleys that went to Troy, and Aelian (*Nat. Anim.* 2.15) of the lubberly element aboard a ship who have not the slightest idea how far they are from land as against the sailors who should know but are often fooled. Artemidorus' statement shows that at some time *perineos* came to signify a low-level official on a vessel, and a passage from Philostratus indicates that he was a kind of cargo clerk (*Vita Ap.* 6.12: καὶ ξυνεμβαίην ἄν σοι τὴν ναῦν περίνεώς τε καὶ μνήμων τοῦ σοῦ φόρτου "And I would gladly embark with you in your ship as clerk [*perineos*]

Big merchantmen carried quartermasters,[85] ship's carpenters (*naupegos* in Greek, perhaps *faber* in Latin),[86] ship's guards (*nauphylakes* in both languages),[87] rowers to man the ship's boats[88] and, though no specific proof is available, no doubt the rest of the specialized personnel found on sailing vessels of all ages.

Sailors generally went naked when aboard ship (Figs. 144, 146-147,

---

and checker of your cargo"). The *toicharchos*, then, as the *perineos'* superior, had responsibility for cargo; for the *toicharchos'* responsibility for passengers, see *Epistula Clementis* 14.2, cited in note 66 above.

A ship's officer called *diopos* is mentioned by Hippocrates (*Epid.* 5.74, 7.36: τῷ ἐκ τοῦ μεγάλου πλοίου διόπῳ "the *diopos* from the big ship"), and in Demosthenes 35.20, 34 the officer "serving as *diopos*" (διοπεύων τὴν ναῦν) executes an affidavit concerning the nature of the cargo along with the captain. This *diopos* may well be the *toicharchos* under another name, the earlier title for the office.

An underofficer called *diaetarius* is mentioned in the Digest, 4.9.1.3: *sunt quidam in navibus, qui custodiae gratia navibus praepronuntur, ut* ναυφύλακες *et diaetarii. si quis igitur ex his receperit, puto in exercitorem dandam actionem, quia is, qui eos huiusmodi officio praeponit, committi eis permittit* "There are certain personnel aboard ship who are in charge of custodial duties in connection with the ship, such as ship's guards [*nauphylakes*] and *diaetarii*. Therefore, if any of these accept receipt, I think the shipowner or charterer must be held legally responsible, inasmuch as he, by placing them in charge of such duties, allows the committing of goods to their care." *Diaetarius* means etymologically "cabin boy" but, to judge by the duties implied in the above passage, the cabin he served was the purser's or supercargo's rather than any passenger's; perhaps he was the equivalent in Latin of the *perineos*.

[84] The *toicharchos'* assistants certainly included the *naustologoi*. *Epistula Clementis* 14.2 (see note 66 above) places them right after the *toicharchoi* and likens them to catechists; they were, then, part of the service personnel aboard. The *Apostolic Constitutions* 2.57 (Migne 1, col. 732) include an instruction to the effect that, after a church service, "The doormen are to stand guard at the men's exits, and the deaconesses at the women's, like *naustologoi*" (στηκέτωσαν δὲ οἱ μὲν πυλωροὶ εἰς τὰς εἰσόδους τῶν ἀνδρῶν φυλάσσοντες αὐτάς, αἱ δὲ διάκονοι εἰς τὰς τῶν γυναικῶν, δίκην ναυστολόγων); this would imply that at least one of their duties was to stand by the gangplanks and check on passengers—very likely as a precaution against that ubiquitous and age-old nuisance, the stowaway.

For the *perineos*, see the previous note.

[85] See the passage from Philostratus cited in note 1 above.

[86] Lucian, *Navig.* 5 (the visitors to the big grain ship *Isis* [cf. NINE, App., pt. 2] learned the dimensions from its *naupegos*).

[87] See note 83 above.

[88] Philostratus (*Vita Ap.* 4.9), describing the activities of the crew of a large merchantman as it prepared to leave harbor, notes that "some, who were rowing personnel, had gotten into the ship's boats" (οἱ μὲν τὰς ἐφολκίδας ἐμβεβήκασιν ἐρετικοὶ ὄντες).

149, 151), while the helmsman and officers wore the common knee-length tunic (Figs. 154, 177). When ashore, sailors wore the tunic but no sandals, going barefoot as they did when on the water.[89]

[89] Dio Chrysostom, *Orat.* 72.1: "Why is it that people, whenever they see someone wearing a tunic and nothing else . . . figure that he is probably a sailor, etc." (διὰ τί ποτε οἱ ἄνθρωποι, ὅταν μέν τινα ἴδωσιν αὐτὸ μόνον χιτῶνα ἔχοντα . . . , λογιζόμενοι τυχὸν ὅτι ναύτης ἐστὶν κτλ.).

APPENDIX

THE USE OF SLAVES

I In Navies

THE expression "galley slave" has two distinct meanings. The student of antiquity, dealing with a slave society, automatically takes it in the sense of a human chattel who, as his assigned duty, helps man the rowing benches of a warship. In common parlance—and certainly in popular literature—a galley slave is something totally different: a criminal condemned to hard labor at the oar of a galley—a *forçat*, to use the convenient term coined by the French, who were partial to this form of penalty. The common impression is that galley slaves are as old as galleys. In point of fact, as a significant naval institution they very likely go back no further than the fifteenth century A.D.[1]

Homer's poems, our earliest source of information, know only the fighter-oarsman, warrior when on land and rower when traveling to an overseas battlefield. Athens, during her heyday as a naval power, relied chiefly on the skill and training of her rowing crews, and these were made up of her citizens, mostly the poor who, unable to afford a suit of armor, could not fight in the army; there was an admixture of foreigners —but no slaves.[2] Only once did she depart from this tradition, during the desperate days before Arginusae in 406 B.C. The fanfare aroused shows how exceptional the step was, and, after the victory, far from continuing the practice, she made free men of the slaves who had served.[3] There is undeniable evidence of the presence of slaves on galleys at times other than emergencies, but these can be explained: they were not rowers but

[1] The traditional view was strongly challenged by W. Westermann, *The Slave Systems of Greek and Roman Antiquity* (Philadelphia 1955) 15-16, who asserted that slaves were used as rowers only at times of absolute emergency. Starr and others (see note 13 below) have demonstrated the total lack of evidence for slaves in the Roman Imperial fleets. Despite this, some still repeat uncritically the traditional view; see L. Wickert, *loc. cit.* (THIRTEEN, note 40) 105, and E. Sander, "Zur Rangordnung des römischen Heeres: Die Flotten," *Historia* 6 (1957) 347-67, esp. 347.

[2] The fundamental article is R. Sargent, "The Use of Slaves by the Athenians in Warfare. II: In Warfare by Sea," *CP* 22 (1927) 264-79. M. Amit, "The Sailors of the Athenian Fleet," *Athenaeum* 40 (1962) 157-78 (repeated, with scant change, as part of Chapter II of his *Athens and the Sea*, Brussels 1965) simply reproduces Sargent with added verbiage and reduced documentation. Jameson has no evidence whatsoever for the suggestion (393-94) that Athens used slaves in 480 B.C.

[3] Sargent 276-77.

322

the personal attendants of the ships' officers or of the hoplites who served as marines, or, when galleys were employed, as they so often were, as troop transports, personal servants of the soldiers transported. Such slaves no doubt assisted at times in the rowing; the point is that they were in no sense regular members of the crew.[4]

In the fourth century B.C., a widespread scarcity of oarsmen developed. Skilled citizen rowers—or even just willing citizen rowers—were in such short supply that navies were forced to turn for the most part to paid professionals, imported if need be.[5] As a matter of fact, the competition was

---

[4] Sargent 273-76. On French galleys in the days of Louis XIV, a captain had no fewer than eight servants aboard over and above his personal domestics, who themselves could number up to five (Masson 239-40). For slave attendants of hoplites, see Sargent, "The Use of Slaves by the Athenians in Warfare. 1: In Warfare by Land," CP 22 (1927) 201-12, esp. 201-207.

Two bits of evidence need more discussion than Sargent (277-78) has given them, Thucydides 1.55.1 and IG ii² 1951.

Thucydides mentions that, in the naval battle in 432 B.C. between Corcyra and Corinth, the Corinthians sank 70 Corcyraean ships (1.54.2) and these yielded no less than 1,000 prisoners (1.54.2), of whom 800 were slave (1.55.1; Thucydides' figure must be a round number, since the nonslave prisoners totaled 250). A. Gomme, A Historical Commentary on Thucydides 1 (Oxford 1945) 196, suggested that the majority of the Corcyraean rowers must have been slave, though admitting in the same breath that this was contrary to Greek practice generally and, indeed, contrary to Corcyraean practice at other times, an admission that leaves little to be said for the suggestion. Westermann (op. cit. note 1 above, 16) would have it that we are dealing here with another Arginusae, that the Corcyraeans, facing a dire emergency, put slaves at the oars. The Corcyraeans went into action with 110 ships (Thucydides 1.47.1); perhaps they manned every hull they could float, including their reserves, in which case they would perforce have taken desperate steps such as enrolling slaves. On the other hand, Thucydides makes particular mention of the large number of fighting men aboard (1.49.1), which may mean as many as 30 or 40 per ship (see THIRTEEN, note 21), i.e., somewhere between 3,300 and 4,400 all told. Of a stricken trireme's crew, those most likely to survive were the officers, marines, and their attendants; stationed on the decks and gangways, all they had to do was jump clear. As a matter of fact, the slaves, with no armor to doff and far less incentive to fight, would be the first to abandon ship—which could explain how so many more of them survived. Of the nonslave captives, almost all came from Corcyra's best families (Thucydides 1.55.1)—more likely a source of hoplites than rowers.

IG ii² 1951 is an inscription, the preserved portion of which lists the names of at least eight crews (cf. W. Pilz, Phil. Woch. 53, 1933, pp. 732-34). Of the nonofficer personnel, many are identifiable as slave. As Pilz has pointed out, the slaves seem to belong to the ships' officers, but their total number on each ship is too great to allow us to explain them as attendants. The inscription obviously formed part of an impressive monument—why not to Arginusae, as A. Körte (among others) has argued (Phil. Woch. 53, 1932, pp. 1027-32)? Pilz' observation would mean that the slaves were recruited through the ships' officers, which would be a natural procedure.

[5] E.g., Demosthenes 50.7-13, 18; 51.6. Cf. T. R. Glover, From Pericles to Philip (London 1917) 328-31.

so keen that Athens, Queen of the Seas, was being outbid by, and losing rowers, to third-rate naval powers like Thasos or Maroneia.[6] In a context such as this, slave rowers simply have no place.

For the Hellenistic Age, our only information concerns Rhodes and Egypt. Rhodes, as proud of her naval tradition as Athens had once been, and, like her, favoring tactics that only crack crews could carry out, manned the benches with her own citizens, eked out with some trained personnel from the nearby coastal cities of Asia Minor.[7] The Ptolemies, with no maritime population to draw on, hired whatever skilled rowers they needed,[8] and conscripted the fellahin for whatever muscle was required.[9] It has frequently been asserted that they relied heavily on slaves, but there turns out to be, on examination, scant support for the statement.[10] Indeed, Hellenistic Egypt was the one locale where slavery played

[6] Demosthenes 50.14-16.

[7] Blinkenberg, op. cit. (THIRTEEN, note 29), no. 88 (= IG XII.1.766, 265-60 B.C.), gives some 70 names of members of the crew of a Rhodian ship. One came from Halicarnassus, while all the others were Rhodian citizens. The inscription carefully notes (lines 286-88) that three of these had foreign-born mothers. Other inscriptions listing rowers who are either Rhodian citizens or subjects are IG XII Suppl. p. 108, no. 210 (2nd-1st B.C.; 4 names of Rhodian citizens) and Segre, "Due nuovi" 459 (1st half of 1st B.C.; a Rhodian citizen serving as oarsman). Syll.³ 1052, 1st half of 1st B.C., lists two Rhodian *nautai*, but these may be from a merchant ship rather than a war galley.

[8] Cf. UPZ II 151.2-4 (259 B.C.): Ἀρεὺς . . . τριημιολίας μισθοφόρου ἐρέτης "Areus . . . oarsman on a *triemiolia misthophoros*." As Wilcken explains in his note to the passage, such a ship was not a chartered galley but one manned by mercenaries. The *triemiolia* by its nature required expert oarsmen; see 129-31 above. P. Grenf. I 9 (cf. *Berichtigungsliste* I and THIRTEEN, note 23), 3rd B.C., mentions mercenary crews, including oarsmen, on duty in the Red Sea. They probably manned the fast craft on antipirate patrol; on the endemic piracy in the Red Sea, see Rostovtzeff, SEHHW 387-88, 924.

[9] For the later period the evidence usually offered is the passage in the Rosetta Stone (196 B.C.) that exempts temple slaves from σύλληψις εἰς τὴν ναυτείαν "seizure for naval service" (*Sammelb.* 8299 [=OGIS 90] 17), which those who cite it take to mean conscription for service in the galleys. There is indirect but more secure evidence for the earlier period. E.g., W. Chrest. 385.30-31 (mid-3rd B.C.) lists two natives who were exempted from compulsory labor on the dikes because they were "of those who had been sequestered for the fleet" (τῶν εἰς τὸ ναυτικὸν κατακεχωρισμένων). PSI 502.24 (257/6 B.C.) refers to "the dispatching of the sailors" (τῇ τῶν ναυτῶν ἀποστολῇ) from upriver to Alexandria, which smacks of the dispatch of a group of conscripts. P. Teb. 703.215-22 (cf. Rostovtzeff's note to the passage) alludes to the desertion of sailors—surely unwilling conscripts rather than well-paid mercenaries. Conscripts would be adequate for helping to fill the benches of the larger units powered by multiple-rower sweeps, which required muscle more than training; see 104 above.

[10] Cf. F. Garofolo, "Sulle armate tolemaiche," *Rendiconti della reale Accademia dei Lincei* 11 (1902) 137-65, esp. 158; J. Lesquier, *Les institutions militaires de l'Égypte sous les Lagides* (Paris 1911) 256-57; Rostovtzeff in CAH VII (1928) 118;

far smaller a role than anywhere else; slaves there were largely the personal servants of the Greek upper stratum of society, a source hardly likely to be tapped for service in the galleys.[11]

The Roman Empire has shared the fate of Athens and Hellenistic

---

Wilcken in note to *UPZ* II 151.2-4. The evidence Garofolo offered was beside the point, as Lesquier himself pointed out (257, note 3). Yet Lesquier's evidence boils down to the passage in the Rosetta Stone mentioned in the previous note, a passage that he, among others, takes to refer to the exemption of temple slaves from conscription into the navy (but see A. Bouché-Leclerq, *Histoire des Lagides* IV, Paris 1907, p. 7 and note 3). Lesquier reasoned that the exemption of one class of slave implied the liability of all others. But Egyptian temple slaves were not slaves in the usual sense; they were, in effect, the peasants of the temple estates (cf. Rostovtzeff, *SEHHW* 322-23). Subsequent writers have been content to cite Lesquier.

Though cogent objections have been raised to Lesquier's view, notably by Westermann (*op. cit.* note 1 above, p. 37) and C. Préaux (*L'économie royale des Lagides*, Brussels 1939, p. 259), a recently published papyrus has brought it to the fore again. P. Hib. II 198, dating mid-3rd B.C., is a copy of a series of royal ordinances. One of these, addressed to the Nile police, concerns deserters from the navy. The police are enjoined (lines 86-88) to apprehend and deliver up [το]ὺς να[ύ]τας τοὺς τὸν χαρακτῆρα ἔχοντας κα[ὶ τ]οὺς [ἀποστάτας] ἐκ τοῦ ναυτικοῦ "those sailors who bear the brand and the deserters from the fleet" (for the restorations in this passage, see N. Lewis, "P. Hibeh, 198, on Recapturing Fugitive Sailors," *AJPh* 89, 1968, pp. 465-69, esp. 465-66). The offenders referred to here would hardly be the well-paid mercenaries from the fleet; they must be unhappy and homesick conscripts. The editors, commenting on the passage, with the support of parallels from Pharaonic times assert that those marked with a brand "will be either slaves, prisoners of war, or impressed criminals," and, of the three alternatives, they vote for slaves, citing Lesquier. Yet are these the only alternatives or, indeed, likely alternatives? We have seen above that there is no evidence for the use of slaves, and we shall see below that there is none for the use of impressed criminals. And prisoners of war were certainly not all branded; as a matter of fact there is a case on record of one who was elevated to a cleruchy, no less (*W. Chrest.* 334, 244/3 B.C.; cf. R. Taubenschlag, *The Law of Greco-Roman Egypt*, New York 1944, p. 53 and note 22). I would suggest a totally different explanation. We know that one of the common purposes of marking a face was to identify runaway slaves (Aristophanes, *Birds* 760; schol. to Aeschines II 83; Lucian, *Timon* 17); why not runaway rowers as well? Could not the branded sailors referred to in the papyrus be second offenders, conscripts who had deserted once before but been recaptured and on recapture were given an immediately identifiable mark to discourage them from another try? Either this explanation or one like it is preferable to assuming a practice in the Ptolemaic navy for which there are no other grounds of support and which is not consonant with the general picture of slavery in Ptolemaic Egypt.

If "the sailors who bear the brand" are second offenders, it follows that the "deserters from the fleet" are first offenders. Deserters, once clear of the base at Alexandria, could head in only one direction—upriver. The Nile police no doubt stopped anyone who looked like a navy man. Those discovered to be branded could be apprehended on sight; others were presumably challenged to produce authorization for being away from their ships.

[11] Cf. Rostovtzeff, *SEHHW* 321-22.

Egypt: historians insist on attributing to it the use of slave rowers.[12]
The facts simply do not bear this out.[13] Sextus Pompey, who could ill
afford to be choosy, used fugitives of all sorts, including slaves,[14] but
these were free volunteers by the time they took their seats on his benches.
Augustus, scraping the bottom of the manpower barrel for his showdown
with Sextus, enlisted slaves—but gave them their freedom before they
stepped on board.[15] From then on, the picture is consistent: the oars that
drove the Imperial fleets were, to be sure, plied chiefly by provincials
rather than citizens, but they were none of them slave.[16]

The next chapter in Mediterranean naval history was written by the
Byzantine and Arab navies. The Byzantine fleet, the successor of Imperial Rome's, carried on the Roman tradition and used only free rowers,
who for the most part were hired.[17] For the Arab navy we are fortunate
enough to have specific information about recruiting practices preserved
in certain papyri of the early eighth century. These show that the Arab

[12] Cf. *IG* v 1.1433 (cf. *SEG* x 1034) and the discussion by A. Wilhelm in *Jahreshefte
des oesterreichischen archäologischen Institutes in Wien* 17 (1914) 1-120, esp. 48-51.
The date is either the end of the 2nd b.c. (Wilhelm 103) or some decades later
(U. Kahrstedt, *Das wirtschaftliche Gesicht Griechenlands in der Kaiserzeit*, Bern
1954, p. 220, note 6). The inscription lists the amount levied by Rome on Messene
as a special war tax, with some indications of who were to pay and how much.
The portion that concerns us (lines 34-39) itemizes a small sum of arrears and those
responsible for it: certain men in military service, deceased freedmen, manual la-
borers assigned to compulsory services, "slave oarsmen" (ἐρέταις δούλοις). Wilhelm
(97), followed by Rostovtzeff (*SEHHW* 751), assumes that the soldiers and rowers
were conscripted for service with the Roman forces, which would imply that these
people had been singled out from the Messenians for a double burden—compulsory
service *and* payment of tax. All we can say is that they were locals who happened to
be away from home when the tax collector came round. The rowers may have been
propelling barges or manning ferries for all we know.

[13] See Starr, Chapter v. Starr's stand has been seconded by Westermann (note 1
above), Kienast (11, 14, 25), and S. Panciera, "Gli schiavi nelle flotte augustee,"
*Atti del Convegno Internazionale di Studi sulle Antichità di Classe, Ravenna, 14-17
ottobre 1967* (Ravenna 1968) 313-30. Panciera argues convincingly that *ILS* 2819,
the tombstone of a certain *Malchio Caesaris trierarchus* "Malchio, trierarch of Caesar," and similar inscriptions—e.g., *ILS* 2820, 2822—of the time of Augustus and
Tiberius involve not slaves, as has been asserted, but provincials (*peregrini*), who,
in these early years of the fleet's existence, when the *esprit de corps* was not so
well established, liked to refer to their connection with the emperor.

[14] Appian, *Bell. Civ.* 2.103. These are the rowers Dio Cassius (49.1.5) refers to as
"warship slaves" (δούλους τριηρίτας).

[15] Suetonius, *Aug.* 16.

[16] Starr 71-77, Panciera, *op. cit.* (note 13 above) 326-30.

[17] See H. Ahrweiler, *Byzance et la mer: la marine de guerre, la politique, et les
institutions maritimes de Byzance aux viie-xve siècles* (Paris 1966), Appendix 1:
"Les équipages," esp. 405.

government met the problem of rowers much as the Ptolemies had—it filled the benches by drafting its Christian subjects.[18] The system does not seem to have been oppressively harsh. The obligation could be met by substituting for personal service a cash payment, to enable the government to hire a replacement, while those who actually served received a money wage according to a regular scale.[19] In short, under normal circumstances the galley crews of this age included no slaves, no more than had those that plied the same waters during the preceding millennium and a half.

The Use of *Forçats* in Ancient Warships

ON 22 January 1443, Charles VII of France gave Jacques Coeur, a canny French shipping magnate who had a private fighting flotilla all his own, the right to impress "personnes oiseuses, vagabonds et autres caïmans" for his crews.[20] For navies bedeviled by a growing shortage of skilled rowers, the act was a veritable signpost to the way to solve their problem: from then on, the vicious practice of putting victims of the law at the oars spread like the plague. By 1550 even Venice, the last to hold out because of stubborn pride in her tradition, centuries old, of using citizen rowers, had given in.[21]

By the seventeenth century the use of *forçats* was nearly universal in the Mediterranean, and it remained so until the death of the oared warship. This fact has profoundly affected our thinking, leaving us with an irresistible inclination to see the same system in effect in antiquity. Yet there is not a shred of conclusive evidence to support such a view;[22] the ancients were as loth to use *forçats* as slaves.

[18] *P. Lond.* IV, Introd. xxxi-xxxv.

[19] Cf. H. Bell, "Two Official Letters of the Arab Period," *JEA* 12 (1926) 265-81. In the second of the two letters, dated A.D. 710, the Arab governor of Egypt for some reason demands personal service instead of an equivalent in cash. The practice of allowing such payments must have been taken over by the Arabs from the later Roman empire; cf. *P. Grenf.* II 80-82, dated A.D. 400-402, three letters dealing with the fulfilling of an hereditary liturgy to pull an oar on the state barge by the payment of money for a substitute. For the evidence for the payment of wages to the crews of Arab galleys, see *P. Lond.* IV, Introd. xxxi.

[20] Masson 80-81.

[21] See Admiral L. Fincati, *Le triremi* (Rome 1881) 29-41, who reproduces verbatim some of the arguments, pro and con, put forth in debate in 1539 as to whether Venice should follow the lead of all her rivals and use *forçats*.

[22] Lesquier, *loc. cit.* (note 10 above) asserts that the Ptolemies used criminals, apparently taking as evidence the presence of naval personnel in mines and quarries; he assumed that only *forçats* would have been put to such tasks. A better explanation is that the sailors were there to handle the barging away of the quarried material

II In Merchant Marines

THOUGH slaves played no role in navies, they played a considerable one in merchant marines. Demosthenes' matter-of-fact allusions to sea-going freighters manned by slaves shows how common the practice must have been in the fourth century B.C.,[23] while various passages from the Digest reveal that, in the Roman Imperial period, it was by no means unusual for officers as well as men, the shipmaster included, to be slave.[24]

---

and the debris; cf. Préaux (note 10 above) 247.

I myself have suggested (see "Galley Slaves," *TAPA* 97, 1966, pp. 35-44, esp. 43-44) that we had one specific instance of the use of *forçats* on the part of the Ptolemies. The passage mentioned earlier that deals with the apprehension of deserters by the Nile police (*P. Hib.* II 198.86-88; see note 10 above) goes on to state that (lines 90-91) "if they fail to deliver [apprehended deserters] and are convicted of this, they are themselves to be sent off to the ships" (ἐὰν δὲ μὴ ἐπαναγάγω[σ]ιν ἐξελεγχθέντες α[ὐτοὶ] ἀποστελλέσθωσαν ἐπὶ τὰς ναῦς). On the surface this certainly looks like the committing of criminals to service in the galleys—a unique example, to be sure, but one which, it could be argued, recommended itself because the punishment so neatly fitted the crime. However, Lewis has shown (*op. cit.*, note 10 above, 467-68) that this is not a unique example of the use of *forçats* but just a common case of what happens to delinquent suretors under Ptolemaic law. The police, once they had apprehended deserters, were suretors for the delivery of these—and, under Ptolemaic law, when a suretor fails to deliver those whose presence he guarantees, he can himself be held personally liable in their place.

[23] 33.8-10, 34.10.

[24] *Dig.* 4.9.7 *pr.*: *debet exercitor omnium nautarum suorum, sive liberi sint sive servi, factum praestare* "The *exercitor* (see THIRTEEN, note 67) is bound to answer for the behavior of all his seamen, whether free or slave"; 9.4.19.2: *si servus tuus navem exercuerit, eiusque vicarius et idem nauta in eadem nave etc.* "If your slave is *exercitor* of a ship and his underslave, being a sailor on the same ship, etc."; 14.1.1.4: *cuius autem condicionis sit magister iste, nihil interest, utrum liber an servus* "Moreover, it is of no significance what status a shipmaster has, whether slave or free." See also 14.1.1.16, 21, 22.

One of the supply ships Caesar used in the Adriatic in 48 B.C. had at least some slaves as well as free men in the crew; cf. *Bell. Civ.* 3.14: "[The ship], because it had no soldiers aboard and was operated under private ownership, . . . was attacked and taken by Bibulus [Pompey's admiral], who exacted punishment of all the slaves and free men down to mere boys, killing them to a man" (*quod erat sine militibus privatoque consilio administrabatur, . . . a Bibulo expugnata est; qui de servis liberisque omnibus ad impuberes supplicium sumit et ad unum interfecit*).

CHAPTER FOURTEEN

Small Craft

AROUND THE shores of the Mediterranean there is still to be found, even in this homogenized age, a bewildering variety of local craft, the product of local needs, conditions, available materials. The same was true in ancient times.

So far as hull is concerned, these small fry, down to the very littlest, were constructed in the same fashion as big craft, that is, of planks carefully edge-joined by means of mortises and tenons.[1] For steering, most used the traditional arrangement, steering oars on both quarters, but not all. Some had only one oar on the starboard quarter.[2] Others steered with a sweep mounted on the stern (Figs. 189, 193). Small sailing vessels naturally differed in rig from their seagoing sisters. A certain number had a fore-and-aft rig, while those which were square-rigged—the great majority—carried only a single sail (cf. 337 below), often set on a retractable mast (Fig. 191).[3]

I SKIFFS

THE ancient world had numerous types of little open boats (Figs. 182, 192, 194) driven mainly by oars, though no doubt most were fitted to step a small auxiliary sail as well. They run the gamut from one-man rowboats on up. They bore various names (compare skiff, dory, dinghy, gig, rowboat, and the like, in English):

[1] See TEN, note 8 and App. 1 (Fiumicino barges).

[2] E.g., a clay model of the 7th or 6th b.c. found on Cyprus; see FOUR, note 124.

[3] The steering arrangement in Fig. 189 is remarkably like that used on Nile boats as early as the third millennium b.c. (see note 63 below).

Fig. 191 shows the mast lowered aft, as was ancient practice (FOUR, note 30), and so does a Roman gravestone of the 2nd-3rd A.D. (G. Tocilesco, *Fouilles et recherches archéologiques en Roumanie*, Bucharest 1900, p. 223, no. 55 and fig. 106). It could also be lowered forward; see Casson, "Harbor Boats," pl. 5.1 (= Moll B IV 98).

kymbe (*cumba* in Latin),[4] *skaphe* (*scapha*),[5] *lembos* (*lembus,* with a diminutive *lenunculus*),[6] *kydaron* (*cydarum*), *mydion* (translated *musculus* in Latin); in Latin alone: *horia, ratiaria.*[7]

Whether the above terms had each a specific designation and, if so, exactly what kind of boat it designated, we do not know. Indeed, most were so flexible in usage that they were not limited to small craft alone. *Kymbe-cumba,* for example, could be used in a generic sense as widely as our word "boat";[8] *skaphe-scapha* was applied to ship's boats (ELEVEN, note 93), which in many cases could accommodate quite a few people, or to harbor barges (336 below), or to sizable rivercraft;[9] *lenunculus* too could be used of harbor barges (p. 336 below) or of good-sized fishing boats, such as the one that ferried Caesar across the Adriatic in midwinter;[10] *lembos-lembus* were standard names for certain types of naval auxiliary galleys (162 above). On the other hand, so far as our available information goes, *horia, mydion-musculus, kydaron, ratiaria* were used only of very small boats.[11]

[4] Often used, e.g., of small fishing boats (Livy 26.45.7; Cicero, *de off.* 3.58-59; Pliny, *NH* 9.33, 35, 145).

[5] *Per. Mar. Eryth.* 3 (cf. 7 and 33): ὁ δὲ τόπος ἀλίμενος καὶ σκάφαις μόνον τὴν ἀπο-δρομὴν ἔχων "The place is harborless and offers refuge only to small boats [*skaphai*]"; Plautus, *Rud.* 162-63: *mulierculas / video sedentis in scapha solas duas* "I see two girls sitting alone in a boat [*scapha*]."

[6] Theocritus, in relating a story about two fishermen, refers (21.12) to their boat as a *lembos.* Lycurgus *in Leoc.* 17) uses *lembos,* and Plautus (*Merc.* 259) *lembus,* of the rowboats that ferried people out to ships anchored in the roads.

[7] Possibly the *placida* and *caupulus* should be added to this list. The first occurs only in the Althiburus Mosaic (Fig. 137), where it is shown as a rowboat with projecting cutwater. References to the second in grammars and glossaries (see *TLL,* s.v.) seem to establish it as the name of a type of skiff.

[8] E.g., Vergil, *Geor.* 4.195-96: *ut cumbae instabiles fluctu iactante saburram / tollunt* "as vessels [*cumbae*] unsteady in a tossing sea take on ballast."

[9] A Nile *skaphe* for hauling wood (σκάφη ξυληγός) was 52½ feet long and 16½ (16 m. x 5) broad (*BGU* 1157, 27/6 B.C.).

[10] Ammianus 16.10.3; *anhelante rabido flatu ventorum lenunculo se commisisse piscantis* "With the sea raging because of high winds, [Caesar] entrusted himself to a fisherman's boat [*lenunculus*]."

[11] Plautus (*Rud.* 910, 1020) uses *horia* of a one-man fishing boat and (*Trin.* 942) *horiola* of a very small rowboat. There is no justification whatsoever for calling the two-masted sailing ship on a mosaic from Thabraka a *horia,* as Duval ("Forme" 140) does, following Gauckler (*DS,* s.v. *horeia*). The *mydion-musculus,* literally "little mouse," outside of mention by grammarians or in glossaries, is attested only

Representations of ancient skiffs show two unusual features. The most striking, appearing in quite a few pictures of Roman Imperial times (e.g., the empty skiff in Fig. 147, the ship's boat in Fig. 144), is a blunt prow.[12] In the boats in question, the stern, unmistakably marked by the steering oars, is well rounded, whereas the prow ends in what looks for all the world like the transom of a modern rowboat. Possibly its purpose was to enable a skiff to moor, head on, flush to a dock. The other unusual feature is the concave prow with projecting cutwater. This is one that is truly ubiquitous: it appears on all sorts of vessels, from the small craft under discussion here (Figs. 177, 182, 192) to ponderous seagoing freighters (174 above), and in all ages, from the second millennium B.C. (35 above) to Roman times.[13]

II RIVERCRAFT

THE documents preserved from Greco-Roman Egypt reveal that all the standard types of merchant galley (EIGHT) saw use on

---

in the Althiburus mosaic (Fig. 137), where it is shown as a rowboat with projecting cutwater. The *kydaron* (*cydarum*) is listed by Pollux (1.82) and Gellius (10.25.5) among various boat types and included in the Althiburus Mosaic, depicted as a two-man fishing boat with projecting cutwater. Three lead seals have been found bearing pictures of *cydara*; two show a boat similar to that in the mosaic, but one a boat with both ends rounded (M. Rostovtzeff, *Tesserarum urbis Romae et suburbi plumbearum sylloge*, St. Petersburg 1903, nos. 944, 945, 948). The papyri attest its use on the Nile: see *P. Oxy.* 1197, A.D. 211 (a *kydaron* with a capacity of 3¾ tons); *P. Oxy.* 1650.12, 1st-2nd A.D.; *P. Oxy.* 1651.15, 3rd A.D. The *ratiaria* is the simplest of all the craft pictured in the Althiburus mosaic.

[12] Other examples: the *horeia* in the Althiburus mosaic (Fig. 137); a boat on the Danube manned by one rower and carrying horses, pictured on Trajan's Column (Lehmann-Hartleben, *Die Trajanssäule*, pl. 19 = Cichorius 34.84 = Moll B VIII 5); a fishing boat in a wall-painting from Herculaneum (Duval, "Forme" pl. 2.11); fishing scene in the mosaic on the floor of the basilica at Aquileia (H. Kähler, *Die Stiftermosaiken in der konstantinischen Suedkirche von Aquileia* [Monumenta artis romanae IV, Köln 1962] pl. 15, upper righthand corner [4th A.D.]). Cf. my note in *MM* 50 (1964) 176.

[13] For an example of a mere one-man skiff with projecting cutwater, see Moll B II 109. For a useful conspectus of all manner and sizes of vessels with this form of prow, see Duval, "Forme" pl. 2. A nice case in point is provided by a series of wall-paintings which picture seven examples of small craft, perhaps types used in local festive ceremonies (see Jacopi, *loc. cit.* [TEN, note 49]); six have rounded prows, one has the projecting cutwater. The date is about the beginning of the 2nd A.D. (Jacopi 89).

the Nile (340 below), and, although specific information is lacking, there is every reason to believe that they were to be found as well on other navigable streams, such as the Rhone or Rhine or Danube. The pictorial evidence furnishes us with illustrations of the heavy oar-driven vessels that carried cargo on the Moselle and other rivers of Gaul,[14] the stout boats with rounded hull that transported supplies on the Danube for Trajan's army,[15] various Tiber craft,[16] the low-walled barges that were towed up the Rhone or Tiber.[17]

The towed boat still exists today in places where labor is cheaper than power.[18] Indeed, until the invention of steam it was practically the only way to get a vessel upstream. The types used for towing up the Tiber are the best documented. The most important was the *caudicaria* (Fig. 179), of which we have a number of pictures that have been identified with a fair degree of certainty.[19] The *caudicaria* had a sturdy, well-rounded hull and carried an efficient sprit-rig (244 above)—clearly it was made for coastal work as well as for being hauled upriver. The mast of a sprit-rig is stepped forward of amidships, and is thereby able to double as a towing mast, which must be set ahead of the center of gravity.[20] The towing mast is an essential element in the towing rigs of all ages, modern as well as ancient; a line is run from the team on the bank over a block on the tip of this mast, then down to the stern, where it is secured.[21] The *caudicariae* seem to have had their towline made fast to a capstan

---

[14] Rostovtzeff, *SEHRE* pl. 39.1 = Moll B IV a 70; Espérandieu 4120.

[15] Pictured on Trajan's Column; see, e.g., Rostovtzeff, *SEHRE* pl. 45; Moll B VIII 3, 5.

[16] See the pictures mentioned in Jacopi's article cited in TEN, note 49.

[17] Casson, "Harbor Boats" pls. 2.1 (= Moll B IV 15-16), 3.2 (= *IH* ill. 67); Rostovtzeff, *SEHRE* pl. 39.5 (= Moll B IV 71).

[18] See *National Geographic*, December 1966, 776 (photograph of a team towing boats in the Nile Delta). I have photographed boats being towed along the canals near Calcutta.

[19] Casson, "Harbor Boats" 36-38 and pls. 2.2 (= *IH* ill. 64), 2.3 (= *IH* ill. 69), 3.1, 4.2, 5.1 (= Moll B IV 98).

[20] See P. Larousse, *Grand dictionnaire universel du XIXᵉ siècle*, s.v. *halage*: "le halage d'un bateau se fait ordinairement en le tirant avec une corde fixée au mât placé dans son axe, en avant du centre de gravité." The mast in Fig. 189, set noticeably forward of amidships and rather short, could well be a towing mast.

[21] Casson, "Harbor Boats" pl. 5.2.

on the poop,[22] very likely to enable these relatively heavy craft to winch themselves upriver at points where the going was particularly hard on the teams.[23] Since the mast was a stout pole that was retractable, it had no ladder and scant standing rigging; so it was sometimes fitted with cleats to allow hands to get aloft.[24]

Other Latin terms specifically connected with rivercraft are *linter*, *stlatta*, and *lusoria*. *Linter* is a name for a general class of small boats, propelled by oars or sail or both, which were frequently found on rivers[25] though they saw use in any shallow waters, such as harbors. The little we know about the *stlatta* indicates it was a small riverboat.[26] *Lusoria*, on the other hand, is rather more specific. Meaning literally "pleasure boat," it was originally used of river houseboats, such as the *thalamegoi* of the Nile (341-42 below), then of working craft.[27] By the fourth A.D. it came to be the official name of the small,

[22] Casson, "Harbor Boats" pls. 2.3 (= *IH* ill. 69), 5.1 (= Moll B ɪᴠ 98).

[23] Cf. Larousse, *loc. cit.* (note 20 above): "le halage à points fixes s'opère en faisant mouvoir des treuils au moyen de machines placées sur le bateau, de manière à enrouler une corde attachée à un point fixe. On peut avoir des points fixes établis d'espace en espace et qui forment autant de stations; mais cela exige que, pendant que le bateau parcourt une station, la corde destinée à lui faire parcourir la station suivante soit portée en avant et déroulée."

[24] Casson, "Harbor Boats" pls. 2.3 (=*IH* ill. 69), 3.1, 4.2, 5.1 (= Moll B ɪᴠ 98).

[25] Propertius (1.14.3-4) speaks of watching "the *lintres* run along [the Tiber] . . . so swiftly . . . and the barges go along on their towlines so slowly" (*tam celeres . . . currere lintres . . . tam tardas funibus ire rates*). Caesar refers to their use on the Saone (*Bell. Gall.* 1.12) and the Seine (*Bell. Gall.* 7.60), and Livy to their use on the Rhone (21.26.8; cf. Polybius 3.42.2, who calls them *lemboi*). Pliny mentions (*NH* 6.105) *lintres* that were dugouts in use in India.

[26] Ausonius (*Epist.* 26.31) speaks of *stlattae* on the Tarn and Garonne. A *stlatta* is pictured on the Althiburus mosaic (Fig. 137), and a line of verse underneath (*CIL* ᴠɪɪɪ 27790.19) refers to use on rivers (*hinc legio stlattis iam transportaverat amne* "From here the legion took care of transport over the stream by means of *stlattae*"). It is represented as a rowboat with projecting cutwater.

[27] Cf. Seneca, *de ben.* 7.20.3: *triremes et aeratas non mitterem, lusorias et cubiculatas et alia ludibria regum in mari lascivientium mittam* "I would not send triremes and bronze-shod craft; I should send pleasure boats [*lusoriae*] and houseboats [*cubiculatae*, i.e., boats with *cubicula* 'bed chambers'] and the other playthings that kings use for desporting themselves on the sea." As working craft, see note 62 below. The *lusoriae* apparently were not limited without exception to rivers. *IGRR* ɪɪɪ 481 (mid-3rd A.D.) mentions an official honored by the people of Termessus because, among other things, "He generously exercised imperial authority on the boat [*lusorion*] during the 9th of November, on which day there was brought a sacred statue of [the Emperor Valerian] (ἀγαγόντα δὲ καὶ ἰνπέριον φιλοτείμως ἐν τῷ λου-

light galleys which the Roman navy at that time had adopted as the standard unit for the forces it maintained on rivers such as the Rhine and Danube.[28]

In Greek, we have the names of several kinds of small craft which, though most of the evidence records them as being used on the Nile, very likely saw service along all navigable streams. The *kontoton*, literally "poled" boat, was surely a punt.[29] The *polykopon* "multi-oared" is self-explanatory. It is attested from the third century A.D. to the sixth,[30] and was used for transporting grain or food,[31] personnel,[32] army supplies.[33] The scant data available (capacity from $12\frac{1}{2}$ to 25 tons; see note 31) indicates that the *polykopon* was of modest size. The *platypegion* "wide-built" was probably a barge.[34] The *halias*, elsewhere attested as an oared coastal craft, appears on the Nile as dispatch boat for the *cursus velox*, the express service of the Roman government post.[35] The *ploion zeugmatikon* "yoked boat"[36] sounds like a catamaran.

σωρίῳ τῇ πρὸ έ εἰδῶν νοεμβρίων ἐν ᾗ ἡμέρᾳ ἐκομίσθη εἰκὼν ἱερὰ κτλ.). The official must have arrived by way of the port of Attalia.

[28] Starr 151-52, Kienast 148-49. They are attested in the 4th A.D. on the Meuse and Rhine (Ammianus 17.2.3, 18.2.12), in the 5th on the Danube (*Cod. Theod.* 7.17.1). They may possibly have been in use as early as the reign of Probus (A.D. 276-282); see *Hist. Aug., Vita Bonosi* 15.1. For their size, see Ammianus 18.2.11-12: Julian used 40 *lusoriae* to ferry 300 men across the Rhine in A.D. 359, an average of 7 to 8 men per·boat.

[29] *PSI* 551.1 (272/1 B.C.), P. Hib. 39.4 (265/4 B.C.), P. Teb. 810.15 (134 B.C.). Alexander had *kontota* built for use in Mesopotamia (Diodorus 19.12.5).

[30] References have been collected by E. Wipszycka in *Chr. d'Eg.* 35 (1960) 219. To her list, add *Stud. Pal.* VIII 774.2 (A.D. 572) and possibly P. Flor. 297.474 (6th A.D.) as emended in *Berichtigungsliste* IV, and P. Bad. II 26.121 (A.D. 292/3) as emended in *Berichtigungsliste* II.

[31] P. Oxy. 2415.46, 51 (3rd A.D.), boats of $12\frac{1}{2}$ and 25 tons respectively; W. Chrest. 46 (A.D. 338), boat of $17\frac{1}{2}$ tons; P. Ross.-Georg. III 5.4, 10 (3rd A.D.); Sammelb. 9563 (4th A.D.).

[32] P. Grenf. II 80, 81, 81a, 82 (ca. A.D. 400); P. Cairo Masp. 67136.7 (6th A.D.).

[33] Stud. Pal. V 119, verso 4.7, 11 (3rd A.D.); P. Lugd.-Bat. II 13, col. 1.6, 12 (4th A.D.), 14.3, 10 (4th A.D.; the editor erroneously calls the vessel a warship).

[34] P. Oxy. 1652 (3rd A.D.); Sammelb. 9614 (A.D. 283); P. Oxy. 2715 (A.D. 386); possibly P. Thead. 59.3 (4th A.D.).

[35] P. Oxy. 2675 (A.D. 318): ἁλιάδων γραμματηφόρων τοῦ ὀξέως δρόμου "the dispatch boats [*haliades*] of the *cursus velox*." Similarly in P. Pan. 1.60-61, 252 (A.D. 298). For the *halias* as a coastal craft, see note 52 below.

[36] P. Oxy. 2415.44, 56 (3rd A.D.), boats with a capacity of 9 to $12\frac{1}{2}$ tons. Another

A riverboat pictured in a relief found in the vicinity of Trier and hence most probably used on the Moselle or Sarre, is unusual (Fig. 195). The boat is short and shallow, and a series of lines crisscross the hull at right angles in a regular pattern. They apparently are intended to portray lashings and, together with the loglike aspect of what they bind, make the craft look very much like a shaped raft, though it boasts such a standard boat appurtenance as a goose-headed sternpost. Even more curious is the sail, a tall oblong square sail which seems to be fitted with two battens running from edge to edge much in the manner of the battens on a Chinese lugsail.[37]

So far as river navigation is concerned, the only information we have is Herodotus' description of the Nile boatmen's technique for drifting downstream. They used a sort of sea anchor consisting of a "doorlike frame made of tamarisk wood over which is lashed reed matting, plus a stone weighing a good 120 pounds with a hole bored in it. The frame, made fast to a line is put out to float off in front of the boat, and the stone is made fast to another line astern. The frame, as the current catches it, drifts along swiftly and tows the *baris* (this is the name given to these boats), and the stone, dragged along behind well below the surface, holds the vessel on course."[38] Similar devices were still in use on the Euphrates up to the last century.[39]

III HARBOR CRAFT

EVERY harbor had its share of rowboats and small sailboats of varying sizes and types—*kymbai (cumbae)*, *skaphai (scaphae)*, *lem-*

---

possibility is a complex of several small boats yoked together like the dugouts of the Dunjec River in Poland; see *MM* 52 (1966) 211-22.

[37] Hennig's drawing (*MM* 48, 1962, p. 315) is misleading. On shaped rafts, see Hornell 61-67, and on their boatlike appurtenances, such as a stempiece, see 64-67. On the Chinese lugsail, see *IH* 176-77.

[38] Herodotus 2.96: ἐκ μυρίκης πεποιημένη θύρη, κατερραμμένη ῥιπὶ καλάμων, καὶ λίθος τετρημένος διτάλαντος μάλιστά κῃ σταθμόν. τούτων τὴν μὲν θύρην δεδεμένην κάλῳ ἔμπροσθε τοῦ πλοίου ἀπίει ἐπιφέρεσθαι, τὸν δὲ λίθον ἄλλῳ κάλῳ ὄπισθε. ἡ μὲν δὴ θύρη τοῦ ῥόου ἐμπίπτοντος χωρέει ταχέως καὶ ἕλκει τὴν βᾶριν (τοῦτο γὰρ δὴ οὔνομά ἐστιν τοῖσι πλοίοισι τούτοισι), ὁ δὲ λίθος ὄπισθε ἐπελκόμενος καὶ ἐὼν ἐν βυσσῷ κατιθύνει τὸν πλόον.

[39] How and Wells, note *ad loc.*

boi (lembi), lintres, lenunculi.[40] But, in addition, there were special craft for two important services: the ancient equivalent of the tug-boat for warping big sailing ships about the narrow confines of a harbor,[41] and barges for unloading them.

The craft for unloading cargo and ferrying it either to harbor warehouses or upriver were called in Greek *hyperetikai skaphai* "service boats,"[42] in Latin *scaphae*,[43] *levamenta* "lighters,"[44] or *lenunculi*. In Portus, the all-important harbor for the city of Rome, both services, the transfer of cargo and the warping of ships, were performed by *lenunculi*. The barges were operated by men called *lenuncularii pleromarii auxiliarii* "boatmen for cargo service."[45] A flat-bottomed craft excavated from the mud of what once was the floor of the harbor may well be a specimen of the sort of craft they used (Fig. 190).[46] A tugboat is illustrated in a relief (Fig. 193) adorning a local tomb, in all probability that of its skipper. We see a heavy skiff rowed by three oarsmen and steered by a helmsman who grasps, not the customary pair of oars on the quarters, but a

[40] Thus the guilds of boatmen found at various ports in Roman Imperial times include *scapharii* (*CIL* II 1180 = *ILS* 1403, 1183, both from Seville; XIV 409 = *ILS* 6146, Ostia; cf. *Dig.* 14.2.4 *pr.*), *lyntrarii* (*CIL* II 1182, Seville; VI 9531, Rome; XIV 4459 = *ILS* 1442, Ostia), *lenuncularii* (there were five guilds of such boatmen at Ostia; see *CIL* XIV 352 = *ILS* 6149 [A.D. 251] and cf. *CIL* XIV 170 = *ILS* 1433 [A.D. 247/8], 4144 = *ILS* 6173 [A.D. 147]). For *lembi* as harbor craft, see the passage from Plautus cited in note 6 above.

[41] Cf. Synesius' description (*Epist.* 147, Migne 66.1545b) of the harbor at Alexandria, where "one vessel is being towed, another carried along by the wind, still another by oars" (καὶ εἵλκετο ναῦς, καὶ ἀνήγετο πρὸς οὖρον αὕτη, κώπαις ἐκείνη). In some harbors, to be sure, the warping was done by a ship's own boat. E.g., Philostratus (*Vita Ap.* 4.9), describing a big freighter getting under way in the harbor of Smyrna, speaks of part of the crew raising anchor, part setting sail, and part "who were rowing personnel had gotten into the ship's boats" (see THIRTEEN, note 88).

[42] Strabo 5.232.

[43] *Dig.* 14.2.4 *pr.*

[44] I.e., the craft operated by the *levamentarii* "lightermen" (*Cod. Theod.* 13.5.1). The term *levamentum* itself is not attested until Mediaeval times; see Du Cange, *Glossarium mediae et infimae latinitatis*, s.v. Since it was indubitably used then to mean barge or small boat, Rougé's suggestion (185) that *levamentarii* were "crane operators" is otiose.

[45] *CIL* XIV 252 (= *ILS* 6176) plus Suppl. I, p. 614 (A.D. 200), 253. Cf. Casson, "Harbor Boats" 34-36.

[46] For another possible example of such barges, see Casson, "Harbor Boats" pl. 2.1 (= Moll B IV 15-16).

single oversized oar mounted on the stern, an arrangement adopted no doubt to give greater leverage for guiding an unwieldy tow. A mast is stepped far up in the bows; its position shows that it must have carried a fore-and-aft sail, most likely a sprit (244 above). The operators of these tugs are probably the men called in inscriptions found about the harbor *lenuncularii tabularii auxiliarii* "boatmen for the service of checking and control"; the tugboat skippers, being the first to greet a new arrival, were a natural choice to charge with making the preliminary check of a ship's papers and assigning it a provisional berth.[47]

IV COASTAL CRAFT

THE average sailing vessel that plied the coasts,[48] being considerably smaller than those used on open water, had only a single-masted rig (Figs. 148, 191).[49] To judge from preserved pictures, in most cases this was a square rig,[50] the best there is for sailing with a favorable wind. However, since coastal craft often cannot afford to wait for the right wind, or must sail in a number of directions, some were fitted with the more flexible fore-and-aft rig, a sprit or lateen

[47] Cf. Casson, "Harbor Boats" 34-35. Five registers of this guild are preserved: *CIL* xiv 250 = *ILS* 6174 (A.D. 152); 4567; 4568 (shortly after A.D. 152); 251 = *ILS* 6175 (A.D. 192); *Not. Sc.* sér. 8, 6 (1953) 278-82, no. 42. *CIL* xiv 341 = *ILS* 6144 is a dedication to one of the patrons.

[48] The generic term was *oraria navis*; see Pliny, *Epist.* 10.15: *nunc destino partim orariis navibus, partim vehiculis provinciam petere* "Now I intend to go to my province partly by coasting vessels [*orariis navibus*], partly by land conveyances."

[49] Thus Synesius, suddenly finding himself out of sight of the coast and rather far out at sea, remarks (*Epist.* 4.161a): μετὰ τῶν ὁλκάδων ἦμεν τῶν διαρμένων (διαρμενίων codd.), αἷς οὐδὲν ἔδει Λιβύης τῆς καθ' ἡμᾶς, ἀλλὰ πλοῦν ἕτερον ἔπλεον "We were among the two-masted merchantmen [*holkades diarmenoi*; see ELEVEN, note 74], who have no need of the Libyan coast which concerned us, but were sailing a totally different course." See the harbor scene on a wall-painting from Stabiae with a row of single-masted vessels at anchor (Rostovtzeff, *SEHRE* pl. 12.1). Among wrecks that have been discovered (TEN, App. 1), the Kyrenia wreck, which measures ca. 15 m. and most probably had a fore-and-aft rig (ELEVEN, note 82), and other vessels of smaller size (e.g., the Marseilles wreck, 17 m. long) were very likely coastal craft.

[50] E.g., S. Reinach, *Répertoire de peintures grecques et romaines* (Paris 1922) 274.4 (Rome), 5 (Sousse), 380.1 and 383.1 (Pompeii); *Répertoire de reliefs grecs et romains* III (Paris 1912) 179.6 (Anzio), 208.3 (Rome), 229.1 (Ostia); Torr, ill. 39 (Ravenna); Moll B x b 76, B xi a 34.

(244-45 above). Aside from the *caudicaria* and possibly the *prosumia*,[51] no other types of sailing craft specifically employed in coastal work are identifiable, though there must have been a good many.

On the other hand, we have abundant evidence for the use along the coasts of the various kinds of merchant galleys—*akatoi* (*actuariae*), *keletes* (*celoces*), *kerkouroi* (*cercuri*), *kybaiai* (*cybaeae*), *lemboi* (*lembi*).[52] Since they were to some extent free of the wind's vagaries, and willy nilly had to stick close to shore, they were a natural choice for coastal work.

V GEOGRAPHICAL TYPES

As a result of varying local sailing conditions or of available boatbuilding materials or even of merely stubborn local adherence to long fixed tradition, boats for performing the same function will differ from place to place in all manner of ways, in shape, size, rig, color, and so on. Our scant evidence, unfortunately, permits us to recognize only very few of such variations.

Thanks to the lucky discovery of some remains of boats in the mud flats off the Thames, we can identify what must have been types of local craft.[53] They differ radically from those of the Mediterranean

---

[51] Gellius refers (10.25.5) to *prosumiae vel geseoretae vel oriolae*. *Geseoretae* are unknown, and (*h*)*oriolae* are very small boats (see note 11 above). However, the Althiburus mosaic (Fig. 137) shows the *prosumia* as sailing vessel, not rowboat, and this seems to be confirmed by the only two instances we have of the use of the term, both in passages from Caecilius preserved in Nonius (p. 536.8 = *ROL* i, pp. 468, 508): *de nocte ad portum sum provectus prosumia* "At night I was brought to the harbor in a *prosumia*"; *Cypro* (*cui pro* codd.) *gubernator propere vertit prosumiam* "The helmsman hurriedly puts the *prosumia* about for Cyprus."

[52] An *akatos* plying between Ascalon and Alexandria, EIGHT, note 7; *actuariolae* for coasting along the west coast of Italy, TWELVE, note 43; *keletes* plying between Syria and Egypt, EIGHT, note 24; a *cercurus* plying between Asia Minor and Athens, Plautus, *Stichus* 366-69; *kybaiai* plying between Syria and Egypt, EIGHT, note 50; a *lembos* plying between Asia Minor and Egypt, EIGHT, note 36.

The *halias* was a coastal craft using sail and oars. Aristotle (*Hist. Anim.* 533b 15-20) refers to its use for fishing with nets and specifically mentions oars. In Polyaenus 6.54 a *halias* sailed from Lemnos to Acanthus, about a 12-hour run with a good wind. Hiero's superfreighter had 34-ton *haliades* as ship's boats (Athenaeus 5.208f, cited in NINE, App., pt. 4). For its use as a government dispatch boat on the Nile, see note 35 above.

[53] See P. Marsden, "A Boat of the Roman Period Discovered on the Site of New

in material or method of construction or both. These Thames boats were made of oak,[54] a wood that southern boatwrights used only sparingly (213 above). Moreover, on some, the planks, instead of being edge-joined one to the other, were nailed to a skeleton of keel and frames[55] and had the seams carefully caulked,[56] as was standard practice in Europe from the Middle Ages on. The hulls were relatively flat-bottomed, as we would expect in craft intended for use in shallow waters. One of the boats was chine-built, i.e., the walls met the floor at a definite angle instead of curving to form rounded bilges.[57]

The information derived from these wrecks is in harmony with what Caesar[58] reports about the coastal craft used just across the Eng-

---

Guy's House, Bermondsey, 1958," *Transactions of the London and Middlesex Archaeological Society* 21 (1965) 118-31; *A Roman Ship from Blackfriars, London* (Guildhall Museum Publication, London 1967). The New Guy's boat was about 50 feet long and 14 broad; it can be dated ca. A.D. 200. The Blackfriars boat was about the same length but wider, ca. 22 feet; it dates to the 2nd century A.D. A third ship, the County Hall boat, though built of northern oak, was put together in traditional Mediterranean fashion, that is with planks edge-joined to each other (202-203 above); see P. Marsden, "The County Hall Ship," *Transactions of the London and Middlesex Archaeological Society* 21 (1965) 109-17. This vessel was about 60 to 70 feet long and 15 wide; it dates ca. A.D. 300.

[54] Marsden, "County Hall" 109, "New Guy's" 119, *Blackfriars* 23.

[55] Marsden, "New Guy's" 123, *Blackfriars* 14, 34-35. The third boat, however, was built in Mediterranean fashion; see "County Hall" 111.

[56] Marsden, "New Guy's" 121 (hazel twigs and also some unidentified root used as caulking material), *Blackfriars* 14 (hazel twigs as caulking material). The County Hall boat, on the other hand, put together with edge-joined planks, was uncaulked ("County Hall" 111).

[57] Marsden, *Blackfriars* figs. 7, 9.

[58] Caesar, *Bell. Gall.* 3.13: *carinae aliquanto planiores quam nostrarum navium quo facilius vada ac decessum aestus excipere possent; prorae admodum erectae atque item puppes ad magnitudinem fluctuum tempestatumque accommodatae; naves totae factae ex robore ad quamvis vim et contumeliam perferendam; transtra ex pedalibus in altitudinem trabibus confixa clavis ferreis digiti pollicis crassitudine; ancorae pro funibus ferreis catenis revinctae; pelles pro velis alutaeque tenuiter confectae, hae sive propter lini inopiam atque eius usus inscientiam, sive eo, quod est magis verisimile, quod tantas tempestates Oceani tantosque impetus ventorum sustineri ac tanta onera navium regi velis non satis commode posse arbitrabantur* "They build their hulls with somewhat flatter bottoms than our craft to make it easier to go through the shallow depths of low tide and over shoals; they build prow and stern as well up rather high to handle the size of the waves when a sea is running; and they use oak throughout to withstand any amount of violence and hard treatment. Beams are of timbers a foot square made fast with iron nails an inch thick, and anchors are held by iron chain instead of rope. Their sails are of hide or of softened leather

lish Channel by the Veneti, a tribe of the Breton coast. These, too, were made of oak and very likely were put together the way the Thames boats were, by nailing strakes to a skeleton of keel and frames.[59] Caesar adds that, to withstand the rigors of the northern climate, the sails were of hide and the anchor cables of chain.

A second area for which we have detailed information about local craft is the Nile. The river exhibited the same multitudinous variety in ancient times as it does today.[60] All the standard types of merchant galley were to be found there—the *keles, kerkouros, kybaia, lembos, phaselion.*[61] The *liburna* and *lusoria*, the former known elsewhere only as a naval craft, from the third A.D. on turn up on the Nile as ordinary cargo boats.[62] The term *ploion Hellenikon* "Greek boat" occurs a number of times; it may very well designate a vessel built in the standard Greek style as against the native Egyptian.[63]

---

instead of canvas, possibly because they have no flax and do not know how to use it, but more likely because they reckon canvas will not stand up to the violence of ocean storms and the force of the winds there and will not drive such heavy vessels efficiently."

[59] This seems a natural deduction from Strabo's remark (cited in TEN, note 38) about the open seams in their ships.

[60] Like seagoing ships (NINE, note 4), Nile craft were sometimes referred to by the cargoes they handled. There was the "sand and manure carrier" (ἀμμοκοπρηγόν *Sammelb.* 423, 3rd A.D.), "manure carrier" (κοπρηγόν *W. Chrest.* 31, A.D. 156), "corpse carrier" (νεκρηγόν *P. Hamb.* 74, A.D. 173), "jar (?) carrier" (κουφηγόν *P. Flor.* 335, 3rd A.D.), "wood carrier" (ξυληγόν *ibid.*), "stone carrier" (λιθηγὸς [βᾶρις] *P. Cairo Zen.* 59745.66, 3rd B.C.), "wine carrier" (οἰνηγόν *PSI* 568, 253 B.C.), "grain carrier" (σιτηγόν *P. Cairo Zen.* 59031, 258 B.C.). And, on the Red Sea, there was even the "elephant carrier" (ἐλεφαντηγὸς [ναῦς] *W. Chrest.* 452, 224 B.C.), boats used to transport the elephants hunted during Hellenistic times for the army's elephant corps. For a picture of such a craft, see *Memoirs of the American Academy in Rome* 13 (1936) pl. 46.3.

The evidence on Nile craft was brought together by M. Merzagora in "La navigazione in Egitto nell' età greco-romana," *Aegyptus* 10 (1929) 105-48. Much new material has been published since.

[61] See EIGHT, notes 24, 40, 50, 36, 59.

[62] A *liburna* is listed in a papyrus of the 3rd A.D. (*P. Ryl.* 223) as carrying miscellaneous cargo. In papyri of the 5th and 6th A.D., a ship called a *libernos* (*Stud. Pal.* VIII 1094, *P. Oxy.* 2042, *Sammelb.* 5953) or *libernion* (*P. Oxy.* 2032.52, 54) is attested as a cargo carrier. For the *lusoria* (or *lusorion*), see *P. Oxy.* 1048.2, 7 (4th/5th A.D.), possibly *P. Oxy.* 1905.21 (4th/5th A.D.), *P. Rend. Harris* 150 (5th A.D.), *Sammelb.* 9563 (4th A.D.).

[63] *P. Lond.* 1164h.6 = *Select Papyri* 38.6 (A.D. 212), *W. Chrest.* 248 (A.D. 220/1), *P. Oxy.* 2136.5 (A.D. 291), *P. Goodsp.* 14.3 (A.D. 343), *P. Mon.* 4.9-11 (A.D. 581). In

Of the craft built in native style, the most common was the *baris*. The *baris* was made in the ancestral Egyptian fashion, of short lengths of acacia wood fastened one to the other by means of keyed joints (Fig. 13) and dowels and reinforced with very little or no framing.[64] It had a square sail amidships for working upriver and oars to help speed the trip downriver. Some of the work boats we see illustrated in the wall-paintings of Pharaonic Egypt must surely be *barides*.[65]

Another native craft was the *thalamegos*. The *thalamegoi*, literally "cabin-carriers," were the *dahabiyehs* of the ancient world, the Nile yachts, such as those that ferried government officials up and down the river.[66] The most grandiose *thalamegos* of all was the wondrous

the first document, the ship is described as having two steering oars with tillers (see ELEVEN, note 135). In the others, details of the tiller are lost or not given. In *P. Mon.* 4.11 the boat is described as *phikopedalos*; this obscure term is also used to describe boats in a papyrus of A.D. 338 (*P. Lugd.-Bat.* XI 1, col. 1.4 and col. 2.3) and another of A.D. 570 (*P. Lond.* 1714.33). Can it be sailors' jargon using the Latin word *ficus* "fig," i.e., a boat with a "fig-[shaped]" steering oar? Perhaps such craft, instead of the standard Greek system of two oars on the quarters, used the Egyptian arrangement of a single oar fixed on the stern (cf. Two, note 39); in the models and pictures from Pharaonic Egypt of boats so rigged, the oar could well be described as "fig-shaped" (Reisner no. 4951 and pl. 26; Moll A IV b 155 and IV c 151, 152). An oar of this shape so mounted appears on a riverboat from Gaul (Fig. 189).

[64] See Two, note 15, and note 38 above. The word *baris* seems to be derived from the Egyptian *br*, a term that included seagoing craft; cf. C. Schaeffer, *Ugaritica* IV (Mission de Ras Shamra XV, Paris 1962) 138-39. A *baris* carries miscellaneous cargo in *P. Hib.* 100.13 (267/66 B.C.), soldiers in *W. Chrest.* 11.1.22 (123 B.C.), fish in *P. Teb.* 701.26 (235 B.C.), stone in *P. Cairo Zen.* 59745.66 (3rd B.C.).

[65] E.g., N. Davies and A. Gardiner, *The Tomb of Huy* (London 1926) pl. 18; Boreux fig. 85.

[66] Strabo (17.800) reports that, on a canal near Alexandria, is "the marina for the *thalamegoi* which officials use for sailing to upper Egypt" (τὸ ναύσταθμον τῶν θαλαμη-γῶν πλοίων, ἐφ' οἷς οἱ ἡγεμόνες εἰς τὴν ἄνω χώραν ἀναπλέουσιν). Cf. *P. Ryl.* 558 (257 B.C.), where a *thalamegos*, used for transport of officials, has a crew of eight; *P. Teb.* 802.7, 9 (135 B.C.), where the *thalamegos* referred to is for the district governor. Ptolemy II's fleet included 800 "*thalamegoi* with gilded sterns and prows" (θαλαμηγά τε χρυσόπρυμνα καὶ χρυσέμβολα Appian, *Praef.* 10). In the mosaic of a Nile scene found at Praeneste (1st B.C.), the roomy boat with cabin amidships depicted there may well be a *thalamegos*; see M. Swindler, *Ancient Painting* (New Haven and London 1929) ill. 510 = Moll B x a 4). The vessel's cabin consists of a roof supported by walls only at each end; along the sides there are merely two columns, and grills partly close in the space between the columns. This corresponds very nicely with a reference in a papyrus (*P. Flor.* 335.2 [A.D. 259, re-edited in *Chr. d'Eg.* 25, 1950, pp. 99-101];

houseboat, a vast floating palace, built by that specialist in show-pieces, Ptolemy IV.[67] Alongside these elegant specimens were much more humble ones that served as ordinary cargo carriers.[68]

Still another native type was the *pakton*. *Paktons* were originally very small craft made of woven branches; they must have been like the coracles of basket work used on the lower Euphrates, and like them must have been liberally pitched over to be made watertight.[69] Later on they were made of wood and were large enough to carry over 10 tons.[70] Alongside *paktons*, the age-old craft made of papyrus reeds were still in use.[71]

---

cf. *P. Fay.* 104.9, 11 [3rd A.D.]) to a καλύβη καὶ τετράστυλον "a shelter with its four-column support" (on καλύβη as the term for a shelter on a boat, see Achilles Tatius cited in NINE, note 64).

[67] Athenaeus 5.204e-206d. It was a catamaran, like Ptolemy's other showpiece, the "forty" (110-11 above). It was over 300 feet long, 45 wide at the broadest point, and towered 60 above the water. The accommodations and decor were the last word in luxury. For a detailed study, see F. Caspari, "Das Nilschiff Ptolemaios IV," *JDI* 31 (1916), 1-74.

[68] *PSI* 332.10, 16 (257/6 B.C.); *BGU* 1882.3 (1st B.C.); *BGU* 802, cols. I.16, IV.11, XII.11, XIV.24 (A.D. 42), probably with a capacity of 2¼ tons (900 artabs; reading ἀγωγῆς for αὐτῆς in XII.13 and restoring it in XIV.26); *P. Oxy.* 1650.20 (1st/2nd A.D.), boat with a capacity of 13 3/4 tons; *P. Oxy.* 1738.2 (3rd A.D.); *P. Ross.-Georg.* V 55.4 (3rd A.D.).

[69] Strabo 17.818: "The pakton is a small boat [*skaphion*] put together with withes so as to resemble woven work; we crossed easily, standing in the water [i.e., that had collected on the boat's floor] or seated on some small boards" (ὁ δὲ πάκτων διὰ σκυταλίδων πεπηγός ἐστι σκάφιον, ὥστ' ἐοικέναι διαπλοκίνῳ· ἑστῶτες δ' ἐν ὕδατι ἢ καὶ σανιδίοις τισὶ προσκαθήμενοι ῥᾳδίως ἐπεραιώθημεν). The crossing was to the island of Philae near the First Cataract.

[70] In *P. Mert.* 19 (A.D. 173) a *pakton* is described which was all of willow wood. It was 18¾ feet long and had a single steering oar and two sculls. Another *pakton* is also described (*P. Oxy.* 2568.14, A.D. 264) as two-oared. *Paktons* were used for transport of goods in *P. Oxy.* 1650 (1st/2nd A.D.), specifically a cargo of 13¾ tons of grain; *P. Oxy.* 1220.12 (3rd A.D.); *BGU* 812 (2nd/3rd A.D.); *PSI* 948.16 (A.D. 345); *P. Ross.-Georg.* II 18.176 (A.D. 140); *P. Oxy.* 2568.14. They were used for transport of people in *P. Ryl.* 225.39 (2nd/3rd A.D.); *P. Oxy.* 2153 (3rd A.D.); *P. Cairo Masp.* 67058 col. VI.5 (6th A.D.). They were in use right up to the early 8th A.D. at least (cf. the mention of πακτωνοπ(οιοί) "*pakton*-makers"—not πακτωνοπ(ράται)—in *P. Lond.* IV 1419.1217).

[71] See Fig. 116 and cf. *UPZ* 81, col. II.6-7 (343 B.C.): "a papyrus boat of the type called *rops* in Egyptian" (πλοῖον παπύρινον, ὃ καλεῖται α⟨ἰ⟩γυπτιστεὶ ῥώψ). The Egyptian word is actually *rms*, and this is attested in Greek; see the Louvre papyrus published in *Archiv* 2 (1903) 515-6: ἐὰν δὲ μὴ ἔχητε πλοῖον, συνεμβήσητε ἅμα ἡμῖν εἰς τὴν ῥῶμσιν "If you have no boat, come with us in the *roms*." As Wilcken has

Other names of local Nile craft are inscrutable. Was the *karis* "shrimp" a boat whose hull was shrimp-shaped?[72] Was the *kasioti-kon* so called because it was used by the people who lived near Mount Kasios at the mouth of the Nile?[73] What kind of craft was the *baioi-elypion*, which could perhaps hold 125 tons?[74] And how are the abbreviations *akor()* and *achor()*, which occur in a list of boats hauling grain,[75] to be resolved?

THE Nile is not alone in furnishing conundrums of nomenclature. What were the *kyknos*, literally "swan," and the *kantharos*, literally "beetle," that were the names of vessels used in the Aegean in the fourth century B.C.?[76] What kind of craft were the *tragoi* "goats," and the *krioi* "rams," native to Lycia?[77] the *vegeiia* (?), possibly native to Gaul?[78] the *gandeia*, possibly a North African type?[79] Until further information is available, none of these questions can be answered.

---

shown (*Mélanges Nicole*, Geneva 1905, p. 587), ῥῶμσιν is to be read instead of the editor's ῥωησιν; and from ῥῶμσ(ιν), a by-form ῥωμψ could easily arise and produce ῥωψ.

[72] *P. Oxy.* 2032.53, 75 and 2480.2, 15, 24, 26, 34, 36 (both 6th A.D.); *P. Iand.* 18.7 (6th/7th A.D.).

[73] *P. Cairo Zen.* 59289.7, 59326.101; *P. Mich. Zen.* 17.4. All are 3rd B.C.

[74] *P. Petrie* III 129.11, where the ship's burden is probably 5,000 artabs. For other instances, see *P. Lille* 25.43, *P. Teb.* 701.260. All are 3rd B.C.

[75] *Sammelb.* 9367, nos. 2.2, 3.2, etc. (163 B.C.).

[76] Nicostratus *ap.* Athenaeus 11.474b: ἡ ναῦς δὲ πότερ' εἰκόσορός ἐστιν, ἢ κύκνος, / ἢ κάνθαρος; "About the ship—is it a 20-oar or a 'swan' or a 'beetle'?" (on the term "20-oar," see NINE, note 5). The *kantharos* is also mentioned by Sosicrates and Menander quoted by Athenaeus in the same passage and by Aristophanes (*Peace* 143).

[77] Pollux 1.83: ἔστι δέ τινα πλοῖα Λύκια λεγόμενα κριοὶ καὶ τράγοι "And there are certain Lycian boats called 'rams' and 'goats.'"

[78] The *vegeiia* (or *vegella*) occurs in the Althiburus mosaic (Fig. 137), where it is pictured as a rowboat with projecting cutwater. It may be identical with the *vetutia* mentioned by Gellius (10.25.5) and the *vehigelorum* (or *vehiegorum* or *veiegorum* or *vegetorum*) listed in the *Corpus Gloss. Lat.* (IV 191.13; V 518.13, 613.12) and defined as *genus fluvialium navium apud Gallos* "a type of rivercraft in use among the Gauls." See *CIL* VIII 27790, commentary to no. 15.

[79] Cf. *TLL*, s.v.

CHAPTER FIFTEEN

Markings and Names

I MARKINGS

SHIPS BORE distinctive devices both fore and aft. It was the device forward that was most important, for this identified the vessel,[1] was its *episemon* or *parasemon* "name-device."[2]

[1] Lucian, *Nav.* 5 (cf. NINE, App., pt. 2): "The prow . . . had an Isis on both sides, the goddess after whom the vessel was named" (ἡ πρῷρα . . . τὴν ἐπώνυμον τῆς νεὼς θεὸν ἔχουσα τὴν Ἶσιν ἑκατέρωθεν); Plutarch, *Mor.* 248a (of a craft of remotely ancient times): "a ship with the name-device [*episemon*] of a lion forward, and a serpent on the stern" (πλοίῳ λέοντα μὲν ἔχοντι πρῴραθεν ἐπίσημον, ἐκ δὲ πρύμνης δράκοντα); Diodorus 13.3.2 (of the triremes at Syracuse in 415 B.C.): "the devices on the prow" (τοῖς ἐπὶ ταῖς πρῴραις ἐπισήμασι [or ἐπιστήμασι]).
A directive recorded in *P. Pan* 2.208-209 (A.D. 300) concerning craft to be built under government auspices for hauling grain down the Nile states that "if desired, shipowners, before launching their craft, may carve on the prows of the ships, as is customary, the devices (*parasema*) of the gods of each city" (κἄν γε βουλόμενον ᾖ τοῖς ναυκλήροις ἐν ταῖς πρῴραις τῶν πλοίων ὡς ἔθος ἐστιν τῶν ἐν ἑκάστῃ πόλει θεῶν τὰ παράσημα ἐγχαραχθήτω τοῖς πλοίοις πρὶν κατασπασθῆναι αὐτά). Having more than one ship with the same name, as no doubt would happen in this instance, apparently raised no problems; cf. notes 66, 71, 72 below.
[2] The term *episemon* can designate any marking on a ship. Strabo, for example, uses it of figureheads (the *hippoi* that characterized a certain class of Phoenician merchantman; see FOUR, note 115). Many writers, however, employ it in the specific sense of the marking that represents a vessel's name: Herodotus 8.88 (σαφέως τὸ ἐπίσημον τῆς νεὸς ἐπισταμένους "[Spectators were aware of Artemisia's prowess] since they knew for certain her ship's name-device"), Plutarch, *Mor.* 248a (cited in the previous note), Hippocrates, *Epist.* 17 (cited in note 7 below). It occurs also in documents; see *Sammelb.* 9223.4, 2 B.C. (πλοίου οὗ ἐπίσημον Αἴγυπτος "vessel with the name-device *Aigyptos*").
The technical term proper, however, was *parasemon*. Cf. Plutarch, *Them.* 15.2: ναῦν . . . ἧς τὰ παράσημα περικόψας ἀνέθηκεν Ἀπόλλωνι "a ship . . . whose name-devices [*parasema*; plural because there was one on each side of the vessel] he cut off and dedicated to Apollo"; Acts 28.11: "[St. Paul embarked] on a ship . . . from Alexandria with the name-device of the Dioscuri" (ἐν πλοίῳ . . . Ἀλεξανδρινῷ παρασήμῳ Διοσκούροις). The *parasemon* was the standard feature by which boats were identified in the business documents from Greco-Roman Egypt: see *P. Pan.* 2.208-209 (cited in note 1 above); *W. Chrest.* 248 (A.D. 220/221), 443 (A.D. 15); the papyri cited in notes 70-73 below. A boat without any name-device was called ἀχάρακτος "uncarved" in the Ptolemaic period (*W. Chrest.* 189 [221 B.C.]; *P. Teb.* 1034.5 and 17, 1035.5 [all 2nd B.C.]) and ἄσημος "unmarked" in the Roman (*M. Chrest.* 341, A.D. 236). Latin simply took over the Greek term; cf. *CIL* III 3 = *ILS* 4395 (2nd A.D.): *navis*

On warships, the name-device was placed on both sides of the bow, either carved (Figs. 125, 127, 129, 131) or painted[3] or in the form of a bronze plaque.[4] Merchantmen carried it on both sides of the stempost (Figs. 144, 151).[5]

The curved stern offered a favored field for additional decoration

---

parasemo Isopharia "ship with Isis of Pharos as device"; *AE* 1951, no. 165b (1st A.D.): *in n(ave)* . . . *vecta Iovis et Iuno(nis) parasemi* "carried in the ship with Jupiter and Juno as device." The general Latin term for device, *insigne*, could be used in the technical sense of ship's device; e.g., the Vulgate renders the identification of St. Paul's ship in Acts 28.11 (see above) as *cui erat insigne Castorum* "[a ship] which bore the device (*insigne*) of the Castores."

A less technical term than *parasemon*, and one that may have included figureheads as well as name-devices, was *antiprosopon*, literally "facing against." See Artemidorus 2.23 (cited in THIRTEEN, note 67), where he explains that, in dreams, the *cheniskos* stands for the captain, and the *antiprosopon* for the bow officer; since the *cheniskos* was the figurehead on the sternpost, the other must be its counterpart forward (not the "prow," as *LSJ* defines it). See also Artemidorus 4.24: τὸν πρῳρέα τις δόξας ἐκφέρειν καὶ κατορύττειν τῆς νεὼς ἀπώλεσε τὸ ἀντιπρόσωπον "someone who dreamed he had given a funeral to and buried his first mate, lost the *antiprosopon* of his ship."

[3] Cf. Aristophanes, *Frogs* 933: the fabulous "tawny cockhorse" is explained as a "device painted . . . on ships" (σημεῖον ἐν ταῖς ναυσίν . . . ἐνεγέγραπτο); Ovid, *Tristia* I.10.2 (cited in note 14 below).

[4] Perhaps the boxlike affair that encloses the name-device in Fig. 130 is supposed to represent a plaque. The Fogg Art Museum of Boston has a bronze plaque, dated early 1st B.C., showing the bust of the goddess of victory, which may well have come from some ship; see G. Hanfmann, "A Roman Victory," *Opus Nobile: Festschrift zum 60. Geburtstag von Ulf Jantzen* (Wiesbaden 1969) 63-67.

[5] For another example, see Alfieri, *op. cit.* (ONE, note 20) fig. 6.

We do not know whether names were also indicated in writing. In an age that had no telescopes, a simple bold device would seem to be more useful, such as that which enabled spectators standing on shore to identify individual galleys during the Battle of Salamis (see Herodotus, cited in note 2 above). Ships named *"Swift"* or *"Foresight"* or the like must have been given some conventionalized emblem that made possible easy recognition. The *stylis* occasionally bore a written name (see note 10 below), but it seems to have referred to a ship's guardian deity rather than its name.

A skiff carved on a gravestone (3rd A.D.; found at Carnuntum) bears the name *Felix Itala*; see A. Schober, *Die römischen Grabsteine von Noricum und Pannonien* (Sonderschriften des österreichischen archäologischen Institutes in Wien, Bd x, Vienna 1923), p. 47, no. 100 and fig. 43 (= Moll B IV 43). This may be just a way of identifying the boat rather than a record of what was actually written on it. The same is true of the plaques bearing names that are sometimes included in pictures of boats; see, e.g., the plaque with the name *Europa* in the graffito cited in NINE, note 34, and the two small boats in a mural on a Roman building that are identified by plaques as the *Nike* "Victory" and the *Lakena* (phonetic spelling for *Lakaina*) "Spartan" (Jacopi, *op. cit.* [TEN, note 49] 61, 63).

(Fig. 146),[6] which could be related in theme to the name-device.[7] The poop, too, was where were placed ornamental representations of deities or other creatures that served as common markings for the galleys belonging to a given state.[8] This practice was carried down to the fifth century B.C. at least; thereafter, Greek galleys either substituted or, in addition, carried the *stylis* (Figs. 108-110, 114, 129). The *stylis* was a pole, generally fitted with a short crosspiece, which was set up alongside the *aphlaston* and which bore either a device symbolizing the guardian deity of the ship[9] or his name written out.[10] The *stylis* found scant favor among the Romans.[11] Republican galleys

[6] Cf. Ovid, *Her.* 16.114: *accipit et pictos puppis adunca deos* "and the curved stern receives pictures of the gods," and Plutarch, *Mor.* 248a (cited in note 1 above).

[7] Cf. Hippocrates, *Epist.* 14: οἶδα . . . τὴν ναῦν ἐκείνην, "Αλιος ἐπιγραφὴ ἦν αὐτῇ "I have seen . . . that ship; Helios [the sun god, identical with Apollo] was the painted device on it." In *Epist.* 17 he refers to the same vessel: ὡς ἀληθέως τὴν 'Ασκληπιάδα νῆα, ᾗ πρόσθες μετὰ τοῦ 'Αλίου ἐπίσημον καὶ 'Υγιείην "truly the Ship of Asklepios—one to which you must add, along with Helios, Hygieia as a device [*episemon*]." The ship, then, bore as device Helios, father of Asklepios, god of medicine, and Hippocrates banteringly suggests the addition of a device in the form of Hygieia, goddess of health. The expression "Ship of Asklepios" may well imply that the latter was somewhere included in the decoration.

[8] Thus Euripides, in describing the Greek galleys at Troy, assigns each contingent a common decoration, mounted on the stern, which served to stamp all within the same contingent. E.g., *IA* 239-41: "Nereids in gilded effigies stood on the tips of the sterns, the mark of Achilles' squadron" (χρυσέαις δ' εἰκόσιν / κατ' ἄκρα Νηρῇδες / ἕστασαν θεαὶ / πρύμναις σῆμ' 'Αχιλλείου στρατοῦ); *IA* 246-50; "the son of Theseus leading sixty ships . . . with the goddess Athena set in a winged chariot with horses" (ἄγων / ἑξήκοντα ναῦς ὁ Θησέως / παῖς . . . θεὰν / Παλλάδ' ἐν μωνύχοις ἔχων πτερωτοῖσιν ἅρμασιν θετόν); see also *IA* 251-58, 273-76. And Aristophanes, in describing the myriad activities connected with the fitting out of a fleet (*Acharnians* 544-54), includes "the gilding of Athena-statues" (547: Παλλαδίων χρυσουμένων). Like the ornaments Euripides describes, these would serve to distinguish all units of Athens' fleet.

[9] The identification of the *stylis* was confirmed by J. Svoronos, "Stylides, ancres hierae, aphlasta, stoloi, acrostolia, embola, proembola et totems marins," *Journal international d'archéologie numismatique* 16 (1914) 81-152, esp. 98-103. Svoronos concludes (100-101) that the symbol was of the tutelary deity of the ship, and (104-20) identifies more or less satisfactorily a good number of types.

[10] A Greek vase-painting of the early 3rd B.C. shows the stern of a galley with a *stylis* on which is inscribed Ζεὺς Σωτήρ "Zeus, the Savior"; see *JDI* 3 (1888) 229 = *GOS*, Clas. 9. As F. v. Duhn, who published the piece, points out (232), until Hellenistic times ships always bore names feminine in form and never of any of the great divinities (cf. 351 below), so the inscription must refer to a guardian deity. Svoronos (91) adds other examples from representations on coins.

[11] Svoronos, *op. cit.* (note 9 above) 84, provides a chart of what he considers

generally carried nothing (Figs. 120-121, 131, but cf. 129), while units of the Imperial navy were given regular standards (Fig. 123), along with a *tutela*,[12] as the guardian deity was called in Latin. Indeed, an image of the *tutela*, prominently placed far aft (see, e.g., the Victory with a wreath in Fig. 146), was an essential feature of every Roman vessel, merchantman as well as war galley.[13] Ships were sometimes named after the guardian deity, sometimes not.[14]

The *tutela* stood near the goose-head, the ancients' preferred motif for finishing off the sternpost. Though the goose-head sometimes appears on galleys (Fig. 139), including war galleys (Fig. 128), it

---

examples of *stylides*. In it he includes a series of square banners, all taken from representations of Roman date. This is a mistake: such banners were in general use, in the Roman army, for example, and on Roman merchantmen, whereas the *stylis* proper—the pole with crosspiece—is the distinctive mark of the oared warship.

[12] A sailor in the Alexandrian squadron identifies his unit with the words *tutela Tauro* "with the Bull-*tutela*" (*CPL* 223). Since it was standard practice for sailors to identify their ship by name, presumably in this case name and *tutela* were identical. A number of ships named *Taurus* are known (357 below).

[13] Cf. Petronius 105: two passengers who had committed an act that was taboo were to be whipped to "appease the guardian deity of the ship" (*ut tutela navis expiaretur*). For the location, see Petronius 108: the helmsman, from his station just forward of the sternpost, persuades two parties locked in conflict to stop fighting and arrange a truce and one holds forth "an olive branch plucked from [sc. the statue of] the vessel's guardian deity" (*ramum oleae a tutela navigii raptum*). Cf. Vergil, *Aen.* 10.171 (*aurato fulgebat Apolline puppis* "The poop gleamed with a gilded Apollo"), Silius Italicus 14.410 (*numen erat celsae puppis Lucrina Dione* "Dione [Venus] of the Lucrine Lake was the deity on the lofty poop"). These tutelary carvings were often quite elaborate; cf. Seneca, *Epist.* 76.13: *navis bona dicitur non quae pretiosis coloribus picta est nec . . . cuius tutela ebore caelata est* "A ship is not said to be good because of the costly colors of its decoration nor . . . because its guardian deity is carved of ivory."

[14] The merchantman in Fig. 144 has a carving of Bacchus as *parasemon* on the stempost and a victory holding a wreath as *tutela* on the stern. This *tutela* must have been a favorite, for it appears on a number of other ships: two mosaics from Tunis of the 3rd A.D. (Foucher, fig. 4; Foucher, *op. cit.* [ELEVEN, note 24] 85, 91) and a wall-painting from Pompeii (R. Hinks, *Catalogue of the Greek, Etruscan, and Roman Paintings and Mosaics in the British Museum*, London 1933, pl. 5).

Ovid, *Tristia* 1.10.1-2, describes a vessel that apparently was named after the *tutela*: *est mihi (sitque precor), flavae tutela Minervae / navis et a picta casside nomen habet* "My guardian deity is yellow-haired Minerva—and may she be such, I pray— and my ship takes its name from a painted helmet [sc. a helmeted bust of Minerva painted on the bows]." On the other hand, Hiero's great freighter, the *Syracusia*, was fitted with a chapel to Aphrodite, presumably the guardian deity (Athenaeus 5.207e, cited in NINE, App., pt. 4).

was the decoration par excellence of merchantmen (Figs. 144-145, 147, 149-151, 156), often gilded for added effect.[15] Plain working boats or small craft could do without this traditional ornament (Figs. 142-143, 148) or could substitute a pennant on a short pole socketed into the top of the sternpost (Fig. 147, center ship).[16]

Other pennants (Figs. 156, 191), or banners on horizontal bars (Figs. 151, 154), could be flown from poles set up near the sternpost.[17]

II NAMES

WARSHIPS: THE EGYPTIAN FLEETS IN NEW KINGDOM TIMES

Egyptian records offer proof—if proof is needed—of how very old is the custom of giving names to ships. We know over three dozen borne by galleys in Egypt's navy, the earliest going back to the first ruler of the New Kingdom, Ahmose, who ascended the throne in 1567 B.C.[18]

In naming their warcraft, the Egyptians, as we might expect, most often sought to honor their god-king. Many boats were named directly after a pharaoh and one of his attributes or virtues (e.g., *Amenhotep II Who Made Strong the Two Lands*,[19] *Ramses II Who Propi-*

---

[15] On the *Isis* (NINE, App., pt. 2) "The sternpost rose in a gradual curve with a gilded goose-head set on the tip of it" (Lucian, *Navig.* 5: ἡ πρύμνα μὲν ἐπανέστηκεν ἠρέμα καμπύλη χρυσοῦν χηνίσκον ἐπικειμένη); cf. Artemidorus 2.23, cited in THIRTEEN, note 67; Lucian, *Jup. Trag.* 47 and *Vera Hist.* 2.41. Apuleius describes (*Met.* 11.16) a ship whose "backward curving stern with its goose-head, clothed in gilded foil, was all agleam" (*puppis intorta chenisco bracteis aureis vestita fulgebat*). Thus the Greek term *cheniskos* was taken over by the Romans in the Latinized form *cheniscus*. The ornament long remained in favor, for a papyrus document of A.D. 581 (*P. Mon.* 4.9-10) describes a Nile boat as ἀγριοχηνοπρύμνης "wild-goose-sterned."

[16] For another example of the short pole, see NINE, note 30.

[17] For other examples of such poles, see Foucher, figs. 2, 4. The ship pictured in Fig. 149 has pennants on the yardarms as well. These may represent signal flags; cf. 246 above.

[18] A preliminary list of Egyptian ship-names was assembled by W. Spiegelberg in *Rechnungen aus der Zeit Setis I mit anderen Rechnungen des Neuen Reiches* (Strassburg 1896) 81-86. Many more are to be found in the list of documentary sources given by A. Schulman in *Military Rank, Title, and Organization in the Egyptian New Kingdom* (Münchner ägyptologische Studien 6, Berlin 1964) 87-169.

[19] Spiegelberg no. 7.

tiates the Aton,[20] *Merenptah Beloved of Sakhmet*[21]). Sometimes a name celebrates just the virtue or attribute (*Appearing in Truth,*[22] *Strong of Appearance,*[23] *Overthrower of the Evil Ones*[24]). Occasionally it refers to a pharaoh's favor in a given locale (*Star in Memphis,*[25] *Star in Both Lands*[26]), more often to his favor with the heavenly gods (*Beloved of Amun,*[27] *Ptah Is Before Him,*[28] *Pacifier of the Aton*[29]).

Now and then the deities themselves were commemorated (*Amun of Front of Beauty,*[30] *Face of Re,*[31] *The Aton*[32]), as well as the land that was their particular concern (*Egypt Is Born,*[33] *Life and Happiness and Health Has Befallen Egypt*[34]). A few boats were named after animals (*The Fishes,*[35] *Bull in Nubia,*[36] *The Cow*[37]). Only exceptionally did the Egyptians follow the practice, so common in later times, of using a name that referred to the craft itself; *Ship of the North*[38] is the sole example I have been able to find.

[20] Spiegelberg no. 16.

[21] Schulman no. 387a. Other examples: *Amenhotep II Appears in the Sun* (?) *Boat* (Schulman no. 23); *Amenhotep II Is Firmly Established* (Spiegelberg no. 7a); *Aakheperure Endures* (Schulman no. 493b); *Thutmose III Tramples Syria* (Schulman no. 495i).

[22] Spiegelberg no. 9; Schulman nos. 321a, 325b, 494f and h, 495b.

[23] Schulman no. 495d.

[24] Schulman no. 495a. Cf. *Powerful Ruler* (Schulman no. 495b), *Repeller of ?-Land* (Schulman no. 495h).

[25] Spiegelberg no. 22; Schulman nos. 325a, c, d, 495b. Others: *Appearing in Memphis* (Spiegelberg no. 3), *Star of Thebes* (Schulman no. 510a).

[26] Spiegelberg no. 4.

[27] Spiegelberg nos. 5, 8, 12, 19; Schulman nos. 323 b-f, 335b, 494d, e, i, 495e, k, l. Cf. *Beloved of Re* (Spiegelberg no. 20, Schulman no. 335a), *Beloved of Amun-Re* (Schulman no. 499c).

[28] Spiegelberg no. 18; Schulman nos. 323a, 493a, 494g.

[29] Schulman no. 202. Cf. *Glittering Like the Aton* (Spiegelberg no. 15, Schulman no. 501a), *Pacifier* (Schulman no. 500c), *Pacifier of the Gods of the Residence* (Schulman no. 120).

[30] Schulman no. 495c. Cf. *Front of Beauty* (Schulman no. 494b).

[31] Schulman no. 495j.

[32] Schulman no. 509b. Also *The Aton Glitters* (Spiegelberg no. 10; Schulman nos. 495b, 509a).

[33] Schulman no. 501b.

[34] Spiegelberg no. 11.

[35] Schulman no. 494a.

[36] Schulman no. 495f.

[37] Säve-Söderbergh, *op. cit.* (Two, note 37) 75.

[38] Spiegelberg no. 2.

Warships: The Athenian Fleet in the Fourth Century b.c.

The earliest name known for a Greek ship is *Argo* "Swift," the galley built and commanded by Jason, whose exploits predate the tale of Troy. Unfortunately it is the only ship-name recorded till practically the end of the fifth century b.c. Then, for the half-century from 377-322 b.c., the Athenian naval lists preserve about 300 names given to galleys in the fleet.[39]

Practically all the ships involved are triremes. A handful of quadriremes are represented, enough to reveal that no attempt was made to distinguish the bigger craft by a distinctive nomenclature.[40] Indeed, the christeners seem to have been more interested in perpetuating favored names than in practical considerations, for there are numerous instances of ships with the same name in the fleet at the

[39] The most complete and detailed study of these names is K. Schmidt's "Die Namen der attischen Kriegsschiffe" (doctoral dissertation, Leipzig 1931). My conclusions are based largely on Schmidt's analyses.

A more easily accessible list is to be found in Miltner 947-52. The following corrections and additions must be made to it:

DELETE

'Αργεία, 'Αρπαλείω, 'Αχιλλεύς, Εὐπρία, Θεαρίς, Κλείω Ποτάμιος. On these deletions, see H. Gaebler, *Die antiken Münzen von Makedonia und Paionia* (*Die antiken Münzen Nord-Griechenlands* III.2, Berlin 1935) 183 note.

ADD

1. Names omitted by inadvertence: Εὐημερία 1611.429, Λευκή 1624.85, Φιλονίκη 1611.433. (Schmidt's no. 214 "Πολυχαρίστη" is *not* to be added; it is a nonexistent name that Schmidt included in error.)

2. Names published subsequently from a fragment, dated to the second half of the 5th b.c. (*IG* II² 1604a, on p. 811): Ἄγλαυρος, 'Αλκμήνη, 'Αρπαγή, 'Αφετή, Γοργοφόνη, 'Ιδαία, Λεοντῆ, Πρωτόπλους. The names Σάλπιγξ, Φήμη, Ναύκρατις, ['Ελευ]-θερία, already attested, also appear on this fragment.

3. Names published subsequently from a fragment of the 4th b.c. (*SEG* x 355): 'Επινομή, 'Εορτή, [Κ]ατα[πλέουσα?]. The name Πανοπλία, already attested, also appears on this fragment.

[40] See *Aktis, Anysis, Achilleia, Eueteria, Eucharis, Hegemonia, Hikane, Kratousa, Nikephoros, Paralia, Petomene, Prote, Salaminia, Salpinx* in Miltner's list; all were borne by both triremes and quadriremes. *Nikosa*, which Miltner lists only as a quadrireme, probably was the name of a trireme in 1612.14. In 357/6 b.c. there was a trireme named *Axionike*; in 325/4 a second *Axionike* appears, this time a horse transport. The *Asklepias* and *Gnome*, after outliving their usefulness as triremes (1628.471, 466), were converted to horse transports without change of name (1629.735, 730; 1631.108, 105).

same time[41]—so many that, to avoid confusion, entries in the records identify units not only by name but by builder as well.[42]

Argo is masculine in gender, but this is exceptional. To the fourth century Greek as to us, ships were female, for he limited himself to names that were feminine in form. In view of the variety he achieved, this obviously had little inhibiting effect. Unlike the Egyptians with their predilection for the names of their god-king and other major deities, he deliberately avoided those of ranking divinities; only adjectival derivatives (e.g., *Artemisia* "Ship of Artemis," *Aphrodisia* "Ship of Aphrodite") or epithets or cult titles (e.g., *Polias* "Guardian of the City," an epithet of Athena, *Tauropole* "Huntress of Bulls," an epithet of Artemis) were allowed.[43] On the other hand, lesser deities or figures from mythology were a rich source, accounting for almost one out of every four names recorded. The selection of these was never haphazard. Names of sea nymphs such as *Nereis, Thetis, Amphitrite* need no explanation. *Aigle*, meaning literally "luster," *Phoibe* "bright," *Phaethousa* "shining," and the like have to do with one aspect of a ship's appearance; *Charis*, meaning literally "grace," *Hebe* "youth," *Aglaia* "beauty," etc., with another; *Atalante, Iris,*[44] *aut sim.* have reference to its speed; *Achilleia, Aianteia,* to its warlike qualities; *Danae, Alkmene, Prokne* and other heroines point to cour-

---

[41] Cf. Schmidt 3-4, 13. E.g., in the year 357/6, two triremes bearing the name *Nike* appear in the records (1611.81, 82).

[42] First attested in 1605.4 (377/6 B.C.). Then, from 1609 (370/69) on, with increasing frequency.

[43] For references for these names and all mentioned subsequently, see Miltner's list.

The proscription against using the names of major gods apparently did not extend to those that were not Greek. The navy had a dispatch galley named *Ammonias* "Ship of Ammon," which probably conveyed missions to Cyrene en route to the shrine of Ammon. It is mentioned by Aristotle in the *Pol. Ath.*, written between 329 and 322 B.C. (61.7: ταμίαν τῆς Παράλου καὶ . . . τῆς τοῦ Ἄμμωνος "a treasurer for the *Paralos* . . . and for the ship of Ammon"; for the form of the name, see Photius, s.v. Πάραλος and Pliny, *NH* 35.101).

The sacred trireme the Corinthians in 344 B.C. equipped and named after the "two goddesses," i.e. Demeter and Persephone (Plutarch, *Timoleon* 8.1), may have borne adjectival forms of the names or may actually have been called *The Two Goddesses*.

[44] Spelled Ἐλπις; see Schmidt 57.

age under hardship; the healing deities such as *Hygieia* and *Iaso*, to beneficial services rendered.

Very much the same considerations lay behind the choice of common nouns and adjectives that served as names. E.g., *Lampra*, literally "gleaming," *Chryse* "golden," *Aktis* "ray" had to do with one aspect of the ship's appearance; *Anthousa* "blooming," *Nea* "new," *Parthenos* "maiden," *Theama* "spectacle," to another; *Agreuousa* "hunting," *Okeia* "quick," *Petomene* "flying" have reference to its speed; *Amynomene* "self-defending," *Sobe* "frightening," *Andreia* "courage," *Thraseia* "boldness," *Dynamis* "power," *Syntaxis* "battle order," to its warlike qualities; *Sozousa* "saving," *Soteria* "safety," *Sosipolis* "saving the city," *Boetheia* "aid"—a favored name[45]—*Hegemonia* "leadership," *Symmachia* "alliance," to its services. And a whole host of names recall the ship's ability, its sureness to succeed. A great favorite was *Nike* "victory";[46] in addition there were *Nikephoros* "bringer of victory," *Nikosa* "winning," *Agathonike* "good victory," *Aristonike* "best victory," as well as *Kratousa* "conquering," *Stephanoumene* "crowned," *Prote* "first," *Protoplous* "first sailer," *Eukleia* "fame." Sometimes the emphasis is on the luck the ship is blessed with, as in names like *Eutyches* "lucky," *Eudaimonia* "prosperousness," *Euploia* "bon voyage," *Apobasis* "disembarcation," *Aura* "fair wind."

Honoring the ships' qualities also explains why many were called

[45] Four names are distinguished by having been given to four different galleys within 55 years or less:

1. *Boetheia*. Four within a period of 35 years: one built by Epicharides, 1615.67 (358/7 B.C.); a second built by Archeneides, 1611.128, 140, 170; a third built by Smicrion, 1631.445; a fourth was one of the two ships handed over by Harpalus, 1632.122 and cf. 1631.170 (323/2).

2. *Demokratia*. One listed in 1604.24 (377/6), 1607.46, 1611.86; a second, with different gear, listed in 1606.59, 1607.87; a third built by Hagnodemus, 1620.32, 60, 63; the fourth was built by Chaerestratus, 1623.326, 1628.107, 1629.603, 1631.193.

3. *Nike*. One listed in 1604.83; a second built by Pistocrates, 1609.93; a third built by Aristocrates, 1624.69; a fourth built by Chaerestratus, 1632.11.

4. *Salaminia*. Four within a period of 35 years: one, a second-class ship, listed in 1611.95, 1612.26 (356/5); a second, a select ship, listed in 1611.164, 183 and in 1612.40; a third, built by Archenides, 1629.750, 853 and 1631.119, 206; the fourth was a quadrireme built by Charetis, 1632.103.

[46] See previous note.

after certain animals, both actual (*Halcyon* "kingfisher," *Dorkas* "gazelle," *Leaina* "lioness," *Lykaina* "she-wolf") and mythological (*Kentaura* "centaur," *Hippokampe* "[life-sized] seahorse"), or after certain weapons (*Aichme* "spear," *Lonche* "lance," *Oistos* "dart," *Sphendone* "sling," *Salpinx* "trumpet").

Rather different in import is a series of abstract nouns which seem to celebrate the qualities of the city and people that the ships in the fleet defended: *Eunomia* "law and order," *Themis* "right," *Demokratia* "democracy" (a particular favorite),[47] *Dikaiosyne* "justice," *Eleutheria* "freedom," *Eirene* "peace," *Pronoia* "foresight," *Sophia* "wisdom," *Mneme* "memory." Others commemorate the city's culture (*Techne* "art," *Tragodia* "tragedy," *Komodia* "comedy," etc.) or the great traditional festivals (*Olympias* "Ship of the Olympic Games," *Nemeas* "Ship of the Nemean Games," *Pompe* "procession").

About ten percent of the ships bore geographical names. The famous dispatch galleys, the *Salaminia*[48] and *Paralia* (as the *Paralos* is called in the documents), recall, the first the island which saw Athens' great naval victory over Persia, the second the shores the ship visited. Other geographical names celebrate the districts of Athens (*Sounias, Eleusis, Pallenis*, etc.) or lands (*Ionike, Hellas, Europe*; these, being derived from mythological figures, offered another dimension as well). Still others were chosen for religious reasons: *Delias* commemorated Delos, the island sacred to Apollo; *Delphis*, Delphi, seat of the famous oracle; *Idaia*, Mount Ida where Cybele lived. Some were mythological: *Aithiopia* was to the Greeks a mystic far-off place, while *Erytheia* was where the three-headed monster Geryon lived.

Certain names clearly had a political import: *Krete* recalled the island with which Athens had important contacts, and *Naukratis* the chief trading city in Egypt; the *Persis* was a reminder of Athens' great defeated enemy, Persia; *Enna*, a key town of Sicily, may be the product of a visit from Sicily's ruler, Dionysius the Elder; *Simaitha*,

[47] See note 45 above.   [48] See note 45 above.

353

named after a stream just north of Syracuse, perhaps recalls Athens' monumental naval venture, the expedition against Syracuse; *Amphipolis* celebrated Athens' rich colony in Thrace.

Lastly, a few galleys were named after their function. E.g., a horse transport was appropriately called *Hipparche* "queen of the horses," and the *Strategis* "flagship" surely performed that role.[49]

HELLENISTIC AND ROMAN WARSHIPS

No galley in the Athenian navy during the period just discussed was named after a person. Yet one entry, dating from the end of the period, is significant. A certain Aristonikos somehow enabled the navy to gain possession of a galley by confiscation. The ship was entered in the records under the name *Aristonike*.[50] This is a proper feminine form, as well as being a name of a familiar type—it means "best victory"; however, the circumstances argue strongly that it was deliberately chosen as a gesture toward the man who was responsible for the ship's presence in the navy. After the death of Alexander in 323 B.C., the Greek world came under the sway of the great Hellenistic rulers, the Antigonids in Greece, the Seleucids in Asia Minor, the Ptolemies in Egypt. Athens had no scruples about naming galleys after these all-powerful personages: the fleet now included units called the *Antigonis*, the *Demetrias*, the *Ptolemais*.[51] Later on, at Actium, Cleopatra put her flag aboard the *Antonias*.[52] By the first century B.C. the aversion to naming ships after important deities had also disappeared; in 69 B.C., one of the *dikrotoi* commandeered by Rome from the Greek maritime cities of Asia Minor was named the *Athena*.[53]

---

[49] Similarly *La Capitana* carried the flag in the fleet of Cosimo di Medici and *La Magistrale* in the fleet of the Knights of Malta.

[50] In 1632.189 the *Aristonike* is listed with no builder's name and with the note ἣν ἔφηνεν ᾿Αριστόνικος Μαραθώνιος "which Aristonikos of Marathon reported to the authorities." Perhaps it was a foreign ship which Athens was able to claim on some legality; cf. Schmidt 51, 92.

[51] Photius, s.v. Πάραλος (*Antigonis, Demetrias*); Schol. in Demosthenes, p. 637 Dindorf (*Ptolemais*).

[52] Plutarch, *Ant.* 60.3.

[53] It was manned by a mixed crew of sailors from Miletus and Smyrna; see *Inscr.*

For the whole of the Hellenistic period, we know hardly more than two dozen names,[54] including those just mentioned. They are enough, however, to reveal that, aside from the new practice of honoring rulers, men-of-war continued to be named in much the same ways. *Cypris* and *Phosphoros* are cult titles of Aphrodite; *Pallas*, of Athena. *Herakles* and *Asklepios* recall two deities honored as benefactors of mankind and enjoying widespread worship at this time. *Parthenos* "maiden" and *Neotes* "newness" celebrate the ship's appearance; *Thera* "hunt," its speed; *Alka* "might," its strength; *Soteira* "savior," its service; while *Eukleia* and *Phama* "fame," *Laonika* "victory of the people," *Nika* "victory," *Prota* "first" proclaim its certain success.[55] *Leontophoros* "lion-bearer"[56] exemplifies an animal name; *Eleutheria* "freedom" and *Eunomia* "law and order" recall desirable qualities of a city; *Isthmia* "Ship of the Isthmian Games" and *Komos* "revel" commemorate the festivals so important in Greek city-state life. There is even an example of a geographical name, *Korkyra*.

FOR THE Roman period, the evidence is much richer. Sailors of the Imperial navy, often recorded their ships on their tombstones, and, since many of these have been found, we know over 80 names of

---

de Délos 1856, a commemorative erected by "the Milesians on campaign together in the *dikrotos* named Athena" (οἱ συνστρατευσάμενοι Μιλησίων ἐν νηὶ δικρότῳ ᾗ ἐπιγραφὴ 'Αθηνᾶ) and 1857, a commemorative with the same wording save for οἱ συνστρατευσάμενοι Ζμυρναίων "the Smyrnaeans on campaign together." Another *dikrotos*, manned entirely by Milesians, was named *Parthenos* (*Inscr. de Délos* 1855).

[54] For *Parthenos*, see the previous note; for *Isthmia*, Plutarch, *Mor.* 676d; for *Herakles* and *Asklepios, Riv. di fil.* 61 (1933) 365-66, 369 (inscription of ca. 200 B.C.). The others all appear, along with warship prows, on coins of Corcyra, dating 300-229 B.C. P. Gardiner, *A Catalogue of the Greek Coins in the British Museum, Thessaly to Aetolia* (London 1883) 129-33 and note on p. xlix, suggested that these were ship-names, probably honoring the winners in rowing races, a suggestion amply confirmed by the similarity of the names to those given to Athenian galleys. *Soteira* also appears as the name of a quadrireme which carried some of Caesar's soldiers to North Africa in 46 B.C. (*Syll.*³ 763).

[55] The names ending in *-a* stem from an area that spoke a dialect different from Athens; the Athenian equivalent would end in *-e*.

[56] See Six, note 52.

galleys, particularly those attached to the two major fleets based at Misenum on the north of the Bay of Naples and at Ravenna.[57]

The names reveal that in this, as in other matters connected with the sea, the Romans for the most part followed in their predecessors' footsteps. Like the Athenians, in assigning names no attempt was

[57] For references for all names mentioned, see Miltner's list, 952-56. The list, unfortunately, is neither complete nor accurate.

The following corrections must be made: s.v. *Aesculapius*, read 2833 for 2836; *Annona*, read 3495 for 3459; *Apollo*, read 3527 for 3572; delete *Arcinice* and *Armenia* (see note 60 below); *Augusta* should read *Augustus* (see x 3450), read 3649 for 3694, and add xi 46 after vi 3151; *Capricornus*, read 3095 for 3059; *Clupeus*, read 1956 for 1965; *Concordia*, read vi 3094 and 3144 for x 3094, 3144; *Fides*, insert x after 1603, 15, and add x 3436; delete *Galeata*; delete *Maia* (the proper reading in x 3507 is *Marte*; see *AE* 1949.210); *Mars*, insert 3507 before 3584; *Mercurius*, after 239, add *Eph. epig.* viii, p. 116, no. 444; *Neptunus*, read 3656 for 3635; *Perseus*, change v to vi; *Salvia*, change xi to vi; after *Triumphus*, add the liburnian *Varvar(ina)* xi 104; *Virtus*, change no. 12 to 126, and the vessel involved is a liburnian, not a trireme.

The following additions must be made:

ships belonging to the fleet at Misenum:
quadriremes: *Dacicus* (Jalabert-Mouterde 1167), *Minerva* (*AE* 1949.206), *Venus* (Jal.-Mout. 1178)
triremes: *Augustus* (Jal.-Mout. 1172), *Fides* (*AE* 1949.208), *Fortuna* (*AE* 1946.145, Jal.-Mout. 1171), *Liber Pater* (*CPL* 120, restored in Jal.-Mout. 1168), *Pax* (Jal.-Mout. 1159), *Pietas* (Jal.-Mout. 1165), *Providentia* (*AE* 1929.142, *CPL* 120), *Salamina* (*AE* 1946.146), *Salus* (*CPL* 120), *Taurus* (*AE* 1929.147, Jal.-Mout. 1162), *Tigris* (*CPL* 120), *Venus* (*AE* 1965.145), *Vesta* (*AE* 1929.146), *Virtus* (*CPL* 120)
liburnians: *Virtus* (Jal.-Mout. 1162)
unclassified: *AE* 1964.103 mentions a *Minervia rate*; Jal.-Mout. 1158 mentions a *Jupiter*.
Ships belonging to the fleet at Ravenna:
liburnians: *Satura* (THIRTEEN, note 59).
Ships belonging to the squadron at Alexandria:
liburnians: *Draco* (*CPL* 210), *Fides* (*CPL* 191), *Lupa* (*BGU* 741.7-8 = M. Chrest. 244, *BGU* 709.23 [reading Λούππας for Τουντας; the ship is the same as in *BGU* 741.7-8]), *Neptunus* (*CPL* 250), *Mercurius* (*CPL* 125), *Sol* (*BGU* 455). Miltner had included only one ship of this squadron, the *Taurus* (*CPL* 223; probably but not certainly a liburnian).
The Syrian squadron: a liburnian *Capricornus* (Jal.-Mout. 1163 and possibly 1174).
The Moesian squadron: liburnians *Armata* (*AE* 1950.175), *Sagita* (*AE* 1967.429).
Fleet or squadron not mentioned: liburnian *Augustus* (*CIL* v 1048, reading *tr(ierarchi) de lib(urna) Aug(usto)*; see Panciera, *op. cit.* [THIRTEEN, App., note 13] 325), liburnian *Sphinx* (Panciera, *op. cit.* 329, note 55), *Triptolemus* (*AE* 1966.97, size unknown).

Of the above names, *Draco, Lupa, Sagit⟨t⟩a* and *Sphinx* had hitherto been unattested.

356

made to distinguish either the type of unit or the units of a fleet: liburnians, triremes, quadriremes, in the same or in different fleets, all bear the same kinds of name and not a few the same name.[58] And half a millennium left many of the grounds for selection unchanged. The Roman navy, too, went in for mythological figures and lesser deities: Venus' son *Cupido*, the sea-nymph *Nereis*, the merman *Triton*, the forest god *Silvanus*, the goat-footed *Satyra*[59] were noted for lightness or quickness of movement; *Castor* and *Pollux, Diomedes, Perseus* were all doughty fighters; *Danae* was a courageous heroine; *Hercules* and *Triptolemus* (who brought agriculture to the earth) performed special services for mankind. Common nouns and adjectives commemorated the ship's looks (*Juventus* "youth," *Radians* "gleaming"), speed (*Pinnata* "winged"), readiness for war (*Armata* "armed"), sureness to achieve success (*Triumphus, Victoria, Fortuna*). The Romans, too, adopted the names of appropriate animals, both real (*Aquila* "eagle," *Crocodilus, Lupa* "she-wolf," *Murena* [a type of Mediterranean fish], *Taurus* "bull," *Taurus Ruber* "red bull") and mythological (*Draco* "dragon," *Gryps* "griffin"), and of instruments of war (*Clypeus* "shield," *Quadriga* "chariot").[60] And numerous ships bore names that celebrated the beneficial qualities of Roman rule: *Annona* "grain-supply," *Clementia, Concordia, Constantia, Fides, Justitia, Libertas, Pax, Pietas* "sense of duty," *Providentia* "foresight," *Salus* "soundness," *Salvia* "saver," *Spes* "hope," *Virtus* "courage." To this same category belong the Greek names *Euandria* or *Euandria Sebasta* (= *Virtus Augusta*)

[58] E.g., *Neptunus* was the name of a trireme and a liburnian in the fleet at Misenum, and of a quadrireme and a trireme in that at Ravenna; *Minerva* was the name of a quadrireme, a trireme, and a liburnian in the fleet at Misenum, and of a trireme in that at Ravenna.

[59] For the feminine form, see Lucretius 4.1169.

[60] A trireme *Arci*( ) occurs in XI 3735 and *Arcin*[ ] in XI 100. Mommsen suggested a Latin-Greek hybrid *"Archinice,"* but such a name, being utterly without parallel, can hardly be right. In XI 102 a trireme is mentioned whose name the editor restores as *Ar*[*me*]*na*, but that cannot be right since the Roman navy used no Greek common words as ship-names. Miltner offers *Armenia*, but with no evidence to support the conjecture. I suggest *Ar*[*ci*]*na*; in other words, all three refer to the same ship-name, *Arcina*, a derivative perhaps of *arx* "citadel" or *arca* "chest."

357

and *Eirena* or *Eirena Sebasta* ($=$ *Pax Augusta*) which were borne by the *triemioliai* that Rome allowed Rhodes to maintain.[61]

Geography also served the Imperial navy but here differences from Greek practice begin to appear. A good many in this category are of bodies of water, such as *Oceanus*, and particularly rivers; practically all the major rivers in the Roman Empire had a ship or ships named after them (*Danuvius* "Danube," *Tigris* and *Eufrates, Nilus, Padus* "Po," *Rhenus* "Rhine," *Tiberis*). The rest are commemorative: *Dacicus* and *Parthicus* must commemorate the wars in Dacia and Parthia; *Phryx* "Phrygian" may refer to the legend that Rome was founded by wanderers from defeated Troy in Phrygia; *Salamina* surely was named after the famous Athenian dispatch galley; *Athenonike* "Athena's victory" must certainly honor Athens' great days on the sea; the liburnian *Varvarina* was a reminder of one of the people from whom this type of craft was adopted.[62]

As we have seen, the Hellenistic world introduced the naming of ships after rulers. The Romans apparently limited themselves on this score to *Augustus*. However, another Hellenistic practice, that of calling ships after major deities, became standard in the Roman navy; indeed about one ship in five was so named. The Roman pantheon is fully represented: *Apollo* (also *Sol* "sun"), *Aesculapius, Ceres, Diana, Juno, Jupiter, Liber Pater, Mars, Mercurius, Minerva, Neptunus, Ops* (goddess of plenty), *Venus* (also *Luciferus*, the equivalent of the Greek *Phosphoros* "light-bringer," one of Venus' titles), *Vesta*. Two Egyptian deities appear, *Ammon*[63] and *Isis*; the latter was at this period particularly revered by sailors and seafarers.

In one key respect the Romans departed completely from Greek practice: no attempt was made to put ships' names into the feminine form; there are as many of the one gender as of the other.

[61] See Six, note 122. All the Rhodian *triemioliai* of the period (1st A.D.) carried these names save one, the *Polias* (a cult title of Athena) in *Eph. Arch.* (1913) 10, no. 9 $=$ Blinkenberg no. 37.

[62] See *RE* s.v. *Varvaria* and *Varvarini* (1955).

[63] Ammon had already qualified for Athens' navy; see note 43 above.

MERCHANTMEN

The monster grain carrier that Hiero II built about 240 B.C. (NINE, App., pt. 4) is almost the first merchant ship we know by name.[64] He called it *Syracusia* in honor of the city he ruled; when he donated it to Ptolemy of Egypt, the name was appropriately changed to *Alexandris*.

Nor do we have very many other examples from all the succeeding centuries. Three or four have been recorded purely by chance; somewhat over a dozen were included in prayers, scratched on rocks, for a bon voyage;[65] about the same number, all of Nile craft, have turned up in the business documents from Roman Egypt.

Few as they are, they indicate that most ships were named after deities. Understandably, those cherished by sailors, *Isis*[66] and *Dioscuri*,[67] were great favorites. A strong runnerup was *Asklepios*,[68] whose worship enjoyed an enormous vogue in the days of the Roman Empire; as the god of healing, he could be expected to keep a vessel sound. Since most of the ships whose names have been preserved belonged to Greeks, the Greek pantheon is well represented: *Apollo, Athena, Artemis, Aphrodite, Demeter, Dionysus, Hermes, Hestia, Nike, Poseidon.*[69] *Herakles* and *Tyche* "Lady Luck,"[70] though not of

[64] The *Halios* (= *Helios*), mentioned by Hippocrates (note 7 above), seems to be the earliest.

[65] Many of these have been collected by Sandberg.

[66] The name of a big freighter on the Alexandria-Rome run (Lucian, *Navig.* 5 [cf. NINE, App., pt. 2]), of a Tiber *caudicaria* (*IH* ill. 64), of a Nile boat (*PSI* 1048, 3rd A.D.). The related form *Isarion* was used for Nile boats (*P. Oxy.* 2415.82, 87 [3rd A.D.], two different boats in the same transport fleet) and *Isopharia*, the "Isis of Pharos," for a seagoing merchantman (see note 2 above). On Isis of Pharos, particularly dear to sailors, see A. Alföldi, "Die alexandrinischen Götter und die Vota Publica am Jahresbeginn," *Jahrbuch für Antike und Christentum* 8/9 (1965/66) 53-87, esp. 64-65.

[67] This was the name of the freighter that took St. Paul from Malta to Italy (see note 2 above). Other examples: *IG* v 1.1548, 1550 (= Sandberg nos. 11, 13).

[68] *IG* xii 8.582, 583 (= Sandberg nos. 40, 41); *SEG* xi 1020 (= Sandberg no. 10), xi 1024 (= Sandberg no. 28); Aristides, *Sermo Sacra B*, p. 303.

[69] Apollo, *P. Oxy.* 2415.25-26, 49 (Aphrodite and Apollo); for Halios, i.e. Helios, see Hippocrates, cited in note 7 above; Athena, *IG* v 1.1553 (= Sandberg no. 16), *SEG* xiv 342 (= Sandberg no. 22); Artemis, *IG* xii 8.585 (= Sandberg no. 43);

the original pantheon, had in this age become important figures. Occasionally the cult titles are used, such as *Pythios* and *Smintheus* for Apollo.[71] The Egyptian deities *Ammon* and *Serapis*,[72] worshipped at this time far beyond the boundaries of Egypt, are also represented, and Nile craft bear the names of lesser native figures, such as *Aigyptos*, the legendary founder of Egypt, and *Ibis*, the sacred bird.[73] From the Christian era we have several examples of ships called *Maria*.[74] There are very few names preserved of vessels belonging to Latin owners: possibly *Fortuna Redux* "good luck that brings one home";[75] *Jupiter* and *Juno*;[76] *Felix Itala*, perhaps the feminine form of *Italus*, legendary founder of Italy.[77] A few mythological names occur: *Aias* (?), *Europa*, *Pantomorphos* "all shapes" (= perhaps Proteus, the Old Man of the Sea).[78] And there are a handful that, appropriate as boat names, fit no particular category: *Eirene* "peace," *Chresmos* "oracle," *Halion Griphos* "sea-fishnet," *Lakaina* "Spartan," *Pontos* "sea," *Thalia* "abundance."[79]

Aphrodite, *SEG* xiv 344 (= Sandberg no. 27), *P. Oxy.* 2415.49 (Aphrodite and Apollo); Demeter, *IG* xii 8.584 (= Sandberg no. 42), *SEG* xi 1012 (= Sandberg no. 23); Dionysus, *SEG* xiv 341 (= Sandberg no. 21), *P. Oxy.* 2415.30; Hermes, *P. Oxy.* 2415.28; Hestia, *SEG* xiv 342 (= Sandberg no. 22); Nike, see note 5 above; Poseidon, *IG* xii 8.582 (= Sandberg no. 40).

[70] Heracles, *IG* xii 5.1.712.26, xii 8.581a (= Sandberg nos. 31, 39); Tyche, *P. Teb.* 486 (2nd/3rd A.D.).

[71] Pythios, *P. Oxy.* 2415.37, 41 (two different boats in the same transport fleet); Smintheus, *IG* xii 8.586 (= Sandberg no. 44).

[72] Ammon, *P. Oxy.* 2415.24, 34, 84 (three different boats in the same transport fleet); Serapis, *IG* xii 8.584 (= Sandberg no. 42). An Egyptian craft named *Nikastachtes* occurs in *Sammelb.* 977 (= Sandberg no. 46), A.D. 14. This may be a mistake for *Nikastarte*, i.e., "Victory of Astarte," the Asia Minor goddess identified with Aphrodite.

[73] Aigyptos, *Sammelb.* 9223.4 (2 B.C.); Ibis, *W. Chrest.* 443 (A.D. 15).

[74] *IG* v 1.1554 (= Sandberg no. 17), xii 5.1.712.56, 65, 75, 78.

[75] *AM* pl. 9b. An inscription over a picture of a ship, it may be the caption for the scene rather than a name.

[76] See note 2 above.

[77] See note 5 above.

[78] A[ía]s, *Sammelb.* 9223.2 (2 B.C.); *Europa*, see NINE, note 34; *Pantomorphos*, *W. Chrest.* 248 (A.D. 220/21).

[79] Eirene, *IG* v 1.1552 (= Sandberg no. 15); Chresmos, *P. Oxy.* 2415.39; Halion Griphos, *P. Teb.* 486; Lakaina, see note 5 above; Pontos, *P. Oxy.* 2415.43, Thalia, *P. Teb.* 486.

CHAPTER SIXTEEN

Harbors

THE HARBORS OF the ancient world are a subject that demands extended treatment. It took a sizable book to provide a satisfactory survey,[1] and whole volumes have been devoted to single harbors.[2] Moreover, in recent years, underwater archaeology has been steadily producing masses of new data.[3] This chapter can do little more than outline the general features of ancient harbors and indicate the bibliography where further information may be found.

I EARLY GREEK HARBORS

EGYPTIANS, Minoans, and the many others who sailed the Mediterranean before the coming of the Greeks no doubt had harbor installations of some sort. Of these, however, only the scantiest of traces have been found, and few which can be identified with complete certainty.[4] Yet, even if we rarely are able to establish the existence of Bronze Age moles or quays, this in no way affects the picture archaeology furnishes of active and far-flung commerce at this time. Readers of Richard Henry Dana's classic, *Two Years Before the Mast*, will recall that, as late as the mid-nineteenth century, a ship large enough to round the Horn could be loaded to the gunwales while lying off an open California beach. Throughout Greco-Roman antiquity, the craft that tramped from coastal town to coastal town

---

[1] K. Lehmann-Hartleben, *Die antiken Hafenanlagen des Mittelmeeres* (Klio, Beiheft 14, Leipzig 1923).

[2] A. Poidebard, *Un grand port disparu, Tyr* (Paris 1939); A. Poidebard and J. Lauffray, *Sidon, Aménagements antiques du port de Saïda* (Beirut 1951), including a short but useful history of the development of harbors by R. Mouterde; R. Bartoccini, *Il porto romano di Leptis Magna* (Bollettino del Centro Studi per la Storia dell' Architettura 13, supplemento al 1958).

[3] Cf. Taylor 160-78; *Archeologia, Tresors des Ages* 17 (July-August 1967) 12-17 (Anthedon), 20-22 (Sabratha), 23-24 (Thapsus), 25-29 (Athlit, just south of Haifa); G. Kapitän, "Sul Lakkios, porto piccolo di Siracusa del periodo greco," *Archivio storico Siracusano* 13-14 (1967-68) 167-80.

[4] Cf. Mouterde 16.

or between small islands often loaded and unloaded off beaches without benefit of harbor installations of any kind (Fig. 191); there is every reason to think that Bronze Age cargo carriers did the same. When Homer describes the harbor of the Phaeacians—who lived, as it were, in Shangri-la—he sings of a protective bay on which the only works of man are the shacks where each shipowner stored his sails and rigging, the stone bollards, probably sunk in the sand, to which ships could tie up, the area where sailmakers and riggers and shipwrights worked, and a shrine to Poseidon.[5]

In the eighth century B.C., man-made harbor arrangements stage a sudden and impressive debut. At Delos there dates from this time a mole 100 m. long made up of massive rough-hewn blocks of local granite.[6] In the next century, other sites join Delos in offering examples of these primitive but mighty moles.[7] And, by the end of the sixth B.C., Polycrates of Samos had built the sophisticated harbor works that Herodotus saw; it included two moles, one 370 m. long and the other 180, and the whole complex was kept safe from enemy attack by being included in the circuit of the city's defense wall.[8]

Thus, by the time Greece entered the Classical Age, her engineers had gone far in adding man's touches to whatever basic elements nature provided at a given port site. A harbor (*limen*) now boasted strong moles (*chomata*) to ensure a quiet and safe anchorage (*hormos*); it was equipped with the quays, open sheds, and warehouses needed for a commercial port (*emporion*) or the boathouses and gear

---

[5] *Od.* 6.263-69: "There was a fine harbor on either side of the city with a narrow entrance. The round-ended ships were drawn up as far as the road. Each and every one [of the owners] has his own gear shed. They have there a meeting place around a fine shrine to Poseidon . . . there they take care of the tackle of their black ships, the ropes and cables, and there they fashion the oars" (καλὸς δὲ λιμὴν ἑκάτερθε πόληος, / λεπτὴ δ' εἰσίθμη· νῆες δ' ὁδὸν ἀμφιέλισσαι / εἰρύαται· πᾶσιν γὰρ ἐπίστιόν ἐστιν ἑκάστῳ. / ἔνθα δέ τέ σφ' ἀγορή, καλὸν Ποσιδήϊον ἀμφίς / . . . ἔνθα δὲ νηῶν ὅπλα μελαινάων ἀλέγουσι, / πείσματα καὶ σπεῖρα, καὶ ἀποξύνουσιν ἐρετμά). See also *Od.* 13.77: πεῖσμα δ' ἔλυσαν ἀπὸ τρητοῖο λίθοιο [on departure, the Phaeacian sailors] "cast off the mooring line from the pierced stone" [sc. a stone bollard with a hole bored in it for making lines fast].

[6] Mouterde 17-18. [7] Lehmann-Hartleben 50-65.

[8] Herodotus 3.60; cf. Lehmann-Hartleben 54-56.

sheds for a naval base (*neorion*); and, for protection, massive defense towers rose at the seaward end of the moles and the whole complex could be brought within the embrace of the town wall (*limen kleistos* "closed harbor").[9]

II GREEK AND ROMAN NAVAL BASES

To ensure an ancient warship's effectiveness, it had to be kept as dry as possible (89-90 above). And so, the principal features of an ancient naval base (*neorion* or *neoria* in Greek, *navalia* in Latin) were the boathouses (*neosoikoi*) where the galleys with their wooden gear were kept under cover, and the sheds (*skeuothekai* or *hoplothekai*) where their sails and rigging were stowed away.[10]

Enough remains of ancient boathouses are extant to give us a fair idea of what they were like.[11] The best preserved are those in the bay of Zea in the Piraeus, built to house the great fleet of triremes Athens maintained during the fifth and fourth centuries B.C. (Fig. 197). They were partly cut out of bedrock, partly built up with blocks of local stone. The essential ingredient was a stone slip, ca. 3 m. (9′ 10″) wide, on which the ship rested. From its landward end each slip sloped downward with a gentle gradient of 1 in 10 for a dry length—that is, the part clear of the water and roofed over —of 37 m. (121′ 5″), a crucial figure since it gives us the length of a trireme. The slip then entered the water to continue submerged for a certain distance. Flanking each slip was a series of stone columns to support a roof of wood and tiles. The clear width of each slip between columns was just under 6 m. (19′ 6″), another crucial figure giving us a trireme's beam. The lines of columns alternated in height, one line higher and the next lower, in order to permit a

[9] Polycrates' harbor at Samos is the earliest of this type that we know of (Mouterde 19). For a full discussion of the *limen kleistos*, see Lehmann-Hartleben 65-74.

[10] See *GOS* 186-89 for references for the Greek terms, and *DS* s.v. *navalia* for the Latin.

[11] For a good summary of what is known about ancient boathouses, see D. Blackman's section in *GOS* 181-86.

pitched roof: the ridge-pole topped a line of higher columns and the roof slanted down over the slips to right and left to the lines of lower columns (Fig. 197, bottom sketch). Building a roof on columns in this fashion instead of on solid walls ensured good ventilation. The landward end of the boathouses was closed off by a continuous stone wall pierced by occasional access doors. The boathouses accommodated not only the ships themselves but the wooden gear—mast, yard, poles, ladders, oars; this was stored either alongside or, more likely, in racks overhead.

Boathouses are known from a few other sites as well, and these show some variations from the type favored by Athens. At Apollonia, the harbor of Cyrene, some of the slips have runners cut down their middle, while at Oeniadae in Acarnania, the slips were beveled to fit the ships' bilges and, instead of being flat all their length, curved upward to cradle the stern.[12] At Apollonia, the gradient was gentler than at Athens, 1 in 14, whereas at Oeniadae it was steeper, roughly 1 in 6, and at Sunium much steeper, 1 in 3.5; here the craft were necessarily drawn up with the aid of winches or the like.

The slips at Zea, made for triremes, had, as mentioned above, a dry length of 37 m. At Apollonia the dry length is just under 40 m. (131' 2") and at Oeniadae 47 m. (154' 2"). The width, however, is about the same as at Zea. It would seem, then, that Apollonia and Oeniadae were built to take larger units, quadriremes and quinqueremes, and that these had roughly the same beam as a trireme. The slips at Sunium, with a dry length of 21 m. (68' 9"), were for smaller craft used as guard ships.

The second key feature in an ancient naval base were the gear sheds in which the galleys' lines and canvas were stored. In the fourth century B.C., Athens built one so elaborate and handsome that it gained a long-lasting reputation as an architectural achievement. It was some 130 m. long (ca. 425'), 18 m. wide (ca. 60'), and 10 m. (ca. 33') high. Two rows of columns divided it into a nave and

---

[12] See the reconstruction in Singer II 517, fig. 470.

aisles; the aisles were two-storied for holding tackle, and the nave provided easy access to the racks.[13]

Of Roman *navalia* we know very little. At Misenum, headquarters of the Empire's major fleet, nothing is left aside from the moles which formed the harbor and the reservoir which ensured the base its drinking water.[14] Yet surely it and every other Roman naval base, just like their Greek predecessors, must have had long lines of boat-houses producing all about a base, as one ancient writer put it,[15] "the look of a continuous colonnade."

III GREEK AND ROMAN COMMERCIAL HARBORS

THE essential elements of a harbor had been worked out by the fifth century B.C. The successive centuries saw chiefly elaboration of facilities and increase in size.

Athens' harbor for merchant shipping (*emporion*) at Piraeus, for example, by the fourth century B.C. boasted stone quays backed by colonnades, which provided an extensive covered area where merchandise could be stocked and business conducted.[16] The Piraeus had only one basin, but certain other ports offered the flexibility of

---

[13] The so-called Arsenal of Philon, minutely described in an inscription (*IG* ii[2] 1668 = *Syl.*[3] 969). For a comprehensive study of the building, see V. Marstrand, *Arsenalet i Piraeus* (Copenhagen 1922), Chapters 2-8 (reconstruction on p. 119; see also Plan iv).

[14] Starr 15-16.

[15] Appian, *Pun.* 96: [The boat houses at Carthage] "have each two Ionic columns in front of them, producing all around the harbor . . . the look of a continuous colonnade" (κίονες δ' ἑκάστου νεωσοίκου προῦχον 'Ιωνικοὶ δύο, ἐς εἰκόνα στοᾶς τὴν ὄψιν τοῦ τε λιμένος . . . περιφέροντες). A number of pictures showing galleys in boathouses that form what can aptly be described as a continuous arcade must surely be illustrations of *navalia*; see F. Coarelli, "Navalia, Tarentum e la topografia del Campo Marzio meridionale," *Studi di topografia romana* (Quaderni dell'Istituto di Topografia Antica della Università di Roma, v, Rome 1968) 27-37, esp. figs. 1, 3-5 (fig. 1 = coin of 47 B.C.; for similar coins see *BM Republic* i, p. 517 and pl. 50.18. Fig. 3 = mosaic of possibly first B.C. though also dated as late as Hadrian; see M. Blake in *Memoirs of the American Academy in Rome* 13, 1936, p. 121 and pl. 28.2. Fig. 5 has been published in a very poor line-drawing in *JDI* 4 [1889] 100 and in an excellent photograph in C. Ragghianti, *Pittori di Pompei* [Milan 1963] pl. 96). For still another example of the motif, see J. Guey and P.-M. Duval, "Les mosaïques de la Grange-du-Bief," *Gallia* 18 (1960) 83-102, esp. 88, 94-95 (mosaic of 2nd A.D.).

[16] Lehmann-Hartleben 120.

two, generally on either side of a promontory, oriented in different directions.[17]

The Hellenistic Age brought to harbor construction the vastness of size and the layout according to an integrated plan that characterized the architecture of the times. In addition, it contributed a feature of the highest practical importance, the lighthouse.[18] The new note was struck with the creation of the greatest harbor yet seen, the one at the mouth of the Nile to serve Ptolemaic Egypt's new capital. The lighthouse, Alexandria's famed *Pharos*, a multiple-level tower at the entrance with a blazing fire at its top able to be seen far at sea, was so striking an achievement that it became one of the Seven Wonders of the Ancient World.[19] (Another port monument of this age was also counted among the seven, the Colossus that beckoned ships into the harbor of Rhodes.[20]) There were two harbors, one facing east and the other west. The arms of the first, the Great Harbor, were formed by two moles, one of which was all of 900 m. long; their tips were 600 m. apart, but reefs and other obstacles in the middle reduced this to two entrances 100 and 200 m. wide respectively. The inner circuit of the Great Harbor was ringed with quays backed up by warehouses as well as with the installations of the naval base. Nearby, palaces and other grandiose buildings added to the general effect.[21]

Greek engineers had consistently turned to stone for their harbor

---

[17] The classic example is Alexandria, with the Great Harbor oriented toward the east and the Eunostos toward the west; see Lehmann-Hartleben, Plan 21. Other examples are Syracuse, Cyzicus, Tyre (but not Sidon; see Poidebard-Lauffrey 83).

[18] On the harbors of the Hellenistic Age, see Lehmann-Hartleben 122-61.

[19] For a suggested reconstruction of the tower, see Singer II 521, fig. 474. For the range of its light, cf. Josephus, *BJ* 4.613: πύργον . . . ἐκπυρσεύοντα . . . ἐπὶ τριακοσίους σταδίους "a tower . . . showing a light visible . . . 300 stades [ca. 33 miles] away."

[20] There is no evidence to support the popular conception of a gigantic figure whose legs straddled the harbor entrance. On the location of the Colossus, see H. Maryon, "The Colossus of Rhodes," *JHS* 76 (1956) 68-86.

[21] Most of our knowledge of Alexandria's harbor comes from Strabo's description (17.791-92, 794-95). See also Lehmann-Hartleben 132-38. For a reconstruction of a typical Hellenistic harbor (Cnidus), see Rostovtzeff, *SEHHW* pl. 112.1. On Carthage's harbor, see J. Baradez, "Le port marchand de Carthage," *CRAI* (1955) 299-300.

works, and taken advantage as much as possible of whatever help nature supplied. The Romans[22] introduced a significant innovation, the use of concrete that would set under water. This powerful and flexible material enabled them to strike out boldly and plant harbors where nature had nothing at all to offer. Claudius, for example, decreed the building of a great port to serve Rome on an open beach just north of the Tiber's mouth; to help make one arm he used the mighty freighter that had brought over the Vatican obelisk as a form and filled it with concrete.[23] Another Roman innovation, made possible by the use of concrete, was the building of moles pierced by arches and of quays resting on arches. These, however, were of

[22] On Roman harbors, see Lehmann-Hartleben 161-217. Portus, the harbor of Rome, and Leptis Magna and several others of Roman date have been investigated since Lehmann-Hartleben wrote. On Portus, see G. Lugli and G. Filibeck, *Il porto di Roma imperiale e l'agro portuense* (Rome 1935); the excellent summary in Meiggs (149-71); the preliminary report on the new excavations in the harbor of Claudius by O. Testaguzza, "The Port of Rome," *Archaeology* 17 (1964) 173-79; and the final report in his *Portus*. On Leptis, see Bartoccini's work cited in note 2 above. On Caesarea, see L. Haefeli, *Cäsarea am Meer: Topographie und Geschichte der Stadt nach Josephus und Apostelgeschichte* (Neutestamentliche Abhandlungen, x Bd, 5. Heft, Münster 1923) 12-16; A. Reifenberg, "Caesarea: A Study in the Decline of a Town," *Israel Exploration Journal* 1 (1950/51) 20-32. On Pompeiopolis see A. Boyce, "The Harbor of Pompeiopolis," *AJA* 62 (1958) 67-78. On Sarepta (between Tyre and Sidon), see *AJA* 74 (1970) 202. On Anthedon, see H. Schläger, D. Blackman, J. Schäfer, "Der Hafen von Anthedon," *AA* 83 (1968) 21-98. For other Italian sites, see the bibliography in L. Crema, *L'architettura romana* (*Enciclopedia Classica*, sezione III: Archeologia e storia dell'arte classica XII.1, 1959) 348. R. Paget, "The Ancient Ports of Cumae," *JRS* 58 (1968) 152-69, gives the history of the ports on this site from their inception through the Roman period. For the ports in Gaul, see A. Grenier, *Manuel d'archéologie gallo-romaine*, Deuxième partie (Paris 1934) 476-529.

[23] Pliny, *NH* 16.202: *longitudo spatium obtinuit magna ex parte Ostiensis portus latere laevo. ibi namque demersa est Claudio principe cum tribus molibus turrium altitudine in ea exaedificatis obiter Puteolano pulvere advectisque* "The length [of the ship] takes up much of the space on the left side of Ostia's port. For, during the reign of Claudius it was sunk there after three masses tall as towers had been built out of Puteoli sand in it and transported [sc. into place]." Cf. 36.70: *in ipsa turribus Puteolis e pulvere exaedificatis perductam Ostiam portus gratia mersit* "after towers had been built out of Puteoli sand in it, it was brought to Ostia and sunk to help make the harbor." "Puteoli sand" is *pozzolana*, the key ingredient in Rome's excellent hydraulic cement. Testaguzza (105-20), on the basis of the recent excavations, offers a reconstruction showing what part of the mole was created by using the ship as a form. See also 72-73 for the possible use of several smaller vessels as forms for other parts of the mole.

limited importance, for they seemed to have been used chiefly along the Campanian coast of Italy during the early Imperial period.[24]

The harbor at Leptis Magna, as improved by Septimius Severus at the end of the second century A.D., has been thoroughly explored and furnishes a good idea of what a medium-sized Roman harbor looked like.[25] At the mouth stood the lighthouse, 21.20 m. (69½′) at the base and at least three levels high,[26] flanking an entrance 80 m. wide.[27] The arms embraced a basin with an expanse of 102,000 sq. m. (25.2 acres); its sides were lined with 1,200 m. of stone quays.[28] Ships made fast nose to, tying up to mooring blocks built into the quays.[29] Back of the quays, approached by flights of steps, were porticoes providing covered space and buildings providing warehousing facilities.

The greatest man-made harbor of antiquity was Portus "the port," the complex mentioned above that served Rome (Fig. 196); it was started by Claudius and expanded by Trajan. Moles jutted into the sea to form a more or less circular expanse of nearly one-third of a square mile in extent, with an entrance 200 m. wide.[30] At one side rose the lighthouse, a massive building of three square levels topped by a cylindrical level which housed the fire (Fig. 144);[31] the inside

[24] Lehmann-Hartleben 165, 167-70. For an example, see the mole that is so prominent a feature in the well-known wall-painting from Herculaneum showing a harbor, probably Puteoli (Singer II 520, fig. 473).

[25] See the reconstructions in Bartoccini pl. A; Crema (op. cit., note 22 above) figs. 398, 399.

[26] Bartoccini 59 and pl. 28.

[27] Bartoccini 11. For a reconstruction of the entrance with the lighthouse, see his pl. B.

[28] Bartoccini 12.

[29] Well indicated in the drawing in Singer II 519, fig. 472. Cf. Fig. 144.

[30] Testaguzza (op. cit., note 22 above) 177, 179; Testaguzza 69, 75.

[31] Testaguzza (121-27), on the basis of a newly discovered painting (125), reconstructs the lighthouse with two square levels and two cylindrical.

The lighthouse became a celebrated building and often appears in paintings and reliefs; G. Stuhlfauth, "Der Leuchtturm von Ostia," Röm. Mitt. 53 (1938) 139-63, provides a large number of examples. It served as model for the bell tower of the church of St. Paul outside the Walls on the road from Ostia to Rome.

The National Museum of Antiquities at Turin has an interesting miniature replica, done in bronze, of a Roman lighthouse; see Bollini, op. cit. (THIRTEEN, note 60) 40, fig. 7. Only the lowermost stage is square; above this rise three round stages.

was ringed with quays where ships could make fast to mooring blocks (Fig. 144, ship to right). Trajan added a hexagonal inner basin,[32] approached from the outer by a channel so narrow most ships surely had to be warped through it. This inner basin, one-eighth of a square mile in size,[33] was surrounded by lines of stone quays and backed up by warehouses and other dockyard facilities.[34] From it a channel enabled barges to go directly to the Tiber for the haul up to Rome.[35]

Big Roman harbors had the equipment and specialized personnel to deal with a high volume of cargo that was often of great bulk. At Portus, for example, by the time of Nero perhaps half a million tons of grain was unloaded each year on the quays.[36] The standard vessels employed by the government as carriers had a capacity of 340 tons;[37] since grain was handled in sacks of a size to make a load for one man,[38] the arrival of each such ship meant 7,500 sacks to be unloaded and sent up the Tiber—and grain ships arrived in fleets not singly (cf. TWELVE, App.). Oil and wine was transported in massive clay jars holding roughly from 20 to 30 liters; each jar with its contents would weigh from 70 to 110 or so pounds, and vessels of just ordinary size carried 2,000 to 3,000 such jars in their holds.[39] The bulkiest cargoes of all were timber and building stone, the timber often in tree-length poles and the stone in chunks weighing dozens of

[32] The arrangement with two basins, an outer and an inner, is also found at Sidon (Poidebard-Lauffray 84; cf. 89 for the Roman date). Achilles Tatius (1.1.1) furnishes a clue to its purpose. He describes the harbor at Sidon as "a wide double harbor in a bight that, with an easy curve, encloses part of the sea. Within the curve of the bight, on the righthand side, a second opening has been dug, and the sea flows in here as well. The harbor also has a subsidiary basin so that merchantmen can lay over here for the winter undisturbed, whereas during the summer they go into the harbor's forward basin" (δίδυμος λιμὴν ἐν κόλπῳ πλατύς, ἠρέμα κλείων τὸ πέλαγος. ᾗ γὰρ ὁ κόλπος κατὰ πλευρὰν ἐπὶ δεξιὰ κοιλαίνεται, στόμα δεύτερον ὀρώρυκται, καὶ τὸ ὕδωρ αὖθις εἰσρεῖ. καὶ γίνεται τοῦ λιμένος ἄλλος λιμήν, ὡς χειμάζειν μὲν ταύτῃ τὰς ὁλκάδας ἐν γαλήνῃ, θερίζειν δὲ τοῦ λιμένος εἰς τὸ προκόλπιον). The description tallies with the extant remains; see Poidebard-Lauffray Plan 2.

[33] Testaguzza (op. cit. note 22 above) 179.

[34] Meiggs 162-66.

[35] Meiggs 159-60.

[36] See ESAR v 139-40.

[37] Dig. 50.5.3, cited in NINE, note 23.

[38] See, e.g., IH ill. 64.

[39] See NINE, note 25 and App., pt. 1.

tons.[40] To transfer it out of the holds to carts or barges, powerful revolving cranes[41] were used, operated by skilled professionals.[42] And, for hauling sacks, amphorae, and other such cargo,[43] there was a veritable army of stevedores[44] backed up by squads of clerks (Fig. 174). Divers were no doubt available for going after any merchandise that went overboard.[45] And there was a special squad of men to handle the sand used as ballast.[46]

[40] Trajan's Column, for example, required 18 cubes of Carrara marble, each weighing 50 tons (*ESAR* v 222). Of two cargoes of building stone that sank off the east coast of Sicily in antiquity and have been investigated by divers, one consisted of 15 blocks totaling 172 tons, of which the biggest single piece weighed 40, and the other of 39 blocks totaling 350 tons, of which the single biggest piece weighed 28½. See G. Kapitän, "Schiffsfrachten antiker Baugesteine und Architekturteile vor den Küsten Ostsiziliens," *Klio* 39 (1961) 276-318, esp. 284, 290.

[41] Vitruvius 10.2.10: *ad onerandas et exonerandas naves sunt paratae, aliae erectae, aliae planae in carchesiis versatilibus conlocatae* "There are [derricks] rigged for loading and unloading ships, some fixed vertically, others set horizontally on revolving platforms"; cf. Rougé 162-64.

[42] The *professionarii de ciconiis* "technicians of the 'storks'"; cf. Rougé, "Ad ciconias nixas," *REA* 59 (1957) 320-28. We prefer to call a derrick a "crane" rather than a "stork."

[43] *IH* ill. 64 shows stevedores handling sacks, while Fig. 144, lower right corner, Fig. 174, and *IH* ill. 69 show them handling amphorae.

[44] Our sources mention *saccarii* "sack-men"; see *CIL* vi 4417 (*ust(rina) sacca-(riorum)* "crematorium of the *saccarii*") and *Cod. Theod.* 14.22, 364 A.D. (*omnia, quaecumque advexerint privati ad Portum . . . , per ipsos saccarios . . . comportari* "[The Prefect is to issue orders that] whatever private shippers convey to The Port be transported by the *saccarii* themselves" [i.e., even nongovernment cargoes are to be handled only by stevedores belonging to the officially recognized guild]). Since a term such as "*amphorarii*" or the like is yet to be attested, possibly the word *saccarius* meant "stevedore," one who handles all forms of cargo.

[45] For the use of divers in salvage operations, see *Dig.* 14.2.4.1: *si navis . . . summersa est et aliquorum mercatorum merces per urinatores extractae sunt* "if a ship sinks . . . and the goods of some merchants have been retrieved by divers [*urinatores*]." There was a guild of *urinatores* at either Ostia or Portus; see *CIL* xiv, suppl. 4620 and note to line 9. For a convenient summary of what is known about ancient divers, see F. Frost, "Scyllias: Diving in Antiquity," *Greece and Rome*, Second Series, 15 (1968) 180-85.

[46] The *saburrarii*, literally "sandmen." Cf. the inscription found at Portus, the harbor of Rome (text in Testaguzza 76): *Sicut coram praecepit v(ir) p(erfectissimus) Messius Extricatus, praefectus annonae, titulus ponetur qui demonstret ex quo loci in quem locum saborrariis saborram tollere liceat* "In accordance with the orders of His Excellency Messius Extricatus, Commissioner of the Grain Supply, a notice shall be posted to show from what point to what point the *saburrarii* may dig sand." The inscription dates 210 A.D. See *CIL* xiv 102 = *ILS* 6177 (156 A.D.) for mention of a guild of "sandmen" (*corpus saburrariorum*).

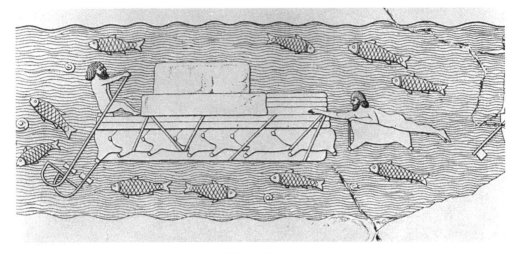

1. Raft of inflated skins, ca. 700 B.C.

2. Hercules on a pot raft, 6th B.C.

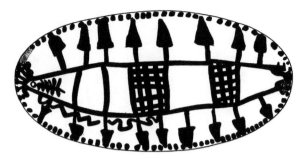

3. Primitive oared riverboat in bird's-eye view,
ca. 3500 B.C.

4. River scene with coracle and fishermen riding inflated floats, ca. 700 B.C.

5. Primitive oared riverboats with cabins, ca. 3200 B.C.

6. The earliest example of a sail, ca. 3100 B.C.

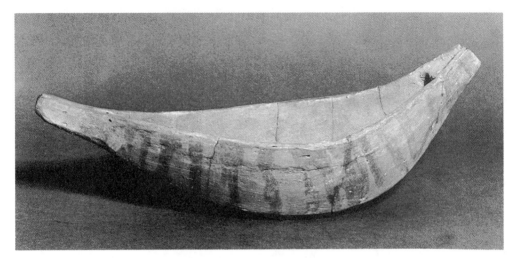

7. Model of a reed boat, ca. 3100 B.C.

8. Models of reed boats, ca. 2000 B.C.

9. Noble hunting in the marshes from a reed canoe, ca. 2500 B.C.

10. Model of a wooden riverboat, ca. 2000 B.C.

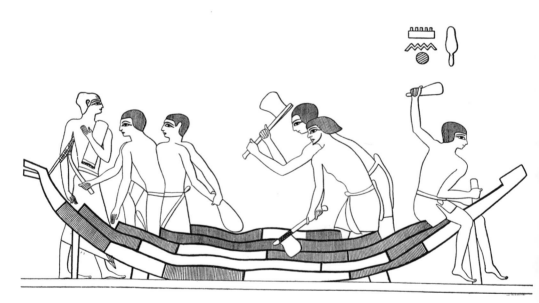

11. Boatwrights putting together a river craft out of short
lengths of planking, ca. 2000 B.C.

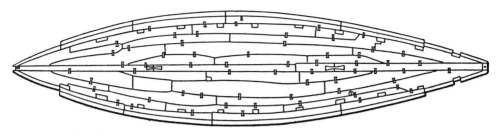

13. Drawing showing the construction of one of the Dahshur boats, ca. 2000 B.C.

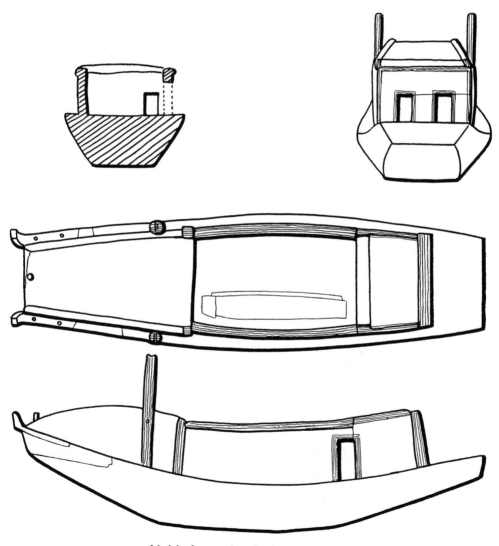

12. Model of a wooden riverboat, ca. 2200 B.C.

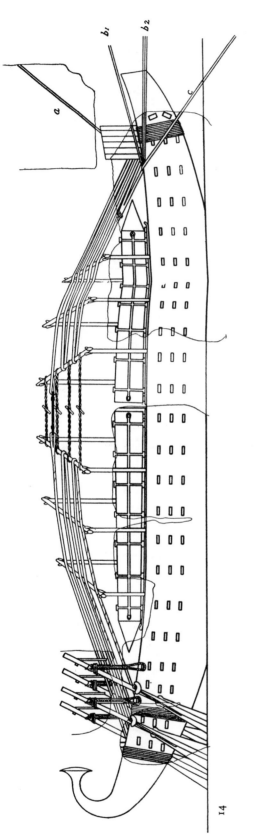

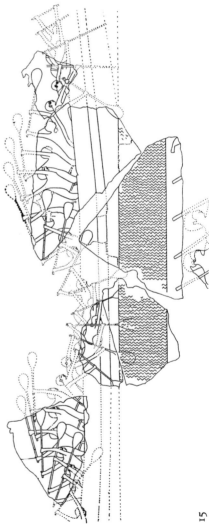

14. Reconstruction of a massive barge carrying obelisks, ca. 1500 B.C. (in actuality probably side by side rather than as shown)

15. The Egyptian method of paddling, ca. 2550 B.C.

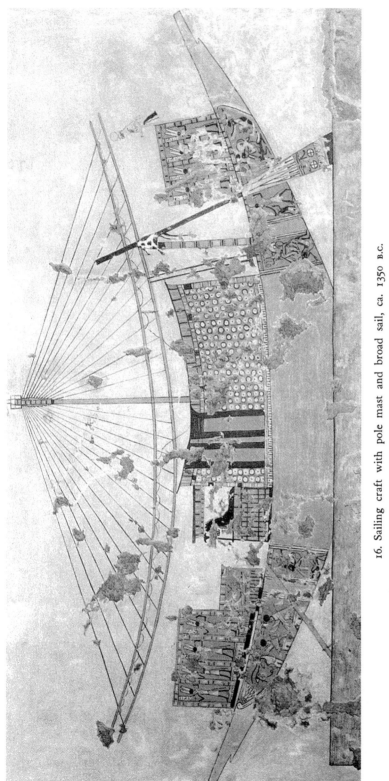

16. Sailing craft with pole mast and broad sail, ca. 1350 B.C.

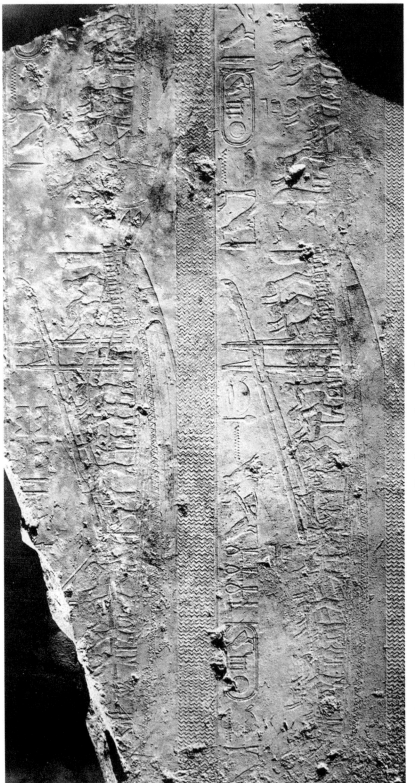

17. Egyptian seagoing vessels, ca. 2450 B.C.

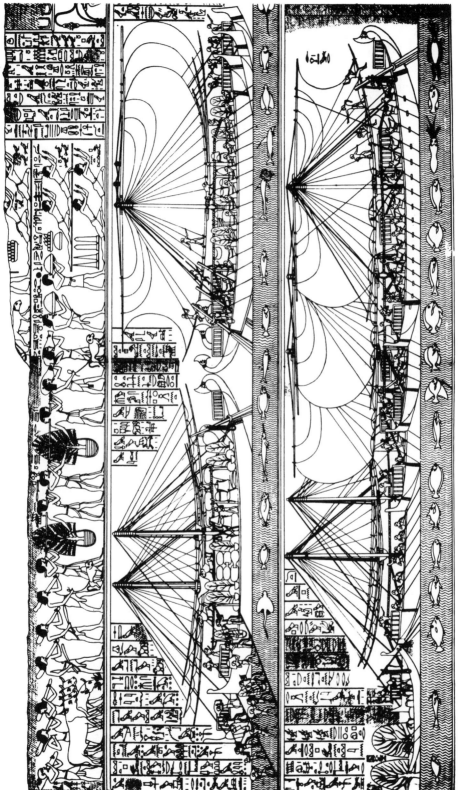

18. Egyptian seagoing vessels, ca. 1500 B.C.

19. Sailing craft with bipod mast and tall, narrow sail, 2400-2300 B.C.

20. Clay model of a Mesopotamian rivercraft, probably a coracle, ca. 3400 B.C.

21. Mesopotamian reed boat, ca. 2300 B.C.

22. Terracotta "frying pan" from Syros, 3rd millennium B.C.

23. Clay model from Palaikastro, Crete, 3rd millennium B.C.

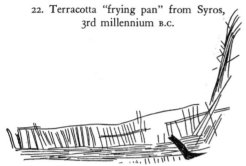

24. Graffiti found on Malta, ca. 1600 B.C.

25. Rock-drawing from Hyria, Greece, ca. 1500 B.C.

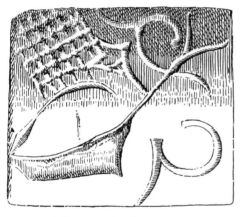

26. Cretan seal, after 1400 B.C.

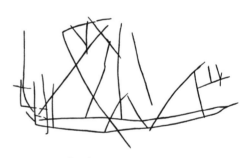

27. Graffito from Cyprus, 1200-1100 B.C.

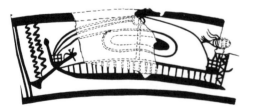

28. Clay box from Pylos, 1200-1100 B.C.

29. Vase from Asine, 1200-1100 B.C.

30. Cup from Eleusis, 850-800 B.C.
(An Aegean galley)

THE AEGEAN MERCHANTMAN

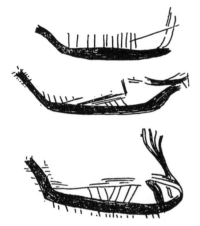

32. Rock-drawing from Hyria,
Greece,
ca. 1500 B.C.

31. Graffiti found on Malta,
ca. 1600 B.C.

33. Detail from a black-figured cup, late 6th B.C.

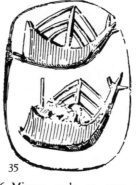

34

35

36

34-36. Minoan seals, ca. 2000 B.C.

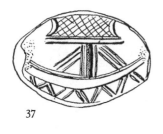

37

38

39

41

42

40

43

44

45

37-45. Minoan seals, ca. 1600-1200 B.C.

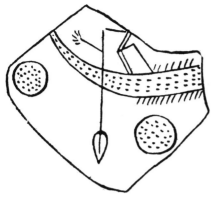

46. Fragment from Phylakopi on
Melos, 3rd millennium B.C.

47. Minoan seal,
ca. 1800 B.C.

48. Minoan seal,
ca. 2000-1600 B.C.

49. Fragment from Mycenae,
ca. 1200-1100 B.C.

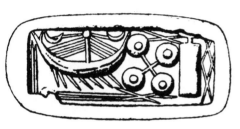

50. Gold ornament from Tiryns,
ca. 1300 B.C.

51. Gold ring from Crete,
after 1400 B.C.

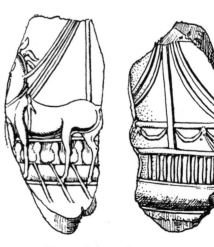

52. Clay seal from Crete, ca. 1400 B.C.
53. Clay seal from Crete, ca. 1400 B.C.

55

54

SMALL CRAFT WITH PROJECTING FOREFOOT

54. Clay model from Mochlos, ca. 2700-2500 B.C.
55. Symbol from the Phaestus disk, ca. 1600 B.C.

ROUND-HULLED MERCHANTMEN

56. Graffito from Malta, ca. 1600 B.C.
57. Syrian ship from an Egyptian wall-painting, ca. 1400 B.C.
58. Syrian ship from an Egyptian wall-painting, ca. 1400 B.C.
59. Decked merchantman pictured on a jar from Cyprus, ca. 1200-1100 B.C.
60. Merchantman on a bowl from Crete, 9th B.C.

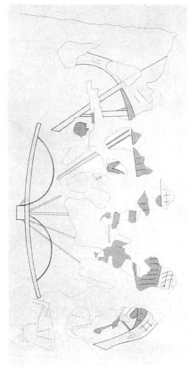

58

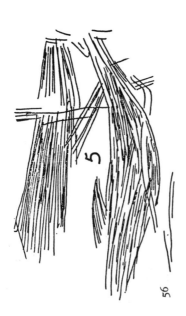

56

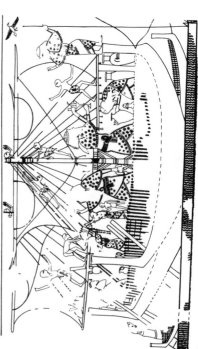

57

60

59

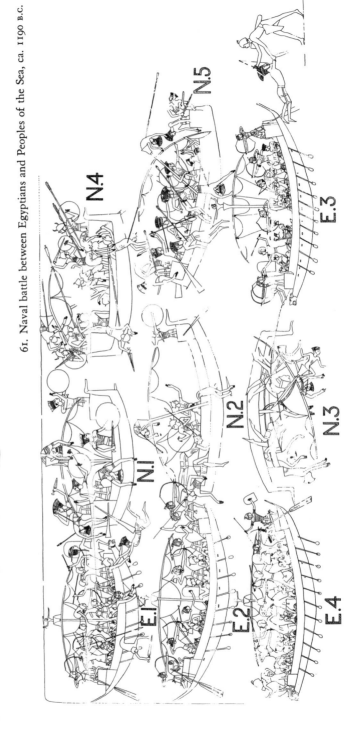

61. Naval battle between Egyptians and Peoples of the Sea, ca. 1190 B.C.

62. Aphract galley, mid-8th B.C.

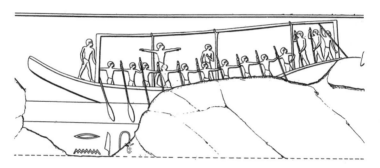

63. Egyptian galley, ca. 2500 B.C.

64. Aphract galley, ca. 725-700 B.C.

65. Warship attacked on shore, first half of 8th B.C.

66. Same ship as in Fig. 65 sailing off.

67. Bow of a warship, mid-8th B.C.

68. Bow of a warship, mid-8th B.C.

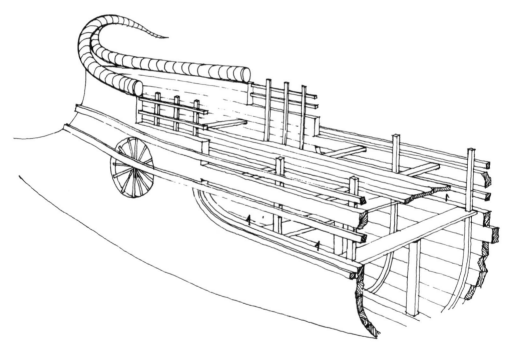

69. Suggested reconstruction of the ship in Fig. 68

70. Two-banked warship with both
levels manned, end of 8th B.C.

71. Two-banked warship with both
levels manned, end of 8th B.C.

72. Warship cruising with upper level manned, mid-8th B.C.

73. Single-banked aphract warship, 6th B.C.

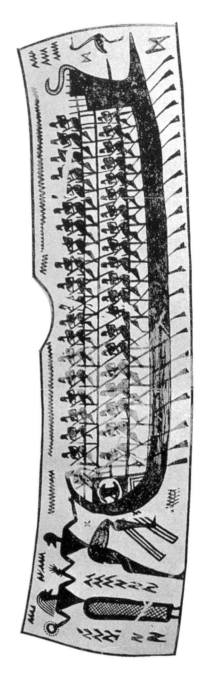

74. Two-banked warship preparing to cruise with upper level manned, second half of 8th B.C.

75. Plan of the fighting deck of the ship in Fig. 76

76. Two-banked Phoenician warship, ca. 700 B.C.

77. Two-banked warship with both levels manned, second half of 8th B.C.

78. Phoenician two-banked warships and transports, ca. 700 B.C.

79. Asia Minor warship, ca. 700 B.C.

80. Merchantman preparing to withstand attack, mid-7th B.C.

81. Pirate craft chasing unsuspecting merchantman, second half of 6th B.C.

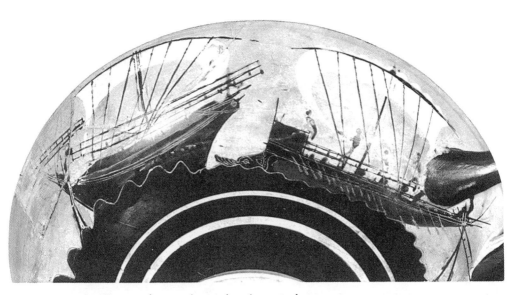

82. Pirate craft preparing to board as merchantman attempts to escape

83. Two-banked warship, with the banks out of time, mid-6th B.C.

84. Decked warship, end of 6th B.C.

85. Two-banked warship, probably a triaconter, late 6th B.C.

86. Merchant galley, 9th-8th B.C.

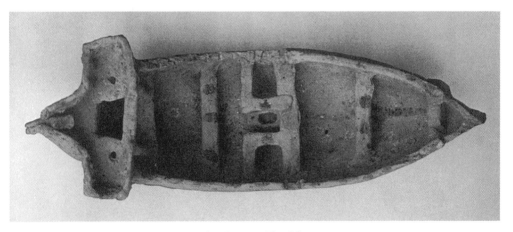

87. Same as Fig. 86

88. Two-banked warships (?) with the upper level manned, third quarter of 6th B.C.

89. Single-banked penteconter, second half of 6th B.C.

90. Aphract warships cruising, second half of 6th B.C.

91. Merchant galley, late 6th B.C.

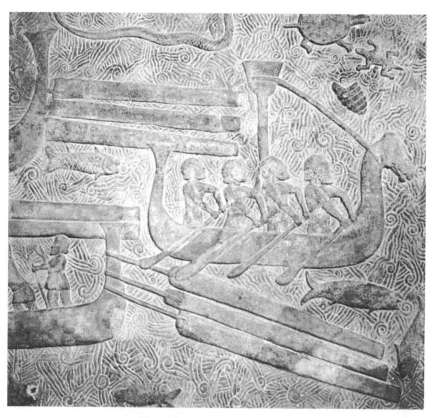

92. Phoenician cargo vessels, ca. 700 B.C.

93. Merchantman on vase from Vulci, 6th B.C.

94. Merchantman from Cyprus, 6th B.C.

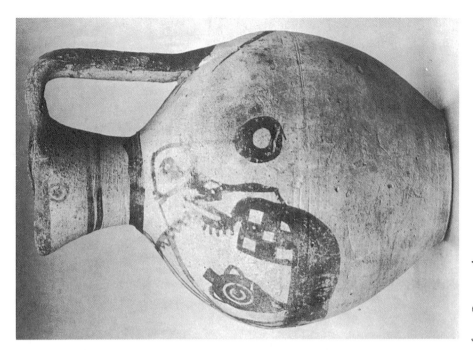

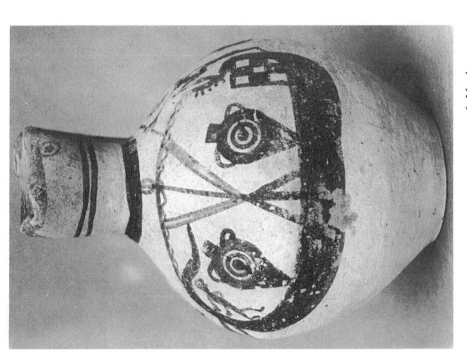

95-96. Merchantman on vase from Cyprus, 7th B.C.

98. Stern of a merchantman, 7th B.C.

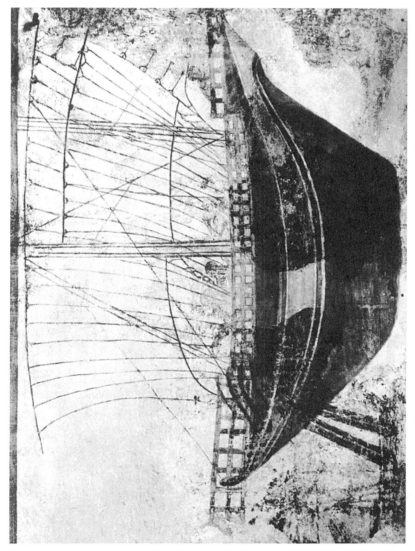

97. Reconstruction of the merchantman pictured in the Tomba della Nave, Tarquinia, early 5th B.C.

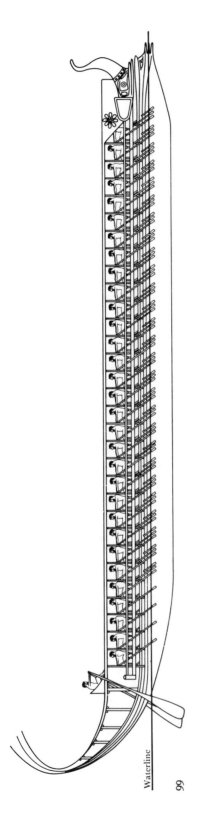

99. Sketch of a trireme of the 5th-4th B.C. in profile view

100. Sketch of the oar arrangements in a trireme of the 5th-4th B.C.

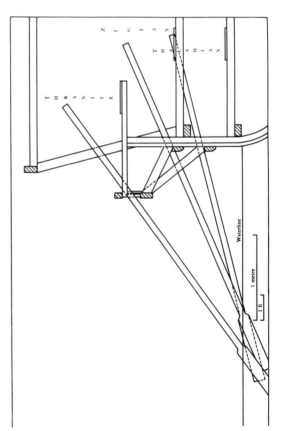

102. Another view

101. View of model showing the oar arrangements in a trireme of the 5th-4th B.C.

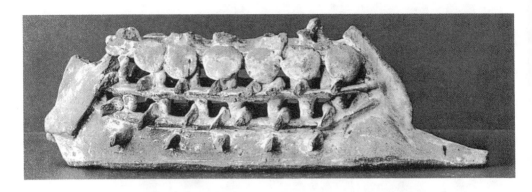

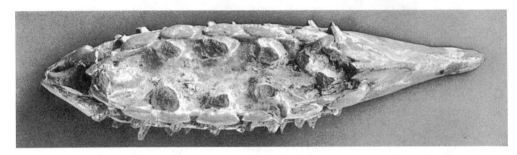

103-104. Model of a Phoenician trireme, 4th-3d B.C.

105. A Persian trireme

106. After portion of a trireme, end of 4th B.C.

107. Prow of a Hellenistic galley, ca. 300 B.C.

108. Reconstruction of a relief showing the stern of
a Rhodian galley, ca. 200 B.C.

109. Graffito from Delos, first half of 1st B.C.

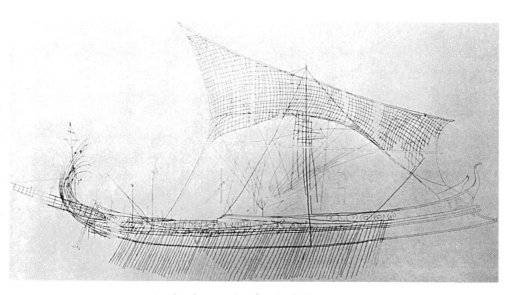

110. Graffito from Delos, first half of 1st B.C.

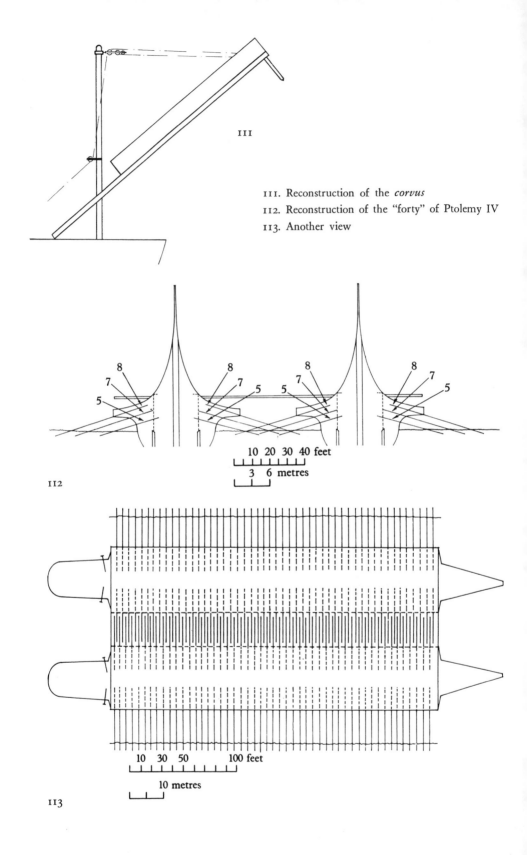

III

111. Reconstruction of the *corvus*
112. Reconstruction of the "forty" of Ptolemy IV
113. Another view

112

113

114. Stern of a galley, 2nd A.D.

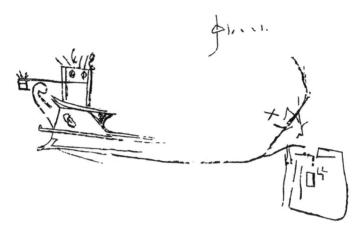

115. Warship equipped with fire pot, 1st B.C.

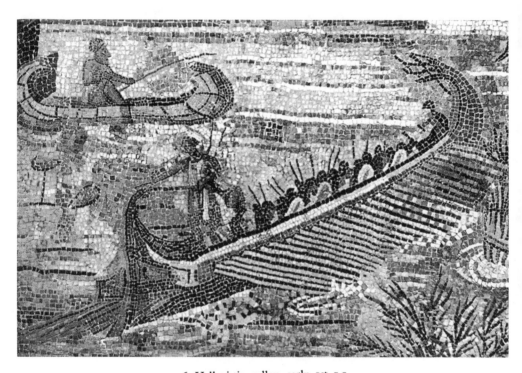

116. Hellenistic galley, early 1st B.C.

117. *Hemiolia*, second half of 6th B.C.

118. Oarport of a Hellenistic galley, 200-180 B.C.

119. Two-banked galley, 2nd-1st B.C.

120

121

120. Galley, probably a quadrireme or quinquereme, 38-36 B.C.
121. Galley, probably larger than a quinquereme, 32-31 B.C.

122

123

122-123. A two-banked and a single-banked galley of the
Roman Imperial Navy, 117-138 B.C.

124. Fore view of a Roman galley, 54-68 A.D.

125. Roman trireme, second half of 1st B.C.

126

NAVISTETRERIS
LONGA

127

126. Graffito of a Roman quadrireme, end of 1st B.C. or beginning of 1st A.D.

127. The prow of Trajan's trireme with artemon raised

128. Emperor Trajan at the helm of a trireme

129. Roman trireme, 1st B.C. to 1st A.D.

130. Two-banked Roman galley, probably a quadrireme or larger,
second half of 1st B.C.

131. Roman trireme, 1st B.C. to 1st A.D.

132. Another view of Figure 130

133. Roman galleys, probably larger than triremes, with a full complement of marines, 1st A.D.

134. Dromon using Greek fire, 14th A.D.

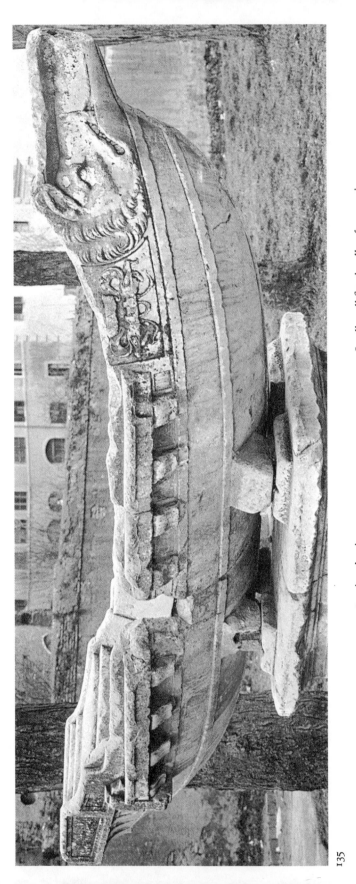

135

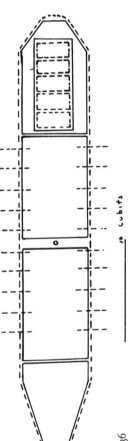

136

135. Small well-fitted galley for carrying passengers,
Renaissance copy of an ancient model

136. Reconstruction of the deck plan of a *kerkouros*

137. Twenty-one different kinds of craft, many of them
merchant galleys, 3rd or 4th A.D.

o cubits

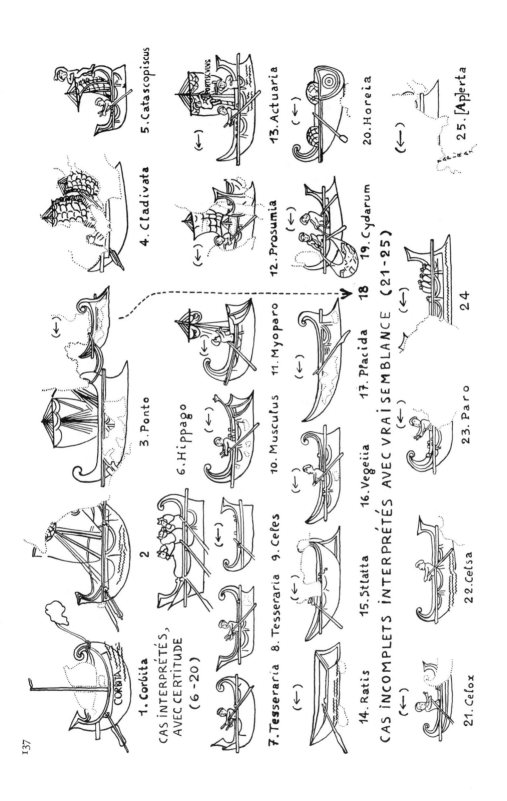

1. Corbita

CAS INTERPRÉTÉS, AVEC CERTITUDE (6-20)

2

3. Ponto

4. Cladivata

5. Catascopiscus

6. Hippago

7. Tesseraria   8. Tesseraria   9. Celes   10. Musculus   11. Myoparo

12. Prosumia

13. Actuaria

14. Ratis   15. Stlatta   16. Vegeia   17. Placida   18   19. Cydarum   20. Horeia

CAS INCOMPLETS INTERPRÉTÉS AVEC VRAISEMBLANCE (21-25)

21. Celox   22. Celsa   23. Paro   24   25. [Ap]erta

137

138. Merchant galley approaching a coast under sail and oars, end of 1st B.C.

139. Roman merchant galley, 1st A.D.

140. Merchant galley loaded with amphorae, 2nd or 3rd A.D.

141. Galley used for transporting beasts for the gladiatorial games, 4th A.D.

142. Sailing vessel, ca. 200 A.D.

143. Sailing vessel, 3rd A.D.

144. Cargo vessel shown entering the harbor of Rome and then tied up at a quay, ca. A.D. 200

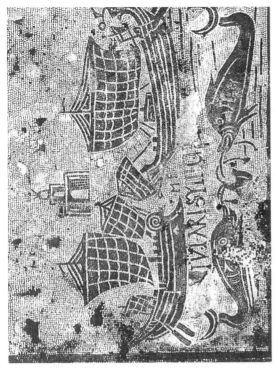

145. Freighter with rounded bow and standard rig and another with projecting cutwater and three-masted rig, ca. A.D. 200

146. Detail of Fig. 144

147. Two vessels with standard rig and one with sprit-rig, 3rd A.D.

148. Two coastal craft at the entrance to Rome's harbor, 3rd A.D.

149. Cargo vessel under full sail, 3rd A.D.

150. Cargo vessel in an Adriatic port, A.D. 98-117

151. Sailing vessel entering port, ca. A.D. 50

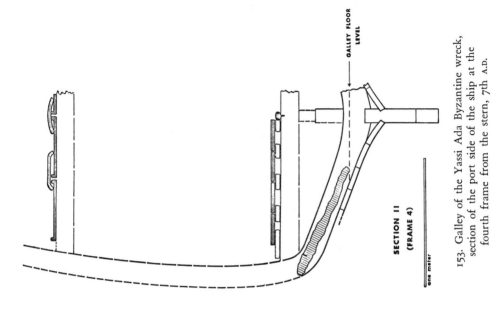

153. Galley of the Yassi Ada Byzantine wreck, section of the port side of the ship at the fourth frame from the stern, 7th A.D.

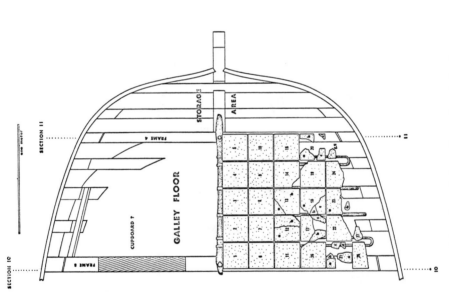

152. Plan view of the galley at hearth level

155. Small cargo vessel with a deck load of amphorae resting on dunnage, 3rd A.D.

154. Ship at the entrance to Rome's harbor, 3rd A.D.

156. Large cargo vessel under full sail, 2nd half of 1st A.D.

157. Detail of the Dramont A wreck showing the bottom of the hull and wine-jars stored in three layers, 1st B.C.

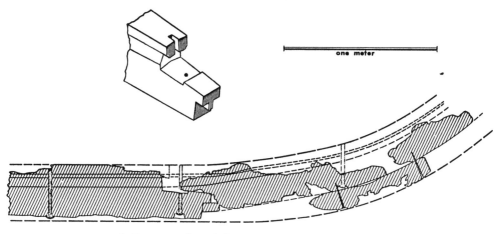

one meter

158. Reconstruction of the scarf joining sternpost to keel
in the Yassi Ada Byzantine wreck

159. Planking from the Grand Congloué wreck: one half of a piece
that had split as if filleted, 2nd B.C.

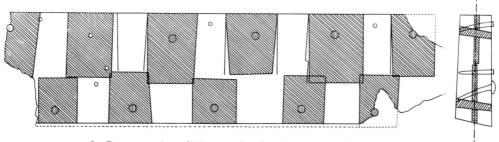

160. Reconstruction of Fig. 159 showing the position of the mortises

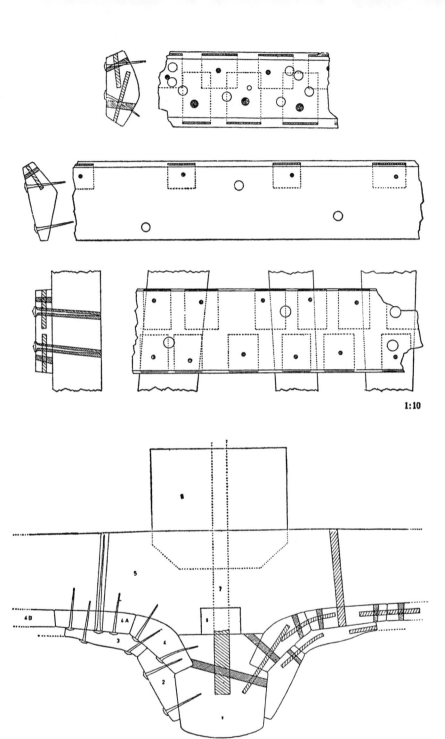

1:10

1:10

161. Reconstruction of elements from the Titan wreck: pieces of planking showing mortises; planking made fast to frame with treenails transfixed by copper nails; keelson, floor timber, keel, and garboard strakes, 1st B.C.

162. The shell-first technique of building a boat: a modern
Swedish boatwright at work, 1929

163. The shell-first technique of building a boat: Roman shipwright
completing a boat, late 2nd or early 3rd A.D.

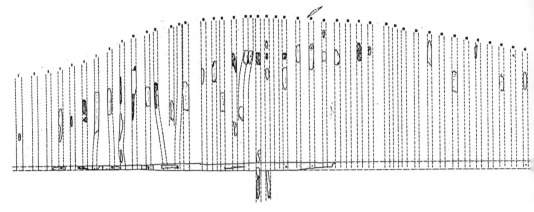

164. Reconstruction showing the keel and frames of the Yassi Adà Byzantine wreck

165. Planking from the Anticythera wreck

166. Planking from the Anticythera wreck, 1st. B.C.

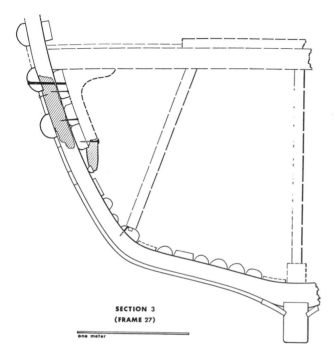

SECTION 3
(FRAME 27)

one meter

167. Reconstruction of a
 midships section of the
 Yassi Ada Byzantine wreck
 showing keel, frame, lining,
 waling pieces, deck beam
 supported by hanging knee
 and diagonal brace, deck
 planking, hatch cover

168. Planking, floor timbers,
 and lining of the
 Chrétienne A wreck,
 1st B.C.

169. Sailing ship with nearly
vertical foremast, A.D. 306

170. Brackets holding a steering oar, 1st B.C. or 1st A.D.

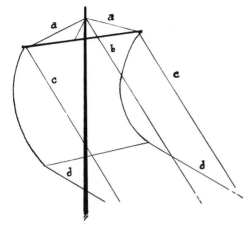

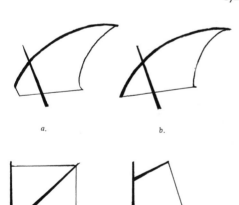

171. Sketch illustrating running rigging:
a lifts, b halyard,
c braces, d sheets

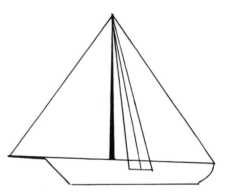

172. Sketch illustrating standing rigging

a.

b.

c.

d.

173. Sketch illustrating fore-and-aft rigs:
a "Arab" lateen,
b Mediterranean lateen,
c sprit rig, d gaff rig

174. Stevedores unloading a cargo of wine jars before shipping clerks on the quay, 3rd A.D.

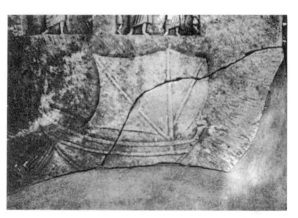

175. Sprit-rigged vessel with two sails traveling "wing and wing," 1st or 2nd A.D.

176. Small vessel with a spritsail, 2nd B.C.

177. Sprit-rigged craft, 2nd or 3rd A.D.

178. Sprit-rigged craft being launched, Roman Imperial period.

179. Sprit-rigged craft, detail of Fig. 147

180. Lateen-rigged craft, date indeterminate

181. Lateen-rigged craft, 2nd A.D.

182. Lateen-rigged craft and small boats off a seaside villa, probably 4th A.D.

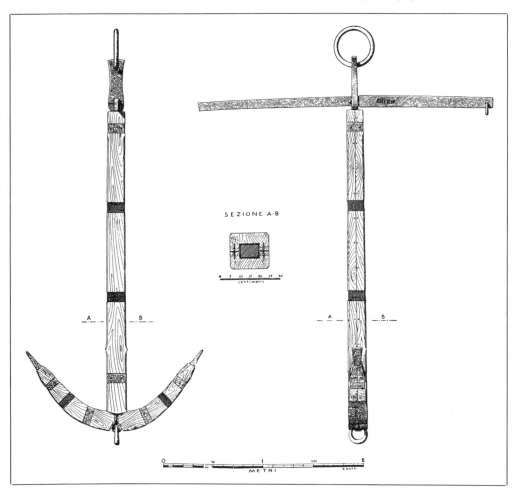

SEZIONE A·B

centimetri

METRI

183. Iron anchor, with removable stock, cased in wood, 1st half of 1st A.D.

184. Wooden anchor with lead stock; behind, hull of one of the Nemi barges covered to the gunwales with a lead sheathing, 1st half of 1st A.D.

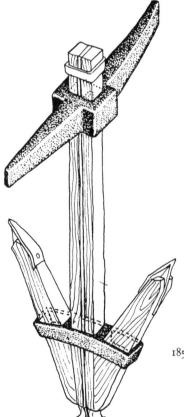

185. Reconstruction of an anchor with arms and shank held together by a lead collar

186. Lead core of a wooden anchor stock accidentally bent double

187. Stone anchors from a wreck of ca. 350 B.C.

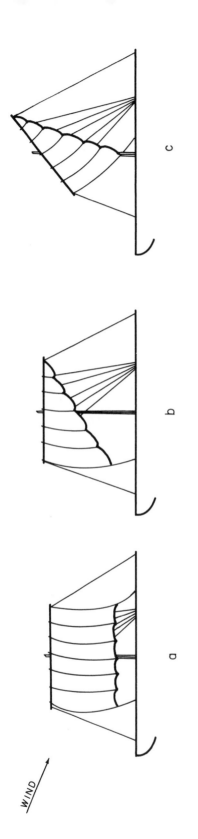

188. Sketch showing method of sailing against the wind with sail brailed up

189. A rivercraft of the Rhine or Moselle, mid-1st A.D.

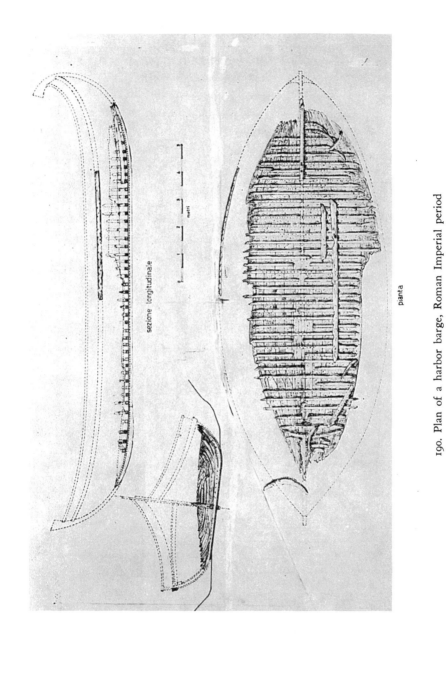

sezione longitudinale

metri

pianta

190. Plan of a harbor barge, Roman Imperial period

191. Coastal craft unloading bars of lead onto a beach, 3rd A.D.

192. Fishing boat, 5th B.C.

193. A harbor tugboat; note the towline that goes from the stern upward, 3rd A.D.

194. Fishing boat, Roman Imperial period

195. River craft of the Saar or Moselle, 2nd-3rd A.D.

196. Model of Portus, the harbor of Rome,
after time of Trajan (A.D. 98-117)

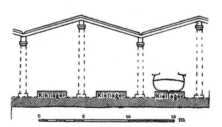

197. Plan and reconstruction of the
boat houses at Zea, 4th B.C.

LIST OF ABBREVIATIONS

AA: Archäologischer Anzeiger (supplement to *JDI*)

AC: l'Antiquité classique

AE: l'Année épigraphique

AJA: American Journal of Archaeology

AJPh: American Journal of Philology

AM: L. Casson, *The Ancient Mariners* (New York 1959)

Anderson: R. Anderson, *Oared Fighting Ships* (London 1962)

Annuario: Annuario della reale scuola archeologica di Atene

Archiv: Archiv für Papyrusforschung

Assmann: A. Assmann, *Seewesen* in A. Baumeisters *Denkmäler des klassischen Altertums* III (Munich and Leipzig 1889) 1593-1639

Assmann, *Segel*: A. Assmann, *Segel* in *RE* (1921)

Ath. Mitt.: Mitteilungen des deutschen archäologischen Instituts, Athenische Abteilung

Babelon: E. Babelon, *Traité des monnaies grecques et romaines* II.2 (Paris 1910)

Bartoccini: R. Bartoccini, *Il porto romano di Leptis Magna* (Bollettino del Centro Studi per la Storia dell' Architettura 13, supplemento al 1958, Rome 1958)

Basch: L. Basch, "Phoenician Oared Ships," *MM* 55 (1969) 139-62, 227-45

Bass, *Gelidonya*: G. Bass, *Cape Gelidonya: A Bronze Age Shipwreck* (Transactions of the American Philosophical Society, New Series 57, Part 8, Philadelphia 1967)

Bass, "Yassi Ada": G. Bass, "Underwater Excavations at Yassi Ada: A Byzantine Shipwreck," *AA* (1962) 537-63

Bauer: A. Bauer, "Seewesen (Diadochenzeit)" in *Die griechischen Altertümer* (Müllers Handbuch der klassischen Altertumswissenschaft 4.1.2, Munich 1893²) 458-66

BCH: Bulletin de correspondance hellénique

Becatti: G. Becatti, *Scavi di Ostia*. IV, *Mosaici e pavimenti marmorei* (Rome 1961)

Benoit: F. Benoit, *l'Épave du Grand Congloué à Marseille* (XIVᵉ supplément à *Gallia*, Paris 1961)

Berichtigungsliste: F. Preisigke and others, *Berichtigungsliste der griechischen Papyrusurkunden aus Ägypten* (Strassburg 1913——)

BGU: Berliner griechische Urkunden (Berlin 1895——)

Blinkenberg: C. Blinkenberg, *Triemiolia* (Det Kgl. Danske Videnska-bernes Selskab. Archaeologisk-kunsthistoriske Meddelelser II.3, Copenhagen 1938)

BM Central Greece: B. Head, *British Museum Catalogue of Greek Coins, Central Greece* (London 1884)

BM Empire: H. Mattingly, *Coins of the Roman Empire in the British Museum* (London 1923——)

BM Republic: H. Grueber, *Coins of the Roman Republic in the British Museum* (London 1910)

Böckh: A. Böckh, *Die Staatshaushaltung der Athener*. III, *Urkunden über das Seewesen des attischen Staates* (Berlin 1840)

Boreux: C. Boreux, *Études de nautique égyptienne: l'art de la navigation en Égypte jusqu'à la fin de l'ancien empire* (Mémoires de l'institut français d'archéologie orientale du Caire 50, Cairo 1925)

Bowen: R. LeBaron Bowen, Jr., "Egypt's Earliest Sailing Ships," *Antiquity* 34 (1960) 117-31

Braemer-Marcadé: F. Braemer and J. Marcadé, "Céramique antique et pièces d'ancres trouvées en mer à la pointe de la Kynosoura (Baie de Marathon)" *BCH* 77 (1953) 139-54

Breusing: A. Breusing, *Die Nautik der Alten* (Bremen 1886)

BSA: Annual of the British School at Athens

Bull. Com.: Bullettino della commissione archeologica comunale di Roma

CAH: Cambridge Ancient History

Casson, "Harbor Boats": L. Casson, "Harbor and River Boats of Ancient Rome," *JRS* 55 (1965) 31-39

Casson, "Isis": L. Casson, "The Isis and her Voyage," *TAPA* 81 (1950) 43-56

Casson, "Speed": L. Casson, "Speed under Sail of Ancient Ships," *TAPA* 82 (1951) 136-48

Casson, "Studies": L. Casson, "Studies in Ancient Sails and Rigging," *Essays in Honor of C. Bradford Welles* (American Studies in Papyrology I, New Haven 1966) 43-58

Chevalier: Y. Chevalier, "La cavité d'emplanture avec monnaie de l'épave antique de l'Anse Gerbal à Port-Vendres (sondage 1963)" *Revue archéologique de Narbonnaise* I (1968) 263-69

Chr. d'Eg.: Chronique d'Égypte

CIAS: Congresso internazionale di archeologia sottomarina

Cichorius: C. Cichorius, *Die Reliefs der Traianssäule* (Berlin 1896-1900)

CIL: Corpus inscriptionum latinarum

CMMS: Corpus der minoischen und mykenischen Siegel

Cohen: H. Cohen, *Description historique des monnaies frappées sous l'Empire romain* (Paris 1880-1892²)

Corp. Gloss. Lat.: G. Goetz, *Corpus glossariorum latinorum* (Leipzig and Berlin, 1888-1923)

CPh: Classical Philology

CPL: R. Cavenaile, *Corpus papyrorum latinarum* (Wiesbaden 1956-1958)

CQ: Classical Quarterly

CR: Classical Review

CRAI: Comptes-rendus de l'académie des inscriptions et belles-lettres de l'Institut de France

Crumlin-Pedersen: O. Crumlin-Pedersen, "Two Danish Side Rudders," *MM* 52 (1966) 251-61.

CVA: Corpus vasorum antiquorum

Dain: A. Dain, *Naumachica* (Paris 1943)

Davison: J. Davison, "The First Greek Triremes," *CQ* 41 (1947) 18-24

de Saint-Denis: E. de Saint-Denis, *Le vocabulaire des manoeuvres nautiques en Latin* (Macon 1935)

Dig.: Digesta, cited from T. Mommsen and P. Krüger, *Corpus iuris civilis* 1 (Berlin 1928¹⁵)

Dolley, "Naval Tactics": R. Dolley, "Naval Tactics in the Heyday of the Byzantine Thalassocracy," *Atti dell' viii congresso di studi byzantini* 1 (Rome 1953) 324-39

Dolley, "Warships": R. Dolley, "The Warships of the Later Roman Empire," *JRS* 38 (1948) 47-53

DS: C. Daremberg and E. Saglio, *Dictionnaire des antiquités grecques et romaines* (Paris 1877-1919)

Dumas: F. Dumas, *Épaves antiques. Introduction à l'archéologie sous-marine méditerranéenne* (Paris 1964)

Duval, "Forme": P. Duval, "La forme des navires romains," *École française de Rome, Mélanges d'archéologie et d'histoire* 61 (1949) 119-49

Ebert, *REV*: M. Ebert, *Reallexikon der Vorgeschichte* (Berlin 1924-1932)

Edmonds: J. Edmonds, *The Fragments of Attic Comedy* (Leiden 1957-1961)

Eph. Arch.: Ephemeris Archaiologike

ESAR: T. Frank and others, *An Economic Survey of Ancient Rome* (Baltimore 1933-1940)

373

Espérandieu: E. Espérandieu, *Recueil général des bas-reliefs de la Gaule romaine* (Paris and Brussels 1907——)

Evans: A. Evans, *The Palace of Minos at Knossos* (London 1921-1935)

FA: Fasti Archaeologici

Fitzwilliam Coll.: S. Grose, *Catalogue of the McClean Collection of Greek Coins in the Fitzwilliam Museum* (Cambridge 1923-1929)

Foucher: L. Foucher, *Navires et barques figurés sur des mosaiques découvertes à Sousse et aux environs* (Institut national d'archéologie et arts. Musée Alaoui, Notes et documents 15, Tunis 1957)

Frost: H. Frost, *Under the Mediterranean* (London 1963)

Gargallo: P. Gargallo, "Anchors of Antiquity," *Archaeology* 14 (1961) 31-35

Goitein: S. Goitein, *A Mediterranean Society*. 1, *Economic Foundations* (Berkeley and Los Angeles 1967)

GOS: J. Morrison and R. Williams, *Greek Oared Ships* (Cambridge 1968)

Guglielmotti: A. Guglielmotti, *Vocabolario marino e militare* (Rome 1889)

Guglielmotti III: A. Guglielmotti, *Storia della marina pontificia* III (Rome 1886)

Hellenica: Hellenica: Recueil d'épigraphie, de numismatique, et d'antiquités grecques, pub. par L. Robert

HO: Hydrographic Office, Washington, D.C. Publications 151, 152, 153, 154A, 154B = *Sailing Directions for the Mediterranean* I-V

Hornell: J. Hornell, *Water Transport* (Cambridge 1946)

Hourani: G. Hourani, *Arab Seafaring in the Indian Ocean in Ancient and Early Medieval Times* (Princeton 1951)

How and Wells: W. How and J. Wells, *A Commentary on Herodotus* (Oxford 1912)

HSCP: Harvard Studies in Classical Philology

Hunterian Coll.: G. Macdonald, *Catalogue of Greek Coins in the Hunterian Collection* (Glasgow 1899-1905)

IG: Inscriptiones Graecae

IGRR: R. Cagnat and others, *Inscriptiones graecae ad res romanas pertinentes* (Paris 1911-1927)

IH: L. Casson, *Illustrated History of Ships and Boats* (New York 1964)

ILS: H. Dessau, *Inscriptiones latinae selectae* (Berlin 1892-1916)

Inscr. de Délos: F. Durrbach and others, *Inscriptions de Délos* (Paris 1926-1937)

Jacoby, *FGH*: F. Jacoby, *Die Fragmente der griechischen Historiker* (Leiden 1957-1958)
Jal, *Arch. nav.*: A. Jal, *Archéologie navale* (Paris 1840)
Jal, *Flotte*: A. Jal, *La flotte de César* (Paris 1861)
Jal, *Gloss. naut.*: A. Jal, *Glossaire nautique* (Paris 1848-1850)
Jalabert-Mouterde: L. Jalabert and R. Mouterde, *Inscriptions grecques et latines de la Syrie* (Paris 1929———)
Jameson: M. Jameson, "The Provisions for Mobilization in the Decree of Themistokles," *Historia* 12 (1963) 385-404
JDI: Jahrbuch des deutschen archäologischen Instituts
JEA: Journal of Egyptian Archaeology
JHS: Journal of Hellenic Studies
JRS: Journal of Roman Studies

Katzev, letters of 8 and 31 March 1970, and copy of the excavation report for 1969, from M. Katzev, Director of the Kyrenia Ship Excavation. See also *National Geographic* June 1970, 841-57
Kienast: D. Kienast, *Untersuchungen zu den Kriegsflotten der römischen Kaiserzeit* (Antiquitas. i, *Abhandlungen zur alten Geschichte*, 13, Bonn 1966)
Kirk: G. Kirk, "Ships on Geometric Vases," *BSA* 44 (1949) 93-153
Köster: A. Köster, *Das antike Seewesen* (Berlin 1923)

Lamboglia: N. Lamboglia, "Il rilevamento totale della nave romana di Albenga," *RSL* 27 (1961) 213-20
Lamboglia, "Punta Scaletta": N. Lamboglia, "La campagna 1963 sul relitto di Punta Scaletta all'Isola di Giannutri," *RSL* 30 (1964) 229-57
Lane: F. Lane, *Venetian Ships and Shipbuilders of the Renaissance* (Baltimore 1934)
Lane, "Tonnages": F. Lane, "Tonnages, Medieval and Modern," *Economic History Review*, Second Series, 17 (1964) 213-33
Lehmann-Hartleben: K. Lehmann-Hartleben, *Die antiken Hafenanlagen des Mittelmeeres* (*Klio*, Beiheft xiv, Leipzig 1923)
Lindos: C. Blinkenberg, *Lindos* ii: *Inscriptions* (Berlin and Copenhagen 1941)
LSJ: H. Liddell, R. Scott, H. Jones, *A Greek-English Lexicon* (Oxford 1940)

375

Marinatos: S. Marinatos, "La marine créto-mycénienne," *BCH* 57 (1933) 170-235

Marsden, *Blackfriars*: P. Marsden, *A Roman Ship from Blackfriars*, London (Guildhall Museum Publication, London 1967)

Marsden, "County Hall": P. Marsden, "The County Hall Ship," *Transactions of the London and Middlesex Archaeological Society* 21 (1965) 109-17

Marsden, "New Guy's": P. Marsden, "A Boat of the Roman Period Discovered on the Site of New Guy's House, Bermondsey, 1958," *Transactions of the London and Middlesex Archaeological Society* 21 (1965) 118-31

Masson: P. Masson, *Les galères de France, 1481-1781* (Annales de la Faculté des Lettres d'Aix 20, 1937, fasc. 1-2, Aix-en-Provence 1938)

M. Chrest: L. Mitteis and U. Wilcken, *Grundzüge und Chrestomathie der Papyruskunde*. Zweiter Band, von L. Mitteis (Berlin and Leipzig 1912)

Meiggs: R. Meiggs, *Roman Ostia* (Oxford 1960)

Migne: J. Migne, *Patrologia Graeca* and *Patrologia Latina*

Miltner: F. Miltner, *Seewesen* in *RE*, Supplementband v, 906-62 (1931)

MM: The Mariner's Mirror

Mohler: S. Mohler, "Sails and Oars in the *Aeneid*," *TAPA* 79 (1948) 46-62

Moll: F. Moll, *Das Schiff in der bildenden Kunst* (Bonn 1929)

Monumenti Antichi: *Monumenti Antichi pubblicati a cura della Accademia Nazionale dei Lincei*, Rome

Monuments Piot: *Fondation Eugène Piot: Monuments et Mémoires*

Morrison, "Nautical Terms": J. Morrison, "Notes on Certain Greek Nautical Terms and on Three Passages in *IG* ii² 1632," *CQ* 41 (1947) 122-35

Morrison, "Trireme": J. Morrison, "The Greek Trireme," *MM* 27 (1941) 14-44

Moschetti: C. Moschetti, *Gubernare navem, gubernare rem publicam* (Quaderni di "Studi Senesi" 16, Milan 1966)

Mouterde: chapter on the history of harbors by R. Mouterde in Poidebard-Lauffray

Müller, *GGM*: C. Müller, *Geographi graeci minores* (Paris 1855-61)

Naumachica: see Dain

Not. Sc.: Accademia dei Lincei, Rome: *Notizie degli scavi di antichità*

Nuova sill.: A Maiuri, *Nuova silloge epigrafica di Rodi e Cos* (Florence 1925)

OGIS: W. Dittenberger, *Orientis graeci inscriptiones selectae* (Leipzig 1903-1905)

Pantera: Pantero Pantera, *L'armata navale* (Rome 1614)

P. Aberdeen: E. Turner, *Catalogue of Greek and Latin Papyri and Ostraca in the Possession of the University of Aberdeen* (Aberdeen 1939)

P. Athen.: G. Petropoulos, *Papyri societatis archaeologicae atheniensis* (Athens 1939)

P. Bad.: F. Bilabel, *Veröffentlichungen aus den badischen Papyrus-Sammlungen*. Heft 2, 4, Griechische Papyri (Heidelberg 1923, 1924)

P. Berol. Frisk: J. Frisk, *Bankakten aus dem Faijûm nebst anderen Berliner Papyri* (Göteborg 1931)

P. Cairo Masp.: J. Maspero, *Papyrus grecs d'époque byzantine* (Catalogue général des antiquités égyptiennes du Musée du Caire 1-3, Cairo 1911-1916)

P. Cairo Zen.: C. Edgar, *Zenon Papyri* (Catalogue général des antiquités égyptiennes du Musée du Caire 79, Cairo 1925———)

P. Col. Zen.: W. Westermann and others, *Zenon Papyri* (Columbia Papyri, Greek Series, 3-4, New York 1934-1940)

P. Fay.: B. Grenfell and others, *Fayum Towns and their Papyri* (London 1900)

P. Flor.: G. Vitelli and D. Comparetti, *Papiri fiorentini* (Milan 1906-1915)

P. Goodsp.: E. Goodspeed, *Greek Papyri from the Cairo Museum* (Chicago 1904)

P. Grenf.: B. Grenfell, *An Alexandrian Erotic Fragment and Other Greek Papyri* (Oxford 1896); B. Grenfell and A. Hunt, *New Classical Fragments and Other Greek and Latin Papyri* (Oxford 1897)

P. Gron.: A. Roos, *Papyri Groninganae, Griechische Papyri der Universitätsbibliothek zu Groningen* (Amsterdam 1933)

P. Hamb.: P. Meyer, *Griechische Papyrusurkunden der Hamburger Staats- und Universitätsbibliothek* (Leipzig 1911-1924)

P. Hib.: B. Grenfell and others, *The Hibeh Papyri* (London 1906, 1955)

P. Iand.: C. Kalbfleisch and others, *Papyri Iandanae* (Berlin and Leipzig 1912———)

P. Lond.: F. Kenyon and H. Bell, *Greek Papyri in the British Museum* (London 1893———)

P. Lille: P. Jouguet, *Institut papyrologique de l'université de Lille: Papyrus grecs* (Paris 1907-1928)

P. Lugd.-Bat.: *Papyrologica lugduno-batava* (Leiden 1941———)

P. Mert.: H. Bell and others, *Catalogue of the Greek Papyri in the Collection of Wilfred Merton* (London and Dublin 1948——)

P. Mich.: H. Youtie and others, *Michigan Papyri* (Ann Arbor 1931——)

P. Mon.: A. Heisenberg and L. Wenger, *Byzantinische Papyri, Veröffentlichungen aus der Papyrus-Sammlung der K. Hof- und Staatsbibliothek zu München* 1 (Leipzig 1914)

P. Oxy.: B. Grenfell, A. Hunt, and others, *Oxyrhynchus Papyri* (London 1898——)

P. Pan.: T. Skeat, *Papyri from Panopolis* (Chester Beatty Monographs 10, Dublin 1964)

P. Petrie: J. Mahaffy and J. Smyly, *The Flinders Petrie Papyri* (Dublin 1891-1905)

P. Rend. Harris: J. Powell, *The Rendel Harris Papyri of Woodbrooke College* (Cambridge 1936)

P. Ross.-Georg.: G. Zereteli, *Papyri russischer und georgischer Sammlungen* (Tiflis 1925-1935)

P. Ryl.: A. Hunt and others, *Catalogue of the Greek Papyri in the John Rylands Library at Manchester* (Manchester 1911——)

PSI: Papiri greci e latini (Florence 1912——)

P. Teb.: B. Grenfell, A. Hunt, and others, *Tebtunis Papyri* (London 1902-1938)

P. Thead.: P. Jouguet, *Papyrus de Théadelphie* (Paris 1911)

Per. Mar. Eryth.: Periplus maris erythraei, cited from H. Frisk, *Le périple de la Mer Érythrée* (Göteborgs Högskolas Årsskrift 33.1, Göteborg 1927)

Phil. Woch.: Philologische Wochenschrift

Poidebard-Lauffray: A. Poidebard and J. Lauffrey, *Sidon, Aménagements antiques du port de Saida* (Beirut 1951)

Prak. Ath.: Praktika tes en Athenais Archaiologikes Hetaireias

RA: Revue archéologique

RE: Paulys Real-Encyclopädie der classischen Altertumswissenschaft

REA: Revue des études anciennes

REG: Revue des études grecques

Reisner: G. Reisner, *Models of Ships and Boats* (Catalogue général des antiquités égyptiennes du Musée du Caire, nos. 4798-4976 et 5034-5200, Cairo 1913)

REL: Revue des études latines

Riv. di fil.: Rivista di filologia e di istruzione classica

378

Rodgers: W. Rodgers, *Greek and Roman Naval Warfare* (Annapolis 1937)

ROL: E. Warmington, *Remains of Old Latin* (Loeb Classical Library 1935-1940)

Röm. Mitt.: *Mitteilungen des deutschen archäologischen Instituts, Römische Abteilung*

RPh: *Revue de philologie*

Rostovtzeff, *SEHHW*: M. Rostovtzeff, *The Social and Economic History of the Hellenistic World* (Oxford 1941)

Rostovtzeff, *SEHRE*: M. Rostovtzeff, *The Social and Economic History of the Roman Empire* (Oxford 1957²)

Rougé: J. Rougé, *Recherches sur l'organisation du commerce maritime en Méditerranée sous l'empire romain* (École pratique des hautes études. vi*e* section, Centre de recherches historiques, Ports, Routes, Trafics 21, Paris 1966)

RSL: Rivista di studi liguri

Salonen, *Wasserfahrzeuge*: A. Salonen, *Die Wasserfahrzeuge in Babylonien* (= *Studia Orientalia* viii.4, Helsinki 1939)

Salonen 1938: A. Salonen, "Zum Verständnis des sumerischen Schiffbautextes AO 5673 mit Berücksichtigung des Textes VAT 7035," *Studia Orientalia* viii.3 (1938) 3-23

Salonen 1939: same as Salonen, *Wasserfahrzeuge*

Salonen 1942: A. Salonen, "Nautica Babyloniaca," *Studia Orientalia* xi.1 (1942) 1-118

Sammelb.: F. Preisigke, F. Bilabel, and others, *Sammelbuch griechischer Urkunden aus Ägypten* (Strassburg, Berlin and Leipzig 1913——)

Sandberg: N. Sandberg, ΕΥΠΛΟΙΑ (Göteborgs Universitets Årsskrift 60 (Göteborg 1954)

Sargent: R. Sargent, "The Use of Slaves by the Athenians in Warfare. II, In Warfare by Sea," *CPh* 22 (1927) 264-79

Schmidt: K. Schmidt, "Die Namen der attischen Kriegsschiffe" (Doctoral dissertation, Leipzig 1931)

SEG: Supplementum epigraphicum graecum

Segre, "Dedica": M. Segre, "Dedica votiva dell' equipaggio di una nave rodia," *Clara Rhodos* 8 (1936) 225-44

Segre, "Due nuovi": M. Segre, "Due nuovi testi storici," *Riv. di fil.* 60 (1932) 446-61

Select Papyri: A. Hunt and C. Edgar, *Select Papyri* (Loeb Classical Library 1932-1934)

Singer: C. Singer and others, *A History of Technology* (Oxford 1954-1958)

Smith: James Smith, *The Voyage and Shipwreck of St. Paul* (London 1848)

SNG: Sylloge nummorum graecorum

Starr: C. Starr, *The Roman Imperial Navy 31 B.C.–A.D. 324* (Cambridge 1960²)

Stud. Pal.: C. Wessely, *Studien zur Paläographie und Papyruskunde* (Leipzig 1901——)

Syll.: W. Dittenberger, *Sylloge inscriptionum graecarum* (Leipzig 1915-1924³)

TAPA: Transactions of the American Philological Association

Tarn, "Alexander's Plans": W. Tarn, "Alexander's Plans," *JHS* 59 (1939) 124-35

Tarn, "Ded. Ship": W. Tarn, "The Dedicated Ship of Antigonus Gonatas," *JHS* 30 (1910) 209-22

Tarn, *Hell. Dev.*: W. Tarn, *Hellenistic Military and Naval Developments* (Cambridge 1930)

Tarn, "Warship": W. Tarn, "The Greek Warship," *JHS* 25 (1905) 137-56, 204-18 (included as an appendix in Torr²)

Taylor: J. du Plat Taylor, *Marine Archaeology* (London 1965)

Testaguzza: O. Testaguzza, *Portus. Illustrazione dei porti di Claudio e Traiano e della città di Porto a Fiumicino* (Rome 1970)

Thiel, *History*: J. Thiel, *A History of Roman Sea-Power Before the Second Punic War* (Amsterdam 1954)

Thiel, *Studies*: J. Thiel, *Studies on the History of Roman Sea-Power in Republican Times* (Amsterdam 1946)

Throckmorton: P. Throckmorton, "The Antikythera Ship," *Transactions of the American Philosophical Society* 55, Part 3 (1965) 40-47

Throckmorton, "Torre Sgarrata": P. Throckmorton, Preliminary report (1969) on the Torre Sgarrata wreck off Taranto (2nd-3rd A.D.), a copy of which was kindly sent to me by the author, who directed the excavation.

Throckmorton-Kapitän: P. Throckmorton and G. Kapitän, "An Ancient Shipwreck at Pantano Longarini," *Archaeology* 21 (1968) 182-87

TLL: Thesaurus Linguae Latinae

Torr: C. Torr, *Ancient Ships* (Cambridge 1895; 2d edn., Chicago 1964)

Ucelli: G. Ucelli, *Le navi di Nemi* (Rome 1950²)

UPZ: U. Wilcken, *Urkunden der Ptolemäerzeit* (Berlin 1922-1937)

van Doorninck: F. van Doorninck, Jr., "The Byzantine Shipwreck at Yassi Ada" (University of Pennsylvania dissertation [unpublished], 1967)

Vegetius: F. Vegetius, *Epitoma rei militaris*, ed. C. Lang (Leipzig 1885)

Wallinga, *Boarding-Bridge*: H. Wallinga, *The Boarding-Bridge of the Romans* (Historische Studies uitgegeven vanwege het Instituut voor Geschiedenis der Rijksuniversiteit te Utrecht, Groningen 1956)

Wallinga, "Nautika": H. Wallinga, "Nautika (1): The Unit of Capacity for Ancient Ships," *Mnemosyne* 17 (1964) 1-40

W. Chrest.: L. Mitteis and U. Wilcken, *Grundzüge und Chrestomathie der Papyruskunde.* Erster Band, von U. Wilcken (Leipzig and Berlin 1912)

Whibley: L. Whibley, *A Companion to Greek Studies* (Cambridge 1931⁴)

abaft: to the rear of

abeam: on either side of the middle area

aft: toward the rear

aloft: in the rigging high above the deck

amidships: in the central part

arms (of an anchor): the lateral extensions, ending in points and often curved, at one end of an anchor; see Fig. 185

backstay: supporting line running from the mast aft; see Fig. 172

ballast: heavy material set deep in a ship's hold to keep the ship stable

beam: the breadth measured at the widest part

beams: horizontal timbers running from side to side that support the decks

belaying pin: fixed wooden pin to which lines are secured

below: under the main deck

bilge: the bottom of the hull, the part the ship would rest on if it ran aground

block: pulley

bollard: post on quay to which mooring lines are secured

bolt-rope: rope along the edge of a sail to give strength there

boom: spar along the foot of a sail

bow: the front part of a ship

bowlines: lines run from the leech of a square sail forward

bow oarsman: the oarsman furthest forward

braces: lines attached to the ends of the yard; see Fig. 171

brails: lines for controlling the area of sail exposed to the wind

bulkhead: wall or partition

bulwark: parapet around an exposed deck

bunt: main body of a sail, belly of a sail

capstan: winch with upright spindle used for moving particularly heavy objects such as the anchor

carlings: short timbers running fore and aft between the beams

carvel-built: said of ships built with planks laid flush, edge to edge

cathead: projecting timber from which the bow anchor can be slung

clew: lower corner of a sail

clinker-built: said of ships whose hull planks overlap each other

companionway: passage, with its ladder, that leads from one deck to another

counter: extension of a vessel's hull aft beyond the sternpost

crutch: wooden prop on which the mast rests when lowered

cutwater: the portion of a vessel's stem that cleaves the water as she moves

deck: a horizontal level in a ship

draft: the depth of water a ship draws, the vertical distance from the bottom of the keel to the waterline

fairleads: rings or eyes or loops to guide a line in a given direction

false keel: an addition to the main keel placed under it

floors, floor timbers: the part of a frame that fits into, or spans, the bottom of the hull

flukes: pointed heart-shaped tips at the ends of an anchor's arms

fore-and-aft rig: a rig whose sails run fore and aft, are aligned with the keel; see Fig. 173

forecastle: the space in the very forward part of the ship

forestay: supporting line running from the mast forward; see Fig. 172

forward: toward the front

frames: the lateral timbers that, appended to the keel, give the vessel its shape, as ribs do in the human torso

freeboard: distance from the waterline to the main deck

futtocks: parts of a frame, particularly the parts that fit along the sides of the hull

galley: the kitchen

gangplank: plank temporarily extended from ship to shore for embarking and disembarking

garboards: the strakes on either side of the keel

gaskets: small cords by which a sail, when furled, is kept bound

gunwale: the uppermost course of planking on a ship's side

halyards: lines for hoisting sail; see Fig. 171

hatch: opening in the deck through which cargo is raised or lowered

helm: the steering apparatus

hold: the space below the main deck where cargo is stored

hull: body of a ship

384

inboard: within the ship

keel: the backbone of a ship; it runs along the lowest part of the hull from stem to stern

knot: nautical mile (6080′) per hour, the standard way of measuring a ship's speed

lapstrake: same as clinker-built

lee: away from the wind

leech: the edge of a square sail that is away from the wind; the after edge of a fore-and-aft sail

leeward: toward the lee side, toward the side away from the wind

lifts: lines running from the masthead to the yard; see Fig. 171

loom: inboard part of an oar

luff: the edge of a square sail that is toward the wind; the forward edge of a fore-and-aft sail

marina: a yacht harbor

moor: secure a ship when it is not in motion

oculus: device in the form of an eye or the like traditionally painted on ships

parral: collar, generally of cord, that holds the yard to the mast

partners: framework which supports the mast at deck level

poop: the rear part of a ship, a deck at the rear

port: the lefthand side facing forward

port (noun): on oared vessels, an opening for working the oar

prow: the front part of the bow

purchase: a block and tackle

quarter: either side of the ship near the stern

quarterdeck: the part of the main deck abaft the mast

quartermaster: petty officer who attends to the helm and assists with navigation

reach: to sail with the wind on the quarter or abeam

rigging: the lines fitted to masts, yards, and sails; see Figs. 171-172

run: to sail with the wind aft

running rigging: the lines that control the movements of sails and spars

GLOSSARY OF NAUTICAL TERMS

scarf: to join two timbers by sloping off the ends of each and fastening
   them together so that they make one piece of uniform size
shank (of an anchor): the main shaft or leg of an anchor; see Fig. 185
sheets: lines attached to the lower corners of a sail; see Fig. 171
shrouds: supporting lines running from a mast to the sides of the hull;
   see Fig. 172
spinnaker: racing sail of immense spread so set as to belly out before
   the wind
sprit: spar that supports a spritsail; it runs from the lower part of the
   mast to the upper after corner of the sail; see Fig. 173
square sail: sail that is set athwartship
standing rigging: rigging that supports a mast
starboard: the righthand side facing forward
stay: line that supports a mast or spar
stem: the foremost timber of the ship, rising up from the forward tip
   of the keel
stern: the rear
stock (of an anchor): lateral bar at one end of an anchor; see Fig. 185
strake: a line of planking extending the length of a vessel
stroke oarsman: the oarsman nearest the stern
superstructure: complex of structures rising above the main deck

tabernacle: housing for the lower part of a retractable mast
tack (verb): to sail at an angle to the wind closer than 90 degrees
tack (noun): the forward lower corner of a sail
tack-rope: line to hold down the tack of a sail
taffrail: bulwark around the stern
tholepin: pin against which the oar is worked
thwarts: cross planks which serve as seats for the oarsmen
tiller: lever fitted to the head of the steering oar or rudder
top: a platform placed at the head of a mast
treenails: wooden pegs used as fastenings

vang: line running from the peak of a sprit, gaff, or lateen yard to the deck

wale: course of planking added on the outside of the ordinary hull planking
waterline: point on the hull that the water reaches when the ship is floating normally

weather (adj.): toward the wind

wing and wing: of a fore-and-aft rigged vessel traveling with sails swung out laterally, one to either side

wooldings: bindings to hold together and stiffen a mast or spar

yard: spar along the head of a sail

GLOSSARY OF GREEK AND
LATIN TERMS

ENTRIES marked (L.) are Latin, all others Greek. Simple numbers refer to pages of the text, compound to chapter and notes. Thus:

gomphos: treenail   223; 4.19; 9 App., n. 26; 10.22

means that the term in question is mentioned on page 223, in note 19 to Chapter Four, in note 26 to the Appendix to Chapter NINE, and in note 22 to Chapter TEN.

acharaktos: "unmarked," of a boat without a name   15.2
achor ( ):   343
actuaria (L.): type of merchant galley   159; 6.102, 8.3, 13.52
actuariola (L.): small actuaria   159; 8.4, 9.8, 12.43, 14.52
adiutor (L.): clerk, a rating in the Roman navy   314
akateion: of the emergency rig in war galleys   236-37, 241
akation: small akatos   154, 159, 264
akatos: type of merchant galley   159-60, 162-63, 338; 11.93, 11.96, 12.19, 13.68
akor ( )   343
akrostolion: sternpost   86, 110; 6.47
akroterion: sternpost   86
ammokopregon: sand and manure carrier   14.60
amphimetrion: ship's floor   222
ancorale (L.): anchor cable   11.101
ankoina: halyard   230, 260-62; 11.23
ankyreion: anchor cable   11.101
anquina (L.): = ankoina
antemna (L.): yard   232
antiprosopon: figurehead   13.67; 15.2
antleterion: bucket for emptying the bilge   9.42
antlia, antlos: bilge   9.39
aperta (navis) (L.): aphract   134
aphlaston: sternpost, ornament atop the sternpost   86, 96, 110, 116, 147, 158, 346; 4.19, 6.47
apobathra: gangplank   11.104, 11.133
archinauphylax: chief ship's guard   13.37
apogaion, apogeion: mooring line   11.101

arbor (L.): mast 231
armenon: sail 233
artemo (L.) = artemon
artemon: bow-sail, foresail 147, 157, 238, 240, 242-43, 264; 11.24, 11.88
asanidon: "unplanked," of open areas on Byzantine galleys 7.44
asemos: "unmarked," of a boat without a name 15.2
askoma: oarport cover 83, 87; 5.52
auchen: loom 11.2, 11.7
auletes: piper 304, 309

baioielypon: type of small craft 333
baris: type of Nile craft 335, 341
beneficarius (L.): administrative officer in the Roman navy 314
bicrota (L.): two-banked war galley 6.126
biremis (L.): two-banked war galley 133; 7.6
bolizo: heave the lead 11.85
bucinator (L.): bugler 312

carchesium (L.) = karchesion
carina (L.): keel 222
castellum (L.): pavesade 154; 7.46
catapirates (L.) = katapeirates
caudicaria (L.): type of coastal and rivercraft 332, 338
caupulus (L.): type of small craft 14.7
celeuma (L.) = keleuma
celeusta (L.): officer on Roman galleys 310
celox (L.): type of merchant galley 160
cercurus (L.) = kerkouros
ceruchus (L.) = kerouchos
chalinos: parral 230, 260-61, 263
chalkoma: bronze sheath for the ram 5.43
cheires siderai: "iron hands," i.e. grapnel 6.87
cheirosiphon: hand gun for Greek fire 7.56
chelandia: type of Byzantine war galley 7.34, 7.39
chelysma: false keel 221; 5.45
cheniscus (L.) = cheniskos
cheniskos: goose-headed sternpost ornament 13.67, 15.2, 15.15
choma: mole 362
clavus (L.): tiller 11.2, 11.10, 11.19

390

constrata (navis) (L.): cataphract 134
constratum (L.): deck 9.54
contus (L.) = kontos
corbita (L.): type of merchantman 169
cornu (L.): yardarm 232; 12.13, 12.14, 12.17, 12.19
coronarius (L.): rating in the Roman navy 314
corvus (L.): "raven," a type of boarding bridge 121
costa (L.): frame 221
cumba (L.) = kymbe
cuneus (L.): tenon 222; 10.7
cursus velox (L.): special service of the Roman public post 334
cybaea (L.) = kybaia
cydarum (L.) = kydaron

decemscalmus (L.): having 10 tholes a side 8.4
dekatarchos: junior officer in Greek navies 309
dekeres: a "ten" 5.5, 6.118; 6 App. 2, n. 5
delphinophoros: "dolphin-bearing," of yards fitted to drop weights 11.67
desmos: joint 222; 10.22
despotes: shipowner 13.67
diaeta (L.) = diaita
diaetarius (L.): type of seaman on a merchant ship 13.83
diaita: cabin 9.63
diarmenos: two-masted 268; 11.74, 14.49
diekplous: type of naval maneuver 280
dieres: two-banked galley 5.5, 7.62
dikrotos, dikroton: two-banked, two-banked galley 133, 135, 354; 4.82,
    4.104, 5.5, 6.55, 6.94-95, 6.124, 6.126-29; 6 App. 1, n. 1; 7.4, 8.15, 8.58,
    12.44
diopos: officer on a merchant ship 13.83
diproros: "double-prowed," of catamaran hulls 6.47, 6.58, 14.67
diprymnos: "double-sterned," of catamaran hulls 6.47, 14.67
diskalmos: having two tholes a side 8.16
dokos: beam 220
dolator (L.): rating in the Roman navy 313
dolon: emergency rig of a war galley 237-38, 264, 267; 13.59
dominus (L.): shipowner 13.67, 13.70
dromon: type of galley in the Byzantine navy 148-54
dromonarion: type of galley in the Arab navy 154

dromonion: small dromon   154
dryochos: stock   223
duplicarius (L.): rating in the Roman navy   312; 13.48

eikoseres: a "twenty"   6 App. 2, n. 17
eikosoros: "twenty-er"   9 App., n. 33; 14.76
elaiochreistes: rating in Greek navies   308, 311
elakate: "spindle," used of the narrow uppermost part of a mast   11.37
elephantegos: elephant carrier   14.60
embolos, embole: ram   85; 5.40, 11.5
emporion: commercial harbor   362, 365
enkoilion: part of a frame   221; 9 App., n. 26
enneres: a "nine"   6 App. 2, n. 5
entorneia: parapet (?)   5.59
epakter: fisherman   8.3
epaktris: type of small boat, merchant galley   8.3, 8.25
epaktrokeles: type of merchant galley   161
epaktron: fishing boat   8.3
epenkenis: gunwale   4.19, 7.49
epholkaion: in Homer, forward through-beam   4.20
epholkion: ship's boat   11.93, 11.95
epholkis: ship's boat   11.93, 13.88
epibates: marine   304, 309; 13.24
epidromos: mizzen   242
epigyon: mooring line   11.101
epikopos: oared   8.16, 8.56
epikrion: in Homer, yard   4.34
epiplous: vice-captain of a war galley   307
episeion: pennant   11.86
episemon: name-device   344; 15.7
episkalmis: strake over the tholes   7.49
epitonos: backstay   230, 260-61; 4.36
epotis: forward face of the outrigger   85-86; 7.48, 11.5
ergates: winch   11.137
ergazomenoi en prora (prymne): "workers in the bows (stern)," deck-
    hands of a war galley   309, 311
eune: in Homer, anchor   4.45
exairetos: "select," of galleys in the Athenian navy   92
exceptor (L.): rating in the Roman navy   314

exercitor (navis) (L.): he who profits from the use of a ship, whether owner or charterer   13.67, 13.80, 13.83; 13 App., n. 24

faber navalis (L.): shipwright, ship's carpenter   311, 320; 10.6, 10.24
forus (L.): deck   9.54
funis (L.): rope   11 App. 1, n. 2

galea: type of galley in the Byzantine navy   148
gandeia (L.): type of small craft   343
gaulos: type of Phoenician merchantman   66, 169; 8.7
geseoreta (L.): type of small craft   14.51
gomphos: treenail   223; 4.19; 9 App., n. 26; 10.22
grammateus: officer in Greek navies   307-308, 314
gubernaculum (L.): helm, steering oar   11.2
gubernator (L.): captain, sailing master, helmsman   310, 316, 318

halias: type of small craft   198, 334; 11.95, 14.52
harmos: joint   222; 10.22
harmonia: joint   222; 4.19
harpax: type of grapnel   122
hegemon ton ergon: "chief of the works," a rating in Greek navies, boatswain (?)   308, 311
hekatozygos: in Homer, "having 100 thwarts," of a very large galley   4.19
hekkaidekeres: a "sixteen"   6.10; 6 App. 2, n. 9
hemiolia: type of fast galley   125, 128-32; 4.95
hendekeres: an "eleven"   6 App. 2, n. 7
hepteres: a "seven"   6 App. 2, nn. 2 and 4
herma: ballast   9.43
hexeres: a "six"   78
himas: line, lift   230, 260-61, 263
hippagogos, hippegos: galley for transport of horses   93
hippos: type of Phoenician merchantman   66; 15.2
histion: sail   233; 4.27
histion lepton: "light sail," special equipment on Athenian galleys   11 App. 2, n. 8
histiokopos: type of small craft   8.3
histodoke: in Homer, mast crutch   4.30
histokeraia: main (?) yard   11.34
histopede: in Homer, tabernacle   4.32-33

histos: mast 231; 4.29
holkas: merchant ship 169, 268; 8.1, 11.74, 14.49
hoplitagogos: galley for transport of troops 93
horia (L.): type of small boat 330
horiola (L.): small horia 14.51
hormos: anchorage 362
hortator (L.): time-beater on a galley 13.45
hypera: brace 230, 259-61; 4.37
hyperesia: "service group," certain personnel of a war galley 13.24
hyperesion: cushion 5.22
hypoblema: type of awning or screen 11.61, 11.99; 11 App. 2, n. 3
hypozoma: undergird 91, 147, 250; 10.45; 11 App. 2, n. 3

iatros: doctor 308, 311
ikria: deck 218, 4.4, 9.58, 11.30
insigne (L.): name-device 11.88, 15.2

kalodion: line 260-61
kalos: brail, line 230, 259; 4.39
kalybe: shelter, cabin 14.66
kamax: tiller, steering bar 11.1-2
kantharos: type of small craft 343
karabos: type of galley in the Arab navy 154
karchesion: mast-top 233; 11.96; 11 App. 1, n. 13
karis: type of small craft 343
kasiotikon: type of small craft 343
kastellatos: having a pavesade 154
katablema: type of awning or screen 11.61, 11.99; 11 App. 2, n. 3
katapeirateria, katapeirates: sounding lead 11.85
katapeltaphetes: catapult operator 309
kataphraktos: cataphract 5.56, 6.63
kataskopos (naus): reconnaissance craft 135
katastroma: deck 87, 145; 5.30, 7.44, 9.54, 9.58
keles: type of merchant galley 157, 159, 160-63, 338, 340; 6.94
keletion: small keles 160
keleuma, keleusma: "beat" (for the rowers of a galley) 12.33
keleustes: officer of a war galley 300-304, 308, 310; 12.32, 13.19
keraia: yard 232; 11 App. 1, n. 13
keras: yardarm 232; 12.24, 13.67

kerkouros: type of merchant galley 157-58, 163-66, 198, 338, 340; 6.98, 11.95

kerkouroskaphe: small kerkouros 163, 166

keroiax: lift 230, 263

kerouchos: lift 230, 263

kleis: in Homer, tholepin 4.22

klimakis: ladder 11.104; 11 App. 2, n. 3

klimax: ladder 11.104

koile naus: hold 9.37

kollema: joining (with scarf or lap joint) 10.21

kollesis: joining 10.18

kontos: boat pole 4.47, 11.92, 11.105, 11.133; 11 App. 2, n. 3

kontotites: "pole man," rating in Greek navies 13.23

kontoton: punt 334

kopeter: plank in which tholepins are set 5.34, 5.51

kopodetes: rating in Greek navies 308

kopregon: manure carrier 14.60

korax: type of grapnel 6.89

kouphegon: jar (?) carrier 14.60

krabatos: berth 7.47

krikos: fairlead 11.42, 11.136

krios: type of small craft 343

kybaia: type of merchant galley 158, 166-67, 338, 340; 8.41, 11.63

kybaidion: small kybaia 166-67

kybernetes: captain, helmsman 300-302, 307, 310, 316; 9 App., n. 47; 13.81

kydaron: type of small boat 330

kyknos: type of small craft 343

kymbe: skiff 330, 335

kyrios: shipowner 13.67

lembos: type of galley used for commerce and war 99, 117, 125-27, 133, 135, 142, 159, 162-63, 295, 330, 335-36, 338, 340; 6.98, 8.20, 11.93, 11.95, 14.25

lembus (L.) = lembos

lenuncularius (L.): operator of a lenunculus 336; 14.40

lenunculus (L.): type of small craft 330, 335, 336-37

levamentarius (L.): lighterman 14.44

levamentum (L.): lighter 336; 14.44

libernion, libernos: type of small craft 14.62

librarius (L.): rating in the Roman navy 314

liburna (L.): liburnian; also a type of small craft 340

limen: harbor 362

limen kleistos: fortified harbor 363

linter (L.): type of small craft 333, 336

lithegos (baris): stone carrier 173; 14.60

lithophoros: "stone bearing," of yards fitted to drop stones 11.67

lusoria, lusorion (L.): type of galley used for commerce and war 333, 340; 14.27, 14.62

lyntrarius (L.): operator of a linter 14.40

magister navis (L.): captain, shipmaster 317; 9.63, 11.10; 13 App., n. 24

malus (L.): mast 231

manganon: block 11.136

manipularis (L.): crew member of a Roman war galley 13.58

medicus (L.): doctor 311

mesodme: in Homer, carlings to hold the mast 4.32

metra: floor timber (?) 222

miles (L.): crew member of a Roman war galley 13.58

moneres: single-banked galley 148; 7.31

monokrotos: single-banked 5.5; 6.126, 12.44

monoxylon: dugout 1.21

musculus (L.): type of small boat 330

mydion: type of small boat 330

myoparo (L.) = myoparon

myoparon: type of fast galley 128, 132, 135; 8.58

myriagogos, myriophoros: "with a capacity of 10,000," descriptive of the size of a merchant ship 9.25

nauarchis: flagship 6.118, 11.88

nauarchos: commodore 13.30

nauclerus (L.) = naukleros

naufylax (L.) = nauphylax

nauklerokybernetes: owner-captain 13.70

naukleros: shipowner or charterer 315; 9.69; 9 App., nn. 35 and 47; 11.40

naumachia: mock naval battle 6.127

naupegion: shipyard 10.6

naupegos: shipwright, ship's carpenter 304, 308, 320; 10.6

nauphylax: ship's guard 309, 312, 320; 13.83

nautes: oarsman 309; 13.23-24, 13.81; 13 App., nn. 7 and 10

naustologos: member of the crew of a merchant ship 13.81, 13.84

navalia (L.): naval base 363, 365

navis longa (L.): war galley 8.7

navicularius (L.): shipowner or charterer 315

neaniskos: marine 9 App., n. 43; 13.33

nekregon: corpse carrier 14.60

neorion: naval base 363

nothos: "bastard," of a small sized sail 11 App. 3, n. 3

neosoikos: ship-shed, boathouse 363

nomeus: frame 222

oiax: tiller 11.2, 11.7

oieion: in Homer, tiller 4.25

oinagogos: wine carrier 9.4

oinegon: wine carrier 14.60

okteres: an "eight" 6.52; 6 App. 2, nn. 12-13

oneraria navis (L.): merchant ship 169; 8.7, 10.50, 11.91

onkos: bracket 11.9

opisthypera: back brace, preventer brace 260-61

optio (L.): officer in the Roman navy 312

ora (L.): mooring line 11.101

oraria navis (L.): coastal craft 14.48

ousia: complement of a galley in the Byzantine navy 149

ousiakos: type of galley in the Byzantine navy 149-50

pakton: type of small Nile boat 342

palus (L.): treenail (?) 223

pamphylos: type of galley in the Byzantine navy 150

parablema: type of awning or screen 5.57, 11.99

pararrhyma: type of awning or screen 5.57, 11.99, 11.61; 11 App. 2, n. 3

paraseion: pennant 11.86

parasema: name-device 344

parastatai: partners 11.59

parexeiresia: outrigger 84; 11.7

pausarius (L.): time-beater, a rating in the Roman navy 310

pedalion: helm, steering oar 4.24, 11.1-2; 11 App. 2, n. 3

pedaliouchos: quartermaster, a rating in Greek navies 308; 13.9

peisma: mooring line 4.43

pentaskalmos: having 5 tholes a side 8.16

pentekaidekeres: a "fifteen" 6 App. 2, n. 9

pentekontarchos: junior officer in Greek navies 302-303, 308-10

penteres: a "five" 78, 97; 7.6

perineos: spare oar for a war galley 5.23

perineos: crew member of a merchant ship 319

periplous: pilot's manual, coast pilot 280; 11.83

pes (L.): sheet 12.13, 12.17-19, 12.43

petasos: "spreader," part of the housing for the steering oar of a gal-
    ley 7.48

phaselion: small phaselos 167-68, 340

phaselos: type of merchant galley 132, 157, 163, 167-68; 6.125, 11.93

phaselus (L.) = phaselos

phikopedalos: "with fig (?) shaped steering oar," used of certain small
    craft 14.63

phoinikis: signal flag 11.89

pistikos: shipmaster 317

pitulus (L.): rating in the Roman navy 311

pitylos: "plash," used of the oars of a galley 12.37

placida (L.): type of small craft 14.7

platypegion: type of small craft 334

pleura: frame 221

ploion: ship, boat 169; 6.128, 8.1, 8.7

ploion Hellenikon: "Greek-style boat," used of certain small craft 340

ploion zeugmatikon: "yoked boat," catamaran (?) 334

polykopon: type of oared small craft 334

polyzygos: "having many thwarts," in Homer descriptive of a large gal-
    ley 4.19

pons (L.): gangplank 11.104

ponto (L.): type of merchantman 169

portisculus (L.): time-beater on a galley 311; 13.45

pous: sheet 230, 259-60; 4.38, 12.13

praetoria navis (L.): flagship 11.88, 11.91

pristis: type of war galley 127; 6.104, 6.108

proem bolion: subsidiary ram 85

professionarius de ciconiis (L.): crane operator 16.42

pronaukleros: representative of a shipowner or charterer 13.69

prorates: bow officer of a war galley 300-302, 307

398

proreta (L.): bow officer of a war galley, first mate of a merchant ship 310

proretes = prorates

proreus: bow officer of a war galley, first mate of a merchant ship 303, 318-19; 9 App., n. 47

propes (L.): tack rope (?) 11 App. 1, n. 1

prosumia (L.): type of small craft 338

protonos: forestay 230, 260-61; 4.31, 11.34, 12.26

prymnesia: mooring lines, especially stern lines 4.44, 11.101

pterna: heel of a mast 7.60, 11.37

pteryx: blade of an oar 11.2

quadrieres (L.): a "four" 7.6

ratiaria (L.): a type of small boat 330

ratis (L.): boat, raft 217

remex (L.): oarsman 6.10

remus (L.): oar 6.10

rhothios: "rush-roaring," used of a galley spurting 12.37

rudens (L.): brail, line 230, 259

saburra (L.): sand, ballast 9.43, 14.8

saburrarius (L.): "sand-man," i.e. handler of ballast 9.43, 16.46

saccarius (L.): "sack-man," i.e. stevedore 16.44

Samaina, Samia: type of Samian war galley 4.104

sanidoma, sanidomata: deck 7.44, 9.54

sanis: plank, strake 223

scala (L.): ladder 11.104

scapha (L.): ship's boat 11.93-94

scapharius (L.): type of boatman 14.40

schedia: boat, raft 217

schiston: "split," part of the housing for the steering oar of a galley 7.48

schoinion: rope 11.101

scriba (L.): officer in the Roman navy 314

secutor (L.): junior administrative officer in the Roman navy 314

selma: thwart, beam, decking 220, 223

semeion: flag, pennant 246; 11.88

sentina (L.): bilge 9.39

sentinaculum (L.): bucket for emptying the bilge 9.42

septesemiodialis (L.): "seven and one-half," used of a rating in the Roman navy 13.53

siparum (L.): topsail 241
sipharos: topsail 241
siphon: cannon for Greek fire 152; 7.50
sitagogos: grain carrier 9.4
sitegos, sitegon: grain carrier 173; 14.60
skalmos: tholepin 86; 7.49
skaphe: skiff, ship's boat, barge 330, 335-36; 11.93, 11.96
skaphos: ship's boat 11.93
skene: shelter, cabin 8.41
skeuophylax: guard of equipment and gear 6 App. 2, n. 17; 13.37
spathe: anchor stock (?) 11.138
speculatoria (navis) (L.): reconnaissance craft 135
stamen (L.): part of a frame, frame 221; 4.19; 9 App., n. 26
statumen (L.): frame 221
steira: in Homer, cutwater 221; 4.19
sterea (tropis): "stiffener keel," fore-and-aft timber running the length
    of a galley 7.45
stereoma: cutwater 221; 5.45
stlatta (L.): type of small riverboat 333
stoichos: file of rowers 6.55
strategis: flagship 354
stratiotis: galley for transporting troops 93
stuppa (L.): tow for caulking 10.39
stuppator (L.): caulker 10.39
stylis: identification device carried at the stern 86, 116, 147, 247, 346; 15.5
suboptio (L.): junior officer in the Roman navy 312
subscus (L.): tenon 222
subunctor (L.): rating in the Roman navy 311
symphoniacus (L.): rating in the Roman navy 312
synnaukleros: co-owner or charterer of a ship 315
synkekrotemenos: "beaten together," used of a trained galley crew 12.29,
    12.32

tabellaria (L.): dispatch boat 11.71
tabula (L.): plank, strake 223; 10.46
tacheia naus: "fast galley," a galley in prime condition 5.81
tarrhos: oarage of a galley 5.37; 11 App. 2, n. 3
tecta (navis) (L.): cataphract 134
tessarakonteres: the "forty" 78

400

tessararios: dispatch galley 135

tessararius (L.) = tessararios

tetreres: a "four" 78, 97; 5.5, 7.6

tetreris (L.) = tetreres

textrinum (L.): shipyard 10.6

thalamegos: type of yacht or houseboat used on the Nile 333, 341-42

thalamia: oarport 83

thalamios: thalamite 5.28

thalamos: cabin 9.64

tholos: shelter, cabin 8.41

thorakion: parapet about a mast-top 233

thranos: "platform," i.e. decking 7.44

threnos: in Homer, after through-beam 4.20, 4.72

threnys: stool 5.33

thyrabathra: companionway ladder 11.137

thyreon: "shield strake," strake with ports for lower bank of oars 7.49

toicharchos: petty officer aboard a war galley, purser or supercargo of a merchant ship 304, 309, 319

toichos: side of a ship 114; 4.19

tonos: device for tightening a hypozoma 92

tonsilla (L.): mooring stake 11.140

topeia: cordage 11.20, 11.101; 11 App. 2, n. 3

toxotes: archer 305, 309

trachelos: "neck," central part of a mast 11.37

tragos: type of small craft 343

transtrum (L.): thwart, beam, hatch 220; 9.47; 11 App. 1, n. 13, 14.58

trapeza: mast step 7.60

trema: boring," i.e. oarport 7.49

triakonteres: a "thirty" 6 App. 2, n. 17

triarmenos: three-masted 115; 6.83; 9 App., n. 29; 11.72, 11.75

triemiolia: type of war galley 128, 129-31, 134, 309, 358; 6.94, 6.128, 13.38; 13 App., n. 8

trierarchos: contributor to the upkeep of a war galley, captain of a war galley 301-302, 304, 307

trierarchus (L.): captain of a galley in the Roman navy 310; 13.63; 13 App., n. 13

trieraules: piper aboard a war galley for piping the time to the rowers 304, 309

trieremiolia: another form of triemiolia 6.120

trieres: trireme, war galley 77-78, 84; 5.3, 5.5, 6.55, 6.118, 7.6
trikrotos: three-banked 5.5, 6.55
triparodos: having 3 gangways 9.57
triremis (L.): trireme 5.3
triskaidekeres: a "thirteen" 6 App. 2, n. 8
triskalmos: having 3 tholes a side 158
trochileia: block and tackle 11.136
trochileion: block 11.136
trochlea (L.) = trochileia
trochos: sheave 11.136; 11 App. 3, n. 2
tropis: keel 222; 4.19
tropos: oar strap 86; 4.23
tropos: beam 220
tropoter: oar strap 86-87; 5.22, 7.49
tutela (L.): image of the patron deity of a ship 347

urinator (L.): diver 16.45
utricularius (L.): "bladder-man," operator of a float, pontoon bridge, or the
    like 1.3

vegeiia (L.): type of small craft 343
vehigelorum (L.): type of river craft 14.78
velarius (L.): deckhand, a rating in the Roman navy 310
velum (L.): sail 233
versoria (L.): part of the rigging 11 App. 1, n. 2
versus (L.): file of rowers 6.10, 6.55
vetutia (L.): type of small craft 14.78
vexillum (L.): flag, pennant 11.88
victimarius (L.): rating in the Roman navy 314

xylegon: wood carrier 14.60
xylokastra: fighting platform on a war galley 7.50-51

zeugle: pennant (rope) 11.8
zeukteria: pennant (rope) 11.17
zoster: wale 223; 5.46
zygios: zygite 5.32
zygon, zygos: thwart, beam 220-21; 4.19, 5.32, 9.64

GENERAL INDEX

For the system of reference, see p. 389.

Abdera 284
Acanthus 14.52
Achaea 6.127
Achradina 196
Actium, Battle of 99, 117, 141, 155, 354; 5.5, 6.88, 6.91-92, 7.17, 9.42, 11.45, 11.54
Adriatic 273, 330; 8.16, 8.54, 8.57, 9.26; 13 App., n. 24
Aegates Islands, Battle of 6.86, 11.53
Aegospotami, Battle of 11.54, 12.44
Aemilius Paulus 119, 292; 6 App. 2, n. 12
Aeschylus 10
Africa 8, 9, 23, 201, 272, 283, 293-94, 297, 343; 1.21, 1.24, 4.115, 6.98, 7.31, 11.63, 11.92, 11.95, 11.114, 12.49
Agathocles 12.108
age of ships 90, 119-20
Agrippa 122; 5.30, 12.32
Agrippa (Herodes I) 12 App., nn. 1, 9
Ahenobarbus, Gn. Domitius 6.127 11.88
Alcaeus 262
Alcibiades 6.87, 11.45, 11.56, 11.89, 12.33, 12.96
Alexander 97-98, 125, 136, 163, 354; 1.1, 5.82, 6.26, 6.73, 6.88, 6.115, 8.38, 14.29
Alexandria 136, 164, 173, 184, 186, 188-89, 199, 268, 283-84, 287, 289, 293, 297-99, 366; 6.121; 6 App. 1, n. 1; 7.2, 7.8, 8.7, 8.17, 8.23, 8.36, 8.50, 11.17, 11.71, 11.75, 12.3, 12.82, 12.88, 13.59, 13.60, 13.68; 13 App., nn. 9-10; 14.41, 14.52, 15.2, 15.12 15.57, 15.66
Alexandria Troas 12.93
Almeria 12.86
Alpheus 286; 12.65
altar 182
Althiburus mosaic 161-62
Amisus 12.93
Anacharsis 11.119
amphora 170-71, 173, 177, 184, 190, 199, 369-70
anchor 48, 198, 250-58; 11.17, 11.96
anchor cable 250, 252, 340; 11.96

animal-shaped figurehead 37-38, 64, 66, 68, 96, 335, 347-48; *see also* goose-head, and cheniskos in Glossary of Greek and Latin Terms
Anquillaria 294
Antigonids 354
Antigonus I 98, 137; 11.25
Antigonus Gonatas 98, 102, 115, 139-40; 6.68, 6.83; 6 App. 2, n. 12
Antioch 12.3
Antiochus III 6.94, 6.102, 6.104, 8.15, 11.39, 12.45
Antium 12.87
Antonius Creticus 6.127
Antony 99, 141, 155; 6.92, 6.125, 6.127, 9.42, 11.54, 11.88
Apamea 6.102
aphract 50, 54-55, 60, 64, 74-76, 130, 134-35, 142-43, 150; 4.65, 6.102, 6.115, 6.119-20, 6.125, 6.130, 12.93, 13.33
Apollonia 364; 6.106
Apollonius 285
Arab craft 201; 10.13, 10.30, 10.37, 11.77; 11 App. 3, n. 4
Arabia 15-16; 1.3
Arab navy, galleys and fleets of 154, 326-27; 8.14
Arabian Gulf 10.37
Aradus 94; 5.93, 6.73
archers 122, 131, 305, 309, 312
Archias 194
Archimedes 172, 176, 185, 194-95, 197-98
Arginusae, Battle of 322; 8.21, 13.8, 13.22; 13 App., n. 4
Arles 6.82
Armenia 1.10
Artemisia 11.87; 15.2
Artemisium, Battle of 124
artistic conventions 71-74, 76; 4.93
Ascalon 288-89, 291; 8.7, 13.68, 14.52
Asia Minor 17, 272, 324, 354; 8.17, 12.45, 14.52
Asine 32
Assyria 4, 6, 56, 65, 163; 1.1, 1.10, 2.39, 2.58
Astypalaea 6.115

403

405

406

harbors (*cont.*)
Caesarea 16.22
Carthage 16.15, 16.17
Cnidos 16.21
Cumae 16.22
Cyzicus 16.17
Delos 362
Leptis Magna 368; 16.2, 16.22
Misenum 356, 365
Oeniadae 364
Piraeus 363-65
Pompeiopolis 16.22
Portus 189, 336, 367-70; 16.22
Puteoli 16.24
Rome: *see* Portus
Sabratha 16.3
Samos 362; 16.9
Sarepta 16.22
Sidon 16.2, 16.17, 16.32
Syracuse 16.3, 16.17
Thapsus 16.3
Tyre 16.2, 16.17
harbor craft 162, 330, 335-37
Harpalus 15.45
Hasdrubal 11.28
hatch 177; 10.32
Hatshepsut 21, 35, 37; 3.7
Hellespont 273
helmsman 4.4, 6.34; *see also* guberna-
tor, kybernetes *in Glossary of Greek
and Latin Terms*
hemp 194, 231
Heraclea Minoa 293
Heracleia Pontica 115; 6.95; 6 App. 2,
n. 11
Herod 12.5, 12.88
Herodotus 6, 14-15, 58-59, 335, 362; 1.3,
1.10, 4.82
Hiero II 181, 185, 194, 198-99, 240, 359;
6.82, 8.40, 9.5, 9.29, 9.40, 9.65; 9
App., nn. 35 and 47; 10.7, 10.31,
10.40, 11.32, 11.75-76, 11.95, 11.115,
13.69, 13.71, 14.52, 15.14
hold 175-78
Holland 1.20
Holland, ships of 11.113
Himilco 12.113
Homer 53-54, 58, 60, 63, 65, 69, 169;
3 App., n. 4
Homeric ships 43-48, 300; 6.66
horse-transport 93-94; 5.81, 8.38, 15.40

houseboat 333, 341-42
hull shape 20-22, 30, 32-33, 35-39, 44-45,
49-50, 52, 64-67, 75, 89, 106-107,
116-19, 158, 332, 336, 339
of merchant galleys 160-62, 166-68
of merchant ships 173-75
of Roman galleys 143-45
Hydaspes 136, 163; 6.115; 6 App. 1

Ibiza 287
Illyria 125; 6.103, 6.125, 6.127
India 5, 7-9, 15, 24, 201, 220-21; 1.2,
1.21, 11.65, 11.77, 13.1, 14.25
Indian Ocean 9, 201
Indonesia 54
Indus 125, 136; 6.115; 6 App. 1
inflated skins 3-4
insulation 9 App., n. 5; 10.18
Ionia 12.3
Iphicrates 6.99, 11.57
Ipsus 6 App. 2, n. 8
Iraq 25
iron fastenings 10.21, 10.32, 10.37
"iron hands" 6.87
Issa 6.104
Isthmus of Corinth 5.62
Italy 67, 119, 194, 221-22, 272, 368; 1.20,
8.54, 11.114, 12.94, 15.59
Italy, galleys of 6.20, 6.34-35, 6.45,
11.31, 12.37, 13.16, 13.26, 13.30, 13.34-
35, 13.53
Ithaca 12.45

Jason 350
Jews 11 App. 3, n. 1
joinery 15, 25-27, 46, 91, 201-202, 204-
206, 217-18, 341
Julian the Apostate 1.1, 1.3, 14.28
junks 2.48, 9.66, 12.22
Justinian 148

keel 8, 10, 13, 15, 25, 27, 45, 47, 187,
201, 203-204, 207, 212, 214-16, 221-
23, 339-40; 2.17
keelson 207
kelek 4, 23; 1.3, 1.10

ladder 251-52, 258, 364
Lade 11.63
Lampsacus 11.54
lapstrake planking: *see* clinker-built

parral 68, 230, 261, 263
partners 69
passengers 157, 161, 167-68, 172, 177,
    180-81, 196-97, 317, 319, 334, 341
Patras 9.8, 11.105, 12.93
patron deity (of a ship) 146
Pausanias 115
Pausistratus 6.93
pavesade 95, 146, 151, 154; 5.53
pay, of rowers 5.35, 13.7
paymaster 303
pedagna 6.34
Pella 6 App. 2, n. 12
Peloponnesian War 5.1
pennant 246-47, 348
penteconter 44, 54, 56, 58-59, 61-63, 77,
    81, 124, 130, 133, 135, 148, 302, 305;
    5.5, 8.11, 8.38, 11.8, 13.2, 13.19
Peoples of the Sea 36, 38, 69; 3 App.,
    n. 4
Perseus 127
Persia 15-16, 201
Persia, galleys and fleets of 163; 5.81,
    5.93, 8.38, 11.90, 13.7, 13.25
Persian Gulf 23, 201; 1.24, 10.13, 10.30
Persian Wars 5.1
Phaeacians 362
Phalara 12.45
Phalerum 294
Pharos 366; 12.63, 12.91, 12.100; 12
    App., n. 9
Pharsalus 12.14, 13.69
Phaselis 5.60
Phasis 8; 1.21, 12.93
Phenice 6.99
Philae 14.69
Phileas 198
Philip V 98, 119, 125, 127; 6.94, 6.104,
    6.106, 6.108, 6.115, 6.118, 6.127; 6
    App. 2, n. 12; 12.99
Philippi 12.93
Philon 16.13
Phocaea 11.53
Phoenicia 43, 56-58, 60, 65-66, 69, 94-96,
    169; 5.19; 6 App. 1; 10.51, 11.25,
    11.120, 13.82, 15.2
Phoenicia, galleys and fleets of 94-97,
    143; 4.58, 5.36, 6.39, 6.73; 6 App. 1
Phoenician trireme 94-96
Phormio 11.105
Phycus 284

pilote 13.16
pilot manuals 245
piloto 13.16
Piraeus 363-65; 6.88
pirates 125, 128-32, 141, 161; 4.118, 6.95,
    6.114, 6.124, 6.127, 8.25, 10.48; 13
    App., n. 8
Pisa 294
pitch 194, 211-12; 6.88; 9 App., n. 5;
    10.37, 10.39
planked boat 8, 13
planking 8-10, 13, 16, 20, 45, 201-204,
    209, 212, 214-18, 223, 329, 339-40
Pliny the Elder 10
Po 7; 1.14, 1.20
Pollux 115
Polycrates 362; 4.104, 16.9
Polycritus 11.88
polyremes 97-123; 6.126-27
    "six" 78, 98-101, 105-106, 137, 140-42,
        156; 13.58
    "seven" 98, 105-106, 116, 137, 140;
        5.92; 6 App. 1
    "eight" 98, 105-107, 113, 138
    "nine" 98, 106, 137-38, 140
    "ten" 98, 117, 137-38, 140-41; 6.118
    "eleven" 98, 117, 138, 140
    "twelve" 140
    "thirteen" 98, 138, 140
    "fifteen" 98, 112, 115, 138, 140
    "sixteen" 98, 100, 106-107, 112-13,
        115, 119, 137-40
    "twenty" 98, 112, 140
    "thirty" 98, 100, 112, 140
    "forty" 98, 108-12, 113, 115, 117, 140;
        6.33
Polyxenidas 11.53, 11.63
Pompeii 144; 8.56, 12.43
Pompey 120; 6.125, 12.14, 13.69
pontoon 1.3, 1.21
portaging 89
Portus 367-70; 11.74; 12 App., n. 5
posticcio 114
pot raft 5
Praeneste relief 145
preventer brace: *see* back brace
Proconnesus 12.96
projecting forefoot 35, 158, 174, 331; 4
    App. 1, n. 16; 14.11, 14.26, 14.78
promotion 303, 309
prow, shape of 12, 14, 23, 28, 31, 33,

411

415

INDEX OF CITATIONS

For the system of reference, see p. 389.

9.1004-1005
10.494-95

Lucian
Amores 6
Charon 3
Dial. Het. 14.3
Dial. Mort. 4
    10.10
Fug. 13
Jup. Trag. 46
    47
    48
    51
Lexiph. 15
Nav. 4

    5

    6
    7
    9
    14
Pseudol. 27
Quomodo Hist. 45
Timon 17
Toxaris 20

Vera Hist. 1.5
    1.42
    2.41

Lucilius
ROL iii, p. 166
    193

    378

Lucretius
4.903-904
4.1169

Lycophron
618

Lycurgus
in Leoc. 17
    70

Marcianus of Heraclea
Epit. Per. Men. 5

12.100
10 App. 3, s.v.
  beams; 11 App.
  1, n. 13

11.35, 11.43
11 App. 1, n. 1
13.20
10.47
11.101, 11.104
11.111
11.72
15.15
9.40, 9.50
11.111
11.72, 11.75
11.41; 11 App.
  1, n. 13
9.54, 9.63; 9
  App., n. 12;
  11.34, 11.65,
  11.86, 11.107;
  11 App. 1, n.
  4; 13.86, 15.1,
  15.15, 15.66
11.1, 11.2, 11.12
12.91, 13.69
12.17, 13.69
11.75
11.75
11.72
13 App., n. 10
9.66, 11.104,
  11.133
8.11
11.101
15.15

9.8
11 App. 1, nn.
  9, 11
11.85

11.12, 11.15
15.59

9.43

8.28, 14.6
12.93

12.81

Marcus Diaconus
Vita Porph. 6
    26
    27
    34
    37
    55
    56-57

Martial
3.67.4; 4.6.21
10.30.12-13

Memnon
13

29
37

Naumachica
5.2.5
5.2.6
5.2.8

5.2.9
5.2.10
5.2.12-13
5.2.13
5.2.67

New Testament
Acts 16.11-12
  20.6, 14-15; 21.1
  27.5
  27.6
  27.9

  27.11
  27.16
  27.17
  27.28
  27.29
  27.30

  27.32
  27.37
  27.40
  28.11

  28.13

Nicander
Ther. 268-70
    823-24

Nonius
151.18
536.8

12.45, 12.80
12.88
12.79
12.90
12.89
12.78
12.85

12.33
8.57

6.52; 6 App.
  2, n. 13
6.95
6.98, 8.39

7.47
7.48
10 App. 3,
  s.v. *frames*
7.60
7.45
7.49
7.46
7.44

12.93
12.93
12.93
12 App., n. 6
12.5; 12 App.,
  n. 6
13.69
11.93, 11.96
5.73, 10.45
11.85
11.129
11.93, 11.96,
  11.129
11.96
9.26
11.17, 11.70
12.93; 12 App.,
  n. 6; 15.2
12.58

8.6, 11.97, 12.19
8.3

13.45, 13.52
14.51

INSCRIPTIONS

M. Chrest. 244	15.57
341	15.2
P. Aberdeen 70.2-5	13.59, 13.60
P. Athen. 63.18	8.13
P. Bad. ii 26.121	14.30
P. Berol. Frisk	
1 col. 22.3-4	9 App., part 5
P. Cairo Masp.	
67058 col. vi.5	14.70
67136.7	14.32
67359 col. ii.2, vi.9	7.63
P. Cairo Zen.	
59002	8.24, 14.52
59012	8.50, 14.52
59015	8.17, 8.36, 14.52
59029	12.5
59031	14.60
59036	6 App. 2, n. 17;
	13.6, 13.29
59053, 59054	8.41
59054.34-35	9 App., n. 34
59110	8.24
59172.6	9.4
59289.7	14.73
59320	8.50, 14.61
59326.101	14.73
59430.12-13	8.13
59445	9 App., n. 34
59548	8.24
59672	8.24, 14.52
59745.66	14.60, 14.64
59754	11 App. 1; 11
	App. 1, n. 4
59755	11.27
59756	11 App. 1; 11
	App. 1, n. 6
P. Col. Zen.	
63	6 App. 2, n. 17
100	11 App. 1; 11
	App. 1, nn. 4-6
P. Fay. 104.9, 11	14.66
P. Flor. 75.8, 29	13.70
297.474	14.30
335	14.60, 14.66
P. Goodsp. 14.3	14.63
P. Grenf. i 9	13.23; 13 App.,
	n. 8
ii 80	14.32
80-82	13 App., n. 19
81, 81a, 82	14.32
P. Gron. 6.3	8.47

P. Hamb. 57	6.121
74	14.60
P. Hib. 39.4	14.29
100.13	14.64
190.86-88	13 App., nn.
	10, 22
198.90-91	13 App., n. 22
P. Iand. 18.7	14.72
P. Lille 22.5, 23.5	8.45
25.43	14.74
P. Lond. 1164h	9.54, 10.33, 11.9,
	11.39, 11.135,
	14.63
1337.2-3	7.63
1341.12	13.74
1419.1217	14.70
1433.23, 27, 28	13.74
1433.64, 129,	
179, 227,	
319	7.62
1434.22	7.63
1434.35	7.62
1434.135;	
1435.10, 95	7.63
1435.98, 103;	
1441.102;	
1449.94	7.62
1464	7.62, 7.63
1714.31-34	9.54, 11.39,
	11.138, 14.63
Inv. 2305	11.63
P. Lugd. Bat.	
ii 1 col. i.3-5, ii.2-4	13.70
13 col. i.6, 12; 14.3, 10	14.33
xi 1 col. i.4, ii.3	14.63
P. Mert. 19	14.70
P. Mich. 467, 468	13.59, 13.60
490	12 App., nn. 4-5
491	12 App., n. 5
Inv. 4607	8.47
P. Mich. Zen. 10	12.5
17.4	14.73
22.2	8.24
P. Mon. 4.9-12	9.54, 11.39,
	14.63, 15.15
P. Oxy. 1048.2, 7	14.62
1197	14.11
1220.12; 1650	14.70
1650.12	14.11
1650.20	14.68
1651.15	14.11

435

ADDENDA AND CORRIGENDA

In addition to those listed on pp. 371-81, the following abbreviations have been used.

Basch: L. Basch, *Le musée imaginaire de la marine antique* (Athens 1987)
Bass-van Doorninck: G. Bass and F. van Doorninck, Jr., *Yassi Ada*. i, *A Seventh-Century Byzantine Shipwreck* (College Station, Tex. 1982)
Casson, *New Findings*: L. Casson, "Greek and Roman Shipbuilding: New Findings," *The American Neptune* 45 (1985) 10-19
Gianfrotta-Pomey: P. Gianfrotta and P. Pomey, *Archeologia subacquea* (Milan 1981)
IJNA: International Journal of Nautical Archaeology
Morrison: John Morrison, ed., *The Age of the Galley: Mediterranean Oared Vessels since pre-classical Times* (London 1995)
Parker: A. Parker, *Ancient Shipwrecks of the Mediterranean & the Roman Provinces*, BAR [British Archaeological Reports] International Series 580 (Oxford 1992)
Steffy: J. R. Steffy, *Wooden Ship Building and the Interpretation of Shipwrecks* (College Station, Tex. 1994)

Page
xvi, Fig. 28: By great good luck the missing portion of the ship-painting on this clay box was later found. It reveals that the lines leading from the mast to the stern were more or less properly reconstructed but that the sail was not: the reconstructed portion abaft the mast is to be eliminated; there is merely blank space there. See *Praktika tês Arkhaiologikês Hetaireias* (1977) 240-41 and pl. 145; Basch 142, fig. 298C.
xxiv, Fig. 137: For an analysis of the types pictured in the mosaic, see I. Pekáry in *Boreas* 7 (1984) 172-92.
5, note 5: For pot rafts pictured on Minoan seals, see L. Basch, *Cahiers d'archéologie subaquatique* 5 (1976) 85-97.
9-10: The discoveries of marine archaeology reveal that sewing as a means of fastening planks to each other continued throughout ancient times for certain types of craft; see Parker 23.
12: On the origin of the Egyptian sail, see Basch 48-49.
14, note 14: On the Dahshur boats, see also C. Haldane, *AJA* 88 (1984)

389, Steffy 33-36. On the Cheops boat, see also N. Jenkins, *The Boat Beneath the Pyramid* (New York 1980); P. Lipke, *The Royal Ship of Cheops*, BAR International Series 225 (Oxford 1984); Steffy 23-29.

14, note 15: C. Haldane and C. Shelmerdine, *CQ* 40 (1990) 535-39, in an attempt to make Herodotus' words agree with archaeological evidence that on some Egyptian craft the planks were bound together with cords (cf. under 15 below), take ἐπάκτωσαν to mean "bound with cord," not "caulked." But the verb πακτόω, whose root meaning is "close," "stop up," cannot have that sense: see L. Casson, *CQ* 42 (1992) 557, note 17. L. Basch, *Archaeonautica* 6 (1986) 191, aware of the proper meaning, suggests that here the verb refers to the caulking of the holes through which sewing twine passed; Herodotus' words, however, contain no reference to sewing of the planks.

15: In some hulls the planks were lashed together as well as secured by mortise and tenon joints; see C. Haldane, *MM* 74 (1988) 141-52.

17: On the nature of the barges that must have been used to transport the Colossi of Memnon, see J. Wehausen et al., *IJNA* 17 (1988) 295-310.

19-20: M. Isler, *Journal of the American Research Center in Egypt* 28 (1991) 155-85, identifies two upright poles with curved ends that appear on a model of a Nile boat as a form of *gnômôn*, i.e., devices to throw a shadow and thereby serve as a sun-compass. He himself cites the obvious objections (e.g., what need would a boat used on the Nile have had for such an instrument?) and attempts to explain them away (185), not very convincingly.

22-24: See also C. Qualls, *Boats of Mesopotamia before 2000 B.C.* (Diss. Columbia University 1981); M. De Graeve, *The Ships of the Ancient Near East (ca. 2000-500 B.C.)* (Orientalia Lovaniensia Analecta 7, Louvain 1981).

30-31: On the ship pictures from Syros, see also Basch 77-88.

31: L. Basch, *Neptunia* 195 (1974) 19-26, publishes a representation of a ship very similar to the graffito from Cyprus (Fig. 27 above) but shown with people aboard. Four of them, clutching models of boats, are pictured walking off the stern to get ashore (the ship, in the usual fashion, is beached stern to). The bows are held in place by an anchor, as indicated by a line slanting outward and downward from the tall stem that presumably represents the cable.

444

32: Important new evidence for Minoan ships is now provided by a painted frieze from about 1600 B.C. found at Thera, which includes a number of marine scenes. One pictures a waterborne procession involving six ships being paddled, a ship being rowed, a ship under sail, and a number of small craft. Another pictures a naval battle. For interpretation of the scenes, see L. Casson, *IJNA* 4 (1975) 6-7; P. Warren, *JHS* 99 (1979) 115-29; L. Morgan, *The Miniature Wall Paintings of Thera* (Cambridge 1988) 121-45; Basch 118-19; *AM²* 19-20. For discussion of the ships, see Casson, *loc. cit.* 3-10; S. Wachsmann, *IJNA* 9 (1980) 287-95 and in Morrison 16; H. Giesecke, *IJNA* 12 (1983) 123-43 (his argument that the six ships were rowed and not paddled is highly dubious); Basch 119-32.

32-33: On the ship pictures on Cretan seals, see also Basch 93-112 for a detailed analysis, profusely illustrated.

34: Basch (107-12) interprets the ships with two or three poles linked by cross-hatching as having a sail or sails stretched between the poles, a rig still in use in various parts of the world.

For a comprehensive listing of the clay models from Cyprus of this period, see K. Westerberg, *Cypriote Ships from the Bronze Age to c. 500 B.C.* (Gothenburg 1983) 7-15.

35: On the Syrian ships pictured in an Egyptian tomb, see also Basch 63-64.

37: S. Vinson, *Journal of the American Research Center in Egypt* 30 (1993) 133-50, cites examples of brails dating a century to a century and a half earlier than the Medinet Habu relief. He rightly holds that brails were not an Egyptian invention but originated in the Mediterranean and came to Egypt from there.

38: For an analysis of the ships of the Peoples of the Sea and the features they share with contemporary and later Aegean craft, see S. Wachsmann, *IJNA* 10 (1981) 187-220, 11 (1982) 297-304.

41: On Bronze Age models, see also P. Johnston, *Ship and Boat Models in Ancient Greece* (Annapolis 1985) 5-34.

41, note 1: Change to read: "For a résumé of opinions, see Miltner 906, Marinatos 182-83, Kirk 125-27, Johnston (*op. cit.* under 41) 7-10, Basch 84-85. All take the low end to be the prow.

41-42: The Thera paintings (see under 32) furnish the clinching proof that the low end with the projection was the stern. What the function of the projection was is still a puzzle. Varying explanations have been offered: see L. Casson, *IJNA* 4 (1975) 7-9 and 7

445

(1978) 232-33; L. Basch, *MM* 69 (1983) 400-403; Basch 127-30; A. Raban, *AJA* 88 (1984) 11-19; S. Wachsmann in Morrison 15-16.

43: On Greek models of watercraft dating to the period covered in this chapter, see Johnston (*op. cit.* under 41) 35-74.

49-60: On the ships pictured on Geometric vases, see also Basch 161-201.

49: F. van Doorninck, Jr., *IJNA* 11 (1982) 277-86, dates the introduction of the ram ca. 900 B.C.

55-59: On the early versions of the two-banked galley, see also J. Morrison and J. Coates, *The Athenian Trireme* (Cambridge 1986) 30-36; Basch 182-87.

56: On the reliefs from the palace of Sennacherib, see also Basch 311-18.

59, note 82: H. Wallinga, *Ships and Sea-Power Before the Great Persian War* (Leiden 1993) 46-52, argues that the term *penteconter* was reserved for the two-banked 50-oared galley and did not include the single-banked version. *AM* 84-86 = *AM²* 77-78.

60: For a group of vases dating to the 7th B.C. that show a number of galleys all of the same type, see J. Biers and S. Humphrey, *IJNA* 6 (1977) 153-55.

60-65: On 6th-century warcraft, see also Basch 202-33.

65: For a comprehensive list of the clay models from Cyprus of this period, see Westerberg (*op. cit.* under 34) 19-43. See also Basch 253-58.

66: On Phoenician merchantmen, see also Basch 306-318.

66, note 115: A Hebrew seal of the 8th-7th B.C. bears a picture of a seagoing merchantman that has a well-rounded tubby hull with a figurehead ending in an animal's head, most likely that of a horse. The rig consists of a single mast carrying a large square sail, which seems to have a boom along the foot, an unexpectedly old-fashioned feature for this time. See N. Avigad, *Bulletin of the American Schools of Oriental Research* 246 (1982) 59-62; *AM²* 79 and plate 26.

67, note 118: For further examples of ships whose bow ends in a massive spur above the waterline, see J. Hagy, *IJNA* 15 (1986) 235, fig. 19; 236, fig. 21.

68, note 120: See also A. Göttlicher, *Materialien für ein Corpus der Schiffsmodelle im Altertum* (Mainz 1978) 70-76, nos. 374-437.

70: New evidence reveals that some Greek war galleys of this period were two-masted; see L. Casson, *IJNA* 9 (1980) 68-69.

73: See S. Wachsmann in Morrison 25-28 for the presentation and analysis of a recently published ship-picture of the late Bronze Age (1200-1100 B.C.) which, as he points out, supports the way I have read the ship-pictures on late Geometric vases (Figs. 70-71 above).

77-80: The long-standing debate about how a trireme was rowed has now been settled once and for all by the construction of a replica based on Morrison's arrangement; in designing the replica Morrison was ably aided by John Coates, a naval engineer. In a series of exacting trials it performed supremely well, sprinting at nine knots, traveling for hours at four knots with half the crew rowing in turns, executing a 180-degree turn in one minute. The crews needed no endlessly long training; they became adept at the oars within weeks. See Morrison-Coates (*op. cit.* under 55-59) 192-228; Morrison-Coates, eds., *An Athenian Trireme Reconstructed: The British sea trials of* Olympias, *1987,* BAR International Series 486 (Oxford 1989); J. Coates, S. Platis, J. T. Shaw, *The Trireme Trials 1988. Report on the Anglo-Hellenic sea trials of* Olympias (Oxford 1990); J. T. Shaw, ed., *The Trireme Project, Operational Experience 1987-90, Lessons Learnt* (Oxford 1993); B. Rankov, *MM* 80 (1994) 131-46. Basch has criticized the replica (Basch 294) and offered a reconstruction of his own (293), but the criticism has been convincingly answered by Morrison and Coates (*MM* 79 [1993] 131-35), who point out (135-37) the flaws that would make his reconstruction unworkable.

80, note 14: *AM* 102 = *AM²* 92.

80-81: The debate over the date of the introduction of the trireme, whether as early as 700 B.C. or as late as 525, continues; for a convenient listing of recent literature, see F. Meijer, *Historia* 37 (1988) 461, note 1. H. Wallinga favors (*op. cit.* under 59, note 82, 15) the latest possible date, 525. His view is in part influenced by his identification (17-19) of the Athenian *naucraria* as the institutional machinery for turning over privately owned armed vessels to the state for naval service. However, the evidence that the *naucraria* had anything to do with ships is debatable; see V. Gabrielsen, *Financing the Athenian Fleet* (Baltimore 1994) 19-24. J. Morrison, who favors the early date, now adds (in Morrison 54-55) to the evidence he offers in support of his view a detail in the relief from the Palace of Sennacherib (Fig. 78 above), which dates ca. 700 B.C. The war galleys pictured there, he points out, have a superstructure above the upper line of rowers that is lacking in the passenger galleys; this, he suggests, was to accommodate a third line of rowers.

82, line 6: For a new estimate of the length of the cubit in nautical contexts, namely .487-.495 m, see J. Morrison, *Antiquity* 65 (1991) 298-305. This would make the 9-cubit oars between 4.38 and 4.46 m long and the 9½-cubit oars between 4.62 and 4.70 m.

82, lines 7-8: The longer oars were used throughout except near the bow and the stern, where the shorter were used; see Morrison-Coates, eds. (*op. cit.* under 77-80) 85.

82, note 27: The ship pictures on the Ficoronian *cista* (Fig. 106) are well reproduced in T. Dohrn, *Die ficoronische Ciste in der Villa Giulia in Rom* (Monumenta Artis Romanae 11, Berlin 1972), pls. 8, 9.

84, lines 9-12: The experience of the trireme replica reveals that the important cluster consists of a thranite, the zygite just alongside of and below him, and the thalamite just sternwards of and below the zygite; see B. Rankov, *MM* 80 (1994) 139-40.

85: Torr (63-64) calculated the weight of a trireme's ram to be a mere 170 lbs. (77 kg) on the basis of an inscription which presumably gave the price of bronze (*CIA* i 319 = *IG* ii^2 371) and two others which presumably gave the price of trireme rams of stated weight (*CIA* ii 809 e 169-72 = *IG* ii^2 1629.1144-47, *CIA* ii 811 c 87-88 = *IG* ii^2 1631.332-34). The figure is invalid. An improved reading of the inscription from which Torr derived his price of bronze (*IG* i^3 472) reveals that the figure given there is for copper and not bronze, and a re-study of the inscriptions that presumably gave the price for rams reveals that the figures probably refer to scrap bronze and not total rams; see W. Murray, *Greek, Roman and Byzantine Studies* 26 (1985) 141-50.

See under 116 for discussion of a ram found off Athlit in Israel that weighs 465 kg. The ram that was fitted to the trireme replica weighs about 200 kg (Morrison-Coates [*op. cit.* under 55-59] 221).

89, note 59: S. Amigues, *RA* (1990) 92-96, takes *entorneia* as a generic term for "frames."

91: There is now incontrovertible proof that warships, like merchantmen, were constructed with the planks edge-joined by means of mortises and tenons, namely, the timbers still in place in the ram found off Athlit (see under 116 below). Cf. also Diodorus 14.72.5, where he reports that, in defense of Syracuse in 396 B.C., Dionysius' ships "with a multitude of ram attacks smashed [the enemy's] jointed-together planks (*syngegomphômenas sanidas*)."

The experience of building and sailing the trireme replica makes it clear that the *hypozôma* must have stretched horizontally within the hull, in a straight line from stem to stern, in order to counter the effects of hogging. In the replica, for various reasons, a steel hawser was substituted for the rope of the original (Morrison-Coates [*op. cit.* under 55-59] 170-72, 206; Morrison-Coates, eds. [*op. cit.* under 77-80] 22, 27-28).

94-96: On the Phoenician trireme, see also Basch 319-34, J. Coates in Morrison 137-38.

95: A Punic wreck (see under 158 below) has a curiously shaped cutwater that, according to L. Basch, *MM* 69 (1983) 120-42, held a long and slender ram, so designed that, after piercing an enemy's hull, it would break off like a bee's stinger. Though very light, it "would be sufficient," Basch asserts (137), "to hole the 'soft underbelly' of the ship under attack before—predictably—breaking off at its base." We know about the "underbelly" of at least one ancient warship and it was hardly "soft"; its planking was almost twice the thickness of the side planking (see under 116 below). After listing the putative advantages of this kind of ram, Basch adds (137), as if it were but a minor consideration, "There was a price to be paid for these advantages: such a ram . . . could only be expected to serve once in the course of a campaign." The price, as a matter of fact, would amount to a logistical absurdity: any vessel equipped with such a ram, after striking a single blow, even a successful one, would immediately have to turn and race for home before becoming an unarmed target for enemies with conventional rams; any navy using such vessels could only win an engagement if its fleet outnumbered the enemy—and not by just a few units but by a great many, since commanders could hardly count on every one of their precious strokes proving fatal. Cf. Casson, *New Findings* 18, note 24.

95, note 94: On the model, see also Johnston (*op. cit.* under 41) 120-22.

97, lines 9-10: J. Morrison, *Classica et Mediaevalia* 41 (1990) 33-41, points out that the inclusion of "fours" in Dionysius' fleet is doubtful, being based on emendations in Diodorus' text. The "four," he argues (37), developed after the "five," probably around the middle of the 4th B.C. (41).

99: For a recently uncovered large (1.2 m long) wall painting, dated ca. 275-250 B.C., that shows a three-level warship, see L. Basch, *MM* 71 (1985) 129-51. The vessel is probably heavier than a trireme but by no means necessarily heavier than an "eight," as Basch (148) claims. It bears the name *Isis.*

100-116: On the oarage of polyremes, see also J. Morrison and J. Coates, *Greek and Roman Oared Warships, 399-31 B.C.* (Oxford 1995), summarized in Morrison 66-77. On the importance of the catapult in their development, see V. Foley and W. Soedel, *Scientific American* (April 1981) 154-63.

100: J. Morrison similarly concludes (in Morrison 68-71) that the "four" was a two-level galley with both levels double-banked, the "five" a three-level galley with two of the three levels double-banked, and the "six" a three-level galley double-banked throughout. See also J. Coates' reconstructions of a "four" and a "five" so arranged (in Morrison 138, 139).

101, lines 18-19: The "five" would more likely have one thranite, two zygites, and two thalamites; see Morrison (*op. cit.* under 97) 37.

101, note 23: *AM* 162-63, 165, 168-72 = *AM²* 146-48, 150, 153-56.

102-103: On the Victory of Samothrace, see also Basch 354-62, who also reconstructs the vessel as a two-level "four," and J. Coates in Morrison 139-40, who reconstructs it as a *triemiolia.* A votive stele from Mysia in Asia Minor of the 2nd-1st century B.C. bears a picture of a galley, identified by the inscription below it as a cataphract, that has an oar arrangement very much like the Victory of Samothrace (I. and T. Pekáry and E. Schwertheim, *Boreas* 2 [1979] 76-86).

102, note 30: *AM* 152, 168-71 = *AM²* 139, 154-56.

104, note 36: *AM* 124-25 = *AM²* 112.

106: J. Morrison (in Morrison 76) takes the "seven" to have been a slightly heavier version of the "six," i.e., with three levels of oars of which the lowermost had three, instead of two, men to the oar. The "nine," he suggests (77), had two levels of oars, with four men to the oar in one and five to the oar in the other.

111-12: For an example of how the two hulls of the "forty" might have been yoked, see Foley-Soedel (*op. cit.* under 100-116) 160-61.

116: Two recent discoveries indicate how big and powerful were the galleys of this age.

The first occurred in 1980 off Athlit, just south of Haifa, where divers came upon an intact bronze ram with remains of the timbers it encased; see L. Casson and J. R. Steffy, eds., *The Athlit Ram* (College Station, Tex. 1991), for a comprehensive study of the find. The ram was a mighty sheath of high grade bronze that weighed 465 kg and measured 2.26 m at its longest point and 95 cm at its highest (17); it had been cast in one piece, an accomplishment testifying to the remarkable skill of the ancient craftsmen (40-50). The surviving timbers reveal that the* vessel was fitted with massive wales that in places reached almost 27 cm in height and over 20 cm in thickness (22). The planks on either side of the wales were edge-joined to them by means of mortises and tenons (21); the bottom planks were about 7.5 cm thick (19), the side planks just under 4 cm (28). A study of the symbols on the ram and comparison with similar symbols on coins make it likely that the ship the ram came from dates to the first half of the 2nd B.C. and belonged to the Ptolemaic navy (51-66, esp. 65-66).

The second discovery arose from the systematic excavation and careful study by W. Murray of the monument that Augustus erected at Nicopolis to celebrate his victory at Actium; see W. Murray and P. Petsas, *Octavian's Campsite Memorial for the Actian War* (Transactions of the American Philosophical Society, vol. 79, part 4, Philadelphia 1989). In this monument, Murray discerned a precious clue to how big the Athlit ship must have been—as well as to how big and massive in general were the polyremes of this age. Part of the monument consisted of a stone retaining wall, into which had been cut a series of curiously shaped sockets; by excavating these sockets so that they were fully laid bare and could be measured at all points, Murray was able to determine their exact shape and size and also that the sizes in the series diminished from huge to quite small (22-59). Several ancient authors mention that part of the monument was given over to a display of rams from vanquished ships (9-10). Murray noted that the sockets reproduced almost exactly the shape of the Athlit ram, and he concluded that they were for holding this display (21); they diminished in size because they held rams from galleys of different sizes. Antony had a heavy fleet ranging from triremes right up to "tens"; the biggest sockets presumably held rams from "tens" and the rest held the rams from the various sizes below that. It is striking that, to judge from the size of the Athlit ram and the point in the series where this would place it, big and heavy though it seems to be, it came from at most a "five," perhaps even a "four" (113-14). The ram that went into the largest socket would have dwarfed it.

121: J. Morrison (in Morrison 76), on the grounds that an "eight" was called *anasteiros*, which he takes to mean "having a high ram," concludes that the bigger polyremes had their rams so placed and that the change arose because in their case ramming was a preliminary to boarding. However, the vessel in Fig. 130 above, which Morrison thinks (77) was probably a "nine," has its ram at the waterline. Moreover, *anasteiros* may very well mean "with the ram elevated," e.g., through the lifting of the bows as a result of a shift aftward of people on deck; see L. Casson, *CQ* 39 (1989) 262-63.

128-31: J. Morrison, in *IJNA* 9 (1980) 121-26, had suggested that a *hemiolia* was a single-level galley with the oars on one side manned by one rower each and those of the other by two each. He now favors (in Morrison 74) an arrangement in which a full file of rowers sat along each gunwale and, in the amidships area, where the beam was broader, alongside sat a half file of rowers. The *triemiolia*, similarly, he reckons (in Morrison 75-76) had two full files and, amidships, one half file. See the reconstruction by J. Coates (139) illustrating this arrangement: on an upper level are two files, one of 21 and another of 24 rowers, with their oars in echelon, and, on a lower level, a half file (loosely speaking; the rowers actually number 15) with its forwardmost oarsman under the fourth in the upper files and the sternmost under the eighteenth.

129, note 115: In *SEG* xxvi 1022.11, the restoration ἡ]μιολίαν offered by E. Badian, *Zeitschrift für Papyrologie und Epigraphik* 23 (1976) 294, is not preferable to τριη]μιολίαν of the ed. pr. Badian's reason for preferring it is that he considers *triêmiolia* to be "a very rare word" (294, note 11). It is by no means rare: cf. 129-31 above.

140, note 17: *SB* 9780 = *P. Lugd.-Bat.* 20.41.

148, note 32: On the ships of Byzantine times, see also J. Pryor, in Morrison 101-10.

151, note 44: J. Pryor takes (in Morrison 104-105) this passage to mean that the *dromon* was fully decked and, of its two levels of rowers, one was above the deck and the other below.

152, note 52: On this passage see now F. van Doorninck, Jr., *MM* 79 (1993) 387-92, whose rendering of it eliminates the implication that a ram attack was involved.

152, note 54: See also, H. E. Davidson, *Byzantinische Zeitschrift* 66 (1973) 61-74; J. Haldon and M. Byrne, *ibid.* 70 (1977) 91-99. Both studies suggest that the mixture was somehow preheated in a vessel attached to the inboard end of the tube and, as it emerged from the muzzle of the tube, was ignited by a marine stationed there.

153: J. Pryor, *Geography, technology, and war. Studies in the maritime history of the Mediterranean, 649-1571* (Cambridge 1988) 59, suggests that the shift from waterline ram to deck-level spur had taken place by the tenth century. On the tenth-century evidence cited for its existence, see under 152, note 52.

158: A pair of Punic wrecks of the 2nd century B.C. has been found off Marsala in western Sicily; see H. Frost, *IJNA* 1 (1972) 113-17, 2 (1973) 33-49, 3 (1974) 35-42, 4 (1975) 219-28, *MM* 67 (1981) 66-75; H. Frost and others, *Not. Sc.*, suppl. to vol. 30 (Rome 1981); Parker 262-64. The excavator has identified them as warships but they are surely merchant galleys; see Casson, *New Findings* 17 and Parker 264. One has a pair of light wooden arms extending forward from the cutwater. This, it is argued, held a pencil-like ram made to hole an enemy and then break off; on this "bee-sting" ram, see under 95.

162: The *lembos* as a merchant galley or naval auxiliary is not attested after the early 2nd century B.C.

163, note 36: The date of *P. Petrie* ii 20 is 252 B.C.

169-82: An excellent idea of what at least a small (about 13.8 m long), one-masted merchantman looked like has been provided by the construction of a full-scale, fully rigged replica of a ship that was wrecked off Kyrenia on the north coast of Cyprus around 300 B.C. See M. Katzev, *Tropis I, 1989* (Proceedings of the 1st Symposium on Ship Construction in Antiquity, Piraeus, 1985) 163-75; Katzev, *INA* [Institute of Nautical Archaeology] *Newsletter* 16.1 (March 1989) 5-10; *AM²* 113-14 and pl. 30; Steffy figs. 3-36, 3-37. See also under 281-91 below.

A scale model of another small (11.25 m long) merchantman, probably of Phoenician origin and dating ca. 400 B.C., that was wrecked off the Kibbutz Ma'agan Micha'el, about 20 miles south of Haifa, provides a good idea of the hull of a vessel very similar to the Kyrenia wreck. See E. Linder, *Biblical Archaeology Review* 18.6 (November/December 1992) 24-35, esp. 30; Parker 247-48.

172-73: Cf. P. Pomey and A. Tchernia, *Archaeonautica* 2 (1978) 233-

51, who arrive at roughly the same estimates of tonnage, i.e., from 330 to 500 or 600 tons for *myriophoroi* (cf. 172, note 25 above) and 1000 for the big grain carriers.

174: Excavation of the wreck of a large merchantman, the Madrague de Giens wreck (see under 214-16 below), has revealed a ramlike cutwater almost perfectly preserved; see P. Pomey, *CRAI* (January-March 1982) 140, 142-45.

175, notes 35, 36: see also Parker 24.

176, note 41: In the fourth line of the note, put a period after "on them") and delete the rest of the note; as J. Oleson, *Greek and Roman Mechanical Water-Lifting Devices. The History of a Technology* (Toronto 1984) 31-32, points out, the word *antlia* in this passage can mean "sump" as well as "bilge," and, in view of the context, probably does, the reference being to pumping water in a mine or the like rather than in a ship.

Whether mechanical pumps other than the screw were used on shipboard is a question. F. Foerster has suggested (*IJNA* 8 [1979] 72-74, cf. *ibid.* 13 [1984] 85-93) that certain remains found on the Los Ullastres wreck (Parker 439) came from a chain-pump, but Oleson (223-26) strongly doubts it, particularly since there is no other evidence for the chain-pump in antiquity. Certain remains found in the barges raised from Lake Nemi have been reconstructed as two types of mechanical pump, a bucket-chain pump and a valved-piston suction pump (Ucelli 181-85). Both reconstructions are to be rejected. The first so transcends the evidence it relies on that it is "archaeologically a fantasy" (Oleson 230-31), while the other not only is based on a sole find dubiously identified as a piston but is a mechanical impossibility (Oleson 231-32). The only type of mechanical pump Oleson thinks was used on ships was the bronze force pump (206-207, 219). He bases this conclusion principally on the discovery of two pair of such pumps in the Dramont D wreck (Parker 167-68) and on some scant remains in the La Tradelière wreck (Parker 433-34); but the pumps on the Dramont D wreck were not in use when the ship went down and could well have belonged to the cargo and not the ship (cf. Parker 167), while the evidence from the La Tradelière wreck (cf. Parker 434) is not conclusive.

176, note 43: For the various forms of ballast found in wrecks, see Parker 28 and 519 (s.v. "ballast," "ballast stones," etc.)

176-77: For an authoritative review of the evidence from marine archaeology for stowage, dunnage, cargo containers, etc., see A. Parker, *IJNA* 21 (1992) 89-100.

177: Marine archaeology has revealed that not a few merchantmen carried cargo in big clay containers—*pithoi* in Greek, *dolia* in Latin—which in some cases are so huge they must have been fixed permanently in the hold. The earliest example found to date is the Kaş (Ulu Burun) wreck of the late 14th century B.C. (see under 214-16); it was carrying ten *pithoi* measuring a meter to over a meter and a half in height and up to a meter at the greatest diameter in at least four of which had been packed pieces of ceramic ware (G. Bass *IJNA* 13 [1984] 273; C. Pulak, *INA Newsletter* 21.4 [Winter 1994] 9). Ships that carried wine in such containers were veritable tankers: e.g., the Le Petit Congloué wreck of the mid-1st A.D. had 15 *dolia* set in three files, each capable of holding 2000 l (Parker 309). For further examples, see Parker Nos. 173, 320, 364 (14 in 3 files), 436, 477, 510, 565 (11-14 in 3 files), 718, 912, 1058, 1103, 1115.

178: On the galley of the Yassi Ada Byzantine wreck, see now Bass–van Doorninck 87-120.

181, note 65: On the Roman wreck found off Ognina, see now Parker 292, where it is dated A.D. 215-230.

182: Many merchantmen carried a *loutêrion*, a wash-basin for cleansing of the hands before performing sacrifice (G. Kapitän, *IJNA* 8 [1979] 97-120).

182, note 69: B. Levy, *Israel Numismatic Journal* 6-7 (1982-83) 111-13, argues convincingly that the figures carrying out the ceremony are Septimius Severus, Julia Domna, and Plautianus, the relief being an ex-voto dedicated by an individual or a group in the wine trade to celebrate the royal family's return from the East (Levy takes the vessel to the left as illustrating the departure from Alexandria and the vessel to the right as illustrating the return to Ostia, but that cannot be correct). She bases her conclusion on the numerous imperial symbols that appear: wolf and twins on the sail, wreath-bearing Victories at masthead and poop, wreath-bearing eagle above, and figures to right and left of the ship holding out wreaths toward it.

189-90: The discoveries of marine archaeology have greatly increased the evidence from wrecks for the dimensions of merchantmen. Here is a new table, including, with revised figures, the wrecks listed on 189-90 above. The large number of small vessels no doubt reflects the fact that the divers, using SCUBA apparatus, are limited to waters more or less near the shore (cf. under 214-16):

Wreck and Reference	Date	Length (meters)	Beam (meters)	Cargo and Estimated Burden
Gela Parker 188-89	6th-5th	20		jars, pottery
Ma'gan Micha'el Parker 247-48	End of 5th	13-15		no cargo
Porticello Parker 332-34	Last qtr. of 5th	17		jars, bronzes, ingots
Kyrenia Parker 231-32	310-300	13.8	4.4	400 jars; 20 tons
Capo Graziano A Parker 117	Mid-2nd			1000-3000 jars
Chrétienne A Parker 140-41	2nd half of 2nd	20?	7?	2000 jars
Porto Ercole B Parker 336-37	2nd half of 2nd	30-33		jars
Spargi Parker 409-11	Last qtr. of 2nd	30		jars, pottery
Bagaud B Parker 64	End of 2nd	12-15		iron bars; 5-10 tons
Mahdia Parker 252-53	End of 2nd			columns, statuary; 250 tons
Pozzino Parker 340	End of 2nd/beg. of 1st	15-18		jars, pottery, metal, glass
Cavalière Parker 133-34	100	13	4.6	jars, pottery; 19-31 tons

456

Grand Congloué B McCann (*op. cit.* under 361-70) 165-66	110-80	20		1200 jars; 40-50 tons
Albenga Parker 49-50	1st qtr. of 1st			jars, pottery; 500-600 tons
Dramont A Parker 165-66	Mid-1st	25	7	1000 jars
Madrague de Giens Parker 249-50	Mid-1st	40	9	6000-7000 jars, pottery; 400 tons
Mal di Ventre Parker 255-56	Mid-1st	36	12	1000 lead ingots
Planier C Parker 316-17	Mid-1st	20	5	jars, pottery
Le Titan Parker 424-25	Mid-1st			jars; 120 tons
Tre Senghe Parker 434-35	25	20-24	5	900 jars
Valle Ponti Parker 443-44	Last qtr. of 1st	25	5.4	jars, kitchenware, wood
Grand Ribaud D Parker 203-4	End of 1st	18		jars, dolia; 45-50 tons
Los Ullastres Parker 439	50 B.C.- A.D. 25	18		jars
Ladispoli A Parker 233	Early 1st	20		dolia, jars
Diano Marina Parker 163	Mid-1st	22-25		16 dolia, 1000 jars
Straton's Tower Parker 413	Mid-1st	40-45	9	no cargo
Chiessi Parker 140	3rd qtr. of 1st			5000-7000 jars

457

Les Roches d'Aurelle Parker 370	End of 1st	12-15		jars, pottery, tiles
St. Tropez A Parker 376	2nd?			marble; 200-230 tons
Procchio Parker 342-43	2nd half of 2nd	18		jars, sulphur ingots
Les Laurons B Parker 236-37	4th qtr. of 2nd	15	5	jars
Marseille Bourse Parker 265-66	Late 2nd	23	4	no cargo; 115-140 tons
Torre Sgarrata Parker 429-30	End of 2nd	30		18 sarcophagi, 23 blocks; 250 tons
Grado Parker 197	200?	18	5	jars, glass
Punta Scifo A Parker 361	Early 3rd	30-35		marble; 150 tons
Giglio Porto Parker 193	1st qtr. of 3rd	30	8	jars
Marzamemi A Parker 266-67	1st half of 3rd			marble; 172 tons
Methone C Parker 275-76	1st half of 3rd			columns; 131.5 tons
Monaco A Parker 279-80	1st half of 3rd	15	4	jars
Punta Ala Parker 345	Mid 3rd	25		jars, dolia
Capo Granitola A Parker 115-16	Mid-3rd			marble blocks; 350 tons
Giardini Parker 190	3rd			24 columns, 13 blocks; 95 tons
Isola delle Correnti Parker 219	3rd-4th			marble blocks; 350 tons

La Luque B Parker 247	1st qtr. of 4th	20	6	jars, lamps
Sobra Parker 408	2nd qtr. of 4th	25		1000 jars
Capo Taormina Parker 125	Rom. Imp. Period			37 columns, 2 blocks; 90-100 tons
Port-Vendres A Parker 329-30	Beg. of 5th	18-20	8	jars; 70-75 tons
Dramont E Parker 168	1st qtr. of 5th	15-18	5-6	jars
St. Gervais B Parker 372-73	1st qtr. of 7th	15-18	6	grain, jars; 40-50 tons
Yassi Ada A Bass-van Doorninck 86	625	20.52	5.22	850-900 jars; 60 tons

198, note 45: See also Oleson (*op. cit.* under 176, note 41) 291-301.

201: As noted above (under 9-10), the sewn boat was well known to the ancient world. A good example is the Bon-Porté wreck of the 6th century B.C. (P. Pomey, *MM* 67 [1981] 225-44; Parker 74-75). For other examples, see Parker 23.

203-209: On Greek and Roman shipbuilding, see also Gianfrotta-Pomey 236-57, Parker 23-25, Steffy 37-91.

203: The Kaş (Ulu Burun) wreck (cf. under 214-16 below) reveals that the practice of edge-joining planks with mortises and tenons goes back at least to the 14th century B.C.; see C. Pulak et al., *INA Newsletter* 20.3 (Fall 1993) 5-8.

On the use of mold frames, see now J. R. Steffy in Bass-van Doorninck 82-83.

208-209: On the gradual shift from shell-first, edge-joined construction to skeleton-first, see J. R. Steffy, *IJNA* 11 (1982) 26-28; Steffy 83-91. By the time of the Serçe Liman wreck of ca. A.D. 1025 (see under 214-16 below), the shift had been completed. The skeleton-first method not only meant savings in shiptimber and labor but also permitted more varied and more commodious hull shapes (Steffy, *IJNA* 26-27, 30-32; Steffy 85).

208, note 35: On the mast-step, see also Gianfrotta-Pomey 250-52. For further illustrations of well-preserved mast-steps, see Pomey et al. (*op. cit.* under 214-16 below) pls. 32, 35.2 (Madrague de Giens wreck); Linder (*op. cit.* under 169-82) 31 (Ma'agan Micha'el wreck).

208, note 37: A wreck very much like the Yassi Ada Byzantine wreck is the Saint Gervais B wreck. Though slightly earlier in date (600-625 A.D.), its construction seems to be even further along toward skeleton-first (Parker 372-73).

209, note 33: See also L. Basch, *Archaeonautica* 6 (1986) 187-98.

210: On lead sheathing, see also Gianfrotta-Pomey 258-59; Parker 27 and 533 (s.v. "lead sheathed," "lead sheathing"). Its use seems to have come to an end toward the end of the 2nd century A.D.

212-13: On shiptimber see also R. Meiggs, *Trees and Timber in the Ancient Mediterranean World* (Oxford 1982) 116-53 and M. Rival, *La charpenterie navale romaine. Matériaux, méthodes, moyens* (Paris 1991). Rival provides a detailed study of all the woods so far identified in wrecks, with an ingenious chart (86) that presents on one page a conspectus of precisely which woods were used for which parts in each ship. He also provides a careful analysis of how the various parts of a tree were exploited for the various cuts of wood that went into a hull. On his somewhat unsure handling of ancient terminology, see my review in *Echos du monde classique/Classical Views* 36 (1992) 238-41.

214-216: By now well over twelve hundred ancient wrecks have been identified and studied to a greater or lesser degree. They are listed, with crisp descriptions and full documentation, in Parker's authoritative and comprehensive compilation. Among them several discovered in the past two decades are of particular note:

Ulu Burun (formerly called Kaş), a wreck found off the southwest coast of Asia Minor that dates toward the end of the 14th century B.C. See G. Bass et al., *IJNA* 13 (1984) 271-79; Bass, *AJA* 90 (1986) 269-96; C. Pulak, *AJA* 92 (1988) 1-37; Bass et al., *AJA* 93 (1989) 1-29; C. Pulak and C. Haldane, *INA Newsletter* 15.1 (March 1988) 2-4; Pulak, *ibid* 15.4 (December 1988) 12-17, 20.3 (Fall 1993) 4-12, 21.4 (Winter 1994) 8-16; Parker 439-40.

Madrague de Giens, a wreck of the mid-first century B.C. found off Toulon in a magnificent state of preservation. Excavation has revealed it to be a big freighter 40 m long, 9 m in beam, 4.5 m deep in the hold, and capable of carrying 375-400 tons. See A. Tchernia, P. Pomey, A. Hesnard, et al., *L'épave ro-*

maine de la Madrague de Giens (xxxive supplément à *Gallia*, Paris 1978); Pomey, *CRAI* (January-March 1982) 133-54; Rival (*op. cit.* under 212-13) 147-244; Parker 249-50.

Serçe Liman, a wreck dating ca. A.D. 1025 found off the southwest coast of Asia Minor. A small coastal craft (15 m long, 5.13 m in beam), it is the earliest known example of skeleton-first construction. See J. R. Steffy, *IJNA* 11 (1982) 13-34; Steffy 85-91; Parker 398-99.

A number of wrecks have been accorded more or less detailed study. E.g.,

Kyrenia: J. R. Steffy, *AJA* 89 (1985) 71-101; Steffy 42-59. Cf. Parker 231-32.

Mahdia: G. Hellenkemper Salies, ed., *Das Wrack, Der antike Schiffsfund von Mahdia*, 2 vols. (Bonn 1994). See esp. 53-81 (O. Höckmann, "Das Schiff"). Cf. Parker 252.

Grand Ribaud D: A. Hesnard et al., *L'épave romaine 'Grand Ribaud D' (Hyères, Var.), Archaeonautica* 8 (1988). Cf. Parker 203-204.

Yassi Ada Byzantine: see Bass-van Doorninck.

All the wrecks discovered so far have been in the general vicinity of the shore, in depths of only up to 60 m or so, since the divers use SCUBA apparatus and this is more or less its effective limit. However, in the future depth may cease to be a limitation, thanks to the development of highly sophisticated vehicles, controlled from shipboard, which do the diving. See A. McCann and J. Freed, *Deep Water Archaeology*, Journal of Roman Archaeology, Supplementary Series 13 (Ann Arbor 1994), for a report on the first ancient wreck so examined, one whose remains lay in the open water between Sicily and Tunisia at a depth of between 750 and 800 m.

217-19: The suggestion by S. Mark, *AJA* 95 (1991) 441-45, that Homer is here describing the building of a sewn boat, one whose planks were laced together with cords, is based on feeble arguments and a mistranslation of the crucial term *harmoniai*; see L. Casson, *IJNA* 21 (1992) 73-74.

218, line 18: C. Kurt, *Seemännische Fachausdrücke bei Homer* (Göttingen 1979) 128-32, holds that *ikria*, although it eventually came to mean superstructure or deck, originally referred to the stanchions of the railings along the sides of vessels, such as are seen in representations of Egyptian and Minoan and other craft, and he asserts (131) that it is to be taken in this sense in *Od.* 5.252. Even if his view of the original meaning is correct, he himself admits (132) that in almost half a dozen other occurrences in the *Odyssey* (12.229, 414; 13.74; 15.283, 552) the term clearly means "deck," and his arguments for shifting to "stanchions" in this one occurrence are not convincing. The decking would have been just a scant amount at prow and stern; cf. 44 above.

220-23: P. Pomey, *Mélanges de l'École française de Rome* (1973) 483-515, demonstrates the accurate use of shipbuilding terms by Plautus and Ovid.

221: On *chelysma* = "false keel," see Amigues (*op. cit.* under 89, note 59) 87-90, where this meaning is defended against recent efforts to throw doubt on it.

229-31: On the terms for various parts of the rigging, see also J. Rougé, *Rph* 50 (1976) 213-20; P. Emanuele, *IJNA* 6 (1977) 181-85; P. Dell' Orco, *IJNA* 10 (1981) 66-71.

230: The ancient world knew the parral fitted with wooden beads (cf., e.g. *IH* ill. 107); see Servius ad *Aen.* 5.487: *dictus est malus . . . quia quasi quibusdam malis ligneis cingitur, quorum volubilitate vela facilius elevantur* "the mast (*malus*) is so called . . . because it is girdled by certain wooden apples (*malis*), as it were, by virtue of whose roundness the sails are more easily raised."

231: On the strength of ropes made of natural fibres, see J. Coates, *IJNA* 16 (1987) 207-11.

231, note 28. The massive anchor cable, 10 cm in diameter, on one of the Nemi barges was of esparto (Ucelli 242 and fig. 275, Gianfrotta-Pomey 285). Numerous ropes of esparto were found in the Marsala wreck (Parker 263).

231, note 29: See also J. Oleson, *IJNA* 12 (1983) 155-70.

233, note 37: A type of Greek cup with vertical looped handles opposite each other, like ears, was called a *karchêsion*. The mast-top of Greek vessels of the 6th and 5th centuries B.C. shows a similar pair of ear-like loops (e.g., *GOS* pl. 21e) for accommodating the twin halyards. J. Boardman, *JHS* 99 (1979) 149-51, suggests that this may have given it its name: it was called after the cup that had so similar a feature.

238, note 63: *P. Lond. Inv.* 2305 has been published as *P. Lond.* vii 2139.

240: On the foresail, see also I. and T. Pekáry, *Acta Archaeologica Academiae Scientiarum Hungarica* 41 (1989) 477-87.

245: There is evidence that the ancients erected markers in the form of poles to warn skippers of shallows or similar dangers. E.g., Arrian, *Indica* 41.2, describes such poles in the Persian Gulf to mark shallows and adds: "Just as in the channel between the island of Leucas and Acarnania signposts have been set forth for those sailing through so that their ships will not go aground on the shallows." Rutilius Namatianus, *De reditu suo* 1.453-62, notes a line of poles topped by tall laurel branches that marked a channel through the shallows along the coast north of Pisa (cf. J. Pryor, *Mediterranean Historical Review* 4 [1989] 274).

245, note 82: Further study of the Kyrenia wreck and the building of a replica of it (see under 169-82), along with the trials the replica was put through, make it certain that the rig consisted of a squaresail; cf. *AM* ² 114 and under 281-91 below.

252: On anchors, see now G. Kapitän, *IJNA* 13 (1984) 33-44, and cf. Parker 29 for figures on the number of wrecks in which anchors of various types have been found.

256: On the anchors of the Yassi Ada Byzantine wreck, see now Bass-van Doorninck 121-43.

258, note 138: A perfectly preserved one-armed anchor was found in the Ma'agan Micha'el wreck; see Linder (*op. cit.* under 169-82) 32.

259, note 1, line 12: The Greek equivalent for the Latin term *propes* is πρόπους (Scholia to Apollonius Rhodius 1.49).

273, note 11: Cf. Polybius 5.5.6: "To sail [from Cephallenia] to Messenia is easy, but to sail back from there, while the Etesians are blowing, is impossible."

277: On the origin of the lateen sail, see L. Casson, *The American Neptune* 31 (1971) 49-51.

278: B. Rankov, *MM* 80 (1994) 131-46, provides an illuminating report on the problems encountered in rowing the trireme replica (see under 77-80) and the various ways in which the sea trials it was put through revealed what was involved in rowing an ancient trireme.

278, note 31: The training, though arduous, need not have taken long. Rankov (*op. cit.* under 278, 133-34), on the basis of his experience with the trireme replica (see under 77-80), concludes that raw Athenian lads who knew how to handle an oar could learn the basics of rowing a trireme in a matter of days.

278, note 32: Raw crews were also trained by rowing on the quiet waters of lakes (Statius, *Theb.* 6.21-22).

279, note 34: Backing water by having the rowers shift to face the prow and pull was common enough to be illustrated in a number of pictures dating from widely separated times; see A. Tilley, *IJNA* 21 (1992) 55-60.

279, note 37: The trireme replica (see under 77-80), in its sea trials, reached a speed of 8.9 knots, and with improvement in design could have done even better; see J. Coates, *IJNA* 20 (1991) 70-71.

281-91: The performance of the replica of the Kyrenia wreck (see under 169-82) provides some indication of how seaworthy and efficient ancient sailing craft could be. Its rig, to be sure, is a reconstruction, but a reconstruction based on meticulous consideration of the available evidence. During a voyage in which the ship encountered all sorts of wind conditions, from calms to a gale with gusts up to 55 knots, it met all demands splendidly. During one 24-hour period marked by consistently favorable winds, it averaged 6 knots, and during one 6-hour segment of that period, when the wind was particularly strong, the average rose to 11 knots. It was even able to advance at a speed of 2 knots sailing into the wind at 50° to 60° off its direction. With no difficulty, it rode out one storm that had sent the local fishing caiques scurrying to the home port for safety. Over the whole voyage of about 500 nautical miles, it averaged 2.85 knots. See Katzev, *INA* (*op. cit.* under 169-82) 8-10.

282, note 47: For a study demonstrating that the winds in ancient times were the same as today, see W. Murray, *Mediterranean Historical Review* 2 (1987) 139-67.

289: Another instance of a voyage made under unfavorable winds is *Hist. Apollon. Reg. Tyri* 7: from Tyre to Antioch, ca. 175 nautical miles, in 3 days, for an average speed of 2.43 knots.

291, note 91: F. Hirth, *China and the Roman Orient* (Shanghai 1885, rep. Chicago 1967) 182-83, cites Chinese sources for a voyage from Alexandria to Antioch (ca. 400 NM) in 6 days, i.e., an average speed of 2.78 knots.

293, note 97: Fleets in a hurry could achieve even higher speeds over short runs. E.g., in 411 B.C. a Peloponnesian fleet, whose commander was anxious to get it from Chios to the Hellespont as soon as possible, covered the leg from Arginusae to Rhoeteium in one long day, starting out just before dawn and arriving just before midnight (Thucydides 8.101.2-3); since the distance is ca.

464

125 NM, if we assume some eighteen hours of travel, the speed works out to an average of ca. 7 knots (cf. Morrison-Coates [*op. cit.* under 55-59] 104-105).

In the summer of 218 B.C., a delegation sent by the Messenians to Philip V at Cephallenia to persuade him to come to the aid of their country pointed out to him that "since the Etesian winds had already set in, the voyage from Cephallenia to Messenia could be made in one day" (Polybius 5.5.2-4). Since the distance is ca. 170 NM, they must have meant a 24-hour day, which works out to an average speed of 7 knots.

301, note 6: After "260-63," add "and Gabrielsen (*op. cit.* under 80-81) index s.v. *trierarchies, trierarchs.*"

302, note 8: Delete "(For the connection of this inscription with the Battle of Arginusae [406 B.C.], see App., note 4)"; see under 323, note 4, last paragraph, below.

305, note 24: J. Morrison, *JHS* 104 (1984) 48-59, a detailed study of the word *hypêresia*, has determined that in the 5th century B.C. the term included the *epibatai* as well as the personnel I have listed.

309: In *SEG* xxxiii 640 (Rhodes, 2nd-1st B.C.) the *ergazomenoi en prymnê* number only four.

309, note 38: See also *SEG* xxxiii 684 (Samos, 324-322 B.C.): a *dekatarchos* of a *dikrotos*; *SEG* xli 717 B and C (Rhodes, 84-67/61 B.C.): *dekatarchoi* aboard anti-pirate patrol craft (*lêstophylakika ploia*).

311-12: A *naupegus aupiciarius* (= *oppiciarius*, from *pix* "pitch") "shipwright who applies pitch" is attested in the Syrian squadron (*AE* 1974.621).

311, note 51: V. Nutton, *Epigraphica* 32 (1970) 66-71, lists six more examples of doctors.

315: On the *nauklêros*, see also J. Vélissaropoulos, *Les nauclères grecs* (Centre de recherches d'histoire et de philologie de la ive Section de l'École Pratique des Hautes Études. iii, Hautes Études du monde Gréco-Romain 9, Paris 1980) 48-56.

315, note 69: The *nauklêros* was aboard the ship that carried Themistocles to Asia Minor after his ostracism in 471 B.C. (Plutarch, *Them.* 25.2).

320: Marine archaeology has contributed some information about the size of crews. The Kyrenia wreck (see under 169-82), a small merchantman some 13.8 m long, had a crew of four (Parker 232).

The Grand Ribaud D wreck, a slightly bigger merchantman (about 18 m long), had a crew of six (Parker 203-204).

320, note 83: On the *diopos*, see also Vélissaropoulos (*op. cit.* under 315) 82-83.

321: Oarsmen while rowing might wear a short tunic reaching to the thighs (Lucian, *Dial. Het.* 14.2).

322-23: The instances in which slaves are attested as being aboard warships are not few. Though in many of such instances, the slaves—as I have argued—were attendants of the marines and officers of the ship, a reappraisal of the evidence makes it undeniable that in some cases they were pulling an oar as regular members of the crew; see A. Graham, *TAPA* 122 (1992) 257-59. They were not, however, owned by the navy using them; their masters were individuals who rented them to it.

When a state found that its need for rowers was greater than could be supplied by its citizens and metics, it did not buy slaves to fill the benches. It first turned to the hiring of free men from other states (cf. 323-24 above). There was good reason for this. Slaves were far from cheap, so their purchase would have involved a sizable capital investment, and, if any were killed in action, their cost was that much state money down the drain. Moreover, the state would have had to pay for their food and shelter all year round and not just during the time they were in service. The foreigner-oarsman, on the other hand, like the mercenary in an army, received a wage only while he served, and, if he was killed in action, that was no loss to his employer. But navies, though they did not own slave-oarsmen, at times found it useful to hire them.

Slaves for hire were readily available, since the owning of slaves to rent out was a standard form of investment in ancient times, even for people of modest means. The slave would earn a wage working as a digger in a mine, a waiter for a caterer, a day laborer, etc., and the earnings—either all or part, depending upon a master's generosity—would end up in the master's pocket; cf. L. Casson, *TAPA* 106 (1976) 39-40 = *Ancient Trade and Society* (Detroit 1984) 41-42. Among the roles filled by rented slaves was that of oarsman on a war galley; see Graham 262. Indeed, it was not uncommon for a master and a slave—or even several of his slaves—to serve as oarsmen on the same ship. This

466

put the master in a position to prevent desertion, something both he and his commanding officer were concerned about—he to avoid loss of a valuable piece of property, the commander to avoid an empty rowing bench (see Graham 266-69).

323, note 4: At the end of the third paragraph, after "more likely a source of hoplites than rowers," add: "K. Welwei, *Unfreie im antiken Kriegsdienst*. ii, *Die kleineren und mittleren griechischen Staaten und die hellenistischen Reiche* (Forschungen zur antiken Sklaverei 8, Wiesbaden 1977) 113-14, overlooks this in arguing that the 800 slave prisoners from the 70 Corcyrean ships which were destroyed must have been rowers, not attendants of the officers and marines. He reasons that, with 30-40 attendants alongside the same number of officers and marines, a ship would have been overloaded and would have lost the necessary maneuverability. That is pure conjecture. We have no idea how many attendants were aboard; very likely far fewer than 30-40, for not every officer and marine need have had one (800 from 70 ships works out to little more than 10 per ship). Moreover, we have no idea how many men a trireme could carry on its decks, particularly the heavy type favored by Corinth. Venetian galleys of the 15th century, which were about the same size as ancient triremes, regularly carried 78 (L. Fincati, *Le Triremi* [Rome 1881] 23-24). French galleys of ca. 1700, whose rowers numbered 202 (against a trireme's 170), carried no fewer than 238 (Masson 313).

323, note 4, last paragraph: Restudy of the evidence indicates that *IG* ii^2 1951 in all probability is not to be connected with the Battle of Arginusae; see Graham (*op. cit.* under 322-23) 263-69. The inscription lists, along with foreigners, many slaves among the crews of the various ships involved; these, like the foreigners, must have been hired, rented to the navy by their masters (who in at least a few cases were rowers in the same crew; see Graham 268-69).

324: During the Second Punic War, Scipio, after his capture of New Carthage in 209 B.C., pressed some of the prisoners of war, both slave and free, into service as oarsmen; see J. Morrison, *CQ* 38 (1988) 251-53.

325, note 10: Welwei (*op. cit.* under 323, note 4) 134-35 takes the deserters characterized as "those sailors who bear the brand" to be slaves, largely on the grounds that there is no evidence for the branding of free rowers who had deserted. There is none for the branding of slave rowers qua rowers, either.

329: On small craft, see also L. Basch, *IJNA* 2 (1973) 329-33.

333-34: O. Höckmann, *IJNA* 22 (1993) 125-35, describes the remains of a number of late Roman rivercraft excavated at Mainz, which he identifies (131) as *lusoriae*.

335, note 37: Change to read: "The sail may have been of leather; cf. *MM* 58 (1972) 105. Henningsen's drawing etc."

336-37: A long, shallow, lightly built small craft has been uncovered at Herculaneum; see J. R. Steffy, *AJA* 89 (1985) 519-21; Steffy 67-71.

337, note 49: The Kyrenia wreck did not have a fore-and-aft rig; see under 245, note 82.

338-40: On boats from northern Europe, see also P. Marsden, *IJNA* 5 (1976) 23-55; L. Lehmann, *IJNA* 7 (1978) 259-67; Steffy 72-77; Parker 15 (where he lists all wrecks found outside the Mediterranean), 25-26.

339: It may have been the Celts who introduced skeleton-first construction in northern Europe; see Marsden (*op. cit.* under 338-40) 42-54; Casson, *New Findings* 18; Nayling et al., *Antiquity* 68 (1994) 596-603 (publication of a recently found "Romano-Celtic boat").

339, note 55: The late Roman rivercraft found at Mainz (see under 333-34) were apparently built by nailing the planks to mold-frames and then removing these and replacing them with permanent frames (Höckmann 126-27).

342: A *skaphopaktôn* is now attested; see *P.Oxy.* 3111.2,6 (A.D. 257); *Zeitschrift für Papyrologie und Epigraphik* 20 (1976) 162, line 6 (A.D. 221). The Greek term *dierama* designates a small boat of some sort; see *CR* 19 (1969) 91-92.

343: The remains of a small craft, some 9 m long and 2.5 m in beam, used on the Lake of Galilee, has been excavated and studied in detail; see S. Wachsmann et al., *The Excavations of an Ancient Boat in the Sea of Galilee (Lake Kinneret)* ('Atiquot, English Series 19, Jerusalem 1990); Steffy 65-67; Parker 193-94.

343, note 75: The term *akôr* (*achôr*) is an abbreviation of *akôritikon* "of Akoris." Boats so called, in other words, came from the port on the Nile called Akoris in Hellenistic times and Achoris in Roman times: see H. Hauben, *Ancient Society* 19 (1988) 210.

348: Add to the fifth line: "Ships bore, as they had for ages, oculi and bowpatches (Fig. 191 above). The oculi were now placed on the stern as well as the prow (e.g., on the sternpost below the goose-head in Fig. 154 above)."

348, note 15: For a list of representations of the *chêniskos*, see I. Pekáry, *Boreas* 5 (1982) 276-79.

348, note 17: J.-M. Gassend et al., *Archaeonautica* 6 (1986) 9-30, publish a graffito, dating to the 1st century A.D., of a merchant-man under full sail. A pole crossed by a horizontal bar rises from the tip of the sternpost, presumably for carrying a banner, as on the ship pictured in Fig. 154 above. The authors call it a *stylis* (28), but that term properly refers to the somewhat similar fitting on a warship (see 346 above).

355: For a Hellenistic warship named *Isis*, see under 99.

356, note 57: The following new names have been attested: in the fleet at Misenum, liburnians *Diana* (*AE* 1975.271), *Juno* (*AE* 1979.167), *Margarita* (*AE* 1974.261); in the fleet at Ravenna, a liburnian *Pinnata* (*AE* 1979.248).

360: An *akatos* named *Ourania* occurs in *P.Oxy.* 3053.19 (A.D. 252) and another named *Antinoos Philosarapis Sôzôn* "Antinous, devoted to Sarapis, the Savior," in *SB* 9571.5 (A.D. 149) = R. Bagnall and W. Harris, eds. *Studies in Roman Law in Memory of A. Arthur Schiller* (Leiden 1986) 11-12. The *Rhedon* and *Medaurus* (*AE* 1979.189) were ships named after Illyrian deities.

361-70: See D. Blackman, *IJNA* 11 (1982) 79-104, 185-211, for a comprehensive review of Mediterranean harbors and in Morrison 224-33 for a review of harbor installations for men-of-war.

A number of individual harbors have received extensive study:
Caesarea: J. Oleson and A. Raban, *The Harbours of Caesarea Maritima: Results of the Caesarea Ancient Harbour Project, 1980-1985*, Vol. I (Oxford 1989)
Cosa: A. M. McCann et al., *The Roman Port and Fishery of Cosa* (Princeton 1987)
Piraeus: R. Garland, *The Piraeus, From the Fifth to the First Century B.C.* (Ithaca 1987)

367: On the technology involved in the building of Roman harbors, see J. Oleson, *IJNA* 17 (1988) 147-57, and, for further bibliography, his *Bronze Age, Greek and Roman Technology. A Select Annotated Bibliography* (New York 1986) 395-405. On the use of concrete at Anzio, see E. Felici, *Archeologia Subacquea* (1993) 71-104.

370, note 44: The dockworkers called *embaenitarii* "putters on board" (*AE* 1971.75) were probably stevedores of some sort.

371: To the entry "*AM*: L. Casson, etc.," add "2nd ed., Princeton 1991."

380: To the entry "Starr: C. Starr, etc.," add "3rd ed., Chicago 1993."

Library of Congress Cataloging-in-Publication Data

Casson, Lionel, 1914-
Ships and seamanship in the ancient world / by Lionel Casson.
    p.   cm.
  Originally published : Princeton, N.J. : Princeton University Press, 1971.
  Includes bibliographical references (p.   ) and indexes.
  ISBN 0-8018-5130-0 (pbk. : alk. paper)
  1. Ships, Ancient.   2. Seamanship—History.   I. Title.
VM16.C233   1995
623.8'093—dc20

95-20034